BIO-INSPIRED AND NANOSCALE INTEGRATED COMPUTING

BIO-INSPIRED AND NANOSCALE INTEGRATED COMPUTING

EDITED BY

Mary Mehrnoosh Eshaghian-Wilner

A JOHN WILEY & SONS, INC., PUBLICATION

Published by John Wiley & Sons, Inc., Hoboken, New Jersey
Published simultaneously in Canada

Library of Congress Cataloging-in-Publication Data:

Bio-inspired and nanoscale integrated computing / edited by Mary Mehrnoosh Eshaghian-Wilner.
p. cm. – (Nature-inspired and nanoscale integrated computing)
Includes bibliographical references and index.
ISBN 978-0-470-11659-3 (cloth)
1. Molecular computers. 2. Biologically-inspired computing. 3. Nanotechnology. 4. Nanomedicine. I. Eshaghian-Wilner, Mary Mehrnoosh.
QA76.887.B56 2009
004.1–dc22

2009013331

Printed in the United States of America

10 9 8 7 6 5 4 3 2 1

CONTENTS

FOREWORD

Bio-inspired and Nanoscale Integrated Computing (edited by Professor Mary Mehrnoosh Eshaghian-Wilner) is the first book in a new and exciting Wiley book series, the *Nature-Inspired Computing Series*. The proliferation of computing devices in every aspect of our lives increases the demand for better understanding of emerging computing paradigms. The new book series seeks to provide an opportunity for researchers to explore the new computational paradigms and their impact on computing in the new millennium.

The series is quite timely since the field of computing as a whole is undergoing many changes. A vast literature exists today on such new paradigms and their implications on a wide range of applications—a number of studies have reported on the success of such techniques in solving difficult problems in all key areas of computing. The list of topics covered by the series is by no means exhaustive but it serves as a guide to the diversity of the topics covered here. It is also hoped that the topics covered will get the readers to think of the implications of such new ideas on the developments in their own fields. These fields are: Quantum Computing, DNA Computing, Genetic Algorithms, Evolutionary Paradigms, Cellular Automata, Neural Networks, Swarm Algorithms, Fuzzy Logic, Computational Synthesis, Machine Learning Techniques, Computational Methods for Biological Systems. The series will also publish books related to technologies that are relevant to nature-inspired computing: Silicon Neuron Processing, Molecular Scale Computers and NanoTech, Optics, Evolvable Hardware, Quantum Hardware, Reconfigurable Hardware, and others.

Bio-inspired and Nanoscale Integrated Computing is a solid inaugural contribution that overviews some of the recent developments in the area of nanocomputing. This is an important emergent discipline which is facilitated by scientific advances in the field of nanotechnology. This book, through a number of chapters covering a wide range of topics, increases the awareness of the potential of nanocomputing as a new paradigm that holds great promise. It is well-positioned to generate more interest and concerted effort in studying the different facets of nanocomputing and applying it to a wider range of problems in science and engineering.

Bio-Inspired and Nanoscale Integrated Computing can be used by scientists, engineers, graduate students, senior undergraduate students, researchers, instructors, and practitioners, who want to learn more about nanocomputing. It is definitely an excellent reference on this topic.

Albert Y. Zomaya

The University of Sydney
Sydney, New South Wales, Australia

PREFACE

The word "bionics" is often described as a compound of "biology" and "electronics." Some say it was coined in the 1950s and that it originates from the Greek word "βíον," meaning "unit of life," and the suffix "-ic," meaning "like" or "in the manner of"–hence "like life." But just like many people of a certain age, I first came to know the term when I was a child (in my case growing up in Tehran) watching the then-famous duo of American TV series, "The Six Million Dollar Man" and "The Bionic Woman." Back then, I had no idea how much of those programs was just fantasy. Not only was I too young to know better, but I was also under the impression that America was so much more advanced than where I lived, and therefore everything must have already been possible in the United States.

As I grew older and eventually moved to the United States during my early teen years, I began to realize more and more just how far removed the bionic fantasy was. The absence of those advancements I had expected to see was certainly a big disappointment, yet it motivated me to begin my undergraduate studies in biomedical and electrical engineering at the University of Southern California (USC).

As part of my studies, it became apparent that computer science and engineering were key disciplines that would make the bionic fantasy a reality. I therefore majored in Computer Engineering in the USC graduate school and received a Master s degree and a Ph.D. My ultimate dream was to make an actual bionic eye. To move one step closer to that dream, I concentrated on designing ultra-dense multiprocessor VLSI chips that had the capability of massive electro-optical intercommunications. I also designed a large number of algorithms that could be mapped onto those chips for image processing and computer vision. Before long, my career took off and my work received recognition in the area of parallel processing. I published, gave talks, taught courses, and organized conferences in various areas within parallel processing. Over time, my bionic dream was diminished to a small side issue.

About two decades later, while employed as a professor and department head at the Rochester Institute of Technology, I learned about the intriguing advancements that were rapidly taking place in the new fields of nanotechnology and bioinformatics. I applied for and received a fellowship at the National Science Foundation's Institute of Pure and Applied Mathematics at the University of California Los Angeles (UCLA). There, I interacted with an interdisciplinary group of scientists who were all interested in the applications of nanotechnology to their fields of expertise. Among them were biologists, chemists, physicists, mathematicians, electrical engineers, mechanical engineers, computer scientists, and so forth. I knew that if I could somehow get all of them to understand each other, we could then also sit and design some of the world's most challenging devices, e.g., those that were part of my "bionic dream." Towards that goal, I began working with a small group of interdisciplinary scientists, and together we wrote a joint article about various ways computing could be done at a nanoscale level. The collaboration also led to the production of some very interesting breakthrough results that were widely disseminated and publicized through various news media. Shortly after that, I was contacted by an editor for the publisher, John Wiley & Sons, who told me Wiley would be interested in a book on the type of collaborative multidisciplinary work in which I was engaged.

"Bio-inspired and Nanoscale Integrated Computing (BioNIC)," the name I had given to my research team at UCLA, was the title that I proposed to Wiley. I proposed a collection of the latest findings of leading scientists who were working in the proposed topic area, with the intent to get one step closer to the creation of actual bionic parts.

Paul Petralia, the Wiley Editor, liked my proposal, then sent it for external review for which it received very positive and high peer ratings. I began working on the book immediately; and after about three years, it is now complete. The quality of the work presented in this book is all to the credit of the outstanding group of scientists with whom I had the distinct privilege and honor to collaborate. I must mention here that this book was initially planned to contain significantly fewer chapters than what you see before you. Only through the remarkable dedication and contributions of the co-authors did this book evolve to its current size. I am immensely thankful and indebted to all of them.

I invite all of you to begin reading with a special emphasis on Chapter 1. That chapter was produced primarily by a group of amazing interdisciplinary UCLA students who took a seminar course on nanocomputing that I taught. Each student studied the application of nanocomputing to their prospective fields and, based on those, they all got together and wrote the chapter. The chapter not only contains a brief but comprehensive background, but it also serves as a guide to various chapters within the book where additional information can be found. Special credit goes to Shawn Singh, who took a superb leading role among all the students that were involved. I also thank Professor Alireza Nojeh for his final review of their work.

There are many additional people to whom I am grateful. First and foremost, I thank Paul Petralia of Wiley, who had the foresight, leadership, and vision to help me produce this book from a simple idea to what it is now. He is also responsible for creating the Nature-Inspired Computing Series that I am co-editing with my dear colleague, Professor Albert Zomaya. This book will serve as the first in the new series and it will be followed by several other interesting titles, some of which are currently in progress. I am extremely grateful and pleased for the opportunity to be working with the renowned Professor Zomaya on this important series.

My undying gratitude also extends to the numerous other wonderful people at Wiley and UCLA who have worked with me nonstop on this book. At Wiley, a special shout-out goes to Lisa Morano Van Horn for her outstanding production work, to George Telecki for seamlessly transitioning into Paul Petralia's role, and to Michael Christian for his routine help with all diverse tasks. Over at UCLA, I would like to thank my administrative assistant, Rose Weaver, and all of my research assistants, including Dr. Shiva Navab, David Shen, Jon Lau, Mike Yip, and Stephen Chu, and Eric Mlinar.

Finally, I don't think this book would have been possible without the various contributions of my family. My parents, Mehdi and Molly, my brothers, Michael and Mark, and my sister, Maggie, have always supported me throughout the various stages of my life and career. I am forever indebted to them. My husband, Arthur, who is an excellent writer himself, has been a great adviser to me. I thank him with all my heart for all his valuable feedback. But most of all, I owe the most credit to my beautiful daughter, Ariana, who has been so patient and understanding about Mommy's crazy schedule. I love her more than words can describe, and I dedicate this book to her and our family.

Thank you for reading this book. Enjoy!

MARY MEHRNOOSH ESHAGHIAN-WILNER

Los Angeles, California
March 2009

CONTRIBUTORS

Sumit Ahuja, Research Assistant, Fermat Lab, Virginia Polytechnic and State University, Blacksburg, Virginia

Lisong Ai, Department of Biomedical Engineering and Division of Cardiovascular Medicine, University of Southern California, Los Angeles, California

Debayan Bhaduri, Research Assistant, Fermat Lab, Virginia Polytechnic and State University, Blacksburg, Virginia

Varun Bhojwani, Analyst, Wachovia Securities, Los Angeles, California

Elaine Ann Ebreo Cara, Department of Electrical Engineering, University of California, Los Angeles, California

R. Chomko, Assistant Research Professor, Department of Electrical Engineering, University of California, Riverside, California

Stephen Chu, Department of Electrical Engineering, University of California, Los Angeles, California

Dr. I. Dumer, Professor, Department of Electrical Engineering, University of California, Riverside, California

Dr. Mary Mehrnoosh Eshaghian-Wilner, Adjunct Professor, Department of Electrical Engineering, University of California, Los Angeles, California

Dr. Robert A. Freitas Jr., Senior Research Fellow, Institute for Molecular Manufacturing, Palo Alto, California

Aaron K. Friesz, Department of Electrical Engineering, University of California, Los Angeles, California

Tzung K. Hsiai, Department of Biomedical Engineering and Cardiovascular Medicine, University of Southern California, Los Angeles, California

Dr. Graham A. Jullien, Professor, Department of Computer and Electrical Engineering, University of Calgary, Calgary, Alberta, Canada

Alex Khitun, Research Engineer, Department of Electrical Engineering, University of California, Los Angeles, California

Dr. Sakhrat Khizroev, Professor, Department of Electrical Engineering, University of California, Riverside, California

Dr. Thomas H. LaBean, Professor, Department of Chemistry and Department of Computer Science, Duke University, Durham, North Carolina

Ling Lau, Hardware Engineer, Cisco Systems, San Jose, California

Dr. Dimitri Litvinov, Director, Center for Nanomagnetic Systems, Associate Professor, Electrical and Computer Engineering and Chemical and Biomolecular Engineering Departments, University of Houston, Houston, Texas

Dr. Tulin Mangir, Professor, Electrical Engineering Department, College of Engineering, Director, Nano Technology Lab, and Director, Wireless and Mobile Security Lab, California State University, Long Beach, California

Dr. Shiva Navab, Staff Scientist, Broadcom Corporation, Irvine, California

Dr. Alireza Nojeh, Assistant Professor, Department of Electrical and Computer Engineering, University of British Columbia, Vancouver, British Columbia, Canada

Dr. Alice C. Parker, Professor, Department of Electrical Engineering, University of California, Los Angeles, California

Dr. John H. Reif, A. Hollis Edens Distinguished Professor of Computer Science, Duke University, Durham, North Carolina

Fady Rofail, Department of Electrical Engineering, University of California, Los Angeles, California

Mahsa Rouhanizadeh, Department of Biomedical Engineering, University of Southern California, Los Angeles, California

Dr. Mario Ruben, Institut fur Nanotechnologie, Karlsruhe Institute of Technology (KIT), Karlsruhe, Germany

Michael M. Safaee, David Geffen School of Medicine, University of California, Los Angeles, California

David D. Shen, School of Law, University of California, Davis, California

Sandeep Shukla, Associate Professor, Department of Electrical and Computer Engineering, Virginia Polytechnic and State University, Blacksburg, Virginia

Gaurav Singh, Fermat Lab, Virginia Polytechnic and State University, Blacksburg, Virginia

Shawn Singh, Department of Electrical Engineering, University of California, Los Angeles, California

Dr. James M. Tour, Chao Professor of Chemistry, Professor of Computer Science, and Professor of Mechanical Engineering and Materials Science, Rice University, Houston, Texas

Ko-Chung Tseng, Department of Electrical Engineering, University of California, Los Angeles, California

Dr. Konrad Walus, Associate Professor, Electrical and Computer Engineering, University of British Columbia, Vancouver, British Columbia, Canada

Dr. Kang L.Wang, Raytheon Chair Professor of Physical Electronics, Director, Marco Focus Center on Functional Engineered Nano Architectonics (FENA), and Director, Western Institute of Nanoelectronics (WIN), University of California, Los Angeles, California

Alexander D. Wissner-Gross, Environmental Fellow, Harvard University, Cambridge, Massachusetts

Daniel Wu, Department of Electrical Engineering, University of California, Los Angeles, California

Chun Wing Mike Yip, Department of Electrical Engineering, University of California, Los Angeles, California

Dr. Hongyu Yu, Assistant Professor, School of Earth and Space Exploration, Department of Electrical Engineering, Arizona State University, Tempe, Arizona

Dr. Lin Zhong, Professor, Department of Electrical and Computer Engineering, Rice University, Houston, Texas

1

AN INTRODUCTION TO NANOCOMPUTING

Elaine Ann Ebreo Cara, Stephen Chu, Mary Mehrnoosh Eshaghian-Wilner, Eric Mlinar, Alireza Nojeh, Fady Rofail, Michael M. Safaee, Shawn Singh, Daniel Wu, and Chun Wing Yip

The continuous shrinking of transistors has made it possible to do amazing feats with computers, and a major theme of the microelectronics era has been "smaller is better." Today, technology has already shrunk to the nanometer scale, causing many practical challenges and motivating the search for new nanoscale materials and designs. In this chapter, we present a brief introduction to the concept of nanocomputing and provide a high level overview of nanocomputing devices and paradigms. We also discuss some applications of nanocomputing such as biomedical engineering and neuroscience.

1.1. INTRODUCTION

In 1959, Nobel laureate Richard Feynman posed this question to his fellow physicists: *"Why cannot we write the entire 24 volumes of the Encyclopedia Britannica on the head of a pin?"* In that lecture, aptly named "There's Plenty of Room at the Bottom," Feynman challenged scientists and engineers to imagine what could be possible using nanoscale structures. He used computing as a prime example, suggesting that wires and circuits could be shrunk to only hundreds of angstroms in size [1], making it possible to combine billions of devices to perform amazing tasks.

The transistor, today's most prevalent modern computing device, can already be manufactured as small as hundreds of atoms across. This could mark the

Bio-Inspired and Nanoscale Integrated Computing. Edited by Mary Mehrnoosh Eshaghian-Wilner

end of the microcomputing era and the beginning of the *nanocomputing era.* Nanocomputing could be fundamentally different from microcomputing: individual particles could play a significant role; quantum effects enter the game much more directly. Even though Feynman's vision has been partially realized with such tiny transistors, researchers are only beginning to explore the fundamental questions: How small can we make computers? How would such computers work? How can we embrace quantum mechanics for computation? What applications are possible with nanocomputing that were not possible with microcomputing? We will explore these questions throughout the book.

In this introductory chapter, we will give a brief overview of devices, paradigms, and applications of nanocomputing. Of course, we cannot include all existing ideas about nanocomputing in the introduction, or even in the entire book. Instead, we aim to inspire the reader with a variety of ideas that appear commonly in nanocomputing research. We will first define computing and nanocomputing and provide some historical context of the microcomputing era. The limitations of today's microcomputers motivate a discussion of nanoscale devices and paradigms, many of which are detailed in later chapters. We then consider two major fields that can greatly benefit from nanocomputing: biology and neurology.

1.2. WHAT IS NANOCOMPUTING?

Computing is the *representation* and *manipulation* of information. Computer games, surfing the Internet, solving complex math equations, and even verbal communication are examples of computing. While we certainly compute using our own thinking power, it is often more useful to create a machine that computes on its own so we can use it to enhance our daily lives. Indeed, our world has become dependent on machines that automatically compute for us. These "computers" are used for entertainment, education, safety, and a vast number of other applications, all of which require manipulation of abstract information.

To build a computing machine, abstract data must eventually be represented by something that physically exists. For example, digital states (such as 1's and 0's) can be represented as high and low voltages on a wire. Similarly, manipulations of abstract data, such as adding two numbers, must eventually be performed by a physical phenomenon that affects the data. Therefore, to explore the world of computing, we must ask the most fundamental question: *How can we use physics to represent and manipulate abstract information?*

Directly or indirectly, researchers from all over the world are exploring this fundamental question. This endeavor spans across several disciplines, including mathematics, computer science, physics, chemistry, and biology. Throughout this book, there are many examples of how physics can be used for computation; some of these ideas may eventually lead to more powerful computers.

Nanocomputing can be interpreted literally as *computing at the nanometer scale*. It is generally agreed that the terms *nanotechnology*, *nanocomputing*, and *nanoscale* are used when considering devices that are at least one dimension smaller than 100 nanometers (nm). Today, devices such as transistors have channels that are well below 100 nanometers in length. Eventually the size of entire computers may also be measured in nanometers.

Just how small is a nanometer? A nanometer is one-billionth of a meter (10^{-9} m). Figure 1.1 shows the approximate size of various physical entities at the nanometer scale. If it were possible to arrange 10 hydrogen atoms side by side, their combined width would measure approximately 1 nanometer. A typical processor found in a modern desktop computer is roughly 10 millimeters wide and 10 millimeters long—in terms of nanometers, this is nearly 10 million nanometers in width and length! Other computers are commonly much smaller; embedded processors used today in cell phones, cars, and many other devices are as small as a fraction of a millimeter squared. Nanocomputers will be even smaller; electrons, atoms, DNA, and proteins are all nanometer scale or smaller and offer a huge variety of ways to represent and manipulate data.

At first this description may seem to be simply a matter of size. For more than 50 years, computers have continued to get smaller and faster, and so it may not be obvious that nanocomputing is drastically different than microcomputing. Looking deeper, however, the nanometer size opens a new world of possibilities. Nanocomputers will be able to fit anywhere, even inside our own bodies. They may be nearly undetectable, certainly invisible to the naked eye. Millions of such computers could work together and intelligently collect data about the world. Quantum physics makes it difficult to continue using old microcomputing techniques, but it also gives us infinitely more possibilities.

With all this in mind, **nanocomputing** can be defined as *the study of devices, paradigms, and applications that surpass the domain of traditional microcomputers by using physical phenomena and objects measuring 100 nm or less*. This definition clarifies that nanocomputing encompasses a large variety of challenges, ranging from effectively fabricating nanoscale devices to creating revolutionary applications for nanocomputers.

In the nanoscale world, it is unlikely that computers will work the same way they work today. Figure 1.2 lists some of the novel paradigms that can be realized with nanoscale physics, roughly estimating how they may be compared to each other given today's understanding of nanocomputing. As the figure shows, there is a vast amount of untapped potential computational power beyond CMOS logic, today's dominant technology.

Another important aspect of nanocomputing is the new set of applications that will be possible with such tiny, powerful computers. In turn, the plethora of applications raises ethical, social, and economic questions that are also of great interest.

The applications and impact of nanocomputing will be discussed later in this chapter, but let us first turn our attention to the fundamental question stated above, that is, exploring how physics can be used for computation. Over the next several sections we will discuss this idea, providing a historical context and generic taxonomy of nanocomputing topics that are detailed in the rest of this book.

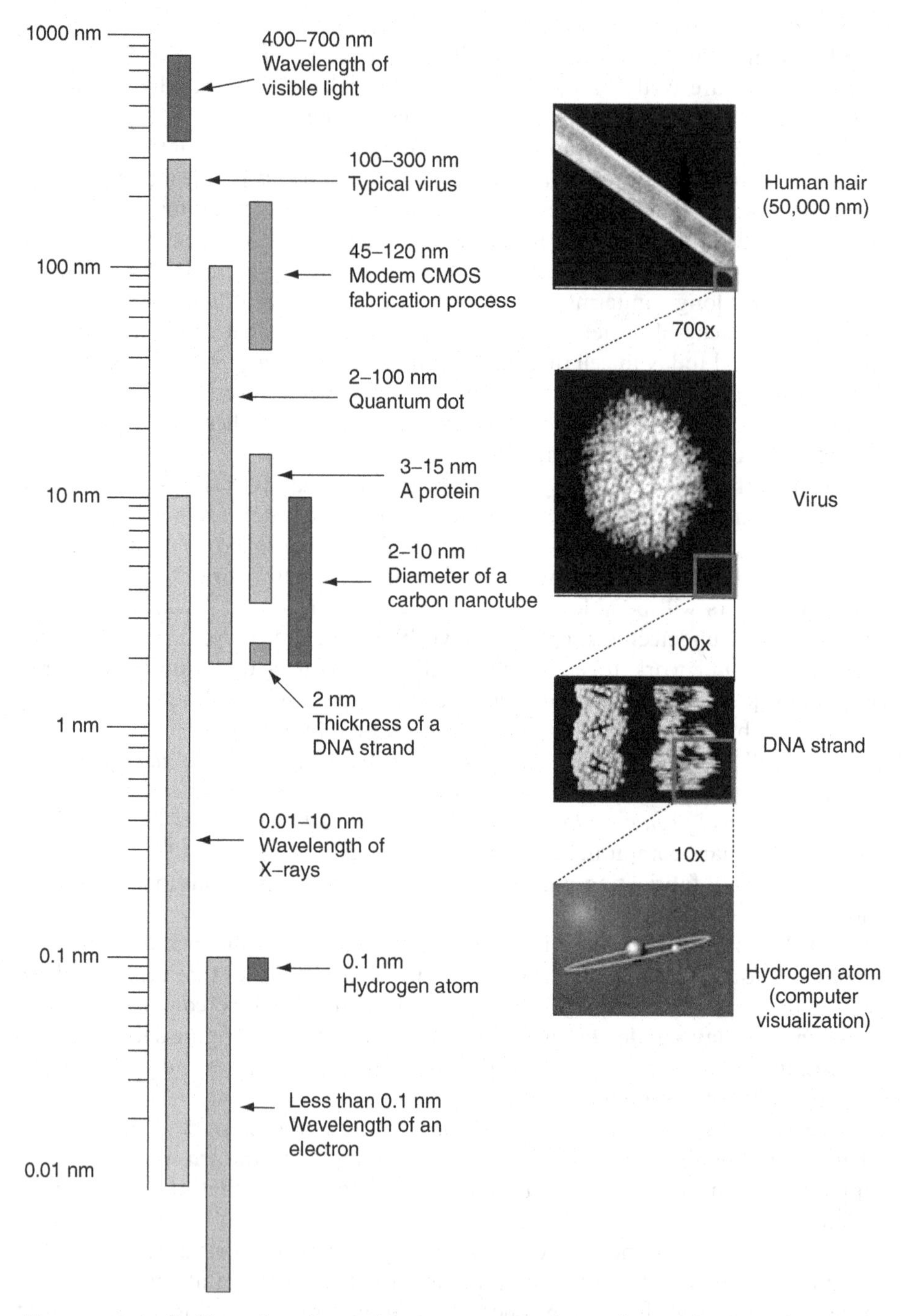

Figure 1.1. Left: Size of various objects, measured on a logarithmic scale. Right: Visual depiction of some of these objects, to compare relative size. Permissions obtained for the DNA strand from CalTech. The human hair and the virus pictures were taken from the Wikipedia public domain.

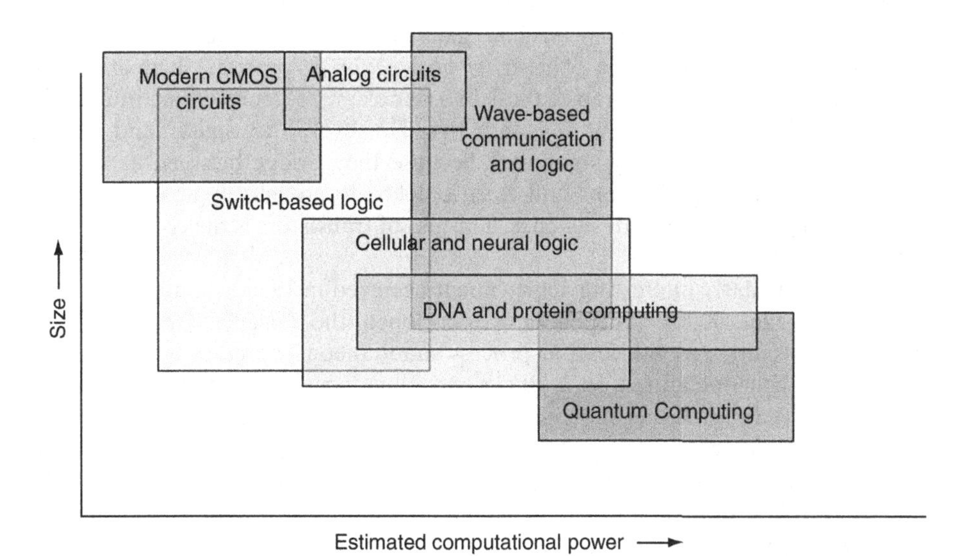

Figure 1.2. Visualization of how future paradigms in nanocomputing may compare to today's CMOS technology.

1.3. THE MICROCOMPUTING ERA: THE TRANSISTOR AS A SWITCH

Traditionally, the most common way to use physics for computation is to cleverly control electricity. Figure 1.3 shows a simplified *transistor*. We can add or remove electrons from the gate. When there is no charge in the gate, the wire can easily transmit its own electrons. If there are electrons in the gate, an electric field of negative charge is created, and this repelling force makes it difficult for electrons to flow through the wire. In a sense, we can control how much current flows through the wire by controlling how much charge we put into the gate.

With this physical device, an abstract 0 or 1 is represented as a low or high current on the wire. This is known as the *digital abstraction.* The transistor's

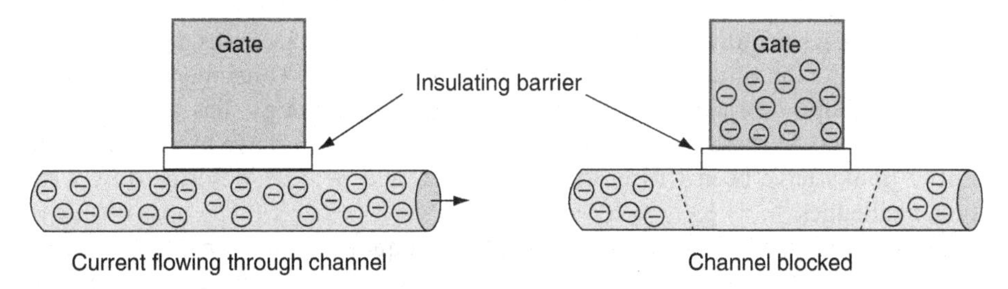

Figure 1.3. A simplified field-effect transistor. Ideally, the gate can "switch" current on or off.

behavior represents a simple *switch*: the gate can allow or prevent current from flowing through the wire. This is the *switch abstraction*. In practice, there are many more abstractions placed on top of these two (for example, representing integers in binary form with a series of 0's and 1's). However, the digital and switch abstractions are particularly significant because they bridge between a physical phenomenon, moving electrons and electric fields, to an entirely abstract world, manipulating 0's and 1's with switches. This use of transistors is the cornerstone of modern computing.

One particularly interesting achievement occurred in 1959, when both Robert Noyce and Jack Kilby independently developed the *integrated circuit*. With integrated circuits, one fabrication process simultaneously creates many transistors, all of them integrated on a single crystalline structure such as silicon. As fabrication techniques began to improve, it became possible to pack more transistors together. By 1965 Gordon Moore, the co-founder of Intel, predicted that the number of transistors that fit into a given area would double every 18 months due to continued improvements in the fabrication process. Following this prediction known as Moore's Law, transistor size, speed, and power consumption have exponentially improved for almost 50 years. Today it is possible to construct hundreds of millions, even billions, of tiny transistors on a small piece of silicon the size of a thumbnail (Fig. 1.4). In turn, it has become practical to create abstract computers that use millions or billions of switches.

Because the fabrication process produces all transistors simultaneously, the cost of fabricating these computers is largely independent of the number of transistors. There is typically a large initial cost, and this initial cost can be amortized over thousands or millions of processors, which can be produced cheaply. The economics of this situation is staggering—with a smaller transistor, performance improves, power consumption decreases, more abstract computation fits onto a single processor, and all this happens as the price of each transistor decreases! With this persistent exponential improvement, it is very easy to manipulate large amounts of abstract information, and computers are used for a prolific number of applications today. All of this has hinged on the fact that transistors continue to get smaller, and this has led to the general trend that "smaller is better."

1.3.1. Difficulties with Transistors at the Nanometer Scale

Transistor sizes are already at the nanometer scale, and this causes many practical difficulties. At the time of publication of this book, many consumer products are using a 45 nm fabrication process, and 32 nm technology has already been demonstrated. At these small sizes, fundamental limitations have to be considered. Entire books have been written on the subject, and here we describe only a few such challenges.

One primary example of these difficulties is a quantum phenomenon known as *tunneling*, visualized in Figure 1.5. Due to the wave nature of particles, electrons can "jump," or tunnel, through barriers with some nonzero probability. This probability increases exponentially as the size of the barrier decreases. The size of

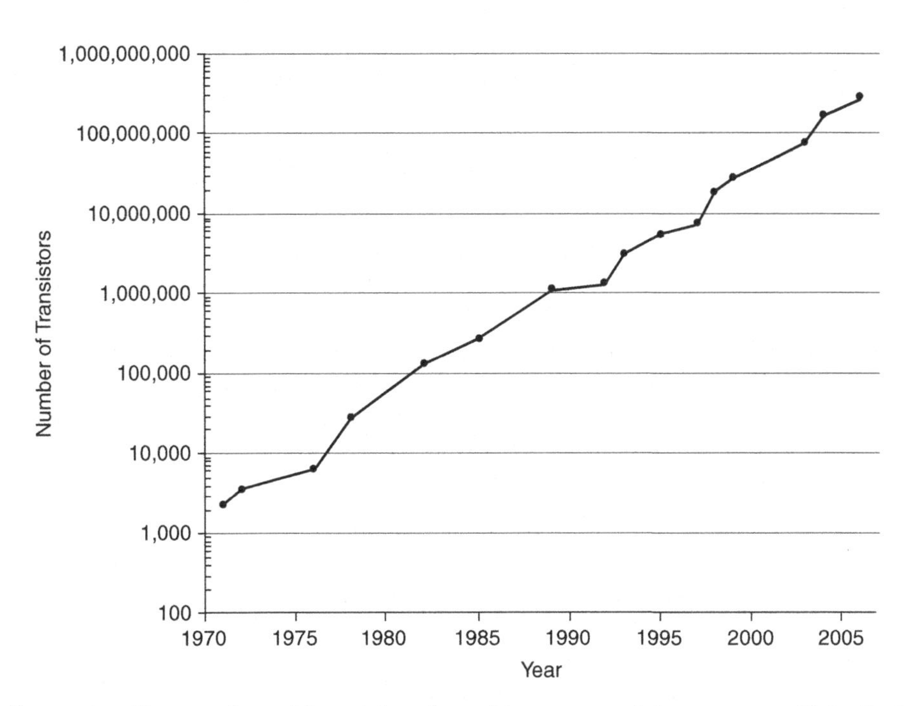

Figure 1.4. The number of transistors found in commercial processors. Note that the y-axis is logarithmic. This exponential trend is mostly due to the decreasing size of transistors. It is expected that very soon transistors will become as small as physically possible, which motivates the exploration of other devices that may be smaller and more powerful. Data acquired from [2].

transistors has decreased so much that in today's tiny transistors, electrons regularly tunnel between the gate and the wire. (Fig. 1.6). Since electrons and charge cannot be controlled as easily at the nanometer scale, the transistor behaves less and less like an ideal switch.

Tunneling has become part of a larger tradeoff between performance and power consumption. The size of transistors has reached the point where traditional models of transistors cannot be applied without a detailed understanding of nonideal characteristics [3]. There are many reasons that electrons can unintentionally leak across the wire, even when the gate tries to block current. Furthermore, the smaller the wires become, the more difficult it becomes for electrons to move through wires; that is, thinner wires have greater resistance. Because of this, even more power is required to push electrons through the wires quickly. Most processors today are limited to about 4 GHz, largely because power requirements beyond this speed are too costly and generate too much heat for a processor to function properly.

Many creative solutions have kept transistors useful despite these limitations. For example, by placing the appropriate stress or strain on the crystalline

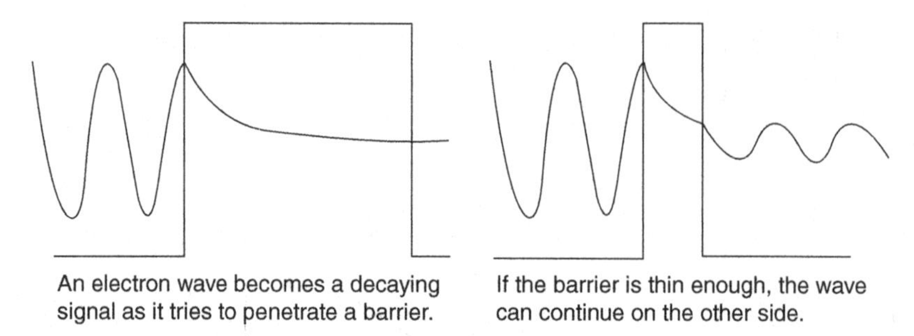

Figure 1.5. Visualization of the tunneling phenomenon. Mathematically, the wave changes into an exponential decay when it enters a region of "high potential"—the barrier—and resumes as a wave after it exits. The probability that an electron will tunnel across the barrier is related to how much amplitude the wave has left once it exits the barrier.

structure of silicon, electrons can move through a transistor more easily. This can be done by adding materials on top of transistors that naturally want to bend, thus pulling or pushing on the silicon. The so-called *strained silicon* [4] has quickly become a standard technique to improve the performance of transistors at 90 nm or less. Another example is the development of better insulating materials, known as *high-K dielectrics* [5]. The right combination of conducting and insulating materials can reduce the amount of undesirable tunneling between the gate and channel, even when the barrier is only a few layers of molecules thick. This advancement has been the key towards 45-nm technology. In the future, it may be necessary to use multiple gates to reliably control the current along a wire. *FinFETs* [6] or *trigate transistors* [7] are two multigate variations of transistors that may take us beyond 45 nm.

There are several more limitations when using tiny transistors that motivate the nanocomputing ideas presented in this book. First, the wiring that interconnects transistors is becoming a very signficant limitation for performance, power, and size of devices. There are even theoretical limitations about how much

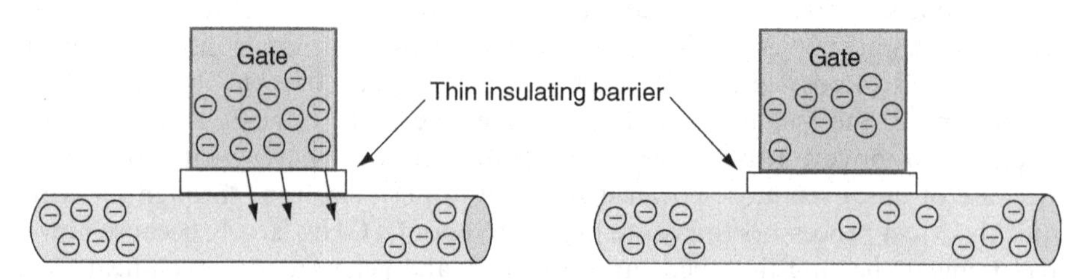

Figure 1.6. One of many nonideal effects in a transistor is that electrons in the gate may tunnel into the wire. This occurs more often as the thickness of the insulating barrier decreases.

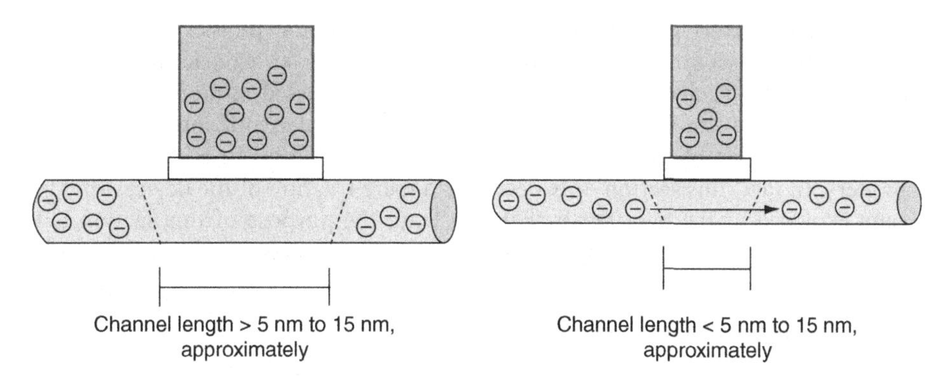

Figure 1.7. When transistors become very tiny, electrons can tunnel across the channel when the gate tries to block current flow. If the channel length is small enough, electrons will regularly tunnel in this way, and the gate would no longer effectively control current flowing through the wire.

area interconnections require as long as we connect transistors with traditional wires [8]. As we will see in the next few sections, there are many ideas in this book that reduce the limitations of wiring.

Second, variations during fabrication are now becoming a very significant problem. Relative to such tiny transistors, variations in geometry or chemical concentrations can easily change or break the behavior of the transistor. This variability decreases the yield and reliability of devices. *Fault-tolerant methodology* (Chapter 10) is desired for computing under unreliable conditions, and new fabrication methods, such as *self-assembly* (Chapter 12), may be better for reliable fabrication at the nanometer scale. Furthermore, *reconfigurability* (Chapter 5) offers a way to keep a device useful by updating or fixing its functionality.

Finally, when transistors become very small (below 5 to 15 nm approximately), electrons will be able to tunnel in a different, much more challenging way: electrons would be able to tunnel through the channel itself, even when the gate tries to block current, defeating the purpose of a gate entirely (Fig. 1.7). It is currently not clear how to overcome this upcoming problem, except to find a better nanoscale device that can behave like a switch [9].

1.4. BEYOND THE TRANSISTOR: NANOSCALE DEVICES

In practice, the use of transistors has been so successful that so far it remains unchallenged as the "best way" to use physics for computation. However, as mentioned above, it is not clear that the transistor will continue to be the best device to use as a switch at the nanometer scale. One major facet of nanocomputing research is finding new devices that exhibit switching or other behaviors that are useful for computing. Unlike the classical transistor, these devices very directly embrace the properties of quantum physics to serve their function. In this section, we briefly describe various nanoscale device technologies, referring to the specific chapters where topics are discussed in more detail.

It should be noted that this introductory material is not intended to be a comprehensive list of nanoscale devices; such information can be found in later chapters. In fact, this section only describes a mere fraction of the devices that are being explored at the nanometer scale. Instead, the purpose of this section is to give an intuitive understanding for several common aspects of nanoscale devices.

1.4.1. Molecular Devices

In general, there are a huge variety of molecules and structures that can be explored (for example DNA, proteins, rotaxanes, nanotubes, and more) [10]. In some sense, atoms and molecules are just highly complicated toy blocks: there are an infinite number of ways to assemble molecules into something useful, limited only by the creativity of future research.

Molecular structures can be used to create very tiny switches, ranging from 1 to 10 nm in size. One possible approach is to control how easily electrons can flow through the molecule, very much like a transistor, but with different underlying physics (e.g., [11]). Another possible approach is to control how light is absorbed or scattered by the molecules (e.g., [12]). These interactions with molecules can be controlled in many different ways, for example, by applying a nearby voltage or by changing the structure of molecules. Molecular switches and molecular computing are discussed further in Chapter 11.

A big challenge with molecular switches—and many nanoscale devices—is to effectively fabricate and interconnect them to perform complex logic functions. In an attempt to circumvent these problems, one proposed molecular device is the *NanoCell* [13]. The NanoCell tolerates defects and variability that occur during self-assembly fabrication. To provide reliability, the NanoCell depends on post-fabrication "training" to create the desired logic function. This approach is interesting for two reasons. First, the logic function of the NanoCell can (ideally) be reconfigured instead of permanently fixed; second, it allows the use of larger and fewer wires to connect between different cells. The function implemented by a single cell would be equivalent to using many transistors, thus simplifying the arrangement of large-scale computations.

1.4.2. Nanotubes

One interesting class of molecular devices is nanotubes, particularly *carbon nanotubes*. Recall that pure carbon has two common crystalline forms: diamond, where carbon atoms form a three-dimensional structure; and graphite, where carbon atoms form flat sheets that can easily slide and peel from each other. A single-wall carbon nanotube (Fig. 1.8) can be visualized as a single sheet of graphite rolled into a tube (though it is not created in this way), with a diameter of only a few nanometers.

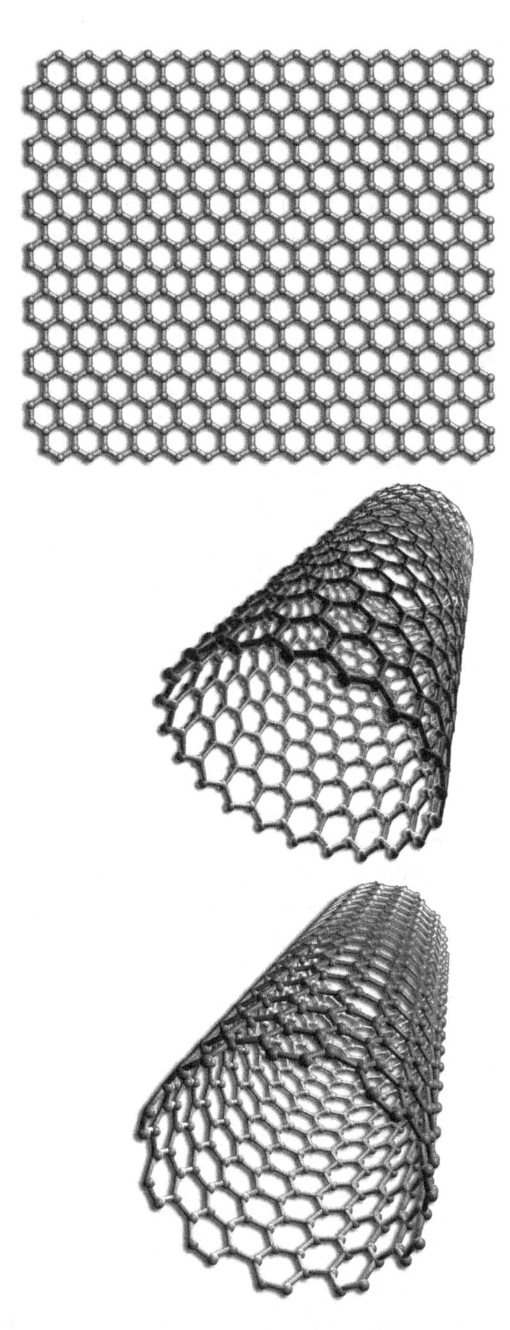

Figure 1.8. Visualization of carbon nanotubes (CNT), an exciting class of molecular devices. CNTs can be visualized as a sheet of graphene (top) rolled into a tube, though they are not created this way. Note that these two nanotubes have slightly different configurations. Various configurations of nanotubes can result in drastically different physical properties. Images created by Gabriele Nateneli.

Carbon nanotubes have very interesting mechanical and electrical properties. Mechanically, they are extremely strong and versatile. The tensile strength and stiffness of carbon nanotubes are extremely high, especially relative to their weight. Also, two tubes arranged with one inside the other can have a very low friction interface, similar to the way two flat sheets of graphite can very easily slide over each other. Therefore, two nanotubes can have very efficient telescoping and rotation motions, useful for nanomachines [14].

Electrically, a nanotube can potentially behave as a *ballistic conductor*. This means that an electron travels through the tube with small, quantized levels of resistance. The typical levels of resistance are much lower than traditional conductors. Nanotubes can exhibit properties of a metal or a semiconductor, depending on how the sheet of graphite is rolled into a tube. It has even been suggested that nanotubes can behave like a *waveguide*, guiding the wave-like properties of an electron similar to the way electromagnetic (optical) waves are guided through a fiber-optic cable [15]. All of these properties are being investigated for future switches and wires. In fact, switches, wires, and support structures have all been demonstrated with carbon nanotubes, but, as with many nanoscale devices, the ability to fabricate a practical nanoscale device with nanotubes and nanowires is still an open challenge. Carbon nanotubes are discussed further in several chapters in this book. See Chapter 2, Chapter 12, and Chapter 18.

1.4.3. Quantum Dots and Tunneling Devices

Many quantum phenomena occur when confining electrons to a very small space, such as the nanoscale range. For example, an electron confined to a small area can only have a select few discrete levels of energy, similar to the discrete levels of energy that an electron may have as part of an atom. When a group of electrons is confined in all axes of movement (i.e., in three dimensions), a *quantum dot* is formed. Similarly, a *quantum wire* is a group of electrons confined along a 1-dimensional line, and a *quantum well* restricts electrons to a 2-dimensional plane. These structures can exhibit properties similar to electrons in atoms or molecules, even if there is no nucleus of protons and neutrons. Their properties can be fine-tuned with more freedom than atoms or molecules, making them very interesting structures to use for computing.

Often, the phenomenon of tunneling, described previously, is combined with quantum dots, wires, and wells to create useful devices. This is in contrast to traditional transistors, where tunneling is very undesirable. Three such nanoscale devices are the *resonant tunneling diode* (RTD), the *single electron transistor* (SET), and *quantum-dot cellular automata* (QCA). An RTD is a device that has a quantum well where electrons can be confined; therefore, electrons in this region can assume only a few possible discrete energy states. When the energy of an incoming electron is close to one of these “resonant” discrete energy states, the electron can tunnel through with high probability. This device can emit extremely high frequencies, in the hundreds of GHz, making it interesting for high-speed applications. RTDs can also be arranged to perform logic functions, digital or otherwise. For example, RTDs have been used to implement cellular automata and cellular nonlinear networks [16, 17], two paradigms described below.

A single-electron transistor, or SET [18], operates on a principle similar to a conventional transistor: a gate can control the flow of electrons through a channel. However, unlike the classical transistor, in a SET the flow of a precise number of electrons is controlled. The device consists of a quantum dot between two barriers. When the gate has a negative electric field, the properties of the quantum dot are changed, effectively preventing electrons from entering (Fig. 1.9). When the gate does not block the flow of electrons, the space between two barriers accepts only a few electrons, typically allowing single electrons to tunnel through one at a time. In addition to quantum phenomena, an important mechanism that is generally dominant in SETs is the so-called coulomb blockade, which is essentially a classical effect, arising from the fact that charge is not continuous, but comes in packets of one electron each. The mutual repulsion of individual electrons, when confined to very small regions, leads to this effect. Currently, the smallest SETs are just as large as transistors, but it is expected that SETs will be able to shrink well beyond the limits of classical transistors. RTDs, SETs, quantum dots, and many other related devices are discussed further in Chapter 2.

Another interesting use of quantum dots is in *quantum-dot cellular automata* (QCA) [19]. A single QCA cell is a container that has several quantum dots. Electrons in this container tunnel between the dots in order to try and find the "ground state," that is, the state where the cell is at its lowest energy. A QCA cell has only two stable states, as shown in Figure 1.10. With two states, a QCA cell can realize the digital abstraction (i.e., logic 0's and 1's). However, there is no switching behavior in this concept. Instead, QCA are based on the property that adjacent cells prefer to have the same ground state. At the inputs, the cells can be constrained. Depending on how cells are arranged, the constrained inputs will propagate in

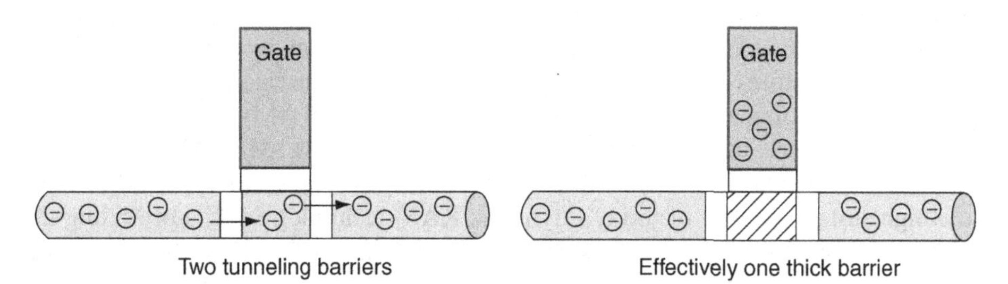

Figure 1.9. Conceptual illustration of a single-electron transistor (SET). A quantum dot (small confined space containing electrons) exists between two tunneling barriers, and electrons can tunnel in and out of the dot, one at a time. When the gate has electrons, the quantum dot changes and no electrons can flow from one side to the other.

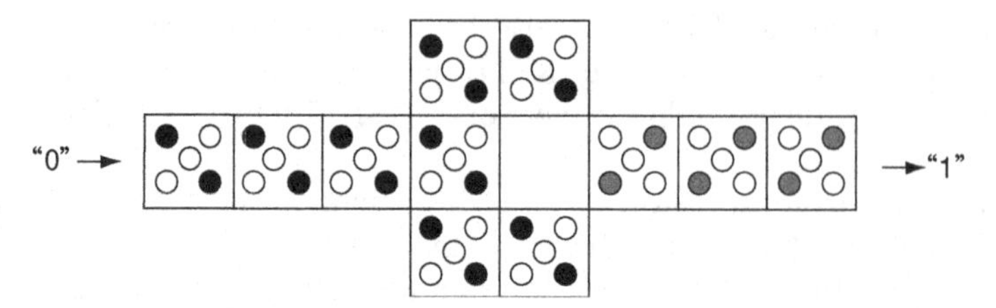

Figure 1.10. Depiction of quantum-dot cellular automata (QCA). Each cell has several quantum dots (in this case five). A cell can have two possible states, where electrons are diagonally oriented. When the input value is changed, the change propagates over time, this particular arrangement of cells represents a logic function that inverts the input.

different ways, allowing for many interesting logic functions. Since there are no wires interconnecting various cells, they can be arranged very compactly, in turn allowing for compact logic functions. Note that quantum dots are only one of serveral ways to implement QCA cells. QCA are discussed in Chapter 4.

1.4.4. Spin Devices

Electron spin is another interesting quantum effect that can be used to create nanoscale switching devices. Various particles can have different types of spin, but electrons in particular can have only two types: *spin up* and *spin down*. This is a natural way to introduce the digital abstraction. Furthermore, electron spin is a main nanoscale property that results in macroscopic magnetic fields. For example, if most of the electrons in a metal object assume a polarized spin state (either all spin up or all spin down), the metal object will be magnetic. One interesting phenomenon that is a result of the relationship between magnetism and electron spin is the *magnetoresistive effect*: the resistance of some materials can change depending on the surrounding magnetic field.

Magnetoresistance is already widely used as the mechanism to read data from a disk drive that stores information magnetically. Manufacturers are also considering the possibility of magnetic random-access memory (MRAM), which would use magnetoresistance to implement nanoscale memory cells. MRAM may have a number of advantages over other types of memory storage. It is expected to have the high storage density of today's dynamic RAM technology, while providing the high speed and power savings of today's static RAM technology. Magnetoresistance can also be used to create a switching device. This sort of switch is called a *spin transistor*, or *transpinnor* [20]. Recall that a classical transistor uses an electric field to make it difficult for electrons to travel across the wire. Similarly, a spin transistor uses magnetoresistance to drastically increase the resistance of the wire, effectively blocking current like a switch. Magnetic storage is discussed in Chapter 6, and spin devices are discussed further in Chapters 7 and 9.

1.5. BEYOND THE SWITCH ABSTRACTION: NANOSCALE PARADIGMS

The use of switches has fundamental limits. At some point, researchers may achieve a switch that is plainly as small as possible. This would mean that the exponential decrease in size of transistors would stop—the end of Moore's Law. It would be desirable to find a new way of continuing Moore's Law, not in the literal sense that transistors could get smaller, but rather that technology could continue to exponentially improve. Furthermore, the complexity of interconnecting switches is a fundamental limitation; even today, the wires are becoming the limiting factor when making high performance nanoscale transistors. In fact, there is no indication that switches, binary logic, or transistors are the best way to compute.

This motivates the search for entirely new paradigms of computing that could eventually replace the digital and switch abstractions. Of course, the common thread in the ideas we consider is that they can be realized at the nanoscale level. Here we briefly describe some of the major paradigms of computing that are part of nanocomputing research, as well as mention the corresponding chapters where they are discussed further. As before, this section is not intended to be comprehensive; instead it aims to inspire the reader by illustrating some novel approaches to computing and how drastically different they are than today's computing technology.

1.5.1. Cellular and Neural Logic

One paradigm is the use of *emergent properties*. The idea is to use a large number of extremely simple processors. Alone, each processing element can be extremely small, and it can do only a limited number of trivial functions. However, powerful computers "emerge" from a group of such processors. Cellular automata, cellular nonlinear networks, and artificial neural networks are major examples of this paradigm. *Cellular automata* (CA) [21, 22] can be understood as a regular grid of identical cells, where each cell changes its state depending on its own state and the state of its neighbors (Fig. 1.11). *Cellular nonlinear networks* (CNN) [23] are similar to cellular automata, the main difference being that each cell in a CNN is an analog processor. *Neural networks* [24], unlike CA or CNN, use simple units that simulate the behavior of biological neurons.

Figure 1.11 shows an example of how a behavior can emerge from a cellular automaton. At each step, every cell changes depending on the previous state of its neighboring cells. The pattern shown in this figure repeats every four steps, but gradually moves diagonally as well. This diagonal movement is a property that emerges from the group of cells. The rules that define how all cells behave are very simple, but the result is a rich, complex world of emergent properties that are being studied by many mathematicians and scientists and can be used for computation.

In nanocomputing, the idea of having a huge number of very simple nanoscale structures is appealing for many reasons. The challenge of how to fabricate such

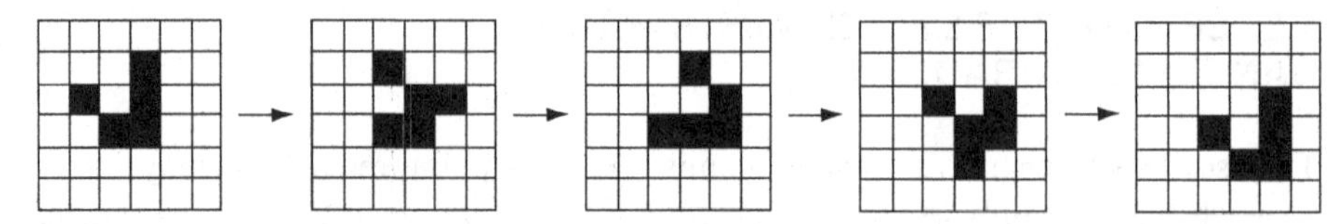

Figure 1.11. Example of a simple cellular automaton. Each cell can be "on" (black) or "off" (white). Depending on its neighbors, a cell may turn on or off in the next time-step. This specific example shows a "glider," a structure that emerges from rules defined by the famous "Game of Life" by John Conway [25]. As time goes forward, the glider moves diagonally. More complex cellular automata can compute anything that a traditional computer can, and possibly more.

a computer becomes much easier because the same nanoscale structure can be repeated in a regular pattern. Most of the ideas in this paradigm are also fault-tolerant, that is, able to handle a few defective cells or neurons. This alleviates the problem of variability when fabricating nanoscale devices. Their regular structure usually implies that they can perform different functions depending on the context, instead of being permanently hard coded with a fixed function; in other words, such computers are highly reconfigurable. Most importantly the potential computing power that is available with emergent properties is immense and only beginning to be explored.

Even with the limited understanding of emergent properties, this paradigm already has many applications. Cellular automata and cellular nonlinear networks have been mostly used for image processing applications due to the highly parallel nature and intuitive correspondence between each simple processor and pixels on an image. Neural networks have been extensively studied for their applications in artificial intelligence and are useful practically anywhere uncertainty is encountered in computation. It has also been shown theoretically that cellular automata and neural networks can do anything that today's computers can do [26, 27]. With all this in mind, one of the main challenges of this paradigm is to find a way to harness emergent properties for a wider variety of applications.

Nanoscale devices that implement cellular automata, cellular nonlinear networks, and neural networks are actively being researched. NanoCells, quantum-dot cellular automata (QCA), and RTDs, described in the previous section, are just a few possible ways to realize this paradigm. Quantom-dot cellular automata are discussed in Chapter 4, and nanoscale neural networks are discussed in Chapter 17.

1.5.2. Wave Computing

Waves are an elegant but complicated way to communicate and manipulate abstract data. One of the most powerful features of waves is the phenomenon of *diffraction*, or the behavior of light as it propagates around objects or through a nonuniform medium. Perhaps the best known example of the power of diffraction is seen—literally—in holography. To display a hologram, light waves are diffracted through patterns that were previously recorded. The diffraction of light actually reproduces all the waves of light that were originally recorded. Because of this, visual holograms are well known to have accurate details and amazing realism. Holography has found many other important engineering applications because diffraction can manipulate light waves in a flexible, powerful way.

Equally important is the *superposition* of waves. Consider two waves that are traveling in opposite directions (Fig. 1.12). The waves continue with the same direction and speed, unaffected by each other. However, if desired, one could measure the *intensity* of the point where both waves cross. The propagation of the waves remains unaffected, but the intensity measured at a point where two waves overlap would be the sum of both waves, a result of constructive or destructive interference.

This provides a convenient way to overcome the significant limitation of wiring: use waves for communication instead of particles. Because of superposition, waves can cross over each other without destroying the information being carried. In addition to guided waves, one could also use waves in free space, removing dependence on wiring. This may seem like a small matter at first, but as shown later in this book, it allows for significant improvements in *theoretical algorithmic performance* of computations. Note that improving algorithmic performance is usually more beneficial than simply making a processor "smaller and faster."

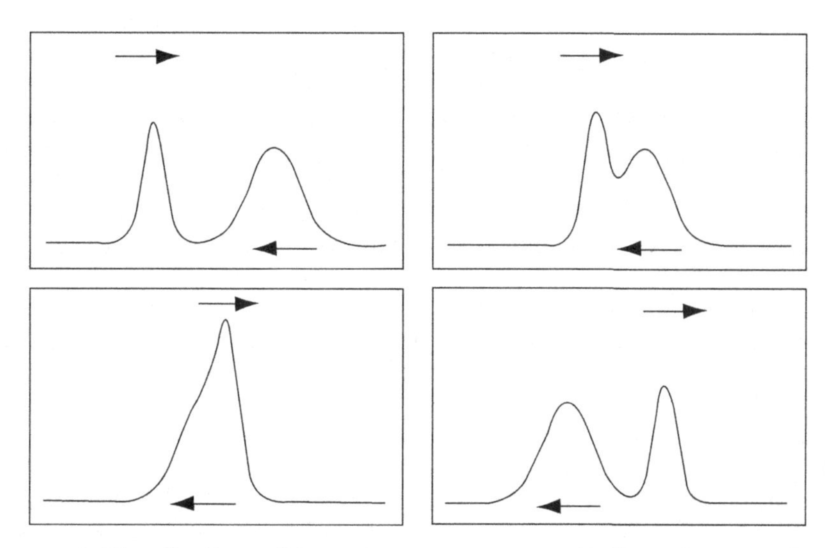

Figure 1.12. Visualization of how two waves cross. Unlike wires carrying an electrical signal, waves can occupy the same space while propagating. However, even though propagation over time remains unaffected, as they cross, the waves do interfere.

In addition to being used for communication, waves can also be used for computation by using interference. For example, constructive and destructive interference can result in either a high intensity or low intensity—introducing the digital abstraction without any switches. Furthermore, several waves can be combined at a single point, an occurence that can be advantageous over traditional logic that requires many transistors to handle many inputs. By combining the benefits of waves over wires with the functionality of waves over transistors, many limitations, theoretical and practical, can be overcome.

The visual portion of the electromagnetic spectrum has wavelengths on the order of hundreds of nanometers, slightly larger than the nanoscale level. The principles of diffraction and interference, however, apply to any type of wave, including X-rays and electrons that have nanoscale or smaller wavelengths (recall from Fig. 1.1). Another useful nanoscale wave, known as a *spin wave*, occurs when the spin state of previous electrons affects the spin state of nearby electrons, causing a propagation of the change in magnetic field (recall that magnetism is the macroscopic property of spin). In addition to the many benefits of wave computing described above, a key benefit of spin waves is that they can conveniently communicate with electronic devices as well. Spin waves for computation are described in Chapters 7, 8, 9, 14, and 19.

1.5.3. DNA and Protein Computing

DNA and protein are nanoscale molecular structures found in almost all existing biological life that we know. Recall that DNA is essentially a sequence of four primitive molecular structures: adenine (A), cytosine (C), guanine (G), and thymine (T), attached to a molecular "backbone." The main property that can be exploited for computing is that adenine bonds only to thymine, and cytosine bonds only to guanine. This means that a sequence of these base pairs has exactly one complementary sequence that will bond to it. By setting up specific sequences of DNA, many clever interlocking "tiles" can be created, and the way these tiles interact is used for computation.

Proteins are complex molecules comprised of a string of amino acids. Biologically, the sequence of amino acids that create a protein is defined by a sequence of DNA. Various proteins can interact to perform useful computations, for example, by exploiting the way a protein structure folds. Proteins are much more complex than DNA, and their use for computation has so far only been simulated [28].

The power of this paradigm is the principle of *constraint satisfaction*. Here, instead of representing abstract operations as physical phenomena, abstract rules are enforced (in this case with molecular structure), and physical phenomena (in this case, bonding between structures) do their best to find the lowest state of energy within the constraints. Performing constraint satisfaction in this way makes it possible to compute many things that are otherwise very difficult. The best example is a demonstration of DNA computing to solve the Hamiltonian Path Problem [29]. This is a well known problem in theoretical computer science for which the best known solutions currently take exponential amounts of time. Because of this high complexity, only very small, trivial inputs can be solved by a traditional computer. To solve a large exponential problem on a traditional computer can take years, or even decades, even with all the computing power available in the world! Many real world problems are also exponentially difficult, and being able to quickly solve these problems would profoundly impact the world. While DNA computing still performs an exponential amount of computation, by using constraint satisfaction of nanoscale physics, this paradigm can quietly bypass many of the limitations of traditional computers, allowing the solution of the Hamiltonian Path Problem to be computed in a tractable amount of time.

Note that constraint satisfaction is implicit in all physical phenomena used for computing. For example, a QCA cell in Figure 1.10 had two stable states because of the way electrons were constrained within quantum dots. The difference here is that the constraint itself is used for computation, rather than to make devices that can be used for computation. Another place that constraint satisfaction is useful is for self-assembly. By specifying tiles of DNA to connect in certain ways, the DNA can automatically assemble itself into structures that satisfy the constraints. DNA self-assembly has been demonstrated as a way to fabricate nanoscale structures [30, 31], which can then be used for other purposes (e.g., as a template to fabricate other integrated devices at the nanometer scale).

The challenge with DNA and protein computing is to manage these complex molecules. Even though DNA computing solves the Hamiltonian Path Problem, extracting the computation from the reactions takes many hours of manual effort, and the constraints to set up do not scale well. For proteins, it is not yet fully understood how they fold to develop complex structures. The number of permutations with sequences of base pairs or amino acids can be astronomical, and finding sequences that result in useful structures is like trying to find a needle in a haystack. As future research overcomes these challenges, DNA and protein computing may become an essential paradigm to solving very difficult problems. Such topics are discussed further in Chapters 13 and 14.

1.5.4. Quantum Computing

Further on the horizon of nanocomputing is the very powerful yet challenging field of quantum computing [32]. Recall that the fundamental question we have been discussing is how physics can be used for computation. All the ideas discussed above, even though they use properties of quantum physics, are only used to represent intuitive, essentially "classical" abstractions, such as digital 1's and 0's. Quantum computing, on the other hand, uses the general principles and mathematics of quantum mechanics as the abstraction itself.

In this paradigm, the leap from physics into an abstract world is accomplished by a *qubit*: the quantum analogy to the digital abstraction. Recall that a classical bit can have one of two possible states, usually denoted as 0 and 1. A qubit also has two such states, often denoted as $|0\rangle$ and $|1\rangle$, but in this case the qubit actually is a combination of both states simultaneously! This combination of two states, however, is something that cannot be measured directly. When trying to observe the state of a qubit, we receive a $|0\rangle$ or $|1\rangle$ with some probability. This probability is the actual information contained by a qubit, and it is almost as if nature teases us by making this information impossible to measure. One of the key tasks of quantum computing is to manipulate this hidden information in such a way that it is meaningful to measure the qubit as only a $|0\rangle$ or $|1\rangle$.

The states $|0\rangle$ and $|1\rangle$ are actually just one "frame of reference" in understanding the actual underlying quantum system. One of the interesting powers of quantum computing is that we can freely decide which frame of reference we want in which to manipulate the qubit. For example, we could view the same qubit as a combination of $|+\rangle$ and $|-\rangle$, two other states that give us a different way of looking at the same qubits. Even more peculiar, two or more qubits may be completely unrelated in one frame of reference, but in another frame of reference, the qubits become *entangled.* This means that a change in one qubit will unavoidably affect the other qubit, and even though this complicates matters, it allows a powerful way to manipulate multiple qubits. Often, a useful quantum circuit first manipulates qubits in one frame of reference, where they are entangled, and then uses another frame of reference where the state of the qubits can be observed.

Perhaps the best known example of the power of quantum computing is its use in finding the prime factors of a given number. Traditional algorithms can take extremely long for large numbers. A quantum algorithm known as Shor's Algorithm [33] uses quantum computing to factor prime numbers. A physical implementation of this algorithm has been demonstrated using 7 qubits, triumphantly factoring the number 15 into the prime numbers 5 and 3 [34]. While the number 15 is not large and it seems like a trivial task that could have been done with any other computer, the real landmark of this result is to demonstrate that quantum computers can indeed work as theoretically proposed.

With every next qubit, the amount of hidden information in a quantum system effectively doubles. With 7 qubits, there are 128 hidden "numbers" that represent a combination of 128 different states. With 20 qubits, there are more than a million such hidden numbers. It might take 30–40 qubits to represent computations that exceed the potential of traditional computers, and each qubit could possibly be represented by a single nanoscale particle! While this is currently a distant dream, the foundations towards realizing this dream are being studied extensively today.

Despite this awe inspiring amount of power that seems possible with quantum computation, there are many daunting challenges to be addressed before quantum computing becomes more practical. First, physically implementing a quantum computer is a tricky task. While it would be ideal to isolate a single quantum system in reality as we can do mathematically, in practice a quantum system also interacts with the rest of the world. Therefore, it is difficult to keep the *coherence* of a quantum system, where coherence is a measure of how long the quantum system can stay intact before it gets disrupted by the surrounding environment. On the other hand, quantum phenomena such as photons of light have a good coherence, but then it becomes difficult to get the photons to interact at all. Another problem is the difficulty of understanding how to develop algorithms for quantum computation. Some algorithms have been proposed that use quantum computation, and a few general computing framework are being proposed for using qubits in a practical setting. Chapter 3 gives a detailed history of the contributions in quantum computing as well as a discussion of its theoretical and practical limitations.

1.6. BIOMEDICAL APPLICATIONS

Recent developments in the field of nanocomputing have laid the groundwork for technology that will revolutionize modern medicine. The most important biomedical research in the latter half of the twentieth century, which culminated in the publication of the human genome, was driven by an understanding of DNA, our genetic code. Similarly, nanotechnology has the potential to usher in an age of nanomedicine, creating a paradigm shift in the way we study and treat disease.

This technology will not come without a heavy price. There are obvious financial impediments and technical challenges, but moral and ethical concerns will also play an important role in the development of this field. An excellent comparison can be drawn with genetically modified foods. A significant amount of the produce and livestock grown in the United States has been subject to genetic engineering. While many consider these modifications to be safe, some are still skeptical. Various nations, particularly some in the European Union, are hesitant to embrace such technology because of a fear that genetically modified foods are inherently unsafe and may damage local ecosystems. Some say nanomedicine may suffer a similar fate. Will governments and health care professionals trust and endorse this technology? Will individuals be comfortable with nanoscale devices circulating through their bodies?

Regardless of the public's willingness to accept nanomedicine, few will dispute its potential to revolutionize the biomedical research. By providing scientists with new techniques for targeting and attacking virtually every human ailment, nanocomputing will usher in an age of medicine in which physicians and scientists can treat disease at a molecular level and attack it in a way never thought possible. As nanotechnology provides more versatile tools, the rules of engagement for diseases will change. Physicians and scientists will no longer be hindered by the small size and tremendous complexity of the human cell, but rather utilize these features to develop therapies that are more specific and effective, producing better outcomes with fewer side effects.

1.6.1. Vaults

The future of nanocomputing knows no bounds and its merger with biomedical research provides unlimited pathways to discovery. It is difficult to imagine an area in which there is more promise and a greater potential to revolutionize the scientific field. It is also important to emphasize the importance of cooperation between scientists, physicians, and engineers to ensure the success of biomedical and nanotechnology related research.

Perhaps the most intriguing development in this field was the discovery of nanoparticles called vaults. Groundbreaking work on vaults was performed at the University of California, Los Angeles, where these nanocapsulses have fascinated scientists since their discovery [35]. Current studies have shown that vaults are found in all eukaryotic cells and are composed of protein and RNA [36]. Through precise genetic manipulation, scientists predict that vaults may be used as structural support for nanomachines as well as integral parts of nanocircuits. Perhaps most appealing is the potential of vaults to serve as vehicles for drug delivery. These nanocapsulses may one day deliver precise amounts of drug to specific cells in the body, increasing their efficacy and eliminating the potential for certain adverse reactions. While this type of technology seems to be straight out of a science fiction film, it is indeed very real and has a tremendous potential to usher in the age of nanomedicine. Vaults may also be used as biological sensors, detoxification centers, and aid in environmental restoration [36]. Equally important will be their contributions to biomedical research as a whole. It is impossible to predict the full potential of vaults but they may revolutionize drug delivery, treatment of disease, and fundamentally change the way we practice medicine.

Imagine for a moment a day in which vital signs, blood chemistries, and even disease progression can be monitored remotely by nanomachines. These safe and affordable nanorobots would be capable of transmitting data to a local physician and may even calculate complete blood counts, cholesterol levels, and search for invading pathogens. Some speculate that such robots could also be used to treat heart attacks and strokes by analyzing and neutralizing blood clots that pose a threat to the patient. Nanotechnology can provide physicians with more accurate and less invasive techniques for treating everything from the common cold to the most severe and debilitating diseases. Medicine would never be the same.

1.6.2. Molecular Motors

In analyzing the problem of providing power for future nanocomputing devices, researchers are exploring the use of molecular motors. These motors, instrumental in the functioning of biological systems such as muscle contraction, will allow for the movement of nanorobots within organisms. All motors consume a form of energy to perform work; in the case of molecular motors, it is some form of chemical energy, such as ATP [37]. Molecular motors are attractive because they are smaller and more efficient than any other man-made motor [37]. Numerous molecular motors exist, the most well-known being the proteins myosin and kinesin. Myosin lies along actin filaments in muscle cells and utilize a single ATP molecule per cycle to perform a power stroke [38]. Kinesin carries cargo in the intercellular space and uses 1 ATP to move 8 nm. The development of effective and reliable molecular motors will be essential for the utilization of nanorobots in biomedical applications.

A focal point of such research is treating motor neuron diseases, such as Amyotrophic Lateral Sclerosis (Lou Gehrig's Disease) and other types of atrophy [38]. Motor neuron diseases typically result in the degradation of neurons that control voluntary muscle, which are critical in performing tasks such as speaking and swallowing. Researchers have discovered a mutation in a molecular motor gene that leads to the buildup of improperly folded proteins in the cell. Many hypothesize that nanorobots and molecular motors may prove useful in preventing such degradation by restoring normal function to the cell and preventing the buildup of protein [37]. Scientists are currently attempting to construct nanoscale devices for this application, but they face many challenges. By studying molecular motors, they hope to discover ways for powering and mobilizing future nanodevices.

1.6.3. Nanorobots

Based on the existence of vaults, it is clear that nature has created its own nanoparticles. But what about man-made nanomachines? As nanocomputing advances, the field of nanorobotics is sure to progress as well. Approximately 10 years ago, the first theoretical design of a nanorobot for medicinal purposes was presented to the scientific community. The device utilized 18 billion precisely arranged atoms to form a diamonoid vessel with active pumping capabilities [41]. This has the potential to deliver over 200 times more oxygen to tissues than red blood cells. Theoretical designs also exist for synthetic white blood cells that are able to digest blood-borne pathogens. These nanorobots would have the ability to operate faster and more reliably than naturally occurring white blood cells.

Other designs include nanorobots with platelet functions that will allow hemostasis in as little as a single second. This complex machine would be invaluable in treating patients with severe hemorrhaging, especially in traumatic injuries. Perhaps most intriguing is the idea of a chromallocyte, a hypothetical mobile repair nanorobot capable of performing chromosome replacement therapy (CRT) [41]. This process involves the replacement of the entire chromatin content of a living cell with a prefabricated set of error-free chromosomes. This may allow for the treatment of entire organs such as the liver or heart and will without a doubt revolutionize the way we treat disease.

These nanomachines will be the core of nanocomputing and nanotechnology's biomedical applications; they represent a fundamental change in the way engineers and doctors will communicate. Indeed, it is essential for researchers on both sides of this fascinating technology to exchange ideas and strategies if they are to fully utilize the potential of nanotechnology. Nanorobotics is discussed in detail in Chapter 15.

1.6.4. Pharmaceuticals

Pharmaceuticals are a multibillion dollar a year industry, evolving daily with the discovery and patenting of new drugs. Developers are constantly seeking stronger and more effective medicines that will also reduce side effects. Nanotechnology and nanocomputing stand as beacons of hope for fulfilling these goals as they encompass several areas of pharmaceuticals, including "discovery, development, delivery and even post-delivery" [39]. Currently, several short- and long-term projects are underway to revolutionize the industry, as well as talks with large pharmaceutical corporations [39]. The National Cancer Institute has also created a nanotechnology branch to allow companies to expedite the processing of their drugs. Presently, many of the benefits achieved by this technology are decreased toxicity and reduction of side effects [39]. Two anti-cancer drugs, Doxil and Abraxane, have had their adverse effects reduced nanoformulation. Other applications include improved targeting of drugs by both oral and parenteral means [40].

These present developments are minuscule compared to the limitless long-term applications of nanocomputing and nanotechnology. Three crucial applications are in the areas of design, delivery, and drug monitoring [40]. Monitoring the efficiency of pharmaceuticals remains an persistent obstacle to both pharmaceutical corporations as well as medical practitioners. The use of nanorobots composed of "diamondoid nanometer-scale nanosensors" may allow imaging after drug delivery [41]. This would enable physicians to consistently monitor patients and evaluate the efficacy of certain medicines over a broad spectrum of individuals. Additionally, by using recently pioneered nanodelivery systems, several drugs may be combined into a single "package." [42].

1.6.5. The Future of Biomedical Nanotechnology

The future of nanotechnology is bright and every day new and important advancements are made in the field of nanocomputing. The merger with biomedical research will bring in a new age of scientific development unlike anything we have ever seen. Like the biotechnology revolution of the 1960s and 1970s, biomedical nanotechnology will revolutionize research and provide an almost infinite supply of techniques for treating the most challenging ailments. Even the simplest tasks may be delegated to nanomachines, which make fewer mistakes and will monitor parts of the human body not possible by physicians. The brain, heart, liver, kidneys, and other vital organs may be under the constant watch of millions of nanorobots that can take precise and accurate measurements in real time. This data can be sent to a local computer where it is processed and transmitted to a healthcare professional for analysis. Doctors will change their approach to treating disease, and computing in nanotechnology will make diagnoses more accurate, treatments more effective, and lives more fulfilling. The technology is real. The potential is real. All we need is time. Chapters 15, 16, and 18 provide additional information.

1.7. NANOCOMPUTING AND NEUROSCIENCE

Nanotech applications in biology and medicine now allow for surgeons to induce desired physiological responses in the human body throughout the central nervous system (CNS). Pioneer work being done in this novel field may one day bring us numerous new therapeutic choices that hold much less risk for patients, as well as prove a more convenient means for surgeons to handle molecular machinery. According to Dr. Gabriel A. Silva, technological advancements must occur alongside clinical neuroscience advancements [43, 44] simply because of the highly interdisciplinary nature. An emerging field of neuroscience nanocomputing is the production of materials and devices designed to interact with neurons at the molecular level. The developing platform technology of nanowires that is to be discussed in this section may prove to have broad applications in neuroscience and, of greater importance, possess the potential to save lives much sooner than expected.

1.7.1. Nanomachinery: Opportunities and Challenges

Imagine wires that were hundreds or thousands of times thinner than the human hair, utilizing blood vessels in the body as conduits towards adjacent individual neurons. These are what we would call nanowires [45]. Dr. Charles M. Leiber, an interested researcher at Harvard University in the field of nanocomputing, invented a nanowire transistor that can detect, stimulate, and inhibit neuronal signals [46]. This gives rise to the question of what a nanowire is. In simple terms, a nanowire is a wire of dimensions in the order of a nanometer. These range in makeup, being either metallic, semiconducting, or insulating. Some previous technology was available in this area but was too large in size. Micropipette electrodes were previously available but were harmful to cells in that they destructively poked cells. By contrast, the tiny nanowire transistors developed by Lieber and colleagues gently touch a neuronal projection to form a hybrid synapse, making them noninvasive and thousands of times smaller than the electronics now used to measure brain activity [46]. In addition to being nonintrusive, nanowires can be biodegradable, biocompatible, and capable of producing diagnostic test results in minutes instead of days.

A great effort has already been invested in this nanomachinery, and a series of promising results have thus been revealed. One such opportunity is the silicon nanowires' precision in its detection of bioterrorism threats. When discussing neuropathological disease processes at a molecular level, scientists can observe that there is potential in a nanowire's ability to limit such disease processes with early detection. Unlike conventional DNA sensors, such nanowire techniques provide much more detailed information on the scale of neurons, as well as give a sharper focus of disease markers in perhaps any bodily fluid in humans [47]. Likewise with degenerative diseases such as Parkinson's and Alzheimer's, nanowires provide hope for treatment rid of damaging side effects (brain tissue scars) by stimulating the affected area of the brain with wires tinier than capillaries themselves. Many researchers envision nanowires connected to a catheter tube and able to be guided throughout the circulatory system to the brain, where nerve-to-nerve interactions will allow neuroscientists to make earlier diagnoses and provide earlier treatment without the cost of time consuming procedures.

Through electrophysiological measurements of brain activity, made possible by nanowires, important signal propagation through individual neurons and neural networks can be understood. Sophisticated networks between the brain and external prosthetic technology can be produced through this revolutionary manipulative technique. Much of this technology has great potential in the field because it can be used to monitor signaling among larger networks of nerve cells, thereby allowing doctors to detect electrical activity going on between neurons, tumors, and brain abnormalities; to localize seizures; and to pinpoint damage caused by injuries and stroke [47]. Eventually, the technology will be used to detect the diverse kinds of neurotransmitters that leap synapses from neuron to neuron. The mystery behind many neural system disabilities such as mental illnesses and certain paralysis diseases could be unraveled with this amazing invention in the scientific community.

Working at the molecular level with nanowires still has its shortcomings and is an incredible challenge in the field of neuroscience and nanocomputing. The extremely intricate composition of the CNS poses obvious challenges to nanocomputing's applications in neuroscience. Specifically, these include cellular heterogeneity and multi-dimensional cellular interactions which explain the basis of its extremely complex information processing [43]. There is also the challenge of guiding nanowire probes to a predetermined location among the thousands of capillary branches in the human brain that reside in the brain's vascular system. And because it is considerably more difficult to manipulate materials on the nanoscale level, it is also difficult to measure the electrical and mechanical properties of the nanomaterials themselves.

Along with developing the functions of engineered machinery to carry out neural regeneration, neuroprotection, and other tasks of the sort, there is an evident need for precise and proper synthesis of such machinery. These "tailored nanotechnologies" cannot provide any solution to neurobiological complications unless they are designed by the most skilled and competent specialists, which in this case is not the role of the neuroscientist [44]. We know that materials scientists, chemists, and specialists of other similar disciplines have, unlike neuroscientists, devoted their careers to the synthesis of such technologies. Neuroscientists in turn contribute to this interdisciplinary science through their wealth of knowledge in neurobiology, neuropathology, and other areas. In this book, an implementation of neural network with nanotechnology is studied in Chapter 17. Evidently, these challenges have the potential to improve what may have otherwise been overlooked in synthesizing machinery. Often such obstacles help us to be more focused on the safety behind clinical neuroscience advancements.

1.7.2. Current Work and Research

Functional nanotechnology, including nanocomputing, is still at its infancy stages, with numerous institutions of various scientific fields finding ways to make nanotechnology as safe and effective as can be. The government has given research grants to scientists from different universities such as Brown, Stanford,

MIT, and CIT in order to facilitate research in such a promising field. Leading researchers such as Dr. Charles M. Lieber of Harvard University have contributed greatly to the advancement of nanocomputing. Dr. Lieber, along with his associate Jong-in Hahm, Ph.D., recently helped start a company called NanoSys, Inc., which is currently in the process of developing nanowire technology and other nanotech products. Other researchers include neuroscientist Rodolfo R. Llinas of the New York University School of Medicine and Masayuki Nakao of the University of Tokyo [45, 47]. As the general public can see, the advent of nanotechnology is one that will affect the lives of people worldwide and not simply arrive to us as a packaged fad whose hype is short lived.

The great deal of research that is currently being done on the neurological applications of nanotechnology was bolstered when six scientists at Brown University were awarded $4.25 million to begin research on such interactions in the mammalian nervous system. Along with many professionals from an array of different fields—surgery, chemistry, physics, and others—these brilliant minds are collaborating to help advance knowledge and further discovery of nanotechnology and nanocomputing. This commitment to research, coupled with a great amount of popular sentiment towards nanotechnology and nanocomputing, may further our own knowledge as observers and students and accelerate the progress of this exciting phenomenon. For more information on funding and patenting issues, see Chapter 20.

1.8. CONCLUSIONS

For the past 50 years, transistors have been shrinking consistently, and we have entered the era of nanocomputing. In this new era, transistors are only a small portion of the technologies that are available at nanoscale. There is a vast landscape of nano devices and paradigms that are currently being studied. In this chapter, we gave a brief introduction to nanocomputing and presented a high level overview of nanocomputing devices and paradigms. Nanocomputing has a potential to provide a remedy for some traditional problems in microelectronics. We also discussed some applications of nanocomputing such as bio medical engineering and neuroscience. The rest of this book will take the reader on a journey from low level device physics to architecture-level, bio-inspired architectures, all of which have the potential to be used for implementing various devices, such as biomedical and biomimetic nanoscale integrated circuits.

REFERENCES

1. R. P. Feynman. There's plenty of room at the bottom: an invitation to enter a new field of physics. *Engineering and Science*, 23(5): 1960.
2. Wikipedia. List of Intel Microprocessors. 2007. http://en.wikipedia.org/w/index.php?title=List_of_Intel_microprocessors&oldid=181278285.

3. N. Weste and D. Harris. *CMOS VLSI Design: A Circuits and Systems Perspective*. Reading, MA: Addison Wesley, 2004.
4. C. K. Maiti, N. B. Chakrabarti, and S. K. Ray. *Strained Silicon Heterostructures: Materials and Devices*. Piscataway, NJ: IEEE, 2001.
5. M. Houssa. *High k Gate Dielectrics*. New York: AIP, 2004.
6. X. Huang, W. C. Lee, C. Kuo, D. Hisamoto, L. Chang, J. Kedzierski, E. Anderson, H. Takeuchi, Y. K. Choi, K. Asano, V. Subramanian, T. J. King, J. Bokor, and C. Hu. Sub 50-nm finFET: PMOS. Electron Devices Meeting, 1999. *IEDM Technical Digest International*: pp 67–70, 1999.
7. B. S. Doyle, S. Datta, M. Doczy, S. Hareland, B. Jin, J. Kavalieros, T. Linton, A. Murthy, R. Rios, and R. Chau. High performance fully depleted tri-gate CMOS transistors. *IEEE Electron Device Letters*, 24(4): pp 263–265, Apr 2003.
8. Harold Abelson and Peter Andreae. Information transfer and area-time tradeoffs for VLSI multiplication. *Communications of the ACM*, 23(1): pp 20–23, 1980.
9. J. Wang and M. Lundstrom. Electron Devices Meeting, 2002. Does source-to-drain tunneling limit the ultimate scaling of MOSFETs? In: *IEDM '02 Digest. International*: pp 707–710, 2002.
10. V. Balzani, A. Credi, and M. Venturi. *Molecular Devices and Machines: A Journey into the Nanoworld*. Weinheim: Wiley-VCH, 2003.
11. Y. Chen, G. Y. Jung, D. A. A. Ohlberg, X. Li, D. R Stewart, J. O. Jeppesen, K. A. Nielsen, J. F. Stoddart, and R. S. Williams. Nanoscale molecular-switch crossbar circuits. *Nanotechnology*, 14(4): pp 462–468, 2003.
12. B. L. Feringa. *Molecular Switches*. Weinheim: Wiley-VCH, 2001.
13. J. M. Tour, W. L. Van Zandt, C. P. Husband, S. M. Husband, L. S. Wilson, P. D. Franzon, and D. P. Nackashi. Nanocell logic gates for molecular vomputing. *Nanotechnology*, 1(2): pp 100–109, Jun 2002.
14. J. Cumings, A. Zettl. Low-friction nanoscale linear bearing realized from multiwall carbon nanotubes. *Science*, 289(5479): pp 602–604, 2000.
15. W. Liang, M. Bockrath, D. Bozovic, J. H. Hafner, M. Tinkham, and H. Park. Fabry–Perot interference in a nanotube electron waveguide. *Nature*, 411: pp 665–669, June 2001.
16. A. Khitun and K. L. Wang. Multifunctional edge driven nanoscale cellular automata based on semiconductor tunneling nanostructure with a self-assembled quantum dot layer. *Superlattices and Microstructures*, 37(1): pp 55–76, 2005.
17. A. Khitun and K. L. Wang. Cellular nonlinear network based on semiconductor tunneling nanostructure. *IEEE Transactions on Electron Devices*, 52(2): pp 183–189, Feb 2005.
18. M. A. Kastner. The single electron transistor and articial atoms. *Annalen der Physik*, 9: pp 885–894, Nov 2000.
19. C. S. Lent, P. D. Tougaw, W. Porod, and G. H. Bernstein. Quantum cellular automata. *Nanotechnology*, 4(1): pp 49–57, 1993.
20. M. Johnson. Bipolar spin switch. *Science*, 260(5106): pp 320–323, 1993.
21. S. Wolfram. *A New Kind of Science*. Wolfram Media, 2002.
22. A. Ilachinski. *Cellular Automata: A Discrete Universe*. Singapore: World Scientific, 2002.
23. L. O. Chua and L. Yang. *Cellular Neural Networks. Circuits and Systems*, 1988.

24. C. M. Bishop. *Neural Networks for Pattern Recognition*. Oxford, UK: Oxford University Press, 1994.
25. M. Gardner. Mathematical games: the fantastic combinations of john conway's new solitaire game 'life.' *Scientific American*, 223(4): pp 120–123, 1970.
26. M. Cook. Universality in elementary cellular automata. *Complex Systems*, 15(1): pp 1–40, 2004.
27. T. Roska and L. O. Chua. The CNN universal machine: an analogic array computer. *IEEE Transactions on Circuits and Systems II: Analog and Digital Signal Processing*, 40(3): pp 163–173, Mar 1993.
28. C. N. Eichelberger and K. Najarian. Simulating protein computing: character recognition via probabilistic transition trees. *IEEE International Conference on Granular Computing, 2006*. pp 101–105, May 2006.
29. L. M. Adleman. Molecular computation of solutions to combinatorial problems. *Science*, 266(11).
30. P. W. Rothemund, N. Papadakis, and E. Winfree. Algorithmic self-assembly of DNA sierpinski triangles. *PLoS Biology*, 2(12): Dec 2004.
31. P. W. Rothemund. Folding DNA to create nanoscale shapes and patterns. *Nature*, 440: pp 297–302, 2006.
32. M. A. Nielsen and Isaac L. Chuang. *Quantum Computation and Quantum Information*. Cambridge: Cambridge University Press, 2000.
33. P. W. Shor. Algorithms for Quantum Computation: Discrete Logarithms and Factoring. *IEEE Symposium on Foundations of Computer Science*: pp 124–134, 1994.
34. L. M. Vandersypen, M. Steffen, G. Breyta, C. S. Yannoni, M. H. Sherwood, and I. L. Chuang. Experimental realization of Shor's quantum factoring algorithm using nuclear magnetic resonance. *Nature*, 414: pp 883–887, Dec 2001.
35. N. L. Kedersha, D. F. Hill, K. E. Kronquist, L. H. Rome. L.H. Subpopulations of liver coated vesicles resolved by preparative agarose gel electrophoresis. *Journal of Cell Biology*, 103: pp 287–297, 1986.
36. National Science Foundation. "Vaults: From Biological Mystery to Nanotech Workhorse?" *Biological Sciences*, Arlington, VA: The National Science Foundation. http://www.nsf.gov/discoveries/disc_summ.jsp?cntn_id=104106&org=MCB, 2007.
37. Regents of the University of Michigan. Molecular Motors. http://www.umich.edu/news/MT/04/Fall04/story.html?molecular. Jan 16, 2007.
38. Y. E. Goldman. Functional studies of individual myosin moleculares. *Annals of the New York Academy of Sciences*, 1(18): 2006.
39. C. McCarthy, Nano Science and Technology Institute. Pharma Explores Business Opportunities for Nanotech. http://www.nsti.org/news/item.html?id=. Mar 13, 2007.
40. Cirrus Pharmaceuticals. Pharmaceutical Nanotechnology. http://www.cirruspharm.com/pdf/brochures/Nanotechnology.pdf. Mar 13, 2007.
41. R. A. Freitas. Computational Tasks in Medical Nanorobotics. Insitute for Molecular Manufacturing, 2007.
42. P. Chapman. Nanotechnology in the pharmaceutical industry. *Expert Opinion on Theraperutic Patients*. 15(3): pp. 249–255, 2005.
43. G. A. Silva. Introduction to nanotechnology and its applications to medicine: applying nanotechnology to medicine. *Surgical Neurology*, 61: pp 216–220, 2004.

44. G. A. Silva. Neurosicence nanotechnology: progress, opportunities, and challenges. *Nature Reviews Neuroscience*, 7: pp 65–74, 2006.
45. National Science Foundation. Nanowires in blood vessels may help monitor,sStimulate neurons in the brain. http://www.nsf.gov/news/newssumm.jsp?cntnid = 104288&org = ENG. July 7, 2005.
46. F. Patolsky, B. P. Timko, G. Yu, Y. Fang, A. B. Greytak, G. Zheng, and C. M. Lieber. Detection, stimulation, and inhibition of neuronal signals with high-density nanowire transistor arrays. *Science*, 313: pp 1100–1104, 2006.
47. Tiny Nanowire Could be Next Big Diangostic Tool for Doctors. *Science Daily*: Dec. 18, 2003.

2

NANOSCALE DEVICES: APPLICATIONS AND MODELING

Alireza Nojeh

This chapter starts with a discussion of the roots of nanotechnology, the distinguishing aspects of the nanoscale, and the definition of nanoscale devices. Some of the common materials and techniques used to fabricate such devices are briefly reviewed. The physics of the operation of nanoscale devices is the subject of the next part. Quantum dots, resonant tunneling diodes, and single-electron transistors, which can be made using traditional fabrication methods, are first studied. Then a number of devices based on newer materials such as carbon nanotubes are introduced, and how the properties of these materials provide opportunities for innovative device design is discussed. The last part of the chapter deals with the theoretical study, modeling, and simulation of nanoscale devices. The challenge of the many-body problem and the difference between first-principles and semi-empirical approaches are highlighted. A brief introduction to some of the simulation methods, such as the Hartree–Fock approximation, density functional theory, molecular dynamics, tight-binding, and Monte Carlo, concludes the chapter.

2.1. INTRODUCTION

2.1.1. The Roots of Nanotechnology

Although one can easily think of many milestones, such as the invention of the scanning tunneling and atomic force microscopes or the discovery of fullerenes in the 1980s, that have pushed us more and more into the nano era, it is not easy to define a single point in time when it all started. For instance, many of the imaging tools relevant to the nanoscale, such as electron microscopes, have been around

Bio-Inspired and Nanoscale Integrated Computing. Edited by Mary Mehrnoosh Eshaghian-Wilner

for a long time. But as we see, they have been called microscopes and not nanoscopes. One tends to think that at least one aspect of what we consider nanotechnology is the extension of microtechnology to smaller sizes, and it is hard to separate the roots of these two, although some clear distinctions can be drawn between them. As discussed in Chapter 1, a particularly interesting early reference to small scale devices and technologies was made by physicist Richard Feynman in a lecture he gave in 1959 [1]. Feynman did not make specific claims about what *would* happen in future, but rather gave a tour of what *could* very logically and naturally be available at small scales from a purely scientific point of view (i.e., without contradicting the laws of physics as we know them). After almost half a century, that lecture can be an insightful introduction for newcomers to the field of nanotechnology and a great inspiration to seasoned nanotechnologists.

2.1.2. Two Ways to Reach the Nanoscale

It could be argued that one of the most dramatic changes in our lifestyles has been brought about by the rapid advance of the microelectronics industry over the last few decades. The steady decrease in the size of transistors, the fundamental building blocks of microelectronic circuits, has lead to today's low cost and high performance computer chips. For instance, the transistors in your computer are about a tenth of a micrometer in size, or maybe even about a twentieth of a micrometer if you have recently bought a new one—that is about 50 nanometers (nm). Thus, for those who work in this industry, nanoelectronics is a completely natural continuation of microelectronics. In fact, many tend to define a device as being a nanodevice if one of its dimensions is smaller than 100 nm. In that sense, the semiconductor industry has already entered the era of nanoelectronics—without really calling itself nanotechnology!

Pushing our traditional microfabrication techniques to make devices much smaller than a micrometer is one way of approaching the nanoscale. One could think of this as the art of "carving" ultra-small objects. However, there are objects that are very small by their nature, such as molecules. They provide another approach to nanotechnology: use small objects already available to us through nature and try to add some level of control to where and how they are arranged so that they can perform the functions we are interested in.

2.1.3. The "Nano" Difference

As pointed out in Chapter 1, there is more to the nanoscale than just the ability to pack a large number of devices in a small area or to manipulate very small objects. There is a fundamental reason that makes "nano" attractive from a scientist's point of view but a nonexpert might have a little more difficulty appreciating it. Let us see what that is.

We know that objects are made of individual units called atoms that are extremely small. This book you are reading, for example, contains on the order of one hundred trillion trillion atoms—that is a 1 with 26 zeros in front of it! When such a large collection of atoms is put together, the general behavior of the ensemble in everyday life experiences does not directly reflect the atomistic nature of the system. It looks like a solid, bulky material for which the behavior can be explained relatively easily: You can describe it in terms of its macroscopic properties, i.e., weight, dimensions, color, etc. You can also use the laws of classical physics, such as Newtonian mechanics, to describe its motion if a force is applied to it.

In fact, atoms behave according to laws discovered much later than Sir Isaac Newton's classical mechanics. The laws of quantum mechanics, developed mainly in the first half of the twentieth century, are rather nonintuitive for someone who is used to everyday phenomena with large objects, but they constitute the best description we have so far for the world of small particles such as atoms. Let us continue with the example of the book. When such a huge collection of atoms is put together, the overall behavior will be a statistical average of the behavior of the individual atoms, subject to external forces and the constraints that keep them together. This average behavior can be explained well with classical physics. The result is that we do not really need quantum mechanics to be able to handle a book! Now, imagine a microelectronic device, such as a transistor, with dimensions on the order of a micrometer. Such a device would contain about one trillion atoms (a 1 with 12 zeros in front of it). That is still a very large number and so, even in such a small device, most of the characteristics are determined by the statistical averages of many interactions and phenomena. Thus, although in order to properly understand these phenomena and the resulting average behaviors one needs to apply quantum mechanics, more phenomenological laws can be derived using statistical analysis that explain the device behavior accurately enough for most engineering purposes. In fact, quantum mechanics is still somewhat masked and its effects are only indirectly visible at the microscale. In other words, you do not need to resort to quantum mechanics every time you want to study or use a microscale transistor. In this sense, even microdevices could be considered somewhat classical.

Now let us imagine devices that have dimensions on the order of only a few nanometers. In matter, atoms are spaced apart by a few angstroms. An angstrom is 10 times smaller than a nanometer. Thus, in the volume of our nanoscale device, there would be only a few hundred or thousand atoms. It turns out that in such small collections of atoms, statistical averages are not always very meaningful. But the individual character of each atom—its quantum mechanical nature—is much more visible. Therefore, when working with nanodevices one observes quantum mechanics very directly and, by the same token, can take advantage of the laws of quantum mechanics to design devices and systems that are not achievable at larger scales (that is, where quantum effects have averaged out to give way to more classical behavior). This direct access to quantum mechanics at the nanoscale makes a fundamental difference that involves more than manipulating small objects or squeezing a large number of devices into a small area. A whole new world of functionalities and possibilities previously unimaginable is opened up. One is tempted to argue that this is the most important aspect of the nano world, but that remains to be determined by the course of progress of this field in the future.

In addition, the surface-to-volume ratio of objects increases as the objects become smaller. For instance, the surface of a cube with side x is $6x^2$ and its volume is x^3, so the surface-to-volume ratio, $6/x$, increases as x decreases. Usually, the surfaces of materials act differently from their bulk in terms of physical properties. As one goes to smaller and smaller scales, surface phenomena become a more important part of device behavior, and this could have very practical implications.

2.1.4. What is a Nanodevice?

The exact definition of what a nanodevice is remains somewhat fuzzy and subjective. There seems to be a general consensus that any device with at least one dimension below 100 nm is a nanodevice. In many cases, depending on how a particular device is made and its principle of operation, some may choose to call it a nanodevice and some may not. What seems to be clear is that, regardless of how we reach the nanoscale and whether or not we manage to make sophisticated nanorobots at some point in future, many interesting phenomena happen at the nanoscale that are worth studying and could be used for new device ideas.

This chapter hopes to reach out to a large audience. For the nontechnical who are intrigued by the word "nano," the chapter provides a general insight. For the nonexpert technical person or those wishing to enter the field of nanodevice research (such as advanced undergraduate or fresh graduate students in science and engineering), it provides an introduction and perspective. The main focus of the chapter is on the electronics aspects of nanodevices, but there exist many other aspects to these devices, as well as different approaches to understanding them and creating new ones. Obviously none of the devices or methods discussed here can be reviewed in full detail in the context of a one-chapter overview, nor can all the valuable references in the field be cited. However, the author hopes that the references provided here will be a good starting point for those interested in learning more about each subject.

2.2. MATERIALS OF NANOTECHNOLOGY

The first question that might come to mind when thinking about materials for nanodevices is such: Can one take any material (even a piece of wood), carve a nanoscale object out of it (assuming one has the means, such as an ultra-sharp carving tip to make something that small!), and hope that it would exhibit some interesting nanodevice characteristics? The answer, in principle, is yes. In practice, however, just as is the case in the microscopic or macroscopic worlds, different materials have different properties at the nanoscale level; depending on the particular application the nanodevice is intended for, one has to spend some effort to find the right material for it.

The well established materials of the microelectronics industry, such as silicon, dielectrics, and metal layers, also represent a very important category of materials necessary in nanodevices. Not only can many nanoscale devices be made using only these materials and standard microfabrication processes, but in many other cases at least part of the structure of a nanodevice involves these materials and processes. We will talk more about this later in the chapter, but for now let us focus on objects that are more "nano" by nature.

2.2.1. Molecules

We know that a molecule is a collection of atoms of either one element or several different elements. Most of the molecules we are familiar with (for example from our high school chemistry) are made of a relatively small number of atoms. Consider the water molecule. It has one oxygen and two hydrogen atoms. However, there is no reason why we could not have much bigger molecules that include hundreds or even thousands of atoms—as a matter of fact, we do! Let us go back and think about our nanodevices for a moment. We estimated that a typical nanodevice would have hundreds or thousands of atoms in it. Can we say a nanodevice is a molecule? In reality, many nanodevices could be considered molecules. Many would be a system of several molecules put together or with other structures in a certain fashion. The term molecular electronics is quite commonly used these days. One difference between these molecules and the more "traditional" molecules that we remember from good old high school chemistry is that there the molecule was a concept describing the smallest building blocks of a much larger substance. When we say a nanodevice can be thought of as a molecule, we are really only emphasizing the fact that it is a stable collection of a number of atoms, and not necessarily a unit that is going to be used to build a macroscopic substance. Also, when one thinks of a molecule, the first examples that usually come to mind are rather small ones that contain only a few atoms (such as the hydrogen or benzene molecules), although much larger molecules are also commonplace (such as many polymer molecules or the DNA that can reach macroscopic length scales). Nanodevices, if one wants to call them molecules, are usually very large molecules.

2.2.2. Nanoparticles

One important set of "objects" that are very useful to the world of nanotechnology is nanoparticles. As the name implies, a nanoparticle is a particle that has dimensions on the order of a few nanometers. This typically includes a few thousands of atoms. Nanoparticles can be made from various materials. Many metals, for instance, form good nanoparticles. Nanoparticles containing more than one element can also be formed. Again, what is really important about a nanoparticle is the fact that it is small enough such that the quantum effects and large surface-to-volume ratio effect discussed before are dominant in it, to the extent that its properties can no longer be described by the same general properties describing a microscopic or macroscopic piece of the same material. In this sense, nanoparticles are at the intermediate level between bulk materials and the atomic world.

2.2.3. Nanowires

We recall from geometry that an object can be zero-dimensional (0D: a point), one-dimensional (1D: a line), two-dimensional (2D: a plane) or three-dimensional (3D: a volume). In analogy, nanoparticles are like points. One important category of nanomaterials is 1D objects or "lines" at the nanoscale, namely nanowires. A typical nanowire has a thickness of a few nanometers to a few tens of nanometers; its length is much more than a nanometer, giving it a very high aspect ratio (like a line). Nanowires have been made of many materials, such as silicon, germanium, zinc oxide, and gallium nitride [2–5]. Nanowires are extremely important due to their one-dimensionality. In particular, the electronic properties in these 1D materials can be drastically different from their bulk counterparts due to the fact that charge carriers (electrons) are confined to a small space (only a few nanometers) in the lateral dimensions. This leads to very strong quantum mechanical effects and can result in outstanding electronic transport characteristics for devices such as transistors and switches needed in ultra-high-performance electronic circuits and chips.

2.2.4. Nanotubes

One particularly important form of nanowires is what we call a nanotube. The name explains it all: It is a hollow cylinder with a diameter on the order of 1 nm. The most common nanotubes are those made of carbon, i.e., carbon nanotubes. Graphite is a well known form of carbon. Pencil lead is made out of graphite. Graphite has an interesting structure, in that it is composed of one-atom-thick layers stacked on top of each other. Each layer, called graphene, is a sheet of carbon atoms arranged in a hexagonal lattice structure, like a chicken wire. A carbon nanotube can be thought of as a single graphene sheet rolled along a certain direction into a nanometer-diameter tube that is hollow inside. Carbon nanotubes can be extremely long, up to several centimeters, and that means they have aspect ratios on the order of 10^7. Thus, they represent highly 1D systems. They can be single-walled (Fig. 2.1), i.e., composed of only one layer, or multi-walled, composed of several co-axial, single-walled nanotubes. Carbon nanotubes were discovered in 1991 by Sumio Iijima [6] and have since attracted much attention in the world of nanotechnology due to characteristics such as outstanding electronic properties arising from their 1D nature, as well as extreme mechanical strength due to the strong carbon-carbon SP^2 bond.

One fundamental difference between nanotubes and other forms of nanowires is that a nanotube is by itself a perfect structure. In other words, most nanowires can be thought of as a very thin wire carved out of a 3D bulk. As such, there could be many unsatisfied and dangling bonds on the surface of these structures. They

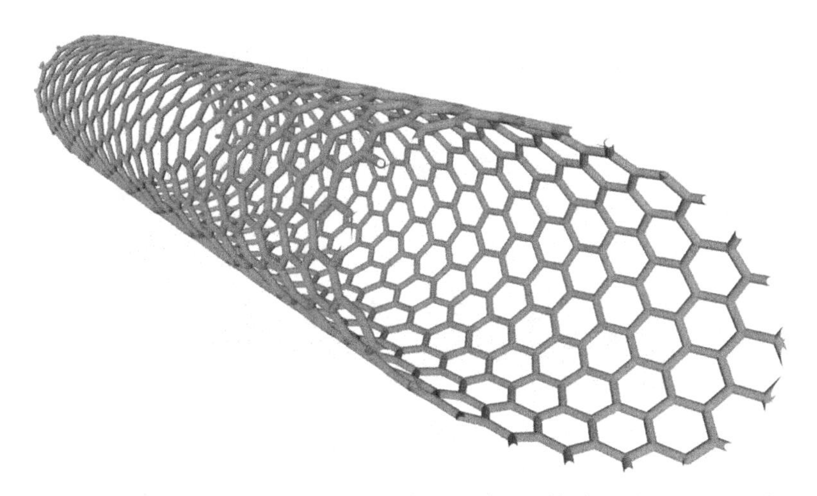

Figure 2.1. 3D view of a section of a single-walled carbon nanotube.

could also be rather brittle. A nanotube is, on the other hand, one of the natural, stable forms of its material. For example, a carbon nanotube is an allotrope of carbon, the same way graphite and diamond are. A nanotube surface does not have dangling bonds and all atoms are in a perfect crystalline structure, contributing to the stability of the overall system. A carbon nanotube is, in fact, a giant molecule. For a detailed study of carbon nanotubes the reader is referred to [7].

2.3. WORKING AT THE NANOSCALE

Nanofabrication, at least so far, has been heavily reliant on the processes used to fabricate microelectronic circuits or micro electro mechanical systems (MEMS). In fact, many of the processes used for nanodevice fabrication are simply microfabrication processes. In most cases, the term "nanofabrication" only implies that we are using them to make nanodevices. For a detailed study of microfabrication and MEMS techniques references [8, 9] can be used. A typical nanodevice fabrication process involves several microfabrication steps, with the addition of one or two new steps to the process. Let us consider a device consisting of a nanotube attached to two metal electrodes lying on an insulating surface (Fig. 2.2). The fabrication of this device starts with thermal oxidation of silicon to create a thin dielectric layer on a silicon wafer. Then lithography and metal deposition are used to create the contact electrodes. Another lithography step helps pattern catalyst islands that will later be used to grow carbon nanotubes. So far everything is standard microfabrication. The last step, i.e., the growth of the nanotube, can be done using chemical vapor deposition (CVD) [10]. Although CVD has long been around for various fabrication purposes,

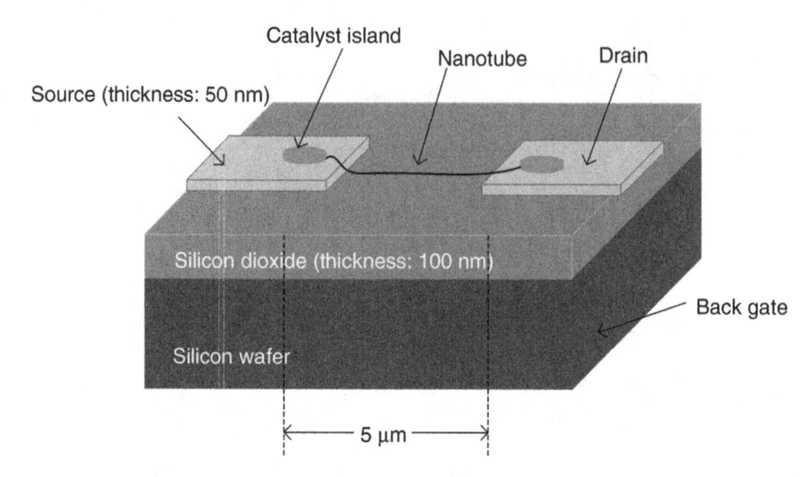

Figure 2.2. Schematic representation of a device including a nanotube attached to two metal electrodes.

nanotube CVD is a relatively recent process that is not standard to the microelectronics or MEMS industry, and can be qualified exclusively as a nanofabrication step. In general, CVD leads to nanotubes growing randomly in all directions; the task of locating the nanotubes in desired locations and orientations after CVD can be quite time consuming and, more importantly, cannot be relied upon for mass production of devices. It has been demonstrated, however, that the direction of the nanotubes grown by CVD can be controlled to a large extent by the application of an electric field [11, 12] or by using the direction of the laminar flow of gases [13, 14] during CVD.

Just as in the case of fabrication, many of the inspection and imaging techniques appropriate for nanotechnology have also existed for a long time. Some of the most commonly used imaging methods with nanoscale (or even subnanoscale) resolution include scanning and transmission electron microscopy (SEM and TEM). These work on principles very similar to optical microscopes; the basic difference is that instead of photons (light particles) they use electrons to obtain an image. The reason is that electrons generated in typical laboratory equipment can easily have associated wavelengths much smaller than photons and thus be focused to a much smaller spot, leading to higher resolution. Another group of techniques for imaging and manipulating nanoscale objects is scanning probe microscopy (SPM). The idea behind these is to have a very small and sharp tip brought in contact with a sample (although the term contact should be used loosely in this context), scan the surface, and record the profile. Based on whether the sample is conductive or insulating and other requirements, various types of SPM methods can be used. Scanning tunneling microscopy (STM) and atomic force microscopy (AFM) are two of the most popular techniques of this kind.

After a device has been fabricated and imaged, various measurements and experiments are performed on it. There are two big challenges. The first is reaching such small devices to perform, say, electrical measurements on them. Usually this can be done by making "interface" structures to connect the nano world to our macroscopic world. For instance, in the case of the device in Figure 2.2, the metal electrodes on the sides can be connected to large pads that one can use for wire bonding to the outside world. Another option is to use high resolution motion probes, such as in an SPM tool, to move, manipulate, and contact objects at the nanoscale. The second challenge is determining whether a measurement or perceived effect is really due to the nanodevice and not other surrounding structures, noise, inaccuracies, etc. This is true in any experiment in general, but it becomes particularly important at the nanoscale since one is dealing with very small signals. Creative techniques need to be developed constantly for such experiments. This is an ongoing challenge in nanotechnology.

2.4. NANODEVICE APPLICATIONS

Although nanotechnology is still very much in a research phase, many applications have been suggested and even demonstrated at the laboratory level for nanoscale devices. It should be noted that many of the more traditional microdevices have also found their way into nanotechnology. Their scaled down versions could well be considered nanodevices, especially since the reduction in scale results in major differences in characteristics and operation, such as the operating temperature. In this section we will first review some of the more established devices in this category, and then present some of the less traditional nanodevice ideas.

2.4.1. "Traditional" Nanodevices

The fundamental building blocks of our computers are field-effect transistors (FETs), which act as switches to perform logic operations. In parallel with efforts to shrink those devices in the well-established silicon technology, as well as in new types of materials (discussed later in this chapter), there have also been many efforts in making alternative devices that could perform such operations. Here we will look at some of the most common devices in this category, namely resonant tunneling diodes (RTDs), single-electron transistors (SETs), and quantum dots (QDs). An interesting overview of these devices can be found in [15]. As discussed, a fundamental aspect of nanodevices is that quantum effects are directly visible in them. We will take a closer look at this here.

2.4.1.1. Resonant Tunneling Diodes. Consider the problem of a "particle in a box." By this we mean a particle, say an electron, confined to a small region in space by a potential energy distribution such as the one shown in Figure 2.3. This represents what is called an infinite potential "well"—namely, a region where the electron is trapped by two barriers on the sides.

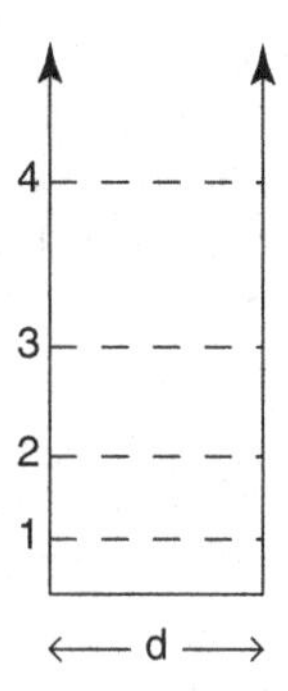

Figure 2.3. A potential well where electrons are confined in space due to energy barriers on the sides. Electrons in such a "box" can only take on discrete energy levels as shown. The horizontal axis is location and vertical axis is energy of the electron.

As will be discussed in more detail later in this chapter, the fundamental equation that governs the behavior of quantum mechanical particles, such as electrons, is the Schrödinger equation. The energy levels that an electron is allowed to have in this one-dimensional potential well can be easily obtained by an analytical solution of the Schrödinger equation. The result is

$$E_N = \frac{\hbar^2 \pi^2 N^2}{2md^2}, \quad N = 1, 2, \ldots$$

Here $\hbar$ is Planck's constant and d is the width of the well. In other words, the electron in the well cannot have just any energy, but must take one of the discrete values given by the above formula. In general, such "quantization" of energy levels also happens in the 3D case. This idea is at the heart of some of the quantum devices that we will discuss in this section. Now consider a region in space where a potential well is connected to two metal electrodes through barriers with finite heights and widths on the sides, such as in Figure 2.4. This is an RTD. An electron can enter this region from outside, leave the region by overcoming the barrier heights (by, for instance, acquiring thermal energy and going to higher energy levels), or move through the barriers by a process called quantum mechanical tunneling. What is interesting is that in the transport characteristics of this device, the effect of these discrete energy levels becomes completely visible.

Now imagine a voltage bias is applied to the structure that leads to a relative shift in the chemical potentials of the two contact electrodes, such as in Figure 2.5. An electron will be able to tunnel through the device from one side to the other only if the biases on the two sides are such that there is an energy level in the well in the range where electrons exist on the left and empty states exist on the right, i.e., when there is a level lower than μ_1 but higher than μ_2. (Remember that in the electrodes all energy states up to the chemical potentials are filled with electrons.)

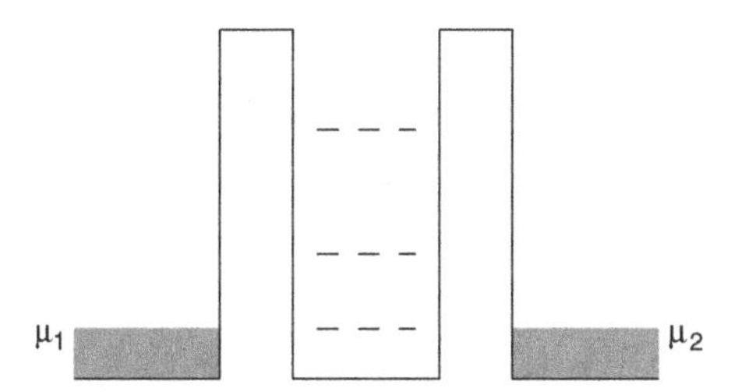

Figure 2.4. The double-barrier structure of an RTD. The gray areas on the sides represent metal contacts or reservoirs of electrons with chemical potentials μ_1 and μ_2. The horizontal axis is location and vertical axis is energy of the electron.

Thus, as the applied bias is increased, every time a new energy level enters the range between the two side chemical potentials, there will be a peak in the device's current versus voltage curve (Fig. 2.6).

If a "gate" electrode is placed below the device to enable us to move the energy levels up and down, then we can use this gate to control which level lies between the side chemical potentials, and therefore we can control the conductance of the device. This is the basis of a three-terminal switching device or resonant tunneling transistor.

Note that the conductance through this device cannot be modeled by simply considering two single barriers in series. In fact, what is essential here is the wave nature of the electrons and the resonance phenomenon in the well that leads to a high transmission probability of electrons from one side to the other, giving rise to the peaks in conductance. This is analogous to the transmission of light through a multilayer structure with layer thicknesses on the order of the wavelength of the incident light or smaller. Resonance phenomena there can lead to high transmission for a given set of layer properties (thicknesses and refractive indices) and incident wavelength. In general, this is the problem of a resonant cavity, which in the case of the RTD is a cavity for electrons.

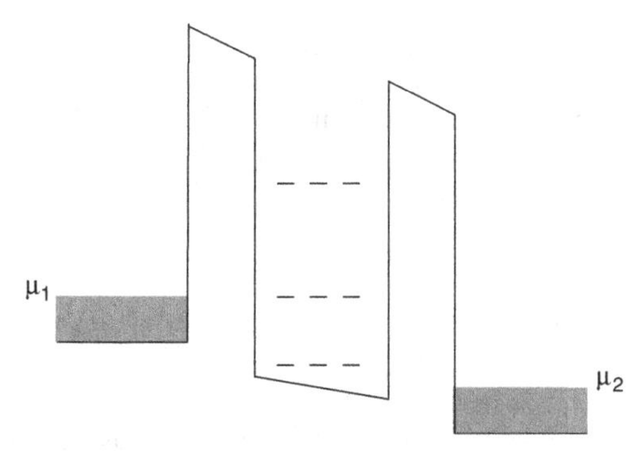

Figure 2.5. A double-barrier under applied bias, which creates a difference between the chemical potentials on the two sides.

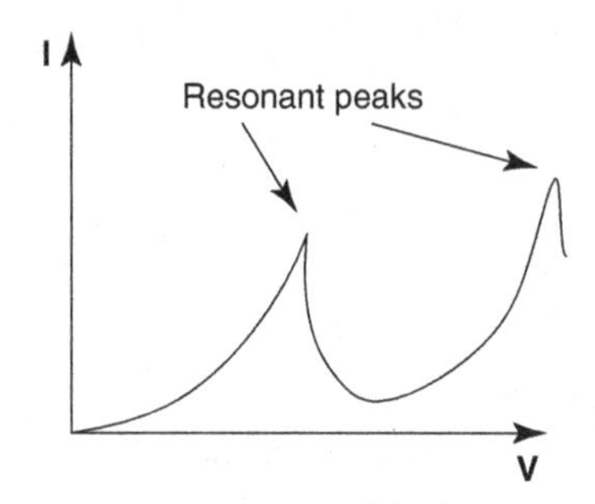

Figure 2.6. Conceptual representation of the current-voltage characteristics of an RTD.

2.4.1.2. Single-Electron Transistors. In the operation of the RTD, we looked at the problem as if there were only one electron in the system. In other words, we neglected electron-electron interactions. Electrons repel each other through Coulomb forces; if an electron is put inside the device, it will be harder for a second one to enter since it will have to face the repulsive force of the first one. This is called Coulomb blockade. In other words, if multiple electrons are to enter the device at a time, each subsequent electron will have to be put at an energy not only determined by the level quantization shown in Figure 2.3, but also by the charging energy of the device due to its capacitance. As a consequence, the difference between the energy of the electrons will be approximately the sum of the two contributions. One effect may be more dominant than the other based on the actual device size and material. In RTDs the device has a relatively large capacitance such that the charging energy e^2/C is much smaller than the level quantization energy discussed above. In an SET, however, the structure is such that the charging energy is much larger than the level quantization energy. Note that although the end result of both effects (resonant tunneling and Coulomb blockade) is that transport happens through discrete energy states and gives rise to unusual device characteristics, the two effects are fundamentally different: one effect is due to the wave nature of particles (quantum mechanical), and the other one due to the fact that the value of charge is quantized (which could be regarded as a somewhat more classical effect). A review of single-electron transistor devices can be found in [16]. Such structures could also be operated under an alternating voltage, and the frequency and amplitude of this signal is such that only one electron is transferred per cycle through the device. Such a device is called a single-electron turnstile [17].

2.4.1.3. Quantum Dots. There seems to be some degree of ambiguity in the literature as to the exact definition of a QD. In particular, the terms SET and QD often seem to be used one instead of the other. A QD could be thought of as a confined region between two potential barriers such as in Figure 2.4, although confinement in all three dimensions is typically assumed. In a sense, RTDs and SETs can be considered special forms of QDs. Both energy level quantization

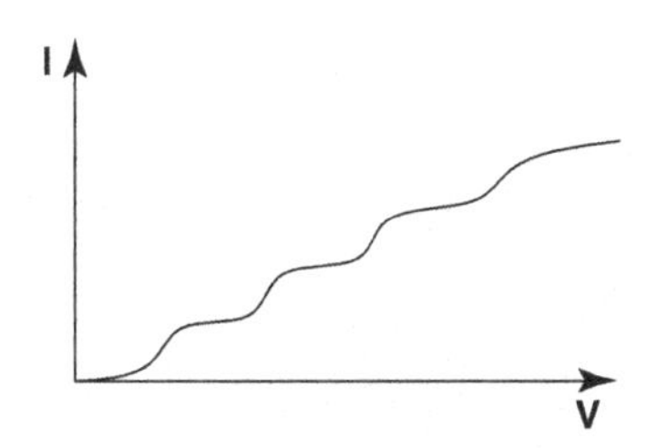

Figure 2.7. The staircase-like current-voltage characteristics of a QD. The width of the plateaus is defined by the separation of the energy levels due to quantum confinement and/or Coulomb blockade.

(due to the wave or quantum mechanical nature of electrons) and Coulomb blockade (due to the discrete nature of charge) are present in the electronic characteristics of a QD. The authors in [15] present the clear distinction that if these two effects are on the same order of strength and significance, the device is a QD; otherwise it is an RTD or SET depending on which effect is dominant. In so doing, they correlate these effects to the number of "small" or confined dimensions in each device.

What is really important in any case is the clear understanding of both these effects and their fundamental difference. Also, note that typically the structure of an RTD is such that as the bias is increased, only one level at a time is in the energy ranges where there are available electrons for transport, giving rise to the peaks in the conductance characteristics of the device and the negative differential conductance after each peak (Fig. 2.6). However, in what we usually think of as QDs or SETs, often a wider energy range can be available for the incoming electrons (compared to the level spacing inside the well); as the bias is increased, first one level enters that range, then a second one, a third one, and so on, without the previous levels leaving that range. Thus, instead of a number of peaks in the current–voltage characteristics, one can have a staircase-like profile: every time a new level enters the allowable energy ranges of the contacts, the current goes up by a finite amount (Fig. 2.7). Because electron energy is quantized in a QD (like the energy levels in an atom) quantum dots are sometimes referred to as artificial atoms. For more detail on the operation of QDs the reader can consult [18, 19].

The above quantum devices can be realized in various ways. One popular method is to create a so called two-dimensional electron gas (2DEG) in the interface region of a heterojunction of III-V semiconductors with different band gaps. A set of electrodes is patterned on top of this 2DEG, and by the application of proper bias to these electrodes, some regions of this 2DEG can be depleted (Fig. 2.8). As a result, an "island" of electrons is created that is connected to the reservoirs on the sides through potential barriers (the depleted regions).

2.4.1.4. Temperature Dependence. Table 2.1 shows the first few allowed energy levels in a potential well with different widths calculated using the formula

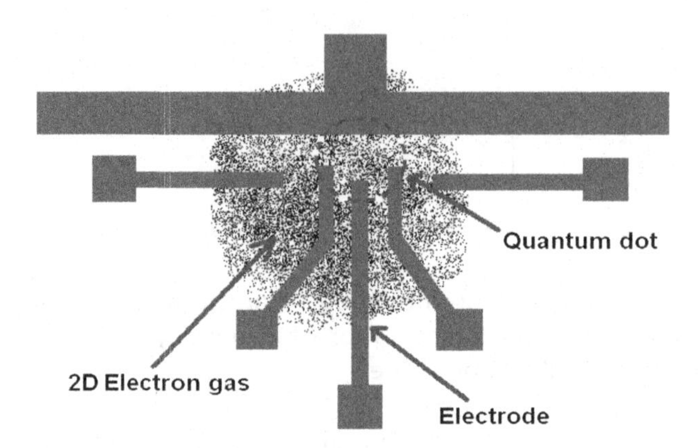

Figure 2.8. Schematic top view of a QD created in a 2DEG.

presented earlier. As can be seen, going from 100 to 1 nm changes the energy level separations significantly.

Note that the thermal energy of electrons at room temperature is kT = 25.9 meV (k is Boltzmann's constant and T is temperature on the absolute or Kelvin scale). If the separation of levels is smaller than this value, the thermal energy of the electrons will be enough to allow them to freely move between these levels; subsequently, the quantization of levels becomes masked at room temperature. Table 2.1 shows that one can hope to see room temperature operation only if the device size is on the order of 10 nm or less. This is in fact the case, and most experiments on QD-type devices are performed at very low temperatures (cryogenic temperatures such as liquid helium temperature). A similar effect also exists with regards to the Coulomb blockade energy e^2/C. This energy will be large enough to have an effect at room temperature only if the device dimensions, and therefore its capacitance, are small enough. This typically happens at the nanoscale. An example of room temperature operation can be seen in [20].

2.4.2. Less Traditional Nanodevices

By making the "island" of electrons smaller and smaller in the devices discussed in the previous section, one goes from the world of microdevices to that of nanodevices. This could be accomplished, for instance, by using lithography

TABLE 2.1. The First Four Energy Levels in an Infinite Potential Well with Various Widths

N	ε_N (meV) for W = 100 nm	ε_N (meV) for W = 10 nm	ε_N (meV) for W = 1 nm
1	0.038	3.8	380
2	0.151	15.1	1510
3	0.339	33.9	3390
4	0.603	60.3	6030

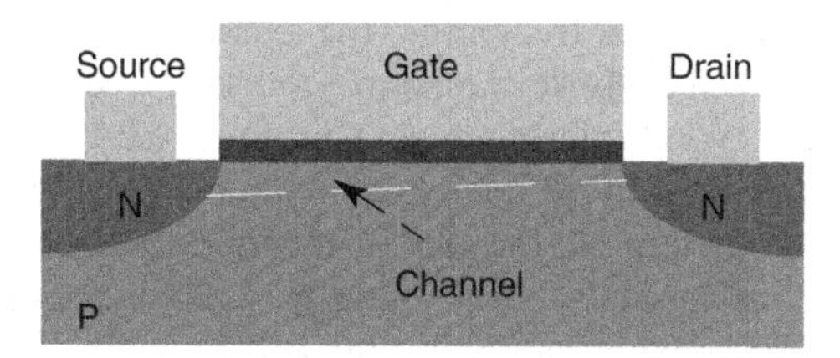

Figure 2.9. Schematic representation (side view) of an MOS FET (N-channel type).

with higher resolution or new methods of fabrication. But there are other types of nanodevices that are "nano" by nature without the need for us to try to make them smaller. This section reviews some of these devices.

2.4.2.1. Nanoscale Field-Effect Transistors. Let us recall from Chapter 1 the operation of a metal-oxide-semiconductor field-effect transistor (MOS FET), shown schematically in Figure 2.9. Conduction in this device happens between source and drain in a channel below the gate. The properties of this channel can be modulated by the voltage applied to the gate (by attracting more charge carriers to the channel or pushing them away from it), and this is the basis of the switching action of the transistor.

Now let us focus more on the channel itself. We introduced nanowires and nanotubes earlier in this chapter. The basic idea behind nanoscale FETs is the use of a nanowire or nanotube as the conducting channel of the transistor. This wire or tube is connected between two electrodes (source and drain), and a third electrode placed in close proximity (above, below, or even on its side) acts as gate (Fig. 2.10). The difference with traditional MOS devices is that here transport does not happen in a 3D bulk material, but rather in a 1D structure, which is radically different. As a result, these nanoscale FETs can have major differences in characteristics from their traditional counterparts, including several advantages.

In these 1D structures or quantum wires, charge carriers are confined in the two lateral directions and behave completely differently compared to 3D materials: not only do they experience strong energy quantization effects, but their movements are strongly correlated with each other. In particular, carriers in these

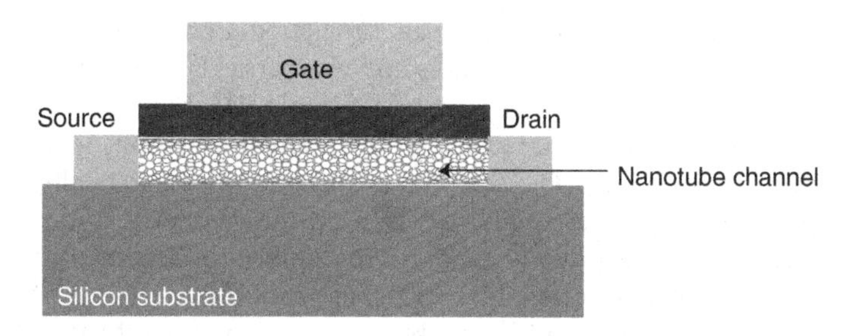

Figure 2.10. Schematic representation (side view) of a nanotube FET.

wires can have very high mean free paths (the average distance they travel before being scattered), and even ballistic transport (collision free transport) can happen over lengths of up to several micrometers. This could lead to very high operation speed. Transistor action in these devices can happen as a result of both direct modulation of the channel due to the gate or the modulation of the tunnel barriers at the source and drain contacts. These nano transistors are currently the subject of active research, both experimentally and theoretically. For instance, silicon nanowire transistors have been reported to have higher on–off ratios and smaller subthreshold slope (approaching the theoretical limit) than planar silicon-on-insulator FETs [21]. FETs based on other types of nanowires, including germanium, zinc oxide, and gallium nitride, have also been investigated [3, 22, 23]. Similarly, high performance FETs based on carbon nanotubes, including ballistic transport devices, have been demonstrated [24–26]. Applications for logic circuits and high frequency operation (GHz range) have also been investigated [27, 28]. For a theoretical study and modeling of nanowire and nanotube transistors, the reader is invited to consult [29–32].

2.4.2.2. Exploiting the Hidden Offerings of Nanomaterials. If the length of the channel in a nanowire or nanotube transistor is made very short, the device will naturally become a quantum dot or single electron transistor. Obviously this can be achieved if high resolution patterning is possible, and it is one way of making nanoelectronic devices with new materials [33, 34]. Note that in this approach we are using only the fact that these materials are one-dimensional or quantum wires, i.e., confinement in two directions is already in place. However, the size of the quantum dot achieved in this way is still determined by our ability to pattern electrodes on the nanotube with very high spatial resolution. Therefore, this kind of quantum dot typically has a length on the order of 100 nm and, as discussed before, room temperature operation is not easy. Here we will discuss briefly how some of the less obvious properties of these nanomaterials can be exploited to make new types of devices that are less dependent on traditional patterning techniques.

Consider a single-walled carbon nanotube. It is a quantum wire and can be made into a quantum dot by creating only one additional degree of confinement in it. Using first-principles calculations, the authors in [35] predict that if the cross section of a nanotube is deformed from circular to elliptical, its electronic structure (both valence and conduction bands) can be altered significantly, changing the nanotube from semiconducting to metallic and vice versa. This property was used in [36] and two kinks were created in a nanotube using AFM manipulation (Fig. 2.11) to induce potential barriers and isolate a dot as small as 20 nm, which exhibited room temperature operation. The device was also studied theoretically in [37].

Another device suggested in [38] is based on the geometrical idea that a dot can be defined by crossing two lines. Here the induced mechanical deformation at the intersection of two nanotubes crossing each other (Fig. 2.12) creates a local potential well with a width of about 1.5 nm, potentially enabling room temperature operation.

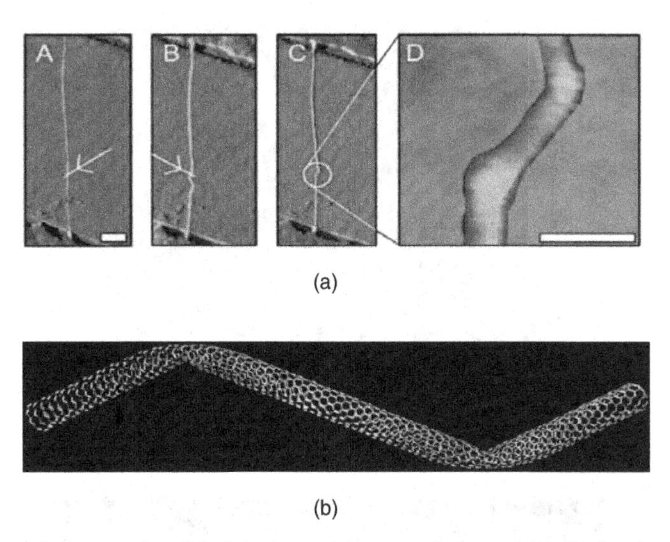

Figure 2.11. (a) Nanotube QD fabricated by creating two kinks in the nanotube shank using AFM manipulation. Scale bar indicates 20 nm.(From [36]. Reprinted with permission from AAAS.) (b) Simulation of the structure Reprinted with permission from [37] (© 2005 American Institute of Physics)

The advantage here is that once the two nanotubes are made to cross one another, the quantum dot is formed by the natural relaxed configuration of the system and its size is defined automatically to be about 1.5 nm, which would be impossible to achieve using the established patterning techniques. The effect of this deformation on the local density of states in a similar structure was also investigated experimentally in [39]. Thus, as can be seen, there are properties in nanomaterials that can be capitalized on for the creation of novel devices. The author believes that this is one of the major advantages of the world of nanoscale devices and that many unexplored opportunities in this regard still exist.

2.4.3. Other Nanodevice Applications

2.4.3.1. Electron Emitters. Electron emitters have long been in use in vacuum tubes and applications such as electron microscopy and electron beam lithography. With the advances in electron optics in these systems, the electron emitter or source remains the performance bottleneck. Parameters such as the brightness, energy spread, and shot noise of the electron beam directly affect resolution, signal-to-noise ratio, and throughput in such applications. Electron emission is the phenomenon of extracting electrons from inside a metal or semiconductor into a vacuum. An electron emitter could be as simple as a tungsten hairpin with a sharp point. Figure 2.13 summarizes various well established electron emission mechanisms including emission by heating the sample to provide the necessary kinetic energy for the electrons to overcome the

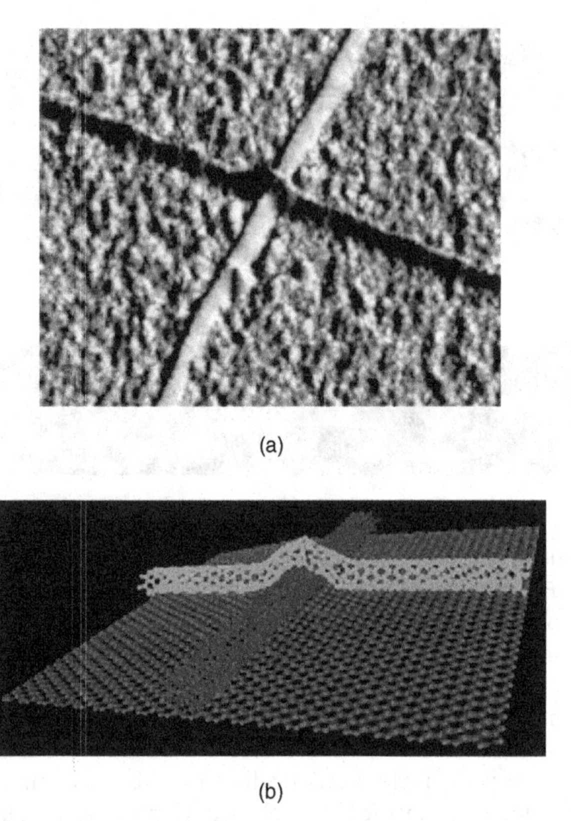

Figure 2.12. Nanotube cross QD. (a) AFM image of structure. (b) Simulated structure using molecular dynamics. Reprinted with permission from [38]. (© 2003 American Chemical Society)

workfunction barrier (thermionic emission), by applying a large external electric field to extract electrons by quantum mechanical tunneling through the potential barrier (field-emission), or by a combination of heat and external field (Schottky emission). Another way to provide the necessary escape energy to the electrons is by using light (photo-emission).

A sharp object enhances an externally applied electric field and makes electron emission easier. With the advent of nanotechnology and the possibility of fabricating very small and sharp tips with high aspect ratios, electron emission research has reached new dimensions and become a very popular subject in the world of nanodevice applications (Fig. 2.14). In particular, carbon nanotubes, with their extremely strong mechanical structure, high current carrying capacity (10^9 A/cm^2—orders of magnitude higher than copper and silver), and high aspect ratios, seem like ideal candidates for field-electron emitters [40–44]. Moreover, their unique structure could lead to interesting electron emission characteristics or be used to make more controllable electron emitters. In one experiment, electron

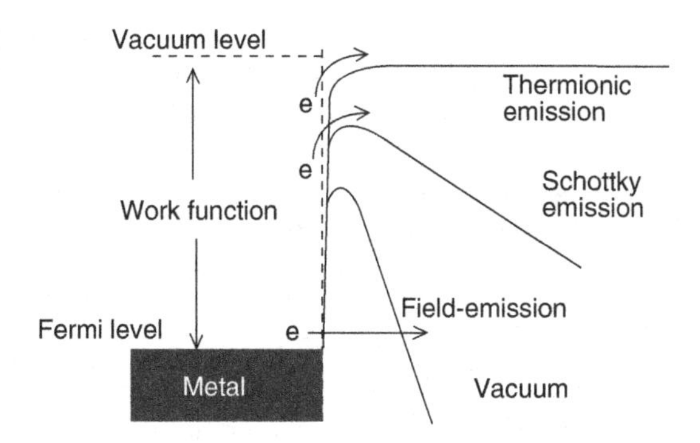

Figure 2.13. Various electron emission mechanisms.

emission from the tip of a nanotube was triggered by a combination of an external field and another electron beam hitting the nanotube with a very high multiplication factor (ratio of emitted electrons to the incoming ones) [45]. This could have many applications in nanoscale vacuum transistors and electron detectors. Other nanowires have also been investigated for electron emitter applications, including silicon, zinc oxide, and tungsten nanowires [46–48]. Controllable nanoscale electron emitters are in demand not only in electron beam lithography and microscopy, but also in vacuum nanoelectronics, flat panel display technologies [49, 50], free-electron analog to digital conversion [51], time-resolved electron holography, and synchrotrons.

2.4.3.2. Sensors and Actuators. In order to sense a molecule or particle, one needs a device that shows a measurable change in characteristics as the result of the proximity or adsorption of that particle. In nanostructures, due to the large surface-to-volume ratio, even a single particle sitting on the nanostructure could potentially have a large effect on the nanostructure properties such as electronic transport characteristics. Sensor applications have, therefore, been investigated in nanotubes and nanowires. The advantages of these sensors include very high sensitivity, low-voltage operation, low power consumption, and portability. Examples include gas sensors [52, 53], biosensors [54–56], and chemical sensors

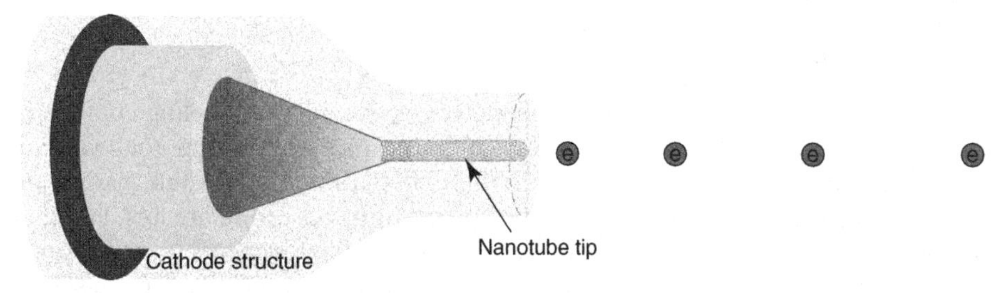

Figure 2.14. Schematic representation of a nanoelectron source for vacuum nanoelectronic applications.

[57, 58]. Similarly, the actuation mechanism in nanostructures is being studied actively. For instance, electromechanical actuation in carbon nanotubes has been demonstrated [59, 60] for applications such as artificial muscles.

2.4.3.3. Interconnect. One of the big challenges facing the integrated circuit industry is the issue of interconnect. As devices shrink and are packed more densely, routing the connections between them becomes more complicated; the interconnect lines themselves can severely affect the operation speed and power consumption of the circuit, to the point of becoming a limiting factor. Various approaches are being pursued to address this problem. One solution is to make 3D integrated circuits (as opposed to current planar or 2D circuits), which would create more ways of connecting devices and make shorter paths possible [61]. So far in this chapter we have been discussing the usage of nanomaterials to make devices. It is noteworthy that several research groups are also working on the possibility of using materials such as carbon nanotubes for interconnecting the devices on a circuit [62, 63]. Since these materials can carry large current densities and operate in the ballistic transport regime, this could lead to potentially faster interconnects with less power consumption.

2.4.3.4. Hydrogen Storage. Alternative energy sources are becoming increasingly important and research in this field is very active. Hydrogen-based energy requires the storage of hydrogen in a safe and efficient manner. Carbon nanotubes are being investigated for their potential in storing hydrogen in high densities [64].

2.4.3.5. Optical Properties of Nanostructures. The focus of this chapter has been mainly on the electronic properties of nanodevices. However, optical properties are closely related to electronic properties; microscale optoelectronic devices are in widespread use. Nanoscale optoelectronic devices are the subject of active research. Examples include lasers based on nanowires [65] and optical emitters [66, 67], detectors [68, 69], and antennas [70] based on carbon nanotubes.

2.5. MODELING AND SIMULATING NANODEVICES

2.5.1. The Art of Modeling

In the world of natural sciences and engineering, the art of modeling consists of describing a physical system in terms of parts and phenomena that we can understand based on our knowledge of nature, in the simplest possible manner, in such a way that none of the fundamental aspects of its operation are lost, and that we can use the model to predict/explain the results of experiments on the system. The model often involves a number of mathematical equations that can be solved to explain or to predict experimental outcomes. As a matter of fact, it could be argued that the most fundamental laws of nature, as we know them, are themselves models at the most basic and general level, and they attempt to describe all of nature and its phenomena in one shot. But as long as these models are not derived from more general models and are at the root of our modeling hierarchy, we call them fundamental laws.

Before the advent of fast calculating machines and computers, researchers had to rely mainly on solving these equations analytically (as much as possible) before attempting to plug in numbers. In the recent times, modeling is often accompanied by heavy numerical simulations; that is, once a good model is chosen for a given system, predictions on the behavior of the system under various circumstances are done by numerically solving the equations using fast computers.

One may ask that if we know the fundamental laws of nature, and if we have a fast computer to solve equations, why can we not always start from those fundamental laws (whether classical or quantum mechanical) and solve the relevant equations for any system in the most general sense? The answer is that this "first-principles" approach usually leads to a complex system of interrelated equations that even the most advanced supercomputers cannot handle numerically for a simple physical system. As a consequence, regardless of how good our computers have become, modeling has remained an integral part of studying natural systems. For example, think about an inert gas. The sciences of statistical mechanics and thermodynamics have provided us with an elegant model that describes an "ideal" gas that results in a very simple equation relating its pressure, volume, and temperature. For many a practical application, this model provides a description that is accurate enough to find those parameters, without the need for heavy calculations. In contrast, imagine the practically impossible task of solving the equations of motion for all the atoms in a given volume of gas (trillions of trillions of atoms) every time one needs to study the gas' expansion as a function of temperature.

With the introduction of every new branch of science and engineering, models start being developed (and gradually improved) to describe systems relevant to the discipline. There are well established models for gases, solids, and fluids that engineers use every day to design new devices and systems. We have been very successful so far in coming up with good physical models to describe systems ranging from giant spacecraft all the way down to microelectronic devices. In other words, when designing a spacecraft, an engineer does not start with the behavior of the individual atoms in it, but rather uses the much more high level models of mechanical engineering that describe the system containing a very large number of atoms. Similarly, even in microelectronic devices, a designer/engineer does not look at individual atoms or electrons, but uses the well established equations of microscopic electronic transport in a system containing a large number of atoms and electrons. These cases (spacecraft and microelectronic device) have in common the fact that in both systems there exists such a large collection of atoms that the details of the exact behavior of each individual atom is of little significance to the overall system, and it is the statistical averages of these behaviors that determine what happens. Thus, the fact that there is a large number of atoms to be accounted for is not only not a threat to modeling, but is actually the very reason why such successful macroscopic models can be developed and used.

2.5.2. The Challenge of Nanoscale Modeling

If you have two planets interacting with each other through gravitational forces, you can use Newtonian mechanics to trace their motion analytically. Add one more planet to the system (the so called three-body problem) and a closed-form analytical solution will generally not exist. This is also true for interacting electrons and protons. There is no analytical solution to the interacting many-body problem and one will need to resort to numerical methods to study such systems. Although simple in theory, the practical difficulty arises from the fact that with the addition of every new particle to the system, the amount of calculations needed grows considerably.

Now let us imagine that we want to study a system that includes only a few electrons. As discussed, we know the fundamental laws that describe the system—that is quantum mechanics. One tends to think that it should not be too hard to analyze such systems with a reasonably good computer. In fact, that might be the case when there are only a few particles involved. However, if there are hundreds or thousands of interacting atoms in the system, such as in nanoscale devices, then the exact solution of the equations of quantum mechanics for the entire system is absurdly out of reach for even the most advanced supercomputers of the modern day. At the same time, hundreds or thousands of atoms do not represent a large enough system where we could neglect the individual character of each atom and simply describe the system in terms of statistical averages. Thus, nanoscale devices present a regime of physical systems—in terms of their size—where we can neither neglect the atomic nature of the structure, nor treat it exactly using numerical methods and fast computers. It is noteworthy that this book focuses on new approaches to making computers with significantly more power than our traditional computers. When such efforts eventually become successful, our method of dealing with nanoscale devices from a modeling and simulation perspective could also change in major ways.

2.5.3. Smart Physical Modeling

So how do we deal with nanoscale devices? The solution lies in a smart approach to modeling. Often, a hybrid, multiscale approach is necessary to treat the system: more classical methods for aspects of the system that are less affected by the atomic nature, and more fundamental, quantum mechanical calculations where needed. Another important factor is that often a given property or experimental outcome could be explained based on a small part of the overall device. The art of modeling nanostructures consists of seeing what is essential to retain in a minimal subsystem (that we can handle with our available computing power in reasonable time) in order to study a certain subject relevant to the nanodevice in question. Nanodevices constitute a very new field and there is not, at least not yet, a general approach to modeling all nanodevices (such as is the case for larger devices like microelectronic transistors). Therefore, every researcher working in the field will, at some point, have to deal with modeling the specific system he/she is studying.

Nonetheless, there are many different levels of theory that have been developed over the years that could be applied based on the level/depth of the model itself. These include methods ranging from semi-classical treatments of atoms to single-electron methods to lower level methods directly relying on the solution of the fundamental equations of quantum mechanics for all the atoms. In the next few sections we shall briefly introduce some of the common methods in this context.

2.5.4. The Methods

Generally speaking, the methods of modeling and simulating nanodevices could be divided into two major categories: the so called *ab initio* (from the beginning) or first-principles approaches, and methods that have more of a phenomenological or empirical aspect to them. In first-principles methods, as the name suggests, the idea is to start with the basic physical equations for the entire system, such as the Schrödinger and Poisson equations, and solve them to find all the desired characteristics like electronic structure or transport properties. The Hartree–Fock theory is one such method that provides a framework for treating the quantum many-body problem by solving the Schrödinger equation. As discussed before, this could be a remarkably difficult task to accomplish (in terms of computational complexity) for a reasonably large system such as a nanodevice with hundreds of atoms. Therefore, other methods have been developed that rely on approximations and fitting parameters that, although not as accurate as first-principles approaches, depending on the problem at hand could still capture most of the underlying physics and handle much larger systems in a practical manner. Molecular dynamics and tight-binding approximations are some important examples in this category.

2.5.4.1. Ab Initio (First-Principles) Approaches. Quantum mechanics and relativity are the theories that describe the world around us (the best theories we have so far, though not necessarily the absolute truth!). In everyday experiences, quantum mechanical and relativistic effects are often masked and simpler theories, such as Newtonian mechanics, provide an accurate enough description. At very small scales, such as the nanoscale, quantum mechanical effects are dominant (although relativistic effects can be negligible in many nanodevice applications). We will begin our discussion here with a quick look at what is at the heart of quantum mechanics.

In one common approach to quantum mechanics, the behavior of a particle can be described by the Schrödinger equation:

$$\mathrm{H}\Psi = \varepsilon\Psi.$$

Here, H is the so called Hamiltonian operator of the system, including both potential and kinetic energy terms and ε represents the allowed energies in the system. Particles are described in terms of waves (Ψ). The amplitude of the wave represents the probability of the particle being at a given location. (You cannot know for sure where a particle is; rather, you have a probability distribution for its existence in various locations in space. To the best of our knowledge so far, this is the reality, but it might sound counterintuitive because this reality reveals itself only at small scales, and so we are not used to it in everyday life.) Note that the Schrödinger equation is, in general, a time-dependent differential equation. Here we have shown the time-independent case just for the sake of introduction. At the root of modeling a nanoscale device is the problem of writing the above equation for the system by including the proper terms in the Hamiltonian and solving it to find Ψ and ε. From Ψ and ε one can proceed to finding other parameters of interest in the system. Thus, for nanoscale modeling, learning quantum mechanics, at least at an introductory level, is the first necessary step. The reader is referred to [71, 72] as good examples of quantum mechanics textbooks. Finding a solution to the above equation without invoking empirical parameters is what *ab initio* approaches try to accomplish.

Slater Determinants and the Hartree–Fock Method. In one simplification, exact instantaneous electron–electron interactions are neglected in studying a system of atoms. The problem is reduced to that of a single electron moving in the "average" field created by the rest of the system. This approximation is very useful in many practical cases and captures the basic physics of the problem while greatly simplifying the computational complexity. When trying to solve the Schrödinger equation in this case, one can represent the overall wave function Ψ of the system as products of individual wave functions for electrons; this is basically the technique of separation of variables in solving differential equations. In addition, Pauli's exclusion principle that no two electrons with the same spin can reside in the same state must be satisfied. This, together with the fact that electrons are indistinguishable particles from one another, requires that the system wave function be antisymmetric with respect to simultaneous change of spin and space coordinates of electrons. A general form for the many-body wave function that satisfies this condition is the so called Slater determinant [73, 74]:

$$\Psi \propto \begin{vmatrix} \phi_1(1) & \phi_1(2) & \cdots & \phi_1(N) \\ \phi_2(1) & \phi_2(2) & \cdots & \phi_2(N) \\ \vdots & \vdots & \ddots & \vdots \\ \phi_N(1) & \phi_N(2) & \cdots & \phi_N(N) \end{vmatrix}$$

In the above each ϕ represents a single-electron wave function (including the spin part). Since swapping two rows or columns in a determinant changes the sign, this form satisfies the antisymmetry condition. In the Hartree–Fock method, the wave function of the system is obtained in the form of a Slater determinant. More on this method can be found in [74, 75]. Although this method neglects electron–electron correlation (i.e., the instantaneous nature of electron–electron interactions as they move in their complex orbitals), it can often provide a good insight into the physics of a problem and even make relatively good quantitative predictions as well. For instance, this method has shown to be quite effective in finding the occupied orbitals of carbon nanotubes and ionization energies of several molecular systems [76, 77].

THE DENSITY FUNCTIONAL THEORY. The density functional theory (DFT) is perhaps the most celebrated of approaches to solving the quantum many-body problem and has become increasingly popular in modeling and simulating solid materials including nanostructures. DFT provides a framework to accurately treat the quantum many-body problem. So, in principle it could be classified as a first-principles approach. However, later we will see that in practical implementations of DFT one needs to use approximate descriptions for the so called exchange–correlation energy. Therefore, DFT is not always necessarily considered an *ab initio* technique. Nonetheless, due to the fact that it does not essentially rely on empirical data, it is included in this section.

A paper published by Hohenberg and Kohn in 1964 [78] demonstrates that in treating an N-body problem, instead of working with the coordinates of all individual particles (electrons and nuclei), one can use a particle density distribution in space. This remarkably simplifies the problem and brings it to a much more manageable level. This is accomplished on the basis of the two following theorems, proven in that seminal paper [75, 78]:

1. The external potential acting on a fully interacting many-particle system in the ground state is determined uniquely by its electron density distribution in space. (Obviously we can add a constant to the potential everywhere without any real effect).
2. The ground state energy of the system (with the correct density) is smaller than the energy in any other configuration (with any other trial density).

The first theorem enables one to transform the many-body Schrödinger equation to an equation for the particle density functional. The second theorem provides a variational principle to solve the equation by finding the density distribution that minimizes the resulting energy for the system, which will be its ground state. A formal approach to finding the solution was proposed by Kohn and Sham in 1965 [79]. The key to this approach is that, using the introduction of an effective potential, the equation for the interacting many-body system is rewritten in such a way that it looks like the equation for a system of noninteracting particles. Then this equation is solved in a self-consistent manner as follows: one makes a guess for the density distribution, uses it to find the effective potential, uses the effective potential to find the orbitals (which is feasible since the equation looks like the one for noninteracting particles), and then calculates the electron density again from these orbitals. This process is continued until the density converges, and then one will have the ground state energy, orbitals and density. This cycle of iterations to find the solution is common to many methods including DFT and Hartree–Fock and therefore these methods are referred to as Self-Consistent Field (SCF) methods.

It should be noted that no miracle is performed in DFT. In following the Kohn–Sham method, a term called the exchange–correlation energy appears in the equations, due to both the quantum mechanical nature of particles and the fact that they are interacting. But this term is not known exactly. Therefore, the final solution will be approximate. There is indeed ongoing work on finding improved descriptions for the exchange–correlation energy in various systems [75]. One of the most widely established methods is the so called local density approximation (LDA). Here, the exchange–correlation energy for the system at each point in space is taken to be equal to that of a system with a uniform density that is the density in the actual system at that location.

Despite the fact that the lack of an exact exchange–correlation energy makes the results obtained by DFT approximate (like other methods), it is important to note that DFT is, in principle, a rigorous approach and is not based on a simplified model for the system. As such, it provides a way not only to understand what approximations one is making (when choosing the exchange–correlation functional) but also to improve those approximations in a systematic manner. In this regard, it also helps to put some of the other methods (that are based on simplified models of the real problem) into perspective and to provide a means of analyzing how accurate they are. For a very well written summary on DFT the reader is encouraged to consult [75]. For more detail [80–82] can be used.

2.5.4.2. Semi-Empirical Methods. As pointed out before, starting from first principles and solving the many-body quantum mechanical problem can often be a formidable task even for the most advanced computers. Therefore, methods have been developed that, in addition to theory, use empirical data to provide simplified descriptions of many-particle systems. Molecular dynamics, tight-binding, and Monte Carlo approaches are some of the popular methods of this genre.

CLASSICAL MOLECULAR DYNAMICS. The basic idea in molecular dynamics is to represent the interaction between each two atoms as a mean field arising from the interaction of all individual electrons and nuclei. For example, in the case of two carbon atoms instead of trying to calculate the force between them as the sum of the attractive and repulsive forces between all the individual electrons and protons (6 protons and 6 electrons in each atom, resulting in a total of $12*12=144$ complex interactions), a semi-empirical model for the force between the two atoms would be used. This force would be expressed as a function of the relative positioning of the two atoms with a number of parameters that are obtained by fitting the predictions of the model to available experimental data.

In very simple terms, the potential energy resulting from the interaction of two atoms could be something like the curve in Figure 2.15. Since force is the negative of the derivative of potential energy with respect to position, it can be seen that if

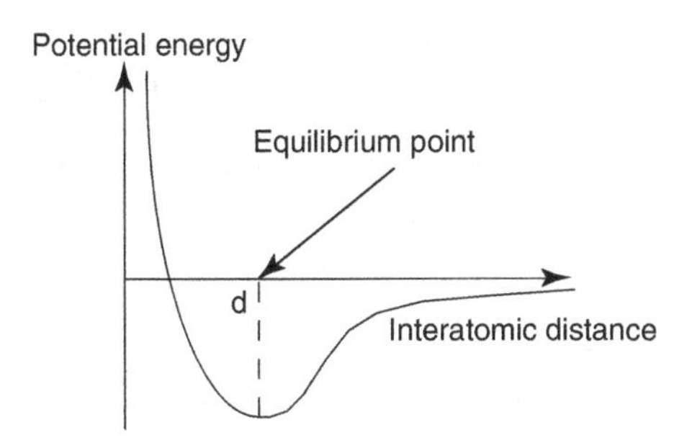

Figure 2.15. A simple representation of the interaction potential energy of two atoms.

the two atoms are a distance d apart, they are in equilibrium. If they are pushed closer, the force between them will be repulsive. If they are pulled farther, the force between them will be attractive. Both cases would lead to the atoms going back to their equilibrium distance.

Here is how molecular dynamics works in principle: consider a system of a number of atoms. We place them in close proximity and would like to find what the equilibrium position of each atom in the "relaxed" structure would be. We can start by calculating the force on each atom resulting from the rest of the atoms, using curves such as the one in Figure 2.15. Then, using Newton's equation of motion $F = ma$, we can calculate the acceleration on each atom due to this force. Now we let all the atoms move under these accelerations for a very small duration of time—a duration short enough that we can assume the force (and, as a result, acceleration) on each atom does not change significantly during that time. After this very short time, we have the new locations of the atoms and their new velocities. From the new locations we can calculate the new accelerations each will be feeling, and then move in time one more short step. If we repeat this process for a large enough number of time steps, at some point we will see that the atoms will not move significantly any longer as time passes. This will be the equilibrium or relaxed configuration of the system. One way to see this is to plot the total potential energy (the sum of all atom–atom interactions) of the system as a function of time. Eventually, this curve will saturate and asymptotically go toward the relaxed energy of the system.

One important issue to consider is that the system could converge to the metastable state closest to where we started from (in terms of initial choice of atomic positions). Therefore, in order to find the global minimum energy configuration of the structure, one may need to repeat the simulation with various choices of initial configurations and compare the end results in terms of total energy, or incorporate mechanisms to "kick" the system out of metastable configurations (using parameters such as thermal energy) with the hope that it will find its global minimum. Many variations of molecular dynamics are in use today and they can handle systems of relatively large numbers of atoms (thousands or tens of thousands of atoms) in reasonable time on present day computers. There have also been many valuable attempts at modeling the interatomic interactions for use in molecular dynamics simulations. One good example is the Brenner–Tersoff potentials that are well established for describing hydrocarbon systems [83, 84]. For a detailed review of molecular dynamics the reader is referred to [85].

TIGHT-BINDING. One way to approach a problem involving multiple atoms is to try to solve the Schrödinger equation by finding an expanded form for the expected solution. It is customary to write the solution in the form of a linear combination of atomic orbitals (LCAO), that is, as a weighted sum of wave functions that have been found for individual, isolated atoms. The parameters involved could be derived from first-principles methods. In many cases, however, these can be found from elsewhere, such as using experimental data. For instance, the core of an atom could be represented by a pseudopotential, instead of explicitly including all the individual electron contributions, just like in molecular dynamics. Thus, as opposed to exact, real orbitals, the atom will be represented by parameters derived by fitting to other available data. In such cases, especially when interactions with close neighbors are considered only, the LCAO method would typically be referred to as the tight-binding (TB) approximation. TB has the advantage of being computationally much less expensive than *ab initio* methods and therefore capable of providing, at least, a qualitative description of the behavior of a relatively large system quickly. For more on LCAO methods and TB, [86–88] can be studied.

MONTE CARLO. Monte Carlo methods are in place in many disciplines. The basic idea is to start from a known point and trace the behavior of a system by following a random-walk process. In each state, the system has the possibility of going to several of subsequent states with different probabilities. The choice of which one of those states the system will go into at each step is decided by a random process that is weighted by those different probabilities. For instance, consider the movement of an atom sitting on a crystalline surface through a loose bond. This atom moves around due to its thermal, kinetic energy. At each point, there exists several sites around it for it to move to. There would be a potential barrier in front of the atom to cross in order to make the transition to any of these sites, but the barriers have different heights and widths for the different sites, which correspond to different transition probabilities. The random walk of this atom could be traced using the Monte Carlo approach. Both semiclassical and quantum Monte Carlo methods are widely in use in nanoscale modeling [89, 90].

2.5.4.3. Use of Green's Functions. Green's functions are powerful tools in solving differential equations. Since the goal in nanosale modeling is to find the solution to the Schrödinger equation, which is a differential equation, Green's functions often come in very handy. In particular, they provide a useful way of dealing with problems involving open boundary systems. Consider, for example, a QD or SET connected to two large metal contacts on the sides. Although the device itself is small (nanoscale), the contacts are usually very large (microscale) and could be treated as semi-infinite systems for the purposes of the analysis. The use of Green's functions greatly simplifies the problem in such cases. In particular, problems involving electronic transport usually have contacts playing a crucial role and thus the so called nonequilibrium Green's function (NEGF) formalism is commonly used. Moreover, Green's functions provide a way of dealing with interparticle interactions in electronic transport modeling. For a study of Green's functions with application to transport problems in nanostructures the reader is referred to [91–93].

2.5.4.4. System-Level Modeling. Although most of the research in the field of nanodevices is currently focused on the device itself, it is noteworthy that valuable efforts are being made on system-level issues, such as designing circuits based on these novel nanodevices and analyzing such systems. Researchers have worked on circuit design paradigms for future computers based on QD-type devices such as quantum dot cellular automata (QCA) [94]. Although this research is very new, even compared to the field of nanodevices itself, there are already examples of computer-aided design (CAD) tools that have been developed for circuit design applications. An example is the software called QCADesigner [95].

2.6. CONCLUSIONS

There are various methods that could be used for the analysis of a given device depending on the aspects being studied. For instance, classical molecular dynamics usually gives an accurate enough result for the mechanical structure of the system, but does not provide much in the way of electronic structure. DFT gives very accurate results for electronic structure and can be used to relax the mechanical structure of the system, although it is extremely expensive computationally. To illustrate how these various methods could be used in practice, let us consider the example of a carbon nanotube with a length of 10 nm connected between two metal electrodes to form a quantum dot. One way to approach the problem of electronic transport in this device is the following: the coordinates of the carbon atoms as well as those of the metal atoms in the contact electrodes are generated based on the lattice structures of the nanotube and the electrode. However, we know that once a short section (10 nm) of the nanotube is cut and put in contact with the electrodes, the atoms, especially at the edges, rearrange themselves slightly. A molecular dynamics simulation is run to determine the exact location of all the atoms in the metal-nanotube-metal structure. Then both DFT and TB are used to determine the electronic structure (orbitals or band structure), while using the NEGF approach to study the transport characteristics. If the current as a function of applied voltage is to be found, it would mean a series of electronic structure calculations with different system Hamiltonians (each reflecting a single bias point). This is very time consuming with DFT. So, one could use TB for this, and at the same time use DFT for some of the bias points to have a benchmark for comparison in order to evaluate the accuracy of the TB results.

The objective of this chapter was to provide an introduction to the world of nanodevices. We briefly reviewed the motivations behind making devices at the small scale, the process of how they are made, some of their applications, and some of the common methods of studying and simulating them. Entering any level of detail in those subjects is obviously beyond the scope of this single chapter, but the author hopes that the chapter shows the big picture of what to expect in this area and how to pursue a given direction of interest.

REFERENCES

1. R. P. Feynman. There's plenty of room at the bottom [data storage]. *Journal of Microelectromechanical Systems*, 1: pp 60–66, 1992.
2. Y. Wu, Y. Cui, L. Huynh, C. J. Barrelet, D. C. Bell, and C. M. Lieber. Controlled growth and structures of molecular-scale silicon nanowires. *Nano Letters*, 4: pp 433–436, 2004.
3. D. Wang, Q. Wang, A. Javey, R. Tu, H. Dai, H. Kim, P. C. McIntyre, T. Krishnamohan, and K. C. Saraswat. Germanium nanowire field-effect transistors with SiO_2 and high-k HfO_2 gate dielectrics. *Applied Physics Letters*, 83: pp 2432–2434, 2003.
4. Y. C. Kong, D. P. Yu, B. Zhang, W. Fang, and S. Q. Feng. Ultraviolet-emitting ZnO nanowires synthesized by a physical vapor deposition approach. *Applied Physics Letters*, 78: pp 407–409, 2001.
5. Y. Huang, X. Duan, Y. Cui, and C. M. Lieber, Gallium nitride nanowire nanodevices. *Nano Letters*, 2: pp 101–104, 2002.
6. S. Iijima. Helical microtubules of graphitic carbon. *Nature*, 354: pp 56–58, 1991.
7. R. Saito, G. Dresselhaus, and M. S. Dresselhaus. *Physical Properties of Carbon Nanotubes*. London: Imperial College Press, 1998.
8. J. D. Plummer, M. Deal, and P. B. Griffin. *Silicon VLSI Technology: Fundamentals, Practice, and Modeling*. Upper-Saddle River, NJ: Prentice Hall, 2000.
9. C. Liu. *Foundations of MEMS*. Upper Saddle River, NJ: Pearson/Prentice Hall, 2006.
10. J. Kong, H. T. Soh, A. M. Cassell, C. F. Quate, and H. Dai. Synthesis of individual single-walled carbon nanotubes on patterned silicon wafers. *Nature*, 395: pp 878–881, 1998.
11. A. Ural, Y. Li, and H. Dai. Electric-field-aligned growth of single-walled carbon nanotubes on surfaces. *Applied Physics Letters*, 81: p 3464, 2002.
12. A. Nojeh, A. Ural, R. F. Pease, and H. Dai. Electric-field-directed growth of carbon nanotubes in two dimensions. *Journal of Vacuum Science and Technology B*, 22: pp 3421–3425, 2004.
13. S. Huang, X. Cai, and J. Liu. Growth of millimeter-long and horizontally aligned single-walled carbon nanotubes on flat substrates. *Journal of the American Chemical Society*, 125: pp 5636–5637, 2003.

14. S. Huang, M. Woodson, R. Smalley, and J. Liu. Growth mechanism of oriented long single walled carbon nanotubes using fast-heating chemical vapor deposition process. *Nano Letters*, 4: pp 1025–1028, 2004.
15. D. Goldhaber-Gordon, M. S. Montemerlo, J. C. Love, G. J. Opiteck, and J. C. Ellenbogen. Overview of nanoelectronic devices. *Proceedings of the IEEE*, 85: pp 521–540, 1997.
16. M. A. Kastner. The single electron transistor and artificial atoms. *Annalen der Physik (Leipzig)*, 9: pp 11–12, 2000.
17. L. J. Geerligs, V. F. Anderegg, P. A. M. Holweg, J. E. Mooij, H. Pothier, D. Esteve, C. Urbina, and M. H. Devoret. Frequency-locked turnstile device for single electrons. *Physical Review Letters*, 64: pp 2691–2694, 1990.
18. C. W. J. Beenakker. Theory of Coulomb-blockade oscillations in the conductance of a quantum dot. *Physical Review B*, 44: pp 1646–1656, 1991.
19. Y. Alhassid. The statistical theory of quantum dots. *Reviews of Modern Physics*, 72: pp 895–968, 2000.
20. K. Matsumoto, M. Ishii, K. Segawa, Y. Oka, B. J. Vartanian, and J. S. Harris. Room temperature operation of a single electron transistor made by the scanning tunneling microscope nanooxidation process for the TiO[sub x]/Ti system. *Applied Physics Letters*, 68: pp 34–36, 1996.
21. Y. Cui, Z. Zhong, D. Wang, W. U. Wang, and C. M. Lieber. High performance silicon nanowire field effect transistors. *Nano Letters*, 3: pp 149–152, 2003.
22. J. Goldberger, D. J. Sirbuly, M. Law, and P. Yang. ZnO nanowire transistors. *Journal of Physical Chemistry B*, 109: pp 9–14, 2005.
23. S. Han, W. Jin, D. Zhang, T. Tang, C. Li, X. Liu, Z. Liu, B. Lei, and C. Zhou. Photoconduction studies on GaN nanowire transistors under UV and polarized UV illumination. *Chemical Physics Letters*, 389: pp 176–180, 2004.
24. J. Appenzeller, J. Knoch, V. Derycke, R. Martel, S. Wind, and P. Avouris. Field-modulated carrier transport in carbon nanotube transistors. *Physical Review Letters*, 89: p 126801, 2002.
25. A. Javey, H. Kim, M. Brink, Q. Wang, A. Ural, J. Guo, P. McIntyre, P. McEuen, M. Lundstrom, and H. Dai. High-k dielectrics for advanced carbon-nanotube transistors and logic gates. *Nature Materials*, 1: pp 241–246, 2002.
26. A. Javey, J. Guo, Q. Wang, M. Lundstrom, and H. Dai. Ballistic carbon nanotube field-effect transistors. *Nature*, 424: pp 654–657, 2003.
27. A. Bachtold, P. Hadley, T. Nakanishi, and C. Dekker. Logic circuits with carbon nanotube transistors. *Science*, 294: p 1317, 2001.
28. S. Li, Z. Yu, S. F. Yen, W. C. Tang, and P. J. Burke. Carbon nanotube transistor operation at 2.6 GHz. *Nano Letters*, 4: pp 753–756, 2004.
29. J. Wang, A. Rahman, A. Ghosh, G. Klimeck, and M. Lundstrom. Performance evaluation of ballistic silicon nanowire transistors with atomic-basis dispersion relations. *Applied Physics Letters*, 86: p 093113, 2005.
30. J. P. Clifford, D. L. John, L. C. Castro, and D. L. Pulfrey. Electrostatics of partially gated carbon nanotube FETs. *IEEE Transactions on Nanotechnology*, 3: pp 281–286, 2004.

31. D. L. John and D. L. Pulfrey. Switching-speed calculations for Schottky-barrier carbon nanotube field-effect transistors. *Journal of Vacuum Science and Technology A: Vacuum, Surfaces, and Films*, 24: p 708, 2006.
32. L. C. Castro and D. L. Pulfrey. Extrapolated fmax for carbon nanotube field-effect transistors. *Nanotechnology*, 17: pp 300–304, 2006.
33. N. Mason, M. J. Biercuk, and C. M. Marcus. Local gate control of a carbon nanotube double quantum dot. *Science (AAAS)*, 303: pp 655–658, 2004.
34. M. T. Woodside and P. L. McEuen. Scanned Probe Imaging of Single-Electron Charge States in Nanotube Quantum Dots. *Science*, 296: pp 1098–1101, 2002.
35. S. Peng and K. Cho. Nano electro mechanics of semiconducting carbon nanotube. *Journal of Applied Mechanics (Transactions of the ASME)*, 69: pp 451–453, 2001.
36. H. W. C. Postma, T. Teepen, Z. Yao, M. Grifoni, and C. Dekker. Carbon Nanotube Single-Electron Transistors at Room Temperature. *Science*, 293: pp 76–79, July 2001.
37. B. Shan, G. W. Lakatos, S. Peng, and K. Cho. First-principles study of band-gap change in deformed nanotubes. *Applied Physics Letters*, 87: p 173109, 2005.
38. A. Nojeh, G. W. Lakatos, S. Peng, K. Cho, and R. F. W. Pease. A carbon nanotube cross structure as a nanoscale quantum device. *Nano Letters*, 3: pp 1187–1190, 2003.
39. L. Vitali, M. Burghard, P. Wahl, M. A. Schneider, and K. Kern. Local pressure-induced metallization of a semiconducting carbon nanotube in a crossed junction. *Physical Review Letters*, 96: p 86804, 2006.
40. K. A. Dean and B. R. Chalamala. Current saturation mechanisms in carbon nanotube field emitters. *Applied Physics Letters*, 76: pp 375–377, 2000.
41. N. de Jonge, M. Allioux, J. T. Oostveen, K. B. K. Teo, and W. I. Milne. Optical performance of carbon-nanotube electron sources. *Physical Review Letters*, 94: pp 186807–186804, 2005.
42. N. de Jonge and J-M Bonard. Carbon nanotube electron sources and applications. *Philosophical Transactions of the Royal Society A: Mathematical, Physical and Engineering Sciences*, 362: pp 2239–2266, 2004.
43. T. Fujieda, K. Hidaka, M. Hayashibara, T. Kamino, H. Matsumoto, Y. Ose, H. Abe, T. Shimizu, and H. Tokumoto. In situ observation of field emissions from an individual carbon nanotube by Lorenz microscopy. *Applied Physics Letters*, 85: pp 5739–5741, 2004.
44. P. G. Collins and A. Zettl. Unique characteristics of cold cathode carbon-nanotube-matrix field emitters. *Physical Review B*, 55: 1997 p 9391, 1997.
45. A. Nojeh, W. K. Wong, E. Yieh, R. F. Pease, and H. Dai. Electron beam stimulated field-emission from single-walled carbon nanotubes. *Journal of Vacuum Science and Technology B: Microelectronics and Nanometer Structures*, 22: pp 3124–3127, 2004.
46. A. I. Klimovskaya, Y. M. Litvin, Y. Y. Moklyak, A. A. Dadykin, T. I. Kamins, and S. Sharma. Field-electron emission at 300 K in self-assembled arrays of silicon nanowires. *Applied Physics Letters*, 89: pp 093122–093123, 2006.
47. C. J. Lee, T. J. Lee, S. C. Lyu, Y. Zhang, H. Ruh, and H. J. Lee. Field emission from well-aligned zinc oxide nanowires grown at low temperature. *Applied Physics Letters*, 81: p 3648, Nov 2002.

48. A. G. Umnov, Y. Shiratori, and H. Hiraoka. Giant field amplification in tungsten nanowires. *Applied Physics A: Materials Science and Processing*, 77: pp 159–161, 2003.
49. J. L. Kwo, M. Yokoyama, W. C. Wang, F. Y. Chuang, and I. N. Lin. Characteristics of flat panel display using carbon nanotubes as electron emitters. *Diamond and Related Materials*, 9: pp 1270–1274, 2000.
50. S. G. Yu, S. Jin, W. Yi, J. Kang, T. Jeong, Y. Choi, J. Lee, J. Heo, N. S. Lee, and J. B. Yoo. Undergate-type Triode Carbon Nanotube Field Emission Display with a Microchannel Plate. *Japanese Journal of Applied Physics*, 40: pp 6088–6091, 2001.
51. K. Ioakeimidi, R. F. Leheny, S. Gradinaru, P. R. Bolton, R. Aldana, K. Ma, J. E. Clendenin, J. S. Harris, and R.F.W. Pease. Photoelectronic analog-to-digital conversion: sampling and quantizing at 100 Gs/s. *IEEE Transactions on Microwave Theory and Techniques*, 53: pp 336–342, 2005.
52. A. Modi, N. Koratkar, E. Lass, B. Wei, and P. M. Ajayyan. Miniaturized gas ionization sensors using carbon nanotubes. *Nature*, 424: pp 171–174, 2003.
53. C. S. Rout, A. Govindaraj, and C. N. R. Rao. High-sensitivity hydrocarbon sensors based on tungsten oxide nanowires. *Journal of Materials Chemistry*, 16: pp 3936–3941, 2006.
54. S. Sotiropoulou and N. A. Chaniotakis. Carbon nanotube array-based biosensor. *Analytical and Bioanalytical Chemistry*, 375: pp 103–105, 2003.
55. K. Besteman, J. O. Lee, F. G. M. Wiertz, H. A. Heering, and C. Dekker. Enzyme-coated carbon nanotubes as single-molecule biosensors. *Nano Letters*, 3: pp 727–730, 2003.
56. G. Zheng, F. Patolsky, Y. I. Cui, W. U. Wang, and C. M. Lieber. Multiplexed electrical detection of cancer markers with nanowire sensor arrays. *Nature Biotechnology*, 23: pp 1294–1301, 2005.
57. J. Kong, N. R. Franklin, C. Zhou, M. G. Chapline, S. Peng, K. Cho, and H. Dai. Nanotube molecular wires as chemical sensors. *Science*, 287: pp 622–625, 2000.
58. X. Y. Xue, Y. J. Chen, Y. G. Wang, and T. H. Wang. Synthesis and ethanol sensing properties of ZnSnO nanowires. *Applied Physics Letters*, 86: p 233101, 2005.
59. R. H. Baughman, C. Cui, A. A. Zakhidov, Z. Iqbal, J. N. Barisci, G. M. Spinks, G. G. Wallace, A. Mazzoldi, D. De Rossi, and A. G. Rinzler. Carbon nanotube actuators. *Science*, 284: pp 1340–1344, 1999.
60. T. Mirfakhrai, J. Oh, M. Kozlov, E. C. W. Fok, M. Zhang, S. Fang, R. H. Baughman, and J. D. W. Madden. Electrochemical actuation of carbon nanotube yarns. *Smart Materials and Structures*, 16: pp S243–S249, 2007.
61. M. Ieong, K. W. Guarini, V. Chan, K. Bernstein, R. Joshi, J. Kedzierski, and W. Haensch. Three-dimensional CMOS devices and integrated circuits, Custom Integrated Circuits Conference, 2003. Proceedings of the IEEE 2003: pp 207–213, 2003.
62. A. Raychowdhury and K. Roy. Modeling of metallic carbon-nanotube interconnects for circuit simulations and a comparison with Cu interconnects for scaled technologies. *IEEE Transactions on Computer-Aided Design of Integrated Circuits and Systems*, 25: pp 58–65, 2006.
63. A. Nieuwoudt and Y. Massoud. Understanding the impact of inductance in carbon nanotube bundles for VLSI interconnect using scalable modeling techniques. *IEEE Transactions on Nanotechnology*, 5: pp 758–765, 2006.

64. F. L. Darkrim, P. Malbrunot, and G. P. Tartaglia. Review of hydrogen storage by adsorption in carbon nanotubes. *International Journal of Hydrogen Energy*, 27: pp 193–202, 2002.
65. M. H. Huang, S. Mao, H. Feick, H. Yan, Y. Wu, H. Kind, E. Weber, R. Russo, and P. Yang. Room-temperature ultraviolet nanowire nanolasers. *Science*, 292: p 1897, 2001.
66. M. Freitag, J. Chen, J. Tersoff, J. C. Tsang, Q. Fu, J. Liu, and P. Avouris. Mobile ambipolar domain in carbon-nanotube infrared emitters. *Physical Review Letters*, 93: pp 076803–076804, 2004.
67. D. L. McGuire and D. L. Pulfrey. A multi-scale model for mobile and localized electroluminescence in carbon nanotube field-effect transistors. *Nanotechnology*, 17: pp 5805–5811, 2006.
68. M. Freitag, Y. Martin, J. A. Misewich, R. Martel, and P. Avouris. Photoconductivity of single carbon nanotubes. *Nano Letters*, 3: pp 1067–1071, 2003.
69. P. Servati, A. Colli, S. Hofmann, Y. Q. Fu, P. Beecher, Z. A. K. Durrani, A. C. Ferrari, A. J. Flewitt, J. Robertson, and W. I. Milne. Scalable silicon nanowire photodetectors. *Physica E: Low-dimensional Systems and Nanostructures*, 3: pp 64–66, 2007.
70. Y. Wang, K. Kempa, B. Kimball, J. B. Carlson, G. Benham, W. Z. Li, T. Kempa, J. Rybczynski, A. Herczynski, and Z. F. Ren. Receiving and transmitting light-like radio waves: antenna effect in arrays of aligned carbon nanotubes. *Applied Physics Letters*, 85: pp 2607–2609, 2004.
71. J. J. Sakurai. In: S. F. Tan, editor. *Modern Quantum Mechanics*. Reading, MA: Addison-Wesley, 1994.
72. W. A. Harrison. *Applied Quantum Mechanics*. Singapore: World Scientific, 2000.
73. J. C. Slater. The Theory of Complex Spectra. *Physical Review*, 34: pp 1293–1322, 1929.
74. A. Szabó and N. S. Ostlund. *Modern Quantum Chemistry: Introduction to Advanced Electronic Structure Theory*. New York: Dover Publications, 1996.
75. A. Gonis. *Theoretical Materials Science: Tracing the Electronic Origins of Materials Behavior*. Warrendale, PA: Materials Research Society, 2000.
76. A. Rosén, D. E. Ellis, H. Adachi, and F. W. Averill. Calculations of molecular ionization energies using a self-consistent-charge Hartree–Fock–Slater method. *The Journal of Chemical Physics*, 65: p 3629, 1976.
77. A. Nojeh, B. Shan, K. Cho, and R. F. W. Pease. Ab initio modeling of the interaction of electron beams and single-walled carbon nanotubes. *Physical Review Letters*, 96: p 56802, 2006
78. P. Hohenberg and W. Kohn. Inhomogeneous electron gas. *Physical Review*, 136: p B864, 1964.
79. W. Kohn and L. J. Sham. Self-consistent equations including exchange and correlation effects. *Physical Review*, 140: p A1133, 1965.
80. R. G. Parr and W. Yang. *Density-Functional Theory of Atoms and Molecules*. Oxford, UK: Oxford University Press, 1989
81. S. Lundqvist and N. H. March. *Theory of The Inhomogeneous Electron Gas*. New York: Plenum Press, 1983.
82. R. M. Dreizler and E. K. U. Gross. *Density Functional Theory*. New York: Springer, 1990.
83. J. Tersoff. Modeling solid-state chemistry: interatomic potentials for multicomponent systems. *Physical Review B*, 39: pp 5566–5568, 1989.

84. D. W. Brenner. Empirical potential for hydrocarbons for use in simulating the chemical vapor deposition of diamond films, *Physical Review B*, vol. 42: pp 9458–9471, 1990.
85. P. B. Balbuena and J. M. Seminario. *Molecular Dynamics: From Classical to Quantum Methods*. Amsterdam: Elsevier, 1999.
86. J. Lowe and K. Peterson. *Quantum Chemistry. Amsterdam*. Boston: Elsevier Academic Press, 2006.
87. W. A. Harrison. *Elementary Electronic Structure*. Singopore: World Scientific, 2004.
88. H. Eschrig. *Optimized LCAO Method and the Electronic Structure of Extended Systems*. Berlin: Springer-Verlag, 1989.
89. M. Suzuki, editor. Quantum Monte Carlo Methods in Equilibrium and Nonequilibrium Systems: *Proceedings of the Ninth Taniguchi International Symposium,* Susono, Japan, November 14–18, 1986. Berlin: Springer-Verlag, 1987.
90. M. H. Kalos, editor. *Monte Carlo Methods in Quantum Problems*. New York: D. Reidel Pub. Co.; sold and distributed in the U.S. and Canada by Kluwer Academic Publishers, 1984.
91. S. Datta. *Electronic Transport in Mesoscopic Systems*. Cambridge: Cambridge University Press, 2003.
92. M. S. Lundstrom and J. Guo. *Nanoscale Transistors: Device Physics, Modeling and Simulation*. New York: Springer, 2006.
93. E. N. Economou. *Green's Functions in Quantum Physics*. Berlin: Springer-Verlag, 2006.
94. W. Porod. Quantum-dot devices and quantum-dot cellular automata. *International Journal of Bifurcation and Chaos*, 7: pp 2199–2218, 1997.
95. K. Walus, T. J. Dysart, G. A. Jullien, and R. A. Budiman. QCADesigner: a rapid design and simulation tool for quantum-dot cellular automata. *IEEE Transactions on Nanotechnology*, 3: pp 26–31, 2004.

3

QUANTUM COMPUTING

John H. Reif

Quantum computation (QC) is a type of computation where unitary and measurement operations are executed on linear superpositions of basis states. This chapter provides a brief introduction to QC. We begin with a discussion of basic models for QC such as quantum TMs, quantum gates, and circuits and related complexity results. We then discuss a number of topics in quantum information theory including bounds for quantum communication and I/O complexity, methods for quantum data compression and quantum error correction (that is, techniques for decreasing decoherence errors in QC), Furthermore, we enumerate a number of methodologies and technologies for doing QC. Finally, we discuss resource bounds for QC including bonds for processing time, energy, and volume, particularly emphasizing challenges in determining volume bounds for observation apparatus.

3.1. INTRODUCTION

3.1.1. Reversible Computations

Reversible computations are computations where each state transformation is a reversible function, so that any computation can be reversed without loss of information. Landauer [1] showed that irreversible computations must generate heat in the computing process, and that reversible computations have the property that if executed slowly enough, they (in the limit) can consume no energy in an adiabatic computation. Bennett [2] (also see Bennett, Landauer [3], Landauer [4], Toffoli [5]) showed that any computing machine (e.g., an abstract machine such as

Bio-Inspired and Nanoscale Integrated Computing. Edited by Mary Mehrnoosh Eshaghian-Wilner

a Turing Machine) can be transformed to do reversible computations. Bennett's reversibility construction required extra space to store information to insure reversibility; Li Vitanyi [6] give tradeoffs between time and space in the resulting reversible machine. An innovative technique due to Bennett [2, 7] can be used to make reversible functions bijective, as required for quantum computations. Given a bijective function f, suppose we can reversibly compute in time $T(x)$ a bijective function f and it's inverse f^{-1} using auxiliary registers for storage of the input. He proves in time $O(T(n))$ we can also reversibly compute the bijective mapping, $(x, 0, 0) \rightarrow (f(x), 0, 0)$, without use of auxiliary registers for storage of the input.

3.1.2. An Introduction to Quantum Computation

Computations and methods that do not make use of quantum mechanics will be termed *classical.* In contrast, *quantum computation* (QC) applies quantum mechanics to do computation. A single molecule (or collection of particles and/or atoms) may have a number n of degrees of freedom known as *qubits.* Associated with each fixed setting X of the n qubits to Boolean values is a *basis state* denoted $|a\rangle$.

Quantum mechanics allows for a linear superposition (also termed an *entangled quantum state*) of these basis states to exist simultaneously. Each basis state $|a\rangle$ of the superposition is assigned a given complex amplitude α; this is denoted $\alpha \; |a\rangle$. *Unitary transformations* are reversible operations on the superpositions which can be represented by unitary matrices A (e.g., permutation matrices, rotation matrices, and the matrices of Fourier transforms) where $AA^T = I$. The sum of the squares of the magnitudes of the amplitudes of all basis states is 1. This sum remains invariant due to the application of a unitary transformations. The Hilbert space H_n is the set of all possible such linear superpositions.

QC is a method of computation where various operations can be executed on these superpositions:

- *unitary operations*, and
- *observation operations*, which allow for the (strong) measurement of each qubit, providing a mapping from the current superposition to a superposition where the measured qubit is assigned a Boolean value with probability given by the square of the amplitude of the qubit in its original superposition.

Elementary unitary operations that suffice for any quantum computation over qubits (see [8] and [9]) include a conditional form of the conditional XOR operation $\oplus$, the Boolean operation NOT, and a constant Boolean operation yielding 0. The time bound for a quantum computations is defined to be the number of such elementary unitary operations.

3.1.3. Surveys of QC

The following are reviews and surveys have been made of QC: Bennett [10], Barenco [11], Benio [12], Brassard [13, 14], Haroche, Raimond [15], Brassard [16], Preskill [17], Scarani [18], Steane [19], and Vedral, Plenio [20]. Also, Taubes [21] and Gershenfeld, Chuang [22] give popular press descriptions of QC. The following are texts on quantum computing.

- Overviews: [23–26]
- Quantum information processing: [27–34]
- Quantum cryptography: [35–37]
- Quantum coding theory: [38, 39]
- Quantum algorithms: [40]
- Experimental implementation of quantum computation: [41–45]

3.1.4. Initial Work in QC

Feynman [46, 47] and Benioff [48] were the first to suggest the use of quantum mechanical principles for doing computation. Deutsch and Jozsa [49] give the first example of a quantum algorithm that gave a rapid solution of an example problem, where the problem (for a given a black box function) is not quickly solvable by any deterministic conventional computing machine. But their problem could be quickly solved using randomization. Bernstein and Vazirani [50] then provided the first example of a fast quantum algorithm for a problem that could not be quickly solved by conventional computing machines even using randomization. (See also Costantini, Smeraldi [51] for a generalization of Deutsch's example and see Collins et al. [52] for a simplified Deutsch–Jozsa algorithm, and Jozsa [53–55] for further work in quantum computation and complexity.)

3.1.5. Organization of this Chapter

In Section 3.1 we introduced QC. In Section 3.2 we introduce formal quantum computing models and in Section 3.3 we discuss quantum complexity classes. Next we overview key topics concerning quantum information processing: Section 3.4 discusses bounds for quantum communication; Section 3.5 discusses methods for quantum errorless compression; Section 3.6 discusses methods for quantum error coding; and Section 3.7 describes methods for quantum cryptography. In Section 3.8 we discuss further algorithmic applications of QC. In Section 3.9 we enumerate various technologies for doing QC. In Section 3.10 we review of the resource bounds of quantum computing as compared with the resources required by classical methods for computation. In Section 3.11 we conclude the chapter. In Appendix 3.12 we discuss the challenge of providing volume bounds for observation apparatus when doing QC.

3.2. QUANTUM COMPUTING MODELS

- *Quantum TMs and other Automata.* Deutsch [56] gave the first formal description of a quantum computer, known as a *quantum TM*. The tape contents of the TM are qubits. *Quantum configurations* of the QTM are superpositions of (classical) TM configurations. A transition of the QTM is a unitary mapping on quantum configurations of the QTM. Thus, a computation of the QTM is a unitary mapping from the initial quantum configuration to the final quantum configuration. Various papers generalize machines and automata to the quantum case. Moore, Crutchfield [57] propose quantum finite-state and push-down automata, and regular and context-free grammars; they generalize several formal language and automata theorems, e.g., pumping lemmas, closure properties, rational and algebraic generating functions, and Greibach normal form. Kondacs and Watrous [58] partially characterize the power of quantum finite state automata. Dunlavey [59] gives a space efficient simulation of a deterministic finite state machine (FSM) on a quantum computer (using Grover's search algorithm discussed below). Watrous [60] investigates quantum cellular automata and Dürr et al. [61, 62] give decision procedures for unitary linear (one dimensional) quantum cellular automata.
- *Quantum Gates.* A set of Boolean gates are *universal* if any Boolean operation on arbitrarily many bits can be expressed as compositions of these gates. Toffoli [5] defined an extended XOR 3-bit gate (which is an XOR gate condition on one of the inputs and is known as the *Toffoli gate*) and showed that this gate, in combination with certain 1-bit gates, is universal. A set of quantum qubit gates are *universal* for Boolean computations for QC if any unitary operation on arbitrarily many qubits can be expressed as compositions of these gates. Deutsch defined the extended quantum XOR 3-qubit gate (known as the Deutsch–Toffoli gate) and proved this gate, in combination with certain one qubit gates, is universal. Barenco [63], Sleator et al. [64], Barenco et al. [65], and DiVincenzo [66] proved the 2-qubit XOR gates with certain 1-qubit gates can implement the Deutsch–Toffoli gate, so are universal for QC (also see Smolin and DiVincenzo [67], DiVincenzo et al. [68], Poyatos et al. [69], and Mozyrsky et al. [70, 71]). Lloyd [72] then proved that almost any 2-qubit quantum logic gate (with certain 1-qubit gates) is universal for QC. Monroe et al. [73], DiVincenz et al. [74] gave experimental demonstrations of quantum gates [75]. Defined a quantum computing model known as a *quantum gate array*, which allows execution of a (possibly cyclic) sequence of quantum gates, each input is a qubit, and each gate computes a unitary transformation.
- *Quantum Circuits.* Yao [76] restricted the concept to (acyclic) *quantum circuits,* which are a generalization of Boolean logic circuits for quantum gates. It suffices that a quantum circuit use only these universal gates. Yao [76] proved that QTM computations are equivalent to uniform quantum

circuit families. Bernstein and Vazirani [50] showed that quantum gates of only logarithmic accuracy suffice for polynomial time quantum circuits. Aharonov et al. [77] discusses a generalization of quantum circuits to allow mixed states, where measurements can be done in the middle of the computation, and showed that such quantum circuits are equivalent in computational power to standard quantum circuits. This generalized an earlier result of Bernstein and Vazirani [50] that showed that all observation operations can be pushed to the end of the computation by repeated use of a quantum XOR gate construction. Aharonov et al. [78] considered an adiabatic model of quantum computation and showed it is equivalent to standard quantum computation.

- *Computer Simulations of QC.* Obenland, Despain [79–81] have given efficient computer simulations of QC, including errors and decoherence, and Cerf, Koonin [82] have given Monte Carlo simulations of QC.

3.3. COMPLEXITY BOUNDS FOR QC

3.3.1. Quantum Complexity Classes and Structural Complexity

Berthiaume, Brassard [83] survey open QC structural complexity problems (also see Berthiaume [84]). QC can clearly execute deterministic and randomized computations with no slow down. P (NP, QP, respectively) are the class of problems solved by deterministic (nondeterministic, quantum, respectively) polynomial time computations. Thus QP is the quantum analog of the time efficient class P. It is not known if QP contains NP, that is, if QC can solve NP search problems in polynomial time. It is also not known whether QP is a superset of P, nor if there are any problems QC can solve in polynomial time that are not in P (but this is true given quantum oracles; see Berthiaume, Brassard [85, 86], Machta [87], and van Dam [88, 89] for complexity bounds for computing with quantum oracles).

3.3.2. Bounded Precision QC

Let *BQP* be the class of polynomial time quantum computations that are computed within bounded error. Most of the algorithms we will mention (such as Shor's) are in the class BQP [50]. Showed that BQP computations can be done using unitary operations with a fixed irrational rotation. Adleman et al. [90] improved this to show that BQP can be computed using only unitary operations with rational rotations, and that BQP is in the class PSPACE of polynomial space computations of (classical) TMs. Practical implementations of QC most likely will need to be done via unitary transitions within some modest amplitude precision. Bernstein, Vazirani [50] proved that BQP computations running in time T can be done with unitary operations specified by only $O(\log T)$ bits of precision.

3.3.3. Quantum Parallel Complexity Classes

Let NC (QNC, respectively) be the class of (quantum, respectively) circuits with polynomial size and polylogorithmic depth. Thus QNC is the quantum analog of the processor efficient parallel class NC. Moore, Nilsson [91] define QNC and show various problems are in QNC; for example, they show that the quantum Fourier transform can be parallelized to linear depth and polynomial size.

3.4. BOUNDS ON MEASUREMENT, SENSING, AND COMMUNICATION

3.4.1. Lower Bounds on Quantum Communication

Cleve et al. [92] prove linear lower bounds for the quantum communication complexity of the inner product function and give a reduction from the quantum information theory problem to the problem of quantum computation of the inner product. Knill, Laflamme [93] characterize the communication complexity of one qubit.

3.4.2. Interaction-Free Quantum Measurement

A method for *(nearly) interaction-free measurement* (IFM) specifies the design of a quantum optical sensing system that is able to determine with arbitrarily high likelihood if an obstructing body has been inserted into the system, without moving or modifying its optical components; moreover, in the case that the obstructing body is present, IFM uses at most an arbitrarily small multiplicative factor of the input intensity to do the sensing. Kwiat et al. [94] (also see [95]) have given a method for IFM which does repeated rounds of measurement to affect small phase changes that eventually determine (via the quantum Zeno effect) whether an obstructing body has been inserted. Kwiat et al. [95] assert their method can be applied to sensing tasks such as photography, but the use of their method for IMF has major practical limitations (i.e., if the obstructing body has not been inserted, then the amount of sensing can be quite large).

3.4.3. Interaction-Free Quantum Sensing

Reif [96] defines *(nearly) interaction-free sensing* (IFS) similarly to IFM, except an upper bound is imposed on both the intensity to do the sensing (which again is an arbitrarily small multiplicative factor of the input intensity) whether or not the obstructing body is present. A quantum optical method for IFS (but not IFM) may be used to do I/O with bandwidth reduced by an arbitrarily small multiplicative factor of the bandwidth required for classical (e.g., conventional optical or electronic) I/O methods. Reif [96] proves there is no method for IFS with unitary transformations and so concludes that I/O bandwidth cannot be significantly reduced by such quantum methods for sensing. (Also see Holevo [97], Fuchs and Caves [98] for proof that quantum methods cannot increase the bandwidth for transmission of classical information.)

3.5. QUANTUM COMPRESSION

As noted above, quantum methods cannot increase the bandwidth for transmission of classical information; still, in certain cases, entangled states can be compressed to fewer qubits. This quantum compression could have important applications in practice, where the number of usable qubits is very limited. Schumacher [99] considered compression and decompression of a noiseless source of n quantum bits (qubits), each sampled independently from a given mixed state quantum ensemble. For such a quantum source, the compression factor obtainable by classical information theory is limited by the Shannon entropy, which in general (except in the case where the quantum ensemble has only orthogonal states) is less than the quantum compression factor given by the von Neumann entropy. In particular, Schumacher [99] proved a *quantum noiseless coding theorem* that states that the source's von Neumann entropy is the number of qubits per source state which is necessary and sufficient to asymptotically (in the limit of large code-block size) encode the output of the source with arbitrarily high fidelity. The quantum noiseless coding of Schumacher has asymptotically optimal fidelity and size; the resulting compressed number of qubits can be far fewer than in the classical case.

Shannon Entropy and the Limitations of Classical Methods for Noiseless Compression. Suppose n characters from a finite alphabet Σ are each sampled independently over some probability distribution p. In classical information theory, the Shannon entropy of each character is $H_S(p) = -\sum_{a \in \Sigma}^{1} p(a) \log p(a)$. A string of these n bits may be compressed without loss to a bit string of mean length $H_S\,(p)n$.

The von Neumann entropy and Quantum Noiseless Compression. Following Schumacher [99], we assume there is a finite quantum state ensemble (S, p) which is a *mixed state* consisting of a finite number of qubit states $S = \{|a_0\rangle, \ldots, |a_{|S|-1}\rangle\}$, where each $|a_i\rangle \in S$ has probability p_i. The compressor is assumed to act on blocks of n qubits (so is a block compressor), and is assumed to know this underlying ensemble (S, p). The *density matrix* of (S, p) is an $|S| \times |S|$ matrix $\rho = \sum_{i=0}^{|S|-1} p_i |a_i\rangle\langle a_i|$. The *von Neumann entropy* (see [99, 100]) corresponding to (S, p) is $H_{VN}(\rho) = -Tr(\rho \log \rho)$. In general, the Shannon entropy $H_S\,(p)$ is greater than or equal to the von Neumann entropy. These entropies are equal only when the states in S are mutually orthogonal.

An Example: Consider a slightly more complex example of a source consisting of a sequence of n photons polarized randomly, with equal probability of phase 0 or phase angle θ (e.g., As a very simple example of a source with low von Neumann entropy, consider N photons polarized randomly, equiprobably at 0 or 1.) In this case, the states are $S = \{|a_0\rangle, |a_1\rangle\}$, where the first state $|a_0\rangle = |0\rangle$, corresponds to phase 0, and the other state $|a_1\rangle = \cos\theta|0\rangle + \sin\theta|1\rangle$ corresponds to phase angle θ, and the probabilities are both $p(0) = p(1) = \frac{1}{2}$. The density matrix is $\rho = \frac{1}{2}|a_0\rangle\langle a_0| + \frac{1}{2}|a_1\rangle\langle a_1| = \frac{1}{2}(|0\rangle\langle 0| + (\cos\theta|0\rangle + \sin\theta|1\rangle)(\cos\theta\langle 0| + \sin\theta\langle 1|)) = \frac{1}{2}((1+\cos^2\theta)|0\rangle\langle 0| + \cos\theta\sin\theta|0\rangle\langle 1| + \cos\theta\sin\theta|1\rangle\langle 0| + \sin^2\theta|1\rangle\langle 1|)$ which has 2×2 matrix form $\frac{1}{2}\begin{bmatrix} 1+\cos^2\theta & \cos\theta\sin\theta \\ \cos\theta\sin\theta & \sin^2\theta \end{bmatrix}$ over the basis vector $\begin{bmatrix} |0\rangle \\ |1\rangle \end{bmatrix}$. Then we can find an appropriate β which gives a change of basis with new basis states $|0'\rangle = |0\rangle$ and $|1'\rangle = \cos\beta|0\rangle + \sin\beta|1\rangle$, providing a diagonal density matrix $\rho' = \frac{1}{2}\begin{bmatrix} (1+\cos^2\theta)+\cos\theta\sin\theta\tan\beta & 0 \\ 0 & \cos\theta\sin\theta+\sin^2\theta\tan\beta \end{bmatrix}$ over the basis vector $\begin{bmatrix} |0'\rangle \\ |1'\rangle \end{bmatrix}$. Although this source has high Shannon entropy $H_S(p)$, it will have low von Neumann entropy $H_{VN}(\rho)$ in the case of a small magnitude phase angle θ However, note that the entropies are the same in the special case where $\theta = \pi/2$, so the states $|a_0\rangle = |0\rangle$, $|a_1\rangle = |1\rangle$ are orthogonal and the density matrix is simply the diagonal matrix $\rho = \frac{1}{2}(|0\rangle\langle 0| + |1\rangle\langle 1|)$which has a diagonal density matrix $\rho = \begin{bmatrix} \frac{1}{2} & 0 \\ 0 & \frac{1}{2} \end{bmatrix}$ over the basis vector $\begin{bmatrix} |0\rangle \\ |1\rangle \end{bmatrix}$.

For technical reasons, the unitary compression and decompression mappings need to preserve the number of bits (some of which are ignored). An *n-to-n' quantum compressor* is a unitary transformation that maps n-qubit strings to n-qubit strings; the first n' qubits that are output by the compressor are taken as the compressed version of its input, and the remaining $n-n'$ qubits are discarded. An *n'-to-n decompressor* is a unitary transformation that maps n-qubit strings to n-qubit strings; the first n' qubits input to the decompressor are the compressed version of the uncompressed n qubits, and the remaining $n-n'$ qubits are all 0. The *source* to the compression scheme is assumed to be a sequence of n qubits sampled independently from (S, p). The *observed output* is the result of first compressing the input qubits, then decompressing them, and measuring the result (over a basis containing the n inputs). The *compression rate* is n/n' and the *compression factor* is n'/n. The *fidelity* of the compression scheme is the probability that the observed output is equal to the original input (that is, the probability that the original qubits are correctly recovered from the compressed qubits). The goal here is a quantum compression with both a high fidelity and a high compression rate.

Example (Continued): Consider again the example of a source consisting of a sequence of n photons polarized randomly, with equal probability of phase 0 or phase angle θ. If θ has small magnitude, then a quantum encoder can compress these photons into an entangled state using just a few photons. Furthermore, a quantum decoder can the recover n photons with the original distribution (with arbitrarily high fidelity for large n) from these compressed photons.

SCHUMACHER QUANTUM NOISELESS COMPRESSION. Schumacher [99] gave a *quantum noiseless coding theorem* which provided asymptotically optimal noiseless compression of a sequence of qubits independently sampled from a finite quantum state ensemble (S, p). The quantum noiseless coding theorem states that for any $\varepsilon,\delta > 0$ and sufficiently large n, (i) there is an n-to-n' quantum compression scheme with fidelity at least $1-\varepsilon$ and compression to length $n' \leq n(H_{VN}(\rho) + \delta)$, and (ii) any n-to-n'' quantum compression scheme which gives compression to length $n'' \leq n(H_{VN}(\rho) - \delta)$, has fidelity $< 1-\varepsilon$. That is, in the limit of large code-block size, the source's von Neumann entropy $H_{VN}(\rho)$ is asymptotically the number of qubits per source state which is necessary and sufficient to encode the output of the source with arbitrarily high fidelity. Given a known finite quantum state ensemble (S, p), Schumacher's compression scheme assumes a known basis for which the density matrix ρ is diagonal, with nonincreasing values along the diagonal.

The proof of the Schumacher quantum noiseless coding theorem and its refinements by Jozsa and Schumacher [101] and H. Szeto [102] make use of the existence of a *typical subspace* Λ (see [101]) within a Hilbert space of n qubits over a source of von Neumann entropy $H_{VN}(\rho)$. The typical subspace Λ has dimension $\leq 2^{nH_{VN}(\rho)}$ and with high probability, a sample of n qubits has an almost unit projection onto Λ. The Schumacher compressor simply transposes (via a permutation mapping) the subspace Λ into the Hilbert space of a smaller block of $nH_{VN}(\rho)$ qubits. These proofs are not completely constructive.

Bennett [103] gave a constructive presentation of the Schumacher compression. He observes that the Schumacher compression can be done by a unitary mapping to a basis for which the density matrix ρ is diagonal (in certain simple cases the density matrix ρ is already diagonal, e.g., when the input is a set of n identical qubits) followed by certain combinatorial computation which we will call the *Schumacher compression function SCHUMACHER*. The Schumacher compression function SCHUMACHER simply orders the basis states first by the number of ones (from smallest to largest) that are in the binary expansion of the bits and then refines this order by a lexical sort of the the binary expansion of the bits. That is, all strings with i ones are mapped before all strings with $i+1$ ones, and those strings with the same number of ones are lexically ordered. Note that for any given value X of the qubits, this transformation SCHUMACHER (X) is simply a deterministic mapping from an n bit sequence to a n' bit sequence defined by a combinatorial computation. In particular, given an n bit binary string X, the transformation SCHUMACHER (X) is the number of n bit strings so ordered before X. It is easy to show that SCHUMACHER(X) is a permutation. Since it is a permutation, it is a bijective function which is uniquely reversible; it is also a unitary transformation. To insure that the overall transformation (for all the states) is a quantum computation, it is essential that the transformation SCHUMACHER (X) be done using only reversible, quantum-coherent elementary operations. (Bennett et al. [104] gave a polynomial time quantum algorithm for the related problem of extraction of only classical information from a quantum noiseless coding.) Cleve, DiVincenzo [105] then developed the first polynomial time algorithm for Schumacher noiseless compression of n qubits. In particular, they explicitly computed the bijective function SCHUMACHER (X) and it's reverse using $O(n^3)$ reversible, elementary unitary operations. This was previously the fastest previous algorithm for the Schumacher encoding and decoding functions. Recently, Reif and Chakraborty [108] gave a time efficient algorithm for asymptotically optimal noiseless quantum compression and decompression, costing only $O(n(\log^4 n)\log\log n)$ elementary quantum operations. This modified Schumacher encoding requires the evaluation of various combinatorial sums, for which Reif and Chakraborty provide efficient recursive, reversible quantum algorithms. The coding of [106] employed a modified Schumacher encoding that was still asymptotically optimal in fidelity and size.

The Schumacher quantum noiseless coding theorem assumes the compressor knows the source. Jozsa et al. [107] recently gave a generalization of the Schumacher compression, that is, the compressor does not know the source, and thus provides the first asymptotically optimal universal algorithm for quantum compression. Also, Braustein et al. [108] have recently given a fast algorithm for an quantum analog of Huffman coding, but do not provide a proof that this coding gives asymptotically optimal noiseless quantum compression (that is, reaches the von Neumann entropy), as provided by Schumacher compression. ([102] assumes the compressor knows the source, but can be extended using the techniques of Jozsa et al. [107] to a asymptotically optimal universal algorithm for quantum compression where the compressor does not know the source.)

3.6. QUANTUM ERROR-CORRECTING CODES

3.6.1. Quantum Coding Theory

The qubit can be defined in quantum information theory as the amount of information that can be carried in a quantum system with two basis states, e.g., the internal degree of freedom of a polarized photon. The qubit is thus a fundamental unit of quantum channel capacity. Nielsen [109] (Svozil [110, 111], Holevo [112], Knill, Laflamme [113, 114], and Ohya [115] develop a theory of quantum error-correcting codes and quantum information theory, e.g., they give the definition of *quantum mutual entropy* for an entangled state. Buhrman et al. [116], and Adami, Cerf [117] contrast quantum information theory with classical information theory. Quantum channel capacity has been investigated for noisy channels (DiVincenzo, et al. [118], Holevo [119], Barnum et al. [120]), very noisy channels (Shor, Smolin [121]), and quantum erasure channels (Bennett et al. [122]). Fuchs [123] showed that nonorthogonal quantum states maximize classical information capacity. (Also, Helstrom [124, 125] defines a quantum theory of information detection, and Fuchs [126] defines a related quantum theory of information distinguishability.)

3.6.2. Decoherence Errors in QC

Quantum decoherence is the gradual introduction of errors of amplitude in the quantum superposition of basis states. All known experimental implementations of QC suffer from the gradual decoherence of entangled states. The rate of decoherence per step of QC depends on the specific technology implementing QC. A significant property of Shor's algorithm is that the precision of the amplitudes in the superpositions need be only a polynomial number of bits. Although the addition of decoherence errors in the amplitudes may at first not have a major effect on the QC, the effect of the errors may accumulate over time and completely destroy the computation. Researchers have dealt with decoherence errors by extending classical error correction techniques to quantum analogs. Generally, there is assumed a decoherence error model where the errors introduced are assumed to be uniform random with bounded magnitude, independently for each qubit.

3.6.3. Quantum Codes

Shor [127] and Steane [128] gave the first techniques for reducing quantum decoherence by the addition of extra qubits which are then projected via observation operations to eliminate errors in the superposition. Calderbank, Shor [129], and Steane [130] then proved that QC can be done with bounded decoherence error, assuming the error correction mechanism is without error itself. Bennett et al. [131], and Laflamme [132] gave the first optimal 5-qubit codes, leading to asymptotically optimal (for large code blocks) quantum error correction codes. Shor [133] and Kitaev [134, 135] extended these techniques to do fault tolerant quantum computation on quantum networks, in the presence of bounded decoherence error, even if the error correction mechanism also suffers from error decoherence errors. A final innovation (Aharonov, Ben-O [136], and Knill et al. [137, 138]) was concatenated versions of the above quantum codes that allow for arbitrarily long QC in the presence of arbitrary (i.e., not necessarily random) decoherence error below a fixed constant threshold. Current bounds on this threshold are very small, and it seems likely (although it is not yet known) that they can be increased to above the decoherance error bounds of experimental techniques for QC.

Also see the texts [38, 39] on quantum coding theory.

3.7. QUANTUM CRYPTOGRAPHY

Here we overview quantum cryptography; also see the following texts: [35–37].

3.7.1. Quantum Keys

Bennett et al. [139] and Bennett and Brassard [140] gave the first methods for quantum cryptography using qubits as keys, which were proved to be secure against certain types of attacks. (However, see the remarks at the end of this subsection.) Surveys of quantum cryptography are given in Bennett, Brassard, Ekert [141], Brassard [142], Bennett, Brassard [143], Brassard [144], and Gisin [145]. Ozhigov [146] gives a protocol for security of information in quantum databases. Hruby [147] discusses further methods for quantum cryptography. Bennett et al. [148], Hughes et al. [149] describes experiments of quantum cryptography, including optical fibers.

Bennett et al. [150] gave a protocol for quantum oblivious transfer. Mayers [151] gives quantum oblivious transfer and key distribution protocols and Mayers [152] extends the protocols to noisy channels. Lo, Chau [153] give a quantum key distribution protocol which is unconditionally secure over arbitrarily long distance.

Brassard, Crpeau [154] gave quantum bit commitment and quantum coin tossing protocols. Brassard et al. [155] gives a quantum bit commitment scheme provably unbreakable by both parties. Yao [156] proved quantum protocols secure against coherent measurements. Brassard et al. [157] shows how to defeat classical bit commitments with a quantum computer. Chau, Lo [158] gives further methods for qubit commitment. Crpeau et al. [159] gives protocols for quantum oblivious mutual identification.

IS QUANTUM CRYPTOGRAPHY ACTUALLY UNBREAKABLE? Unfortunately, some of the methods for quantum cryptography that claim to be unbreakable can in fact be broken by sidestepping assumptions assumed in the proofs of their security. For example, to break the well known quantum cryptography method of Bennett and Brassard [140], Brandt [160] provided a method (experimentally demonstrated by Kim et al. [161]) that exploited entanglement of momentum with the phase of photons, making observations of the momentum portions to infer transmitted phases. It is not clear what other prior results in quantum cryptography could be broken by similar techniques, which places the field of quantum cryptography in some doubt. (Also, Lo, Chau [162] have recently argued that quantum bit commitment and ideal quantum coin tossing are impossible in certain cases that are not covered in the preceeding results.)

3.7.2. Distributed Quantum Networks

Future hardware will have to be fast, scalable, and highly parallelizable. A *quantum network* is a network of QCs executing over a spatially distributed network, where quantum entanglement is distributed among distant nodes in the quantum network. Thus, using *distributed entanglement*, a quantum network distributes the parts of an entangled state to various processors, which can act on the parts independently. Pellizzari [163] proposes quantum networks using optical fibers; Cirac, Zoller et al. [164] and Bose, Vedral [165] show state transfer distribution can be done among distant nodes. For example, [164] uses a cavity QED device that traps atoms in multiple cavities and exchanges photons between the cavities to establish the distributed entanglement. Various basic difficulties were overcome. They are explained below.

How can one do state transfer distribution? Bennett et al. [166, 167] and Brassard [168] developed a technique known as *teleportation* to transmit arbitrary input states with perfect fidelity. It does this by separating the input state into classical and quantum components. The input can then be reconstructed from these components with perfect fidelity.

How can one cope with communication errors and attenuation in a quantum network? Wootters, Zurek [169] proved that a single quantum cannot be cloned. (Note: Buzek, Hillery [170] recently claimed a universal optimal cloning of qubits and quantum registers in a distributed quantum network, but this seems inconsistent with the no-cloning theorem.) That no-cloning theorem implies that once a signal becomes attenuated in an optical fiber communication channel, then it cannot (generally) be amplified. It would at first appear that communication and quantum network links may be limited to distances of the order of the attenuation length in the fiber. However, the range of quantum communication could be extended using *quantum repeaters* that do quantum error correction, restoring the quantum signal without reading the quantum information. Ekert, Huelga et al. [171] extend the techniques of distributed quantum computation to noisy channels, and showed that for quantum memories and quantum communication, a state can be transmitted over arbitrary distances with bounded error, provided a minimum gate accuracy can be achieved—a constant factor of this error.

3.8. OTHER ALGORITHMIC APPLICATIONS OF QC

The early literature in QC provided some examples of QC algorithms for problems constructed for the reasonable purpose of showing that QC can solve some problems more efficiently than conventional sequential computing models. Later, quantum algorithms were developed for variety of useful applications. See the texts on quantum algorithms: [27–34, 40].

- *Quantum Fourier Transforms.* Drutsch, Jozsa [49] gave an $O(n)$ time quantum algorithm for creating a uniform superposition of all possible values of n bits, which is a *quantum Fourier transform* over the finite field of size 2. Simon [172] used this quantum Fourier transform to give an efficient time quantum algorithm for determining whether a function over a finite domain is invariant under some XOR-mask. This provided one of the first examples of a quantum algorithm which efficiently solves an interesting problem that is costly for classical computation. Brassard, Hoyer [173] gave improvements to Simon's algorithm. There have been a number of efficient quantum algorithms for extensions of the quantum Fourier transform: to the approximate quantum Fourier transform (Coppersmith [174]), over various domains (Griffiths, Niu [175], Hoyer [176]), over symmetric groups (Beals [177]), and over certain nonabelian groups (Pueschel, Roetteler, Bet [178]). Vedral, Barenco, Ekert [179] give efficient quantum networks for elementary arithmetic operations using the quantum Fourier transform. Grigoriev [180] used the quantum Fourier transform to test shift-equivalence of polynomials.

- *Quantum Factoring.* The most notable algorithmic result in QC to date is the quantum algorithm of Shor [181, 182] for discrete logarithm and integer factorization in polynomial time (with modest amplitude precision). (Also see a review of the algorithm by Ekert and Jozsa [183].) Shor's algorithm uses an efficient reduction (due to Miller [184]) from integer factoring to the problem of approximately computing the period (length of a orbit) within an integer ring. Shor approximates the period by repeated use of a quantum Fourier transform over an integer ring and greatest common divisor computations. There has been further considerable work on Shor's quantum factoring algorithm: Zalka [185] improved the time complexity; Beckman et al. [186] describe its execution on quantum networks with small size and depth; Obenland, Despain [187], Plenio, Knight [188] consider the feasibility of executing Shor's quantum factoring algorithm on various quantum computer architectures. (The latter provide somewhat pessimistic lower bounds for the factorization time of large numbers on a quantum computer in the presence of decoherance errors.) [189] describes a 7-qubit demonstration of Shor's factorization algorithm using nuclear magnetic resonance. Kitaev [190] gave an independent derivation of Shor's factoring result using a reduction to find an abelian stabilizer.
- *Quantum Search.* Another significant efficient QC algorithmic result is the algorithm of Grover [191], which searches within a database of size N in time $\sqrt{N}$. (An interesting property of Grover's algorithm for search is its similarity to the quantum Zeno affect technique for quantum measurement [94, 95]. In particular, the algorithm also uses $O(\sqrt{N})$ stages of unitary operations, each quite similar to a stage of the quantum Zeno sensing method.) Grover refined his result to require only a single query [192] and to use almost any unitary transformation [193]; Zalka [194] showed Grover's algorithm cannot be further asymptotically sped up and so is optimal for database search; and Pati [195] gave further improvements to the bounds. Biron et al. [196] extended Grover's algorithm to arbitrary initial amplitude distribution. Cockshott [197] gave fast quantum algorithms for executing more general operations on relational databases, and Benjamin, Johnson [198] discuss the use of Grover's algorithm and related quantum algorithms for other data processing problems. Farhi et al. [199] showed that Grover's algorithm could not be extend to quickly determine

parity of N bits; in particular, they showed that any quantum algorithm for parity takes at least $N/2$ steps. Meyer [200] and Terhal, Smolin [201] propose quantum search algorithms that do not to require entanglement. Brassard et al. [202, 203] combine the algorithmic techniques of Grover and Shor to give a fast quantum algorithm for approximately counting (i.e., finding the number of matches in a database).

While Grover's algorithm is clearly an improvement over linear sequential search in a database, it appears less impressive in the case of an explicitly defined database which needs to be stored in volume N. Parallel computation can do search in a database of size N in time at most polylogarithmic with N (that is, in time $O(\log^{O(1)} N)$) by relatively straightforward use of parallel search Moreover, Grover's algorithm may not have a clear advantage even in the case of an implicitly defined database, which does not need to be stored, but instead can be constructed on the fly (e.g., that arising from NP search methods). In this case, Grover's search algorithm can be used to speed up combinatorial search within a domain of size N to a time bound of $O(\sqrt{N})$. (Hogg [204], and Hogg, Yanik [205] investigate similar quantum search techniques for local and other combinatorial search problems.) In this case Grover's algorithm appears to require only volume logarithmic in the search space size N. In contrast, parallel computation takes volume linear in the combinatorial search space, but takes just time polylogorithmic in the search space.

- *Quantum Simulations in Physics.* The first application proposed for QC (Feynman [46]) was for simulating quantum physics. In principle, quantum computers provide universal quantum simulation of any quantum mechanical physical system (Lloyd [206], Zalka [207], Boghosian [208])). Proposed QC simulations of quantum mechanical systems include many-body systems (Wiesner [209]); many-body Fermi systems (Abrams, Lloyd [210]); multiparticle (ballistic) evolution (Benioff [211]); quantum lattice-gas models (Boghosian, Taylor [212]), Meyer [213, 214]); Ising spin glasses (Lidar, Biham [215]); the thermal rate constant (Lidar, Wang [216]; and quantum chaos (Schack [217]).
- *Quantum Learning.* QC may have some interesting applications in learning theory and related problems. Bshouty, Jackson [218] describe learning Boolean formulas in disjunctive normal form (DNF) over the uniform distribution of inputs, using a quantum example oracle; Ventura, Martinez [219] describe a QC learning algorithm for learning DNF using a classical example oracle. Also, Yu, Vlasov [220] describe image recognition using QC, Tucci [221] investigates quantum Bayesian networks, and Ventura, Martinez [222] describe a quantum associative memory,
- *Quantum Robotics.* Benioff [223] considers a distributed QC system with mobile *quantum robots* that can carry out carrying out measurements and physical experiments on the environment; as an example, Benioff gives an

algorithm for the problem of measuring the distance between a quantum robot and a particle on a 1D space lattice. Hogg [207] proposes the use of distributed QC to allow small-scale sensors and actuators to be controlled in a distributed manner. Further discussion of the applications of QC are given by Landauer [224, 225].

- *Winding Up Quantum Clocks.* The precision of atomic clocks are limited by the spontaneous decay lifetimes of excited atomic states. An interesting application of QC proposed by Huelga [226] (also see Bollinger et al. [227]) is to extend these lifetimes by using quantum error correcting codes to inhibit the spontaneous decay. A similar idea can be used for improving the precision of frequency standards and interferometers.
- *Quantum Strategies.* Meyer [228, 229] has proposed a class of generalized games that allow for quantum strategies and he proves that they provide an improvement over conventional mixed (randomized) strategies for certain games.

3.9. POSSIBLE TECHNOLOGIES FOR DOING QC

Here we provide an overview of various experimental implementations of quantum computation; also see the texts [41–45]. As noted previously, any QC can be realized by a *universal* set of gates consisting of the 2-qubit XOR operation along with some 1-qubit operations. There are two basic approaches known to do QC.

(A) *Micromolecular QC.* Here QC on n qubits is executed using n individual atoms, ions, or photons, and each qubit is generally encoded using the quantized states of each individual atom, ion, or photon. The readout (observation operation) is by measurement of the (eigen) state of each individual atom, ion, or photon. We enumerate a number of proposed micromolecular QC methods as follows:

- *Quantum Dots.* Burkard et al. [230], Loss, Di Vincenzo. [231], and Meekhof et al. [232] describe the use of coupled quantum dots to do QC. ([78] proposes quantum computation using Cooper pairs.)
- *Ion Trap QC.* Cirac, Zoller [233], Pellizzari et al. [234], and James [235] proposed using a linear array of cold trapped ions (the ions are trapped by electromagnetic fields) whose energy states are used to store the qubits; vibrational modes between consecutive ions also can be used to store states of qubits. The coupling of the qubits is by electrostatic repulsion between the ions. Unitary transitions on superpositions can be executed via an associated array of lasers, each of which pulses a distinct ion; these induce electric dipole moments that determine the transitions. A group at the National Institute of Standards at Boulder, Colorado (Wineland et al.

[236, 237] King et al. [238], and Turchette et al. [239]) and a group at Los Alamos (Hughes [240], Hughes et al. [241], and James [242]) have experimentally demonstrated trapped ion QC. These and other researchers have addressed various key issues associated with quantum computation with trapped ions.

- Deterministic entanglement of two trapped ions (Turchette et al. [239])
- Decoherence bounds (Hughes et al. [240] and Plenio, Knight [244])
- Measurement and state preparation, i.e., initialization of the collective motion of the trapped ions (Schneider et al. [245] and King et al. [238]),
- Coherent quantum-state manipulation of trapped atomic ions (Wineland et al. [236, 237]),
- Heating of the quantum ground state of trapped ions (James [246]) and quantum computation with "hot" trapped ions (Schneider et al. [247]).

- *Cavity QED.* A group at Cal Tech (Turchette [248]) have experimentally demonstrated the use of trapped photons in a cavity QED system to execute 2-qubit XOR gates and thus in principle can do universal QC. The qubits are encoded by the circular polarization of photons interacting. The XOR unitary transitions on superpositions can be executed by resonance between interacting photons in the cavity; the coupling of qubits is via resonance between interacting photons using a Cesium atom also in the cavity, and the coupling is tuned by the spacing of mirrors in the cavity.
- *Photonics.* Various groups (i.e., Chuang et al. [249, 250] and Torma, Stenholm [251]) have experimentally demonstrated QC using optical systems where qubits are encoded by photon phases, and universal quantum gates are implemented by optical components consisting of beamsplitters and phase shifters as well as nonlinear media (also see the linear optics QC proposed by Adami, Cerf [252]).
- *Heteropolymer.* This is a polymer consisting of a linear array of atoms, each of which can be either in a ground or excited energy state. Teich et al. [253] first proposed classical (without quantium superpositions) molecular computations using heteropolymer. Later Lloyd [254] extended the use of heteropolymers to QC using the energy states to store the state of the qubits. The coupling of qubits may be via electric dipole moments that cause energy shifts on adjacent atoms. Unitary transitions on superpositions can be executed via pulses of a laser at particular frequencies; these induce electric dipole moments that determine the transitions.
- *Nuclear Spin.* DiVincenzo [255] and Wei et al. [256, 257] proposed the use of nuclear spin to do QC; see the remarks following the discussion of bulk QC.
- *Quantum Propagation Delays.* Castagnoli [258] proposed a way to do QC using retarded and advanced propagation of particles through various media.

Of these, ion trap QC, cavity QED QC, and photonics have been experimentally demonstrated up to a very small number of qubits (about three bits). The apparent intention of such micromolecular methods for QC is to have an apparatus for storing qubits and executing unitary operations (but not necessarily executing observation operations) which requires only volume linear in the number of qubits. One difficulty (addressed by Kak [259], and Murao et al. [260]) is *purification of the initial state*: If the state of a QC is initially in an entangled state, and each of the quantum gate transformations introduces phase uncertainty during the QC, then the effect of these perturbations may accumulate to make the output to the QC incorrect. A more basic difficulty for these micromolecular methods is that they all use experimental technology that is not well established; in particular, their approaches each involve containment of atomic size objects (such as individual atoms, ions, or photons) and manipulations of their states. A further difficulty of the micromolecular methods for QC is that apparatus for the observation operation—even if observation is approximated—seems to require volume growing exponential with the number of qubits, as described earlier in this paper.

(B) *Bulk (or NMR) QC.* Nuclear magnetic resonance (NMR) spectroscopy is an imaging technology using the spin of the nuclei of a large collection of atoms. Bulk QC is executed on a macroscopic volume containing a large number of identical molecules in solution, each of which encodes all the qubits. The molecule can be chosen so that it has n distinct quantized spins modes (e.g., each of the n nuclei may have a distinct quantized spins). Each of the n qubits is encoded by one of these spin modes of the molecule. The coupling of qubits is via spin-spin coupling between pairs of distinct nuclei. Unitary operations such as XOR can be executed by radio frequency (RF) pulses at resonance frequencies determined in part by this spin-spin coupling between pairs of nuclei (and also by the chemical structure of the molecule). Bulk QC was independently proposed by Cory, Fahmy, Havel [261] and Gershenfeld, Chuang [22, 262]. Also see Berman et al. [266] and the proposal of Wei et al. [257] for doing NMR QC on doped crystals rather than in solutions, and see Kane [264] for another solid state NMR architecture for quantum computing using silicon.

- *Bulk QC.* was experimentally tested (Jones et al. [265, 266]) and applied to demonstrate the following tasks: quantum search (Jones [267]), approximate quantum counting (Jones, Mosca [268]), Deutsch's problem (Jones, Mosca [269]), and the Deutsch–Jozsa algorithm on 3 qubits (Linden, Barjat, Freeman [270]).
- *Advantages of Bulk QC.* (i) it can use well established NMR technology and in particular macroscopic devices, The main advantages are (ii) the long time duration until decoherence (due to a low coupling with the environment) and (iii) scaling to more qubits than other proposed technologies for QC.

- *Disadvantages of Bulk QC.* A possible disadvantage of bulk QC is that it appears to allow only a *weak* measurement of the ensemble average that does not provide a quantum state reduction; that is, the weak measurement does not alter (at least by much) the superposition of states. Another disadvantage of bulk QC is that it may require (for a variety of reasons) macroscopic volumes and volumes which grow exponentially with the number of qubits. Macroscopic volumes may be required for measurement via conventional means. However, known quantum algorithms can still be executed even in this case (e.g., see Gershenfeld, Chuang [22, 262]). So the lack of strong measurement is not a major disadvantage.

Also, bulk QC requires the initialization close to a pure state. If bulk QC is done at room temperature, the initialization methods of Cory, Fahmy, Havel [261] (using logical labeling) and Gershenfeld, Chuang [22, 262] (using spatial averaging) yield a pseudo-pure state, where the number of molecules actually in the pure state drops exponentially as $1/c^n$ with the number n of qubits, for some constant c (as noted by Warren [271]). If we approximate the resulting measurement error by a normal distribution, the measurement error is (with high likelihood) at least a multiplicative factor of $1 - c'/\sqrt{N}$, for some constant c'. To overcome this measurement error, we need $1/c^n > c'/\sqrt{N}$, and so we require that the volume be at least $N > (c^n/c')^2$. Hence, for the output of the bulk QC to be (weakly) measured, the volume (the number N molecules) of bulk QC needs to grow exponentially with the number n of qubits. Recently, there have been various other proposed methods for initialization to a pure state:

- Barnes [272] proposes the use of very low temperatures.
- Gershenfeld, Chuang [22, 262] suggest the use of gradient fields.
- Knill et al. [273] suggest a randomization technique they call temporal averaging.
- Recent work of Schulman, Vazirani [274] provides polynomial volume for initialization, with the assumption of an exponential decrease in spin-spin correlations with the distance between the nuclei located within a molecule. (In particular, they assume that the statistical correlation between and two bits on a molecule falls of exponentially with the distance between these bits). Although their methods may provide a solution in practice, known inter-atomic interactions such as the spin-spin correlations are generally considered to be governed by potential force laws which decrease by inverse polynomial powers rather than by an exponential decrease.

It has not yet been experimentally established which of these pure state initialization methods scale to a large number of qubits without large volume.

(Note: Some physicists feel that it has not been clearly established whether (a) NMR is actually a quantum phenomenon with quantum superposition of basis states, or (b) if NMR just mimics a quantum phenomenon and is actually just classical parallelism, where the quantum superposition of basis states is encoded using multiple molecules where each molecule is in a distinct basis state. If the latter is true (each molecule is in a distinct basis state) then the volume may grow grow exponentially with the number n of qubits, since each basis state may need to be stored by at least one molecule, and the number of basis states can be 2^n. (See Williams, Clearwater [23]) Also, even if each molecule is in some partially mixed quantum state (see Zyczkowski et al. [275]), the volume may still need to grow very large.)

In summary, some possible disadvantages of bulk QC that may make it difficult to scale are (i) the inability to do observation (strong measurement with quantum state reduction); (ii) the difficulty to do even a weak measurement without the use of exponential volume; (iii) difficulty (possibly now resolved) to obtain pure initial states without the use of exponential volume; and (iv) the possibility that bulk QC is not a quantum phenomena at all (an unresolved controversy within physics), and so may require use of exponential volume.

It is interesting to consider whether NNR can be scaled down from the macroscopic to molecular level. DiVincenzo [255] and Wei et al. [256, 257] propose doing QC using the nuclear spins of atoms or electrons in a single molecule. The main advantages are (i) small volume and (ii) the long time duration until decoherence (an advantage shared with NMR). The key difficulty of this approach is the measurement of the state of each spin, which does not appear to be feasible by the mechanical techniques for detection of magnetic resonance usually used in NMR (which can only do detection of the spin for large ensembles of atoms).

3.10. RESOURCE BOUNDS

In this chapter, we have discussed many applications of quantum computation that provide advantages over classical methods of computation. Certain applications of QC (e.g., quantum cryptography) require only a small or constant number of qubits, whereas other applications (e.g., factoring and database search) require a large number of qubits and moreover require an observation operation at least as the final step of the QC. For these advantages to be practical, we need to determine that there are no unfeasible, large resources required by QC. Hence we complete the chapter with a review of the resource bounds of quantum computing compared to the resources required by classical methods for computation. In particular, we will conclude that for the advantages of QC (with a large number of qubits) to be practical for applications requiring a large number of qubits, there needs to be determined (theoretical and practically) bounds on the volume required of observation operations. This seems to us a major missing element in the field of QC.

The energy consumption, processing rate, and volume are all important resources to consider in computing devices. Conventional (classical) electronic supercomputers of the size of a work station operate in the range of 10^{-9} Joules per operation, at up to about 50 giga-ops per second, with memory of about 10 to 100 giga-bytes, and in a volume of about 10 cm^2. The volume scales as the number of bits of storage.

- *Energy Bounds for QC.* The conventional linear model of QC allows only unitary state transformations and so by definition is reversible (with the possible exception of the observation operation that does quantum state reduction). Benioff [48] noted that as a consequence of the reversibility of the unitary state transformations of QC, these transformations dissipate no energy. But this does not consider (i) the precision of the amplitudes to be preserved nor (ii) the expected time duration required to drive the operation to completion. Gea-Banacloche [276] and Ozawa [277] independently derived lower bounds on the energy needed to execute within a given precision of the amplitudes and in a given time, an elementary qubit logical operation on a quantum computer. They derived energy lower bounds depending inversely on the time duration for the operation and on the precision of the amplitudes to be preserved. Hence, for polynomial time quantum computations requiring polynomial relative precision, the lower bounds on energy are polynomial, though the constant factors could be a limitation for practical implementations. (Recall that Bernstein, Vazirani [50] proved that BQP computations can be done with unitary operations specified by only logarithmic bits of precision, which corresponds to relative precision ε where $\varepsilon > 1/n^{O(1)}$.) These energy lower bound results were stated to be independent of the nature of the physical system encoding the qubits and under what the authors claimed normal circumstances in a wide variety of conditions for implementations of quantum computers. Nevertheless, the matter still appears to not be completely resolved, since there may be physical implementations of quantum computers that do not abide by their assumptions. Energy bounds for the quantum qubit logical operations require better understanding and study, particularly with respect to their dependence on the technology used.
- *Processing Rate of QC.* In QC, the rate of execution unitary operations largely depends on the implementation technology (see Section 3.9); certain techniques can execute unitary operations in microseconds (e.g., bulk NMR) and some might execute at microsecond or even picosecond rates (e.g., photonic techniques for NMR). The time duration to do observation can also be very short, but it may be highly dependent on the size of the measuring apparatus and the required precision. (See the following discussion on the observation operation and its volume).
- *Volume Bounds for QC.* We now consider (perhaps more closely than usual in the quantum literature) the volume bounds of QC. Potentially, the modest volume bounds of QC may be the one significant advantage over classical methods for computation. Due to the *quantum parallelism* (i.e., the superposition of the basis states allows each basis state to exist in parallel),

the volume would *appear* to be no more that the number of qubits. This may be true, but there are a number of substantial issues that need to be carefully considered. Recall the observation operation both provides a measurement of a qubit with a resulting state reduction. However, the QC literature has not yet carefully considered the volume bounds for the observation operation and as we shall see, it is not yet at all clear what the volume required. In spite of major works on the mathematical and physical foundations of quantum observation, the precise nature of quantum state reduction via a strong quantum measurement remains somewhat of a mystery. Two distinct approaches to the mathematical and physical foundations of observation have been developed:

(a) the *Copenhagen Formulation,* where the observation is simply *an assumed basic operation* and is considered to be done by a macroscopic measuring device, and
(b) the *von Neumann Formulation* [100, Chapter 4: Macroscopic Measurement], which views the measuring apparatus as well as the quantum system measured as both part of a quantum system. Hence the evolution of the system (and resulting experimental predictions) can be distinct from that predicted by the Copenhagen formulation of observation (which does not take this into account since the measuring apparatus is assumed in their formulation to be very large).

See Cerf and Adami [18] for a comparison the Copenhagen and von Neumann formulations and see Hay, Peres [278] for an example of this difference. In summary, the Copenhagen and the von Neumann formulations for observation differ in the assumed context (macroscopic or microscopic measurement apparatus). (Note: Attempts to rectify the difference between the Copenhagen and the von Neumann formulation for observation are given in Hay, Peres [278] and in Zurek [279], but it appears not yet resolved.) The Copenhagen formulation for observation is generally used in the context of quantum physics experiments which use macroscopic measurement apparatus. However, the Copenhagen formulation does not seem to be applicable in the context of a microscopic measurement apparatus, which is so small that it is subject to quantum effects (and thus is within a unitary quantum system). So the Copenhagen formulation for observation may not appropriate for molecular size QC. Although the von Neumann formulation of observation is not relevant to the vast majority of physics experiments (since their experiments generally use large measuring apparatus and small number of degrees of freedom (qubits)), it appears to be appropriate for molecular size QC. It is possible that the volume for quantum observation apparatus grows very quickly with the number of qubits in the von Neumann formulation. In particular, no one has proved a upper bound on the volume for quantum observation (as a function of the number of qubits) assuming the von Neumann formulation. See the Appendix for a further discussion of the problem of determining volume bounds for the observation operation in QC.

3.11. SUMMARY AND ACKNOWLEDGMENTS

We have have provided an overview of the field of quantum computing and surveyed its major algorithmic results and applications as well as physical implementations and their limitations. Some issues such as energy costs as well as volume of observation apparatus for quantum computing are still unresolved, and the later are further addressed in the Appendix. We would like to thank G. Brassard for his clear explanation of numerous results in the field of QC. Also, I would like to thank P. Shor and U. Vazirani for references and for illuminating discussions on quantum computation, particularly on the issue of volume bounds for quantum observation.

3.12. APPENDIX: VOLUME OF OBSERVATION APPARATUS FOR QUANTUM COMPUTING

Here we discuss the challenge of providing volume bounds for observation apparatus when doing QC.

3.12.1. A Potentially Fallacious Proof of Small Volume

First we note that one might be tempted to give a constructive proof that observation can be done on n qubits in small volume, along the following lines:

(i) *Basis Step.* We begin with a simple, well established experimental method for observation of a single qubit in small quantum system with say n_0 qubits for a constant n_0. There are many other examples of experimentally verified methods for observation using macroscopic measurement apparatus. (For example, a number of proposed QC architectures (e.g., the Cirac and Zoller [233] 3, and Pellizzari et al. [234] proposed ion trap QC and Kane's [264] silicon-based NMR QC) give specific descriptions of measuring apparatus that have been experimentally verified for observation of a single qubit within a quantum computing systems with a constant number of qubits. While their measuring apparatus is macroscopic, it still must have just some finite volume.

(ii) *Inductive Step.* However, then we just scale up by using the same experimental apparatus to do observation on each of n qubits (that is, repeating the observation for each of the other qubits). This seems to result in a small volume (perhaps even linear size) apparatus for observation.

The potential fallacy of this line of argument is:

(a) In the basis step, the experiments of [230, 260] did not provide bounds on the errors (or fidelity) of the measurement as a function of the volume of the measuring apparatus.

(b) The inductive step fails to take into account quantum effects involving both the measuring apparatus and the n qubits, as might be predicted by the von Neumann formulation of quantum measurement in the case where the measuring apparatus is so small that it is subject to quantum effects.

That is, there needs to be given, in addition to the experimental description (which is only established for n_0 qubits):

(iii) *A mathematical analysis of the quantum effects (in the context of a closed unitary system) involving the measuring apparatus as the number n of qubits grows large.* In particular, bounds on the errors (or fidelity) of the measurement need to be determined as a function of the size of the measuring apparatus.

Without this crucial final element, the proof is certainly not complete. Since the observation operation is not reversible, such a proof (in the context of a closed unitary system) seem unlikely to be obtainable.

3.12.2. Possible Experimental Demonstrations of Measurement

Another approach would be to test a proposed small-volume apparatus for observation on n qubits for moderate size n (say, in the range of a few hundred, which is required for a nontrivial factoring computation). But the experimental evidence of the volume bounds for observation is unclear, since the QC experiments have not yet been scaled to large or even moderate numbers (say dozens) of qubits, and there are few if any physics experiments for this case. (Shnirman, Schoen [120] describe the use of a single-electron transistor to perform quantum measurements; D'Helon, Milburn [280] describe quantum measurements with quantum computers; and Ozawa [281] describes methods for nondestructive (known as *nondemolition*) quantum measurements of certain quantum computations.)

Hence, at this time that there appears to be *neither a mathematical proof nor an experimental demonstration* (for even a moderately large number of qubits n) *that observation can be done in small volume* (in a closed quantum system). Thus there is no evidence (either mathematical or experimental) at this time that QC using measurement scales to large numbers of qubits with small volume.

We first consider a number of related questions concerning measurement and quantum state reduction:

3.12.3. Is a Quantum Observation Instantaneous?

It appears not. Brune et al. [282] describe the progressive decoherence of the meter in a quantum measurement.

3.12.4. Is an Observation Always Reversible?

It appears the answer is no (in a narrow mathematical sense of a state reduction), yes (for small closed state spaces), and no (in a practical sense for entanglements in a large state space).

- By the strict mathematical definition of the state reduction due to observation, an observation is not reversible in general. Under what conditions is a measurement reversible in the strict mathematical sense? That is, when can we measure classical information from a quantum source (yielding a set of pure states with their probabilities with a reduction of quantum entropy), but later be able to reverse this process to regenerate the entangled source state? Bennett et al. [104] show that this is possible in the very special case where the source states can be partitioned into two or more mutually orthogonal subsets. (Other necessary and sufficient conditions for measurements to be reversible have been proved in Bennett et al. [104]; Chuang, Yamamoto [283] describe how to regenerate a qubit if it has observable error.)
- There is experimental evidence that the physical execution of some reductions via measurement are in fact reversible (at least in very small closed systems). Mabuchi, Zoller [284] have observed inversions of quantum jumps in very small quantum-optical systems under continuous observation, and Ueda [285] compares the notions of mathematical and physical reversibility.
- On the other hand, in the case of entanglements in a large state space, even if a measurement is in principle reversible in a closed system due the reversible nature of the diffusion process, the likelihood of such a reverse to the original state, within a moderate (say polynomial in n) time duration, appears to drop exponentially with the number of qubits n. Gottfield [286] and Diosi, Lukacs [287] explain quantum state vector reduction via strong measurement as a physical process, e.g, state diffusion into the atoms of the measurement apparatus. (Also see Pearle [288, 289].) This diffusion due to reduction may be modeled by a system similar to a rapidly mixing Markov system in probability theory, which seems to provide a very low (dropping exponentially with n) likelihood for reversibility within a polynomial time duration. (Others have modeled measurement by a nonlinear interactions with the environment, which are irreversible.)

3.12.5. Should You Avoid Observation Operations?

An alternative approach is to completely avoid observation operations on the basis that the observation operation is not actually essential to many quantum computations. (This seems somewhat surprising, given the extensive use of the observation operation in the QC literature for both algorithms and quantum error correction.) Bernstein, Vazirani [50] (by showing that any given observation operation can be delayed to future steps by use of the using XOR operation) proved that all observation operations can be delayed to the final step of a quantum computation. For a small $\varepsilon > 0$, let some particular qubit (of the linear superposition of basis states) be *ε-near classic* if had the qubit been observed, the measured value would be a fixed value (either be 0 or 1) with ε probability. Suppose the output of a QC consists of the observation of a subset S of the qubits; the resulting reduced superposition will be termed the *output superposition.*

Bernstein and Vazirani [50] and Brassard et al. [173, 202] observe that any QC can be repeated to insure the output qubits are ε-near classic in the final output superposition after the repetitions. Note that if a QC with bounded amplitude precision is reduced by an observation, the output qubits yield the correct value with high likelihood. Hence we may consider simply not doing the observation reduction to a basis state in the final step; in place of this (reduced) output superposition we simply output the nonreduced quantum state superposition of the QC that exists just prior to the final observation step. This alternative approach can eliminate entirely the observation operation from many quantum computations, and so provides small volume, but has the drawback of providing a nonclassic output consisting of a nonreduced quantum state superposition. The potential difficulty with this approach is as follows: If this (nonreduced quantum state superposition) output is then processed by a classical computing machine, it may propagate unwanted quantum effects to the classical computing machine.

3.12.6. What about Approximate Observation Operations?

An approach to this difficulty is to only do the observation operation approximately within accuracy ε; this may suffice for many QC applications. However, even if the observation operation is done ε-approximately by unitary operations, it appears to require a number of additional qubits n' growing exponentially with the n, the original number of qubits of the QC. In fact, we know of no upper bound on n' better than $2^n \log(1/\varepsilon)$.

3.12.7. Why May the Volume Required by Observation Apparatus Not Be Small?

We next consider whether it is reasonable to expect that a mathematical proof (or such experimental demonstrations) of small volume quantum observation will ever be done. We provide an informal argument that even an ε-approximate observation cannot be done in polynomial time using small volume, where ε is the inverse of a polynomial. (It should be emphasized that the following is not a formal proof in any sense.) Since for n qubits, the size of the basis state space grows as 2^n in the general case, it seems reasonable to assume (e.g., where the physics of the strong measurement is modeled by a diffusion process [286–288] that is rapidly mixing) that the likelihood of reversibility within polynomial time bounds drops exponentially with the number n of qubits. Thus, in the context of polynomial time computations, the ε-approximate observation is assumed irreversible with high likelihood.

The argument will hinge on the assumption, made by conventional formulations of quantum computation, that quantum computation (including both unitary qubit operations as well as the nonunitary observation (or projection) operation) can be exectuted for any given number n qubits, which makes an implicit assumption that both unitary and nonunitary operations can be executed at any scale.

Let us also assume that the number of qubits n is small (at most a few hundreds). For sake of contradiction, let us for the moment suppose that (i) quantum computing scales to at least moderate size (say a few tens of thousands of qubits), and (ii) an ε-approximate observation operation can be done on one of n qubits by a microscopic measuring device of size $n' = n^c$, for a constant c, and operating within time polynomial in n. Since n is small, the measuring device is surely of sufficiently small size so that its physics is consistent with established quantum physics. (Observe that if quantum computing is to scale to at least moderate size n', then surely quantum effects need to hold for molecules of size n'). This implies that we need to view the apparatus for the observation as executing polynomial time unitary quantum computation, which is reversible, so the reverse of the observation also executes in quantum polynomial time. Hence we have an apparent contradiction, since we have assumed the ε-approximate observation is not reversible in polynomial time. (Note: This argument does not require that the governing physical laws shift at some definite size from a quantum-mechanical paradigm to a classical paradigm; instead, the argument requires that if the quantum-mechanical paradigm is valid at size n then it also is valid at some what larger size $n' = n^c$.)

Due to informal nature of this argument, it only provides partial evidence that (with the above assumption), QC with the observation operation does not scale to a large number of qubits within small volumes, and in particular that a polynomial time ε-approximate observation operation requires very large volume and cannot be done at the micromolecular scale for moderate large n. We feel our above argument is far too informal to provide a resolution of the issue. It remains a major open problem in QC to *provide a formal proof that either (i) there is large volume required for observation or (ii) there is not.*

3.13. CONCLUSIONS

This chapter provided an overview of the field of QC, including abstract models for QC. We have surveyed major potential applications of QC, for example, cryptography, speeding up combinatorial search and integer factorization were surveyed. Major technologies for experimental implementation of QC have been enumerated. Even after two decades of major research efforts, all prior QC demonstrations to date have been limited to very small number of qubits. Various key technical challenges and fundamental resource bounds for scaling QC to larger problem sizes have been reviewed. It remains to be demonstrated if these challenges can be overcome to allow QC to scale to problem instances of practical importance to combinatorial search and integer factorization.

REFERENCES

Note: QCQC 98 is an acronym for Proceedings of 1st NASA Workshop on Quantum Computation and Quantum Communication (QCQC 98), Springer-Verlag, Feb 1998.

1. R. Landauer. Irreversability and heat generation in the computing process. *IBM Journal of Research Development*, 5(183): 1961.
2. C. H. Bennett. Logical reversibility of computations. *IBM Journal of Research and Development*, 17: pp 525–532, 1973.
3. C. H. Bennett and R. Landauer. Physical limits of computation. *Scientific American*, p 48, July 1985.
4. R. Landauer. The physical nature of information. *Physics Letters A*, 217: p 188, 1996.
5. T. Toffoli. In: J. W. de Bakker, J. van Leeuwen editors. *Automata, Languages and Programming*. New York: Springer-Verlag, p 632, 1980.
6. M. Li and P. Vitanyi. Reversibility and adiabatic computation: trading time and space for energy. *Proceedings of the Royal Society of London, Series A*, 452: pp 769–789, 1996. (Online preprint quant-ph/9703022)
7. C. H. Bennett. Time/space trade-offs for reversible computation. *Siam Journal on Computing*, 18 p 766, 1989.
8. A. Barenco, C. H. Bennett, R. Cleve, and D. P. DiVincenzo. *Physical Review A*, 51: p 1015, 1995.
9. C. H. Bennett and D. P. DiVincenzo. Progress towards quantum computation. *Nature*, Oct 1995.
10. C. H. Bennett and D. P. DiVincenzo. Quantum computing: towards an engineering era? *Nature*, 377: 1995.
11. A. Barenco. Quantum physics and computers. *Contemporary Physics*, 37: p 375, 1996. (Online preprint quant-ph/9612014)
12. P. Benioff. Quantum ballistic evolution in quantum mechanics: application to quantum computers. To be published in *Physical Review A*, 1996. (Online preprint quant-ph/9605022)
13. G. Brassard. New Trends in Quantum Computing. 13th Symposium on Theoretical Aspects of Computer Science, Grenoble. *Lecture Notes in Computer Science*: New York: Springer-Verlag, Feb 1996. (Online preprint quant-ph/9602014)
14. G. Brassard. New horizons in quantum information processing. *Proceedings of the 25th Colloquim on Automata, Languages, and Programming,* Aalborg, Denmark, May 1998.
15. S. Haroche and J. M. Raimond. Quantum computing: dream or nightmare? *Physics Today*, 49(8): p 51, 1996.
16. G. Brassard. Quantum information processing: the good, the bad and the ugly. In: S. Burton, Kaliski Jr., editor. Advances in Cryptology: CRYPTO '97, Volume 1294 of *Lecture Notes in Computer Science*: pp 337–341, Aug 1997.
17. J. Preskill. Quantum computing: pro and con. Submitted to *Proceedings of the Royal Society London, A*, 1997. (Online preprint quant-ph/9705032)
18. V. Scarani. Quantum computing. Accepted for publication in *American Journal of Physics*, 1998). (Online preprint quant-ph/9804044)
19. A. Steane. Quantum computing. Reports on Progress in Physics, 1998. (Online preprint quant-ph/9708022)
20. V. Vedral and M. B. Plenio. Basics of quantum computation. Invited basic review article for *Progress in Quantum Electronics*, 1998. (Online preprint quant-ph/9802065)
21. Taubes. All together for quantum computing. *Science*, 273: 1996.

22. N. Gershenfeld and I. L. Chuang. Quantum computing with molecules. *Scientific American*, 278(6): pp 66–71 June 1998.
23. C. P. Williams and S. H. Clearwater. *Explorations in Quantum Computing*. New York: Springer-Verlag, 1997.
24. R. Brylinski and G. Chen. *Mathematics of Quantum Computation* (Computational Mathematics Series). Boca Raton, FL: CRC Press, Feb 2002.
25. J. Gruska. *Quantum Computing*. (Advanced Topics in Computer Science Series). New York: Osborne/McGraw-Hill, Mar 1999.
26. M. Hirvensalo. *Quantum Computing* (Natural Computing Series). Berlin: Springer-Verlag GmbH and Co., Dec 2003.
27. T. Beth and G. Leuchs. *Quantum Information Processing*, 2nd ed. Weinheim: Wiley-VCH, Mar 2005.
28. M. Brooks. *Quantum Computing and Communications*. London: Springer-Verlag., May 1999.
29. S. Imre and F. Balazs. *Quantum Computing for Communications*. Chichester, UK: Wiley, Jan 2005.
30. G. Leuchs. *Quantum Information Technology*. Weinheim: Wiley-VCH, Mar 2003.
31. C. Macchiavello, G. M. Palma, and A. Zeilinger. *Quantum Computation and Quantum Information Theory*. Singapore: World Scientific Publishing, Jan 2001.
32. M. L. A. Nielsen and I. L. Chuang. *Quantum Computation and Quantum Information* (Cambridge Series on Information and the Natural Sciences). Cambridge, UK: Cambridge University Press, Oct 2000.
33. W. P. Schleich and H. Walther. *Elements of Quantum Information*. Weinheim: Wiley-VCH, Jan 2007.
34. V. Vedral. *Introduction to Quantum Information Science* (Oxford Graduate Texts Series). Oxford, UK: Oxford University Press, Sep 2006.
35. G. Van Assche. *Quantum Cryptography and Secret-Key Distillation*. Cambridge, UK: Cambridge University Press, July 2006.
36. D. Bouwmeester, A. K. Ekert, and A. Zeilinger. *The Physics of Quantum Information: Quantum Cryptography, Quantum Teleportation, Quantum Computation*. Berlin: Springer Verlag, Jan 2007.
37. A. V. Sergienko (Eds.), *Quantum Communications and Cryptography*. American Physical Society. Boca Raton, FL: CRC Press, Nov 2005.
38. D. Evans, J. J. Holt, C. Jones, and K. Klintworth. *Coding Theory and Quantum Computing*. (Contemporary Mathematics Series). Newport, RI: American Mathematical Society, Sep 2005.
39. G. Frank. *Quantum Error Correction and Fault Tolerant Quantum Computing*. Boca Raton, FL: CRC Press, Mar 2008.
40. A. O. Pittenger. *An Introduction to Quantum Computing Algorithms* (Progress in Computer Science and Applied Logic (PCS)). Amsterdam: Birkhauser, Nov 1999.
41. G. Chen, D. A. Church, B. Englert, and C. Henkel. *Quantum Computing Devices: Principles, Designs, and Analysis* (Chapman and Hall/CRC Applied Mathematics and Nonlinear Science Series). Boca Raton, FL: Chapman & Hall/CRC Press, Sep 2006.
42. R. Clark (eds.), *Experimental Implementation of Quantum Computation*. Fairfax, VA: IOS Press, Dec 2002.

43. T. Metodi and F. Chong. Quantum Computing for Computer Architects (Synthesis Lectures on Computer Architecture). San Rafael, CA: Morgan and Claypool Publishers, Nov 2006.

44. B. E. Kane and T. W. Sigmon. *Quantum Dot Devices and Computing*. SPIE Society of Photo-Optical Instrumentation Engineering, (Apr 2002).

45. A. Leggett, B. Ruggiero, and P. Silvestrini. *Quantum Computing and Quantum Bits in Mesoscopic Systems*. Norwell, MA: Kluwer Academic/Plenum Publishers, Dec 2003.

46. R. P. Feynman. Simulating physics with computers. *International Journal of Theoretical Physics*, 21(6–7): pp 467–488, 1982.

47. R. P. Feynman. Quantum mechanical computers. *Foundation of Physics*, 16(6): pp 507–531, 1986.

48. P. Benioff. Quantum mechanical models of Turing machines that dissipate no energy. *Physical Review Letters*, 48: p 1581, 1982.

49. D. Deutsch and R. Jozsa. Rapid solutions of problems by quantum computation. In: Proceedings of the Royal Society, London A, 439: pp 553–558, 1992.

50. E. Burnstein and U. Vazirani. Quantum Complexity Theory. Proceedings of the 25th Annual ACM Symposium on Theory of Computing. Association for Computing Machinery, New York, 1993, pp 11–20. Published in: *SIAM Journal on Computing*, 26(5): pp 1411–1473, Oct 1997.

51. G. Costantini and F. Smeraldi. A Generalization of Deutsch's Example. 1997. (Online preprint quant-ph/9702020)

52. D. Collins, K. W. Kim, and W. C. Holton. Deutsch–Jozsa algorithm as a test of quantum computation. Approved for publication in *Physical Review A*, 1998. (Online preprint quant-ph/9807012)

53. R. Jozsa. *Proceedings of Royal Society London, A*, 435: p 563, 1996.

54. R. Jozsa. Entanglement and quantum computation. In: S. Huggett, et al., editors. *Geometric Issues in the Foundations of Science*. 1997. (Online preprint quant-ph/9707034).

55. R. Jozsa. Quantum effects in algorithms. QCQC 98, Feb 1998. (Online preprint quant-ph/9805086)

56. D. Deutsch. Quantum theory: the Church–Turing principle and the universal quantum computer. *Proceedings of the Royal Society, London, A*, 400: pp 97–117, 1985.

57. C. Moore and J. P. Crutchfield. Quantum automata and quantum grammars. 1997. (Online preprint quant-ph/9707031)

58. A. Kondacs and J. Watrous. On the power of quantum finite state automata. In: 38th Annual Symposium on Foundations of Computer Science, Miami Beach, Florida, Oct 1997. IEEE, pp 66–75.

59. M. R. Dunlavey. Simulation of finite state machines in a quantum computer. 1998. (Online preprint quant-ph/9807026)

60. J. Watrous. On one-dimensional quantum cellular automata. In: 36th Annual Symposium on Foundations of Computer Science, Milwaukee, Wisconsin, Oct 1995. IEEE, pp 528–537.

61. C. Dürr, H. L. Thanh, and M. Santha. A decision procedure for well-formed linear quantum cellular automata. In: 13th Annual Symposium on Theoretical Aspects of

Computer Science, Volume 1046 of *Lecture Notes in Computer Science*, Grenoble, France, Feb 22–24, 1996. New York: Springer, pp 281–292.

62. C. Dürr and M. Santha. A decision procedure for unitary linear quantum cellular automata. In 37th Annual Symposium on Foundations of Computer Science, Burlington, Vermont, Oct 1996. Piscataway, NJ: IEEE Press, pp 38–45. (Online preprint quant-ph/9604007)
63. A. Barenco. A universal two–bit gate for quantum computation. 1995. (Online preprint quant-ph/9505016)
64. D. P. DiVincenzo, N. Margolus, P. Shor, T. Sleator, J. Smolin, and H. Weinfurter. Elementary gates for quantum computation. Submitted to *Physical Review A*, 1995.
65. A. Barenco, C. H. Bennett, R. Cleve, D. P. DiVincenzo, N. Margolus, P. Shor, T. Sleator, J. Smolin, and H. Weinfurter. Elementary gates for quantum computation. *Physical Review A*, 52: p 3457, 1995. (Online preprint quant-ph/9503016)
66. D. P. DiVincenzo. Two-bit gates are universal for quantum computation. *Physical Review Letters A*, 50(1015): 1995.
67. J. A. Smolin and D. P. DiVincenzo. Five two-bit quantum gates are sufficient to implement the quantum fredkin gate. *Physical Review A*, 53: p 2855, 1995.
68. D. P. DiVincenzo. Quantum gates and circuits. Proceedings of the ITP Conference on Quantum Coherence and Decoherence, Dec 1996. Submitted to *Proceedings of Royal Society London, A*. (Online preprint quant-ph/9705009)
69. J. F. Poyatos, J. I. Cirac, and P. Zoller. Complete characterization of a quantum process: the two-bit quantum gate. *Physical Review Letters*, 08: Nov 1996. (Online preprint quant-ph/9611013)
70. D. Mozyrsky, V. Privman, and S. P. Hotaling. Extended Quantum XOR gate in terms of two-spin interactions 1996. (Online preprint quant-ph/9610008)
71. D. Mozyrsky, V. Privman, and S. P. Hotaling. Design of gates for quantum computation: the three-spin XOR gate in terms of two-spin interactions. *International Journal of Modern Physics B*, 12: pp 591–600, 1998. (Online preprint uant-ph/9612029)
72. S. Lloyd. Almost any quantum logic gate is universal. Los Alamos National Laboratory preprint (1997c).
73. C. Monroe, D. M. Meekhof, B. E. King, W. M. Itano, and D. J. Wineland. Demonstration of a fundamental quantum logic gate. *Physical Review Letters*, 75: p 4714, 1995.
74. D. P. DiVincenzo and D. Loss. Quantum Information is physical. To be published in *Superlattices and Microstructures*, Special issue on the cccasion of Rolf Landauer's 70th Birthday, 1998. (Online preprint cond-mat/9710259)
75. D. Deutsch. Quantum computational network. *Proceedings of the Royal Society London, A*, 425: pp 73–90, 1989.
76. A. C. C. Yao. Quantum circuit complexity. In: Proceedings of the 34th IEEE Symposium on Foundations of Computer Science, Palo Alto, California, Nov 1993: pp 352–361.
77. D. Aharonov, A. Kitaev, and N. Nisan. Quantum circuits with mixed States, Proceedings of the Thirtieth Annual ACM Symposium on Theory of Computation (STOC): pp 20–30, 1997. (Online preprint quant-ph/9806029); (1998).

78. D. Aharonov, W. van Dam, J. Kempe, Z. Landau, S. Lloyd, and O. Regev. Adiabatic quantum computation is equivalent to standard quantum computation. In: Proceedings of 45th Annual IEEE Symp. on Foundations of Computer Science (FOCS): pp 42–51, 2004.
79. K. M. Obenland and A. M. Despain. Models to reduce the complexity of simulating a quantum computer. 1997. (Online preprint quant-ph/9712004)
80. K. M. Obenland and A. M. Despain. Simulating the effect of decoherence and inaccuracies on a quantum computer. QCQC 98, Feb 1998. (Online preprint quant-ph/9804038)
81. K. M. Obenland and A. M. Despain. A parallel quantum computer simulator. *High Performance Computing*: 1998. (Online preprint quant-ph/9804039)
82. N. J. Cerf and S. E. Koonin. Monte Carlo simulation of quantum computation. *Mathematics and Computers in Simulation*, 47: pp 143–152, 1998. (Online preprint quant-ph/9703050).
83. A. Berthiaume and G. Brassard. The quantum challenge to structural complexity. In: Proceedings of the 7th Annual IEEE Conference on Structure in Complexity: pp 132–137, 1992.
84. A. Berthiaume. L'ordinateur quantique: complexit'e et stabilisation des calculs. Ph.D. thesis, Dept. d'informatique et de Recherche Operationelle, Universite de Montreal, 1995.
85. A. Berthiaume and G. Brassard. Oracle quantum computing, In: Proceedings of the Workshop on Physics and Computation–Physcomp '92. Piscataway, NJ: IEEE Press, pp 195–199, Oct 1992.
86. A. Berthiaume and G. Brassard. Oracle quantum computing. *Journal of Modern Optics*, 41(12): pp 2521–2535, 1994.
87. J. Machta. Phase information in quantum oracle computing. 1998. (Online preprint quant-ph/9805022)
88. W. van Dam. Two classical queries versus one quantum query. 1998. (Online preprint quant-ph/9806090)
89. W. van Dam. Oracle interrogation: Getting all information for almost half the price CWI, University of Oxford, 1998. (Online preprint quant-ph/9805006)
90. L. M. Adleman, J. Demarrais, and M. D. A. Huang. Quantum computability. *SIAM Journal on Computing*, 26(5): pp 1524–1540, Oct 1997.
91. C. Moore and M. Nilsson. Parallel Quantum computation and quantum codes. 1998. (Online preprint quant-ph/9808027)
92. R. Cleve, W. van Dam, M. Nielsen, and A. Tapp. Quantum entanglement and the communication complexity of the inner product function. QCQC 98, Feb 1998. (Online preprint quant-ph/970801)
93. E. Knill and R. Laflamme. On the power of one bit of quantum information. 1998. (Online preprint quant-ph/9802037)
94. P. G. Kwiat, H. Weinfurter, T. Herzog, A. Zeilinger, and M. A. Kasevich. Interaction-free measurement. *Physical Review Letters*, 74: pp 4763–4766, 1995.
95. P. G. Kwiat, H. Weinfurter, and A. Zeilinger. Quantum seeing in the dark. *Scientific American*: Nov 1996.
96. J. H. Reif. On the impossibility of interaction-free quantum sensing for small I/O bandwidth. Information and Computation: pp 1–20, Jan 2000. (Online preprint at http://www.cs.duke.edu/ reif/paper/qsense/qsense.pdf)

97. A. S. Holevo. Some estimates of the information transmitted by quantum communication channels. *Problems of Information Transmission* (USSR), 9: pp 177–183, 1973.
98. C. Fuchs and C. Caves. Ensemble-dependent bounds for accessible information in quantum mechanics. *Physics Review Letters*, 73: pp 3047–3050, 1994.
99. B. Schumacher. On quantum coding. *Physical Review Letters A*, 51: p 2738, 1995.
100. J. von Neumann. *Mathematical Foundations of Quantum Mechanics.* New York: Springer Verlag, 1932. Reprinted in: *Princeton Landmarks in Mathematics.* Princeton, NJ: Princeton University Press, 1996.
101. R. Jozsa and B. Schumacher. A new proof of the quantum noiseless coding theorem. *Journal of Modern Optics*, 41: p 2343–2349, 1994.
102. K. Y. Szeto. Data compression of quantum code. July 1996. (Preprint quant-ph/9607010)
103. C. H. Bennett. Quantum information and computation. *Physics Today*, pp 24–30, Oct 1995.
104. C. H. Bennett, G. Brassard, Jozsa, Mayers, Peres, Schumacher, and Wootters. Reduction of quantum entropy by reversible extraction of classical information. *Journal of Modern Optics*, 1994.
105. R. Cleve and D. P. DiVincenzo. Schumacher's quantum data compression as a quantum computation. 1996. (Online preprint quant-ph/9603009)
106. J. H. Reif and S. Chakraborty. Efficient and exact quantum compression. *Journal of Information and Computation*, 205: pp 967–981, 2007.
107. R. Jozsa, M. Horodecki, P. Horodecki, and R. Horodecki. May 1998. Universal quantum data compression. (Preprint quant-ph/9805017)
108. S. L. Braustein, C. A. Fuchs, D. Gottesman, and H. K. Lo. A quantum analog of Huffman coding. May 1998. (Report no. quant-ph/9805080)
109. M. A. Nielsen. The entanglement fidelity and quantum error correction, 1996. (Online preprint quant-ph/9606012)
110. K. Svozil. Quantum algorithmic information theory. Lectures given at the summer school, Chaitin Complexity and Applications, Mangalia, Romania, June 1995. (Online preprint quant-ph/9510005)
111. K. Svozil. Quantum information theory. *The Journal of Universal Computer Science*, 2(5): pp 311–346, May 1996.
112. A. S. Holevo. Coding theorems for quantum communication channels. Steklov Mathematical Institute, 1997. (Online preprint quant-ph/9708046)
113. E. Knill and R. Laflamme. Concatenated quantum codes 1996. (Online preprint quant-ph/9608012)
114. E. Knill and R. Laflamme. A theory of quantum error-correcting codes 1996 (Online preprint quant-ph/960403)
115. M. Ohya. A mathematical foundation of quantum information and quantum computers: on quantum mutual entropy and entanglement. 1998. (Online preprint quant-ph/9808051)
116. H. Buhrman, R. Cleve, and A. Wigderson. Quantum vs. classical communication and computation. 1998. (Online preprint quant-ph/9802040)
117. C. Adami and N. J. Cerf. What information theory can tell us about quantum reality. QCQC 98, Feb 1998. (Online preprint quant-ph/9806047)

118. D. P. DiVincenzo, P. W. Shor, and J. A. Smolin. Quantum channel capacity of very noisy channels. To appear 1998. (Online preprint quant-ph/9706061)

119. A. S. Holevo. The capacity of quantum channel with general signal states. 1996. (Online preprint quant-ph/9611023)

120. H. Barnum, M. A. Nielsen, and B. Schumacher. Information transmission through a noisy quantum channel. 1997. (Online preprint quant-ph/9702049)

121. A. Shnirman and G. Schoen. Quantum measurements performed with a single-electron transistor. Submitted to Physical Review B, 1998. (Online preprint cond-mat/9801125)

122. C. H. Bennett, D. P. DiVincenzo, and J. A. Smolin. 1997. Capacities of quantum erasure channels. (Online preprint quant-ph/9701015)

123. C. Fuchs. Nonorthogonal quantum states maximize classical information capacity. 1997. (Online preprint quant-ph/9703043)

124. C. W. Helstrom. Detection theory and quantum mechanics. Information and Control, 10(3): pp 254–291, Mar 1967.

125. C. W. Helstrom. Detection theory and quantum mechanics (II). Information and Control, 13(2): pp 156–171, Aug 1968.

126. C. A. Fuchs. Distinguishability and accessible information in quantum theory. 1996. (Online preprint quant-ph/9601020)

127. P. Shor. Scheme for reducing decoherence in quantum memory. *Physical Review A*, 52: pp 2493–2496, 1995.

128. A. M. Steane. Error correcting codes in quantum theory. *Physical Review Letters*, 77: p 793, 1996.

129. A. R. Calderbank and P. W. Shor. Good quantum error-correcting codes exist. *Physical Review A*, 54: p 1098, Dec 1995. (Online preprint quant-ph/9512032)

130. A. Steane. Multiple particle interference and quantum error correction. *Proceedings of the Royal Society of London A*, 452: p 2551, 1996. (Online preprint quant-ph/9601029)

131. C. Bennett, D. DiVincenzo, J. Smolin, and W. Wootters. Mixed state entanglement and quantum error correction. *Physical Review A*, 54: p 3824, 1996. (Online preprint quant-ph/9604024).

132. R. Laflamme, C. Miquel, J. P. Paz, and W. H. Zurek. Perfect quantum error correction code. *Physical Review Letters*, 77: p 198, 1996. (Online preprint quant-ph/9602019)

133. P. W. Shor. Fault-tolerant quantum computation. 37th Symposium on Foundations of Computing, Los AlamitosCA: IEEE Computer Society Press, pp 56–65, 1996. (Online preprint quant-ph/9605011)

134. A. Y. Kitaev. Quantum computing: algorithms and error correction. 1996. (Preprint in Russian.)

135. A. Y. Kitaev. Fault-tolerant quantum computation by anyons, 1997. (Online preprint quant-ph/9707021)

136. D. Aharonov and M. Ben-Or. Fault Tolerant Quantum Computation with Constant Error. In: Proceedings of the Twenty-Ninth Annual ACM Symposium on Theory of Computing, El Paso, Texas, May 1997, pp 176–188. (Online preprint quant-ph/9611025)

137. E. Knill, R. Laflamme, and W. Zurek. Threshold accuracy for quantum computation. 1996. (Online preprint quant-ph/9610011)
138. E. Knill, R. Laflamme, and W. H. Zurek. Resilient quantum computation: error models and thresholds. 1997. (Online preprint quant-ph/9702058)
139. C. H. Bennett, G. Brassard, S. Breidbart, and S. Wiesner. Quantum cryptography, or unforgeable subway tokens In: D. Chaum, R. L. Rivest and A. T. Sherman (eds.), *Advances in Cryptology: Proceedings of Crypto, 82*: pp 267–275, 1982. New York: Plenum Press, 1983.
140. C. H. Bennett and G. Brassard. Quantum cryptography: public key distribution and coin tossing. In: Proceedings of the IEEE International Conference on Computers, Systems, and Signal Processing, Bangalore, India. New York: IEEE, 1984, pp 175179.
141. C. H. Bennett, G. Brassard, and A. Ekert. Quantum cryptography. *Scientific American*, Oct 1992 pp 50–57.
142. G. Brassard. Cryptology column—quantum cryptography: a bibliography. *Sigact News*, 24(3): pp 16–20, 1993.
143. C. H. Bennett and G. Brassard. An update on quantum cryptography. In: G. R. Blakley, D. Chaum, editors. Advances in Cryptology: Proceedings of CRYPTO 84, Volume 196 of *Lecture Notes in Computer Science*, Aug 1984, pp 475–480. New York: Springer-Verlag, 1995.
144. G. Brassard. Cryptology column: quantum computing: the end of classical cryptography? SIGACTN: *SIGACT News* (ACM Special Interest Group on Automata and Computability Theory), 25: 1994.
145. N. Gisin, G. Ribordy, W. Tittel, and H. Zbinden. Quantum cryptography. *Review of Modern Physics*, 74: p 145, 2002.
146. Y. Ozhigov. Protection of information in quantum qatabases, 1997. (Online preprint quant-ph/9712016)
147. J. Hruby. Q-deformed quantum cryptography. In: A. De Santis, editor. Advances in Cryptology: EUROCRYPT 94, Volume 950 of *Lecture Notes in Computer Science*. Springer-Verlag, 1995, pp 468–472.
148. C. H. Bennett, F. Bessette, G. Brassard, L. Salvail, and J. Smolin. Experimental quantum cryptography. *Journal of Cryptology*, 5(1): pp 3–28, 1992.
149. R. J. Hughes, G. G. Luther, G. L. Morgan, C. G. Peterson, and C. Simmons. Quantum cryptography over underground optical fibers. In: Neal Koblitz, editor. Advances in Cryptology: CRYPTO '96, Volume 1109 of *Lecture Notes in Computer Science*. Berlux: Springer-Verlag, pp 329–342, Aug 1996.
150. C. H. Bennett, G. Brassard, C. Crpeau, and M. H. Skubiszewska. Practical quantum oblivious transfer. In: J. Feigenbaum, editor. Advances in Cryptology: CRYPTO '91, Volume 576 of *Lecture Notes in Computer Science*. New York: Springer-Verlag, pp 351–366, Aug 1992.
151. D. Mayers. On the security of the quantum oblivious transfer and key distribution protocols. In: Don Coppersmith, editor. Advances in Cryptology: CRYPTO '95, Volume 963 of *Lecture Notes in Computer Science*. New York: Springer-Verlag, pp 124–135, Aug 1995.
152. D. Mayers. Quantum key distribution and string oblivious transfer in noisy channels. In: Neal Koblitz, editor. Advances in Cryptology: CRYPTO '96, Volume 1109 of

Lecture Notes in Computer Science. New York: Springer-Verlag, pp 343–357, Aug 1996.

153. H. K. Lo and H. F. Chau. Quantum computers render quantum key distribution unconditionally secure over arbitrarily long distance. 1998. (Online preprint quant-ph/9803006)
154. G. Brassard and C. Crpeau. Quantum bit commitment and coin tossing protocols. In: A. J. Menezes, S. A. Vanstone, editors. Advances in Cryptology: CRYPTO '90, Volume 537 of *Lecture Notes in Computer Science*. New York: Springer-Verlag, pp 49–61, (Aug 1990)
155. G. Brassard, C. Crpeau, R. Jozsa, and D. Langlois. A quantum bit commitment scheme provably unbreakable by both parties. In: 34th Annual Symposium on Foundations of Computer Science, Palo Alto, California, pp 362–371, Nov 1993.
156. A. C. C. Yao. Security of quantum protocols against coherent measurements. In: Proceedings of the Twenty-Seventh Annual ACM Symposium on the Theory of Computing, Las Vegas, Nevada, pp 67–75, May–June 1995.
157. G. Brassard, C. Crpeau, D. Mayers, and L. Salvail. 1998. Defeating classical bit commitments with a quantum computer. (Online preprint quant-ph/9806031)
158. H. F. Chau and H. K. Lo. An empty promise with a quantum computer. *Fortsch Phys*, 46: pp 507–520, 1998. (Online preprint quant-ph/9709053)
159. C. Crpeau and L. Salvail. Quantum oblivious mutual identification. In: L. C. Guillou, J.-J. Quisquater, editors. Advances in Cryptology: EUROCRYPT 95, Volume 921 of *Lecture Notes in Computer Science*. New York: Springer-Verlag, pp 133–146, May 1995.
160. H. E. E. Brandt. Quantum-cryptographic entangling probe. *Physical Review A*, 71: 042312(14), 2005.
161. T. Kim, I. Stork genannt Wersborg, F. N. C. Wong, and J. H. Shapiro. Complete physical simulation of the entangling-probe attack on the Bennett–Brassard 1984 protocol. *Physical Review A*, 75: 042327, 2007.
162. H. K. Lo and H. F. Chau. Why quantum bit commitment and ideal quantum coin tossing are impossible. Accepted for publication in a special issue of *Physica D*: 177–187, 1998. (Online preprint quant-ph/9711065)
163. T. Pellizzari. Quantum Networking with Optical Fibres. Submitted to *Physical Review Letters*, 1997. (Online preprint quant-ph/9707001)
164. J. I. Cirac, P. Zoller, H. J. Kimble, and H. Mabuchi. Quantum state transfer and entanglement distribution among distant nodes in a quantum network. *Physical Review Letters*, 78: p 3221, 1997.
165. S. Bose, V. Vedral, and P. L. Knight. A multiparticle generalization of entanglement swapping. 1997. (Online preprint quant-ph/9708004)
166. C. H. Bennett, Brassard, Crepeau, Jozsa, Peres, and Wootters. Teleporting an unknown quantum state via dual classical and Einstein–Podolsky–Rosen Channels. *Physical Review Letters*, 70: p 1895, 1993.
167. C. H. Bennett, Brassard, Popescu, Schumacher, Smolin, and Wootters. Purification of noisy entanglement and faithful teleportation via noisy channels. *Physical Review Letters*, 76: p 722, 1996.
168. G. Brassard. Teleportation as a quantum computation. *Physica D*, 120: pp 43–47, 1998. (Online preprint quant-ph/9605035)

169. W. K. Wootters and W. H. Zurek. A single quantum cannot be cloned. *Nature*, 299: p 802, 1982.
170. V. Buzek and M. Hillery. Universal optimal cloning of qubits and quantum registers. QCQC 98, Feb 1998. (Online preprint quant-ph/9801009)
171. A. Ekert, S F. Huelga, C. Macchiavello, and J. I. Cirac. Distributed quantum computation over noisy channels. 1998. (Online preprint quant-ph/9803017)
172. D. R. Simon. On the power of quantum computation. In: Proceedings of the 35th Annual Symposium on Fundamentals of Computer Science. Piscataway, NJ: IEEE Press, pp 116–123, Nov. 1994.
173. G. Brassard and P. Hoyer. An exact quantum polynomial-time algorithm for Simon's problem. In: Proceedings of the Fifth Israeli Symposium on Theory of Computing and Systems (ISTCS'97). (Online preprint quant-ph/9704027)
174. D. Coppersmith. An approximate fourier transform useful in quantum computing. IBM Research Report RC19642: 1994.
175. R. B. Griffiths and C. Niu. Semiclassical Fourier transform for quantum computation. *Physical Review Letters*, 76: pp 3228–3231, 1996. (Online preprint quant-ph/9511007)
176. P. Hoyer. Efficient quantum transforms. 1997. (Online preprint quant-ph/9702028)
177. R. Beals. Quantum computation of Fourier transforms over symmetric groups. In: Proceedings of the Twenty-Ninth Annual ACM Symposium on Theory of Computing, El Paso, Texas, May 4–6: pp 48–53, 1997.
178. M. Pueschel, M. Roetteler, and T. Bet. Fast quantum Fourier transforms for a class of non-abelian groups. Universitaet Karlsruhe, 1998. (Online preprint quant-ph/9807064)
179. V. Vedral, A. Barenco, and A. Ekert. Quantum networks for elementary arithmetic operations. 1996. (Online preprint quant-ph/9511018)
180. D. Grigoriev. Testing shift-equivalence of polynomials by deterministic. *probabilistic and quantum machines, Theoretical Computer Science*, 180(1–2): pp 217–228, June 1997.
181. P. W. Shor. Algorithms for quantum computation: discrete logarithms and factoring, Proceedings of the 35th Annual Symposium on Foundations of Computer Science, Santa Fe, New Mexico, Nov 1994.
182. P. W. Shor. Polynomial-Time Algorithms for Prime Factorization and Discrete Logarithms on a Quantum Computer. *SIAM Journal of Computing*, 26: p 1484, 1997. (Online preprint quant-ph/9508027)
183. A. Ekert and R. Jozsa. Shor's quantum algorithm for factoring numbers. *Review of Modern Physics*, 1996.
184. G. L. Miller. Reimann's hypothesis and test for primality. *Journal of Computer Systems Science*, 12: pp 300–317, 1976.
185. C. Zalka. Fast versions of Shor's quantum factoring algorithm. 1998. (Online preprint quant-ph/9806084)
186. D. Beckman, A. N. Chari, S. Devabhaktuni, and J. Preskill. Efficient networks for quantum factoring. 1996. (Online preprint quant-ph/9602016)
187. K. Obenland and A. M. Despain. Simulation of factoring on a quantum computer architecture. In: Proceedings of the 4th Workshop on Physics and Computation, Boston, Massachusetts, Nov 1996. Boston: New England Complex Systems Institute.

188. M. B. Plenio and P. L. Knight. Realistic lower bounds for the factorization time of large numbers on a quantum computer. *Physical Review A*, 53: p 2986, 1996. (Online preprint quant-ph/9512001)
189. L. M. K. Vandersypen, M. Steffen, G. Breyta, C. S. Yannoni, M. H. Sherwood, and I. L. Chuang. *Nature*, 414: pp 883–887, 2001. (Online preprint doi:10.1038/414883a).
190. A. Y. Kitaev. Quantum measurements and the Abelian Stabilizer Problem. 1995. Preprint.
191. L. K. Grover. A fast quantum mechanical algorithm for database search. Proceedings of the 28th Annual ACM Symposium on Theory of Computing, Philadelphia: pp 212–216, May, 1996.
192. L. K. Grover. Quantum computers can search arbitrarily large databases by a single query. 1997. (Online preprint quant-ph/9706005)
193. L. K. Grover. Quantum computers can search rapidly by using almost any transformation. *Physical Review Letters*, 80: pp 4329–4332, 1998. (Online preprint quant-ph/9712011)
194. C. Zalka. Grover's quantum searching algorithm is optimal. 1997. (Online preprint quant-ph/9711070)
195. A. K. Pati. Fast quantum search algorithm and bounds on it. Theory Div. BARC. Mumbai, India, 1998. (Online preprint quant-ph/9807067)
196. D. Biron, O. Biham, E. Biham, M. Grassl, and D. A. Lidar. Generalized Grover search algorithm for arbitrary initial amplitude distribution. QCQC 98, Feb 1998. (Online preprint quant-ph/9801066)
197. P. Cockshott. Quantum relational databases. 1997. (Online preprint quant-ph/9712025)
198. S. C. Benjamin and N. F. Johnson. Structures for data processing in the quantum regime. University of Oxford, 1998. (Online preprint cond-mat/9802127)
199. E. Farhi, J. Goldstone, S. Gutmann, and M. Sipser. A limit on the speed of quantum computation in determining parity. 1998. (Online preprint quant-ph/9802045)
200. D. A. Meyer. Sophisticated quantum search without entanglement. UCSD preprint, 2000.
201. B. M. Terhal and J. A. Smolin. Single quantum querying of a database. *Physical Review A*, 58 pp 1822–1826.
202. G. Brassard, P. Hoyer and A. Tapp. Quantum counting. 1996. (Online preprint quant-ph/9805082)
203. G. Brassard, P. Hoyer, and A. Tapp. Quantum counting. In: Proceedings of the 25th Colloquim on Automata, Languages, and Programming, Aalborg, Denmark, May 1998.
204. T. Hogg. Quantum computing and phase transitions in combinatorial search. *Journal of Artificial Intelligence Research*, 4: pp 91–128, 1996. (Online preprint quant-ph/9508012)
205. T. Hogg and M. Yanik. Local search methods for quantum computers. 1998. (Online preprint quant-ph/9802043)
206. S. Lloyd. Universal quantum simulators. *Science*, 273: p 1073, 1996.
207. C. Zalka. Efficient simulation of quantum systems by quantum computers. 1996. (Online preprint quant-ph/9603026)

208. B. M. Boghosian. Simulating quantum mechanics on a quantum computer. *Physica D*, 120: pp 30–42, 1998. (Online preprint quant-ph/9701019)
209. S. Wiesner. Simulations of many-body quantum systems by a quantum computer. 1996. (Online preprint quant-ph/9603028)
210. D. S. Abrams and S. Lloyd. Simulation of many-body fermi systems on a universal quantum computer. 1997. (Online preprint quant-ph/9703054)
211. P. A. Benio. Review of quantum computation. *Trends in Statistical Physics,* Trivandrum, India: Council of Scientific Information, 1996.
212. B. M. Boghosian and W. Taylor. Quantum lattice-gas models for the many-body Schrodinger equation. Sixth International Conference on Discrete Fluid Mechanics, Boston University, Boston, Aug 1996. (Online preprint quant-ph/9701016)
213. D. A. Meyer. Quantum mechanics of lattice gas automata I: one particle plane waves and potentials. 1996. (Online preprint quant-ph/9611005)
214. D. A. Meyer. From quantum cellular automata to quantum lattice gases. *Journal of Statistical Physics*, 85: pp 551–574, 1996. (Online preprint quant-ph/9604003)
215. D. A. Lidar and O. Biham. Simulating using spin glasses on a quantum computer. *Physical Review E*, 56: p 3661, 1997. (Online preprint quant-ph/9611038)
216. D. A. Lidar and H. Wang. Calculating the thermal rate constant with exponential speed-up on a quantum computer. University of California, Berkeley, 1998. (Online preprint quant-ph/9807009)
217. R. Schack. Using a quantum computer to investigate quantum chaos. 1997. (Online preprint quant-ph/9705016)
218. N. H. Bshouty and J. C. Jackson. Learning DNF over the uniform distribution using a quantum example oracle. In: Proceedings of the Eighth Annual Conference on Computational Learning Theory, Santa Cruz, California, July 1995. New York: ACM Press, pp 118–127.
219. D. Ventura and T. Martinez. A quantum computational learning algorithm. (Previously entitled: Quantum harmonic sieve: learning DNF using a classical example oracle.) 1998. (Online preprint ph/9807052)
220. A. Y. Vlasov. Quantum computations and images recognition. QCM'96, Japan, Sep 1996. (Online preprint quant-ph/9703010)
221. R. R. Tucci. How to compile a quantum Bayesian net. 1998. (Online preprint quant-ph/9805016)
222. D. Ventura and T. Martinez. Quantum associative memory. 1998. (Online preprint quant-ph/9807053)
223. P. Benioff. Tight binding Hamiltonians and quantum Turing machines. *Physical Review Letters*, 78: pp 590–593, 1997. (Online preprint quant-ph/9610026)
224. R. Landauer. Is quantum mechanics useful?. *Philosophical Transactions of the Royal Society London*, 353, 367:1–2, 1995.
225. R. Landauer. Is quantum mechanically coherent computation useful? In: D. H. Feng, B. L. Hu, editors. Proceedings of Drexel-4 Symposium on Quantum Nonintegrability-Quantum-Classical Correspondence, Philadelphia, Sep 8, 1994. Boston: International Press, 1997.
226. S. F. Huelga, C. Macchiavello, T. Pellizzari, A. K. Ekert, M. B. Plenio, and J. I. Cirac. On the improvement of frequency standards with quantum entanglement. 1997. (Online preprint quantph/9707014)

227. J. J. Bollinger, W. M. Itano, D. J. Wineland, and D. J. Heinzen. Optical frequency measurements with maximally correlated states. *Physical Review A*, 54:R4649, 1997.
228. D. A. Meyer. Quantum strategies. *Physical Review Letters*, 82: pp 1052–1055, 1999.
229. D. A. Meyer. *Quantum Games and Quantum Algorithms, Quantum Computing and Quantum Information Science*. New York: ACM Press, pp 11–20, 1993.
230. G. Burkard, D. Loss, and D. P. DiVincenzo. Coupled quantum dots as quantum gates. 1998. (Online preprint cond-mat/9808026)
231. D. Loss (Basel) and D. P. DiVincenzo. Quantum computation with quantum dots. 1997. (Online preprint cond-mat/9701055)
232. D. M. Meekhof, C. Monroe, B. E. King, W. M. Itano, and D. J. Wineland. Generation of nonclassical motional states of a trapped atom. *Physical Review Letters*, 76: p 1796, 1996.
233. J. I. Cirac and P. Zoller. Quantum computations with cold trapped ions. *Physical Review Letters*, 74: p 4091, 1995.
234. T. Pellizzari, S. A. Gardiner, J. I. Cirac, and P. Zoller. Decoherence, continuous observation and quantum computing: a cavity QED model. Submitted to *Physical Review Letters*, 1995.
235. D. F. V. James. Quantum dynamics of cold trapped ions, with application to quantum computation. 1997. (Online preprint quant-ph/9702053)
236. D. J. Wineland, C. Monroe, D. M. Meekhof, B. E. King, D. Leibfried, W. M. Itano, J. C. Bergquist, D. Berkeland, J. J. Bollinger, and J. Miller. Quantum state manipulation of trapped atomic ions. Proceedings of Workshop on Quantum Computing, Santa Barbara, California, Dec 1996. Submitted to *Proceedings of Royal Society of London A*. (Online preprint quant-ph/9705022)
237. D. J. Wineland, C. Monroe, W. M. Itano, D. Leibfried, B. E. King, and D. M. Meekhof. Experimental issues in coherent quantum-state manipulation of trapped atomic ions. *Journal of Research of the National Institute of Standards and Technology*, 03: pp 259, 1998. (Online preprint quant-ph/9710025)
238. B. E. King, C. S. Wood, C. J. Myatt, Q. A. Turchette, D. Leibfried, W. M. Itano, C. Monroe, and D. J. Wineland. Initializing the collective motion of trapped ions for quantum logic. 1998. (Online preprint quant-ph/9803023)
239. Q. A. Turchette, C. S. Wood, B. E. King, C. J. Myatt, D. Leibfried, W. M. Itano, C. Monroe, and D. J. Wineland. Deterministic entanglement of two trapped ions. Time and Frequency Division, National Institute of Standards and Technology, Boulder, 1998. (Online preprint quant-ph/9806012)
240. R. J. Hughes. Cryptography, quantum computation and trapped Ions. Submitted to Philosophical Transactions of the Royal Society, Proceedings of the Royal Society Discussion meeting on Quantum Computation: Theory and Experiment, London, England, Nov 1997. (Online preprint quant-ph/9712054)
241. R. J. Hughes, D. F. V. James, J. J. Gomez, M. S. Gulley, M. H. Holzscheiter, P. G. Kwiat, S. K. Lamoreaux, C. G. Peterson, V. D. Sandberg, M. M. Schauer, C. M. Simmons, C. E. Thorburn, D. Tupa, and P. Z. The Los Alamos trapped ion quantum computer experiment. *Fortsch. Phys.*, 46: pp 329–362, 1998. (Online preprint quant-ph/9708050)
242. D. F. V. James, M. S. Gulley, M. H. Holzscheiter, R. J. Hughes, P. G. Kwiat, S. K. Lamoreaux, C. G. Peterson, V. D. Sandberg, M. M. Schauer, C. M. Simmons,

D. Tupa, P. Z. Wang, and A. G. White. Trapped ion quantum computer research at Los Alamos, Los Alamos National Laboratory. QCQC 98, Feb 1998. (Online preprint quant-ph/9807071)

243. R. J. Hughes, D. F. V. James, E. H. Knill, R. Laflamme, and A. G. Petschek. Decoherence bounds on quantum computation with trapped ions. 1996. (Online preprint quant-ph/9604026)
244. M. B. Plenio and PL. Knight. Decoherence limits to quantum computation using trapped ions. *Proceedings of the Royal Society of London, A*, 453: pp 2017–2041, 1997. (Online preprint quant-ph/9610015)
245. S. Schneider, H. M. Wiseman, W. J. Munro, and G. J. Milburn. Measurement and state preparation via ion trap quantum computing. *Fortsch. Phys.*, 46: pp 391–400, 1998. (Online preprint quant-ph/9709042)
246. D. F. V. James. The theory of heating of the quantum ground state of trapped ions. *Physical Review Letters*, 81: pp 317–320, 1998. (Online preprint quant-ph/9804048)
247. S. Schneider, D. FV. James, and G. J. Milburn. Method of quantum computation with "hot" trapped ions. Submitted to *Physical Review Letters*, 1998. (Online preprint quant-ph/9808012)
248. Q. A. Turchette, C. J. Hood, W. Lange, H. Mabuchi, and H. J. Kimble. Measurement of conditional phase shifts for quantum logic. *Physical Review Letters*, 75: p 4710, 1995.
249. I. L. Chuang and Y. Yamamoto. A simple quantum computer. Submitted to *Physical Review A*, 1995. (Online preprint quant-ph/9505011)
250. I. L. Chuang, L. M. K. Vandersypen, X. Zhou, D. W. Leung, and S. Lloyd. Experimental realization of a quantum algorithm. *Nature*, 393: pp 143–146, 1998. (Online preprint quant-ph/9801037)
251. P. Torma and S. Stenholm. Polarization in quantum computations. 1996. (Online preprint quant-ph/9602021)
252. C. Adami and N. J. Cerf. Quantum computation with linear optics. QCQC 98: Feb1998. (Online preprint quant-ph/9806048)
253. W. Teich, K. Obermeyer, and G. Mehler. Structural basis of multistationary quantum systems, II. Effective few-particle dynamics. *Physical Review B*, 37(14): pp 8111–8120, 1988.
254. S. Lloyd. A potentially realizable quantum computer. *Science*, 261: p 1569, 1993.
255. D. P. DiVincenzo. Quantum computation and spin physics. Proceedings of the Annual MMM Meeting, Nov, 1996; to be published in *Journal of Applied Physics*, 1997. (Online preprint cond-mat/9612125)
256. H. Wei, X. Xue, and S. D. Morgera. Single molecule magnetic resonance and quantum computation. 1998. (Online preprint quant-ph/9807057)
257. H. Wei, X. Xue, and S. D. Morgera. NMR quantum automata in doped crystals.1998. (Online preprint quant-ph/9805059)
258. G. Castagnoli. Quantum computation based on retarded and advanced propagation. 1997. (Online preprint quant-ph/9706019)
259. S. Kak. On initializing quantum registers and quantum gates. Louisiana State University, 1998. (Online preprint quant-ph/9805002)
260. M. Murao, M. B. Plenio, S. Popescu, V. Vedral, and P. L. Knight. Multi-particle entanglement purification protocols. 1997. (Online preprint quant-ph/9712045)

261. D. G. Cory, A. F. Fahmy, and T. F. Havel. Nuclear magnetic resonance spectroscopy: an experimentally accessible paradigm for quantum computing. In: Proceedings of the 4th Workshop on Physics and Computation. Boston: New England Complex Systems Institute, 1996.
262. N. Gershenfeld and I. Chuang. Bulk spin resonance quantum computation. *Science*, 275: p 350, 1997.
263. G. P. Berman, G. D. Doolen, G. V. Lopez, and V. I. Tsifrinovich. Quantum entangled states and quasiclassical dynamics in macroscopic spin systems. 1998. (Online preprint quant-ph/9802015)
264. B. E. Kane. A silicon-based nuclear spin quantum computer. *Nature*, 393: 133–137, 1998.
265. J. A. Jones, R. H. Hansen and M. Mosca. Logic gates and nuclear magnetic resonance pulse sequences. 1998. Submitted to Journal of Magnetic Resonance. (Online preprint quant-ph/9805070)
266. J. A. Jones, M. Mosca, and R. H. Hansen. Implementation of a quantum search algorithm on a nuclear magnetic resonance quantum computer. *Nature*, 393: pp 344–346, 1998. (Online preprint quant-ph/9805069)
267. J. A. Jones. Fast Searches with nuclear magnetic resonance computers. *Science*, 280: Apr 1998 p 229.
268. J. A. Jones and M. Mosca. Approximate quantum counting on an NMR ensemble quantum computer. Submitted to *Physical Review Letters*, 1998.
269. J. A. Jones and M. Mosca. Implementation of a quantum algorithm to solve Deutsch's problem on a nuclear magnetic resonance quantum computer. *Journal of Chemical Physics*: in press, Aug 1998. (Online preprint quant-ph/9801027)
270. N. Linden, H. Barjat, and R. Freeman. An implementation of the Deutsch–Jozsa algorithm on a three-qubit NMR quantum computer. 1998. (Online preprint quant-ph/9808039)
271. W. S. Warren. The usefulness if NMR quantum computing. *Science*, 277: pp 1688–1689. (See also response by N. Gerenfeld and I. Chuang, ibid, pp 1689–90, 1997)
272. S. E. Barnes. Efficient quantum computing on low temperature spin ensembles. 1998. (Online preprint quant-ph/9804065)
273. E. Knill, I. Chuang, and R. Laflamme. Effective pure states for bulk quantum computation. 1997. (Online preprint quant-ph/9706053)
274. L. J. Schulman and U. Vazirani. Scalable NMR quantum computation. 1998. (Online preprint quant-ph/9804060)
275. Zyczkowski, Horodecki, Sanpera, Lewenstein. On the volume of the set of mixed entangled states. 1998. (Online preprint quant-ph/9804024)
276. J. Gea-Banacloche. Minimum energy requirements for quantum computation. *The American Physical Society*, 89(12): Nov 18, 2002 pp 217901-1–217901-4.
277. M. Ozawa. Conservative Quantum Computing. *Physical Review Letters*, 89: 057902, 2002.
278. O. Hay and A. Peres. Quantum and classical descriptions of a measuring apparatus. 1997. (Online preprint quant-ph/9712044)
279. W. H. Zurek. Decoherence and the transition from quantum to classical. *Physics Today*, 44: pp 36–44, 1991.

280. C. D'Helon and G. J. Milburn. Quantum measurements with a quantum computer, 1997. (Online preprint quant-ph/9705014)

281. M. Ozawa. Quantum nondemolition monitoring of universal quantum computers. *Physical Review Letters*, 80: p 631, 1998. (Online preprint quant-ph/9704028)

282. M. Brune, E. Hagley, J. Dreyer, X. Maitre, A. Maali, C. Wunerlich, J. M. Raimond, and S. Haroche. Observing the progressive decoherence of the meter in a quantum measurement. *Physical Review Letters*, 77: p 4887, 1996.

283. I. L. Chuang and Y. Yamamoto. Quantum bit regeneration. *Physical Review Letters*, May 13: 1996. (Online preprint quant-ph/9604031)

284. H. Mabuchi and P. Zoller. Inversion of quantum jumps in quantum-optical systems under continuous observation. *Physical Review Letters*, 76: p 3108, 1996.

285. M. Ueda. Logical Reversibility and Physical Reversibility in Quantum Measurement. International Conference on Frontiers in Quantum Physics, Kuala Lumpur, Malaysia, New York: Springer-Verlag, July, 1997. (Online preprint quant-ph/9709045)

286. K. Gottfield. *Quantum Mechanics*. London: Benjamin-Cummings, Ch. 4, pp 165–190, 1966. (Reprinted by Addison-Wesley, 1989)

287. L. Diosi and B. Lukacs (Eds.), *Stochastic Evolution of Quantum States in Open Quantum Systems and in the Measurement Process*. London: World Scientific, 1994.

288. P. Pearle. State vector reduction as a dynamical process. In: *Proceedings of SUNY-Albany Conference on Fundamental Questions in Quantum Mechanics*. A. Inomata, J. Kimball, L. Roth, editors. 1984.

289. P. Pearle. Models for reduction. In: *Quantum concepts in Space and Time,* R. Penrose and C. J. Isham, editors, London: Clarendon Press, pp 84–108, 1985.

4

COMPUTING WITH QUANTUM-DOT CELLULAR AUTOMATA

Konrad Walus and Graham A. Jullien

This chapter describes a promising emerging technology for nanoscale computing based on cellular automata. Rather than relying on the on/off state of a current switch to encode information, this paradigm, called quantum-dot cellular automata (QCA), represents binary information in the electronic configuration of a set of coupled quantum dots, molecules, or magnetic nanoparticles. QCA logic and circuits are implemented by connecting various logic gates using arrays of QCA cells, which also make up the interconnect network. Control over information flow and synchronization, as well as true power gain, is possible by clocking the cells. Using the available building blocks, several circuit examples are provided, together with a description of available computer-aided design tools. This chapter also presents a description of recent experiments and several promising implementations of QCA.

4.1. INTRODUCTION

The seemingly endless progress of microelectronics is the result of the ability of the semiconductor industry to continuously scale down the transistor. However, there are challenges to this scaling as the lateral dimensions of transistors approach a few tens of nanometers, and the gate oxides are reduced to just a handful of atoms. These challenges include leakage currents through the gate oxide, the discreteness of impurities, low on/off ratios due to drain induced barrier lowering (DIBL), and very high power dissipation. When scaled down to the level of single molecules, several additional problems emerge. A molecular transistor would require a channel, implemented with a single molecule, and three contacts, most likely implemented with metal. Given the significant size difference between the metallic

Bio-Inspired and Nanoscale Integrated Computing. Edited by Mary Mehrnoosh Eshaghian-Wilner

contacts and the molecular channel, it is very difficult to achieve high coupling from the three contacts to the molecular channel. If the source and drain are tightly coupled to the molecule, the gate remains very weakly coupled. As a result, the on/off ratio of the device is very poor. Remember that the current success of the metal-oxide-semiconductor transistor has been due to its nearly ideal properties, which have been maintained in a succession of highly controllable silicon technology advances. Finding a replacement that has all of these features and is also faster and/or cheaper is a formidable task.

There has been great research interest in finding the "next switch" that will replace the current transistor [1], but there have also been alternative proposals to replace the underlying paradigm on which computers operate. One of these paradigms was introduced to the research community in 1993 by C. S. Lent; he published the first work on a novel paradigm for computing at the nanoscale that is based on cellular automata, which he called quantum-dot cellular automata (QCA) [2]. Rather than encoding information in the on/off state of a current switch, the QCA paradigm encodes information in the electronic configuration of coupled quantum dots [3–5], metallic islands [6–26], molecules [27–41], or alternatively the magnetization in anisotropic nanomagnets [42, 43]. Information is transmitted and processed via the electrostatic or magnetic exchange interaction between individual QCA cells [44]. The potential advantages of QCA include high device density, high operating speed, and low power dissipation; the degree to which these advantages can be realized depends on the particular choice of physical implementation. The lack of physical connection (at the molecular scale) to each molecule and the fact that no current passes through each molecule makes the paradigm more suitable as a molecular implementation of the switch [27].

The cellular automata computing architecture paradigm has been studied for many years; in fact, John von Neumann, who is credited with inventing the stored program concept for building computers—the concept on which most of today's computer processors are still based—conceived the first cellular automaton in the late 1940s [45]. Cellular automata (CA) are essentially discrete dynamic systems whose behavior is completely specified in terms of local interactions with adjacent cells. The cell is usually assumed to be a finite-state machine driven by a central clock that synchronously advances time in all cells in the automata array. The outputs of neighboring cells at a given clock pulse are used, by each cell, to compute the cell's state following the next clock pulse. Most research has looked at fairly simple cells that can be implemented with little hardware cost. The QCA cell originally introduced by Professor Lent and his colleagues has a very simple structure with only two states provided by the two energetically low-lying electronic configurations of non-bonding electrons; the process by which electrons move about in the cell is quantum mechanical tunneling. The cell is designed such that these two states are energetically degenerate in the absence of any external influence. The configuration becomes deterministic under the perturbing influence of the electronic configuration of adjacent cells. A major feature of the technology (besides having to be built at nanometer scales) is that the logic state of the cell is determined by the position of extra charge in a fixed location on the substrate or molecule, rather than by the flow of charge in and out of the cell, which is how logic states are changed in standard CMOS-based logic. Since the movement of charge during logic state switching gives rise to most of the power dissipation in modern CMOS logic, QCA technology promises very low power dissipation.

4.2. THE CAD TOOL: QCADesigner

When commercial fabrication processes have reached a stage of advancement where any emerging technology is an economic reality, the potential of that technology should already have been explored and uses for the technology found. In fact, this vertical approach may even change the amount of effort spent on developing a technology based on, for example, the uses (or lack of them) found and the fabrication tolerances required. The best way to evaluate a technology is to provide tools for design and simulation and then to make these available to a large group of future designers within that technology. These tools, with sufficiently accurate simulation models and computational techniques, are able to predict the level of success of nanocircuit designs, even though such nanotechnology is not available. Throughout this chapter we will be illustrating the design concepts and showing simulation results using the QCA CAD tool, QCADesigner [46]. QCADesigner provides a comprehensive design and layout tool, along with a simulation engines that uses two different quantum mechanical approximations. The tool is currently available as a free download from http://www.qcadesigner.ca/ and provides an excellent vehicle for exploring the concepts discussed here. The various elements of QCADesigner will be introduced, where required, later in the chapter.

4.3. THE QCA CELL

QCA computing involves the interaction of many QCA cells, each constructed from a square pattern arrangement of four quantum dots. It is important to note that although initial work on QCA focused on semiconductor quantum dots, recent research has included cells implemented from single molecules [27–41] as well as magnetic nanoparticles [42, 43]. The electronic QCA cell is charged with two mobile nonbonding electrons that repel each other as a result of their mutual Coulombic repulsion, and, occupy one of the two diagonals of the cell in the ground state. These two states are referred to as the "ACTIVE" states and are used to encode binary information as depicted in Figure 4.1. Other higher energy states can be ignored if the energy of those states is sufficiently high.

The quantum dots can be coupled through a quantum mechanical tunneling barrier that enables the controlled transfer of electrons from one site to another. The cells are typically illustrated in the literature as having a bounding box (shown in Fig. 4.1). This bounding box does not necessarily have a physical structure; it is, rather, used for illustration and design purposes to isolate one cell from the next. Some implementations are based on double dots, two of which are required to

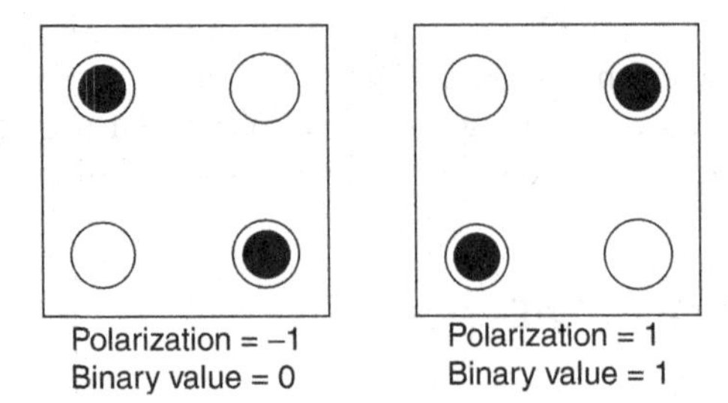

Figure 4.1. QCA cell is constructed from four quantum dots arranged in a square pattern. The two mobile electrons in the cell tend to repel each other and occupy the diagonal sites of the cell. The electronic configuration of the cell is used to encode binary information.

make a single cell, while others employ the cell in its entirety. Figure 4.2 illustrates the numeric indexing normally used to refer to individual dots in a cell.

Using this indexing convention, cell *polarization* is defined as

$$P = \frac{(\rho_1 + \rho_3) - (\rho_2 + \rho_4)}{\rho_1 + \rho_2 + \rho_3 + \rho_4},$$

where ρ_i is the charge density at dot i. The value of ρ_i represents the average number of electrons occupying dot i [44]. In an isolated cell, the two ACTIVE states will be energetically equivalent and, quantum mechanically, the state of this isolated cell can form a superposition of the two diagonal states. In such a superimposed state, the cell will have no net polarization. An unpolarized cell is said to be in a "NULL" state and the influence of such a cell does break the energy degeneracy in the neighboring cells; essentially the cell is said to be "off". However, in a four-dot cell such superimposed states only exist if the cell is maintained in coherent isolation; real devices will naturally drop to a random ACTIVE state due to the loss of quantum mechanical coherence to the environment. Therefore, maintaining a stable NULL state in four-dot cells is a difficult technical challenge. In order to overcome the challenges associated with coherent NULL states, recent proposals have included two additional sites that introduce a

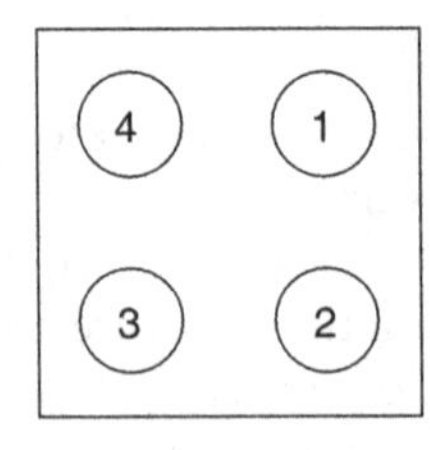

Figure 4.2. Cell indexing used throughout this chapter.

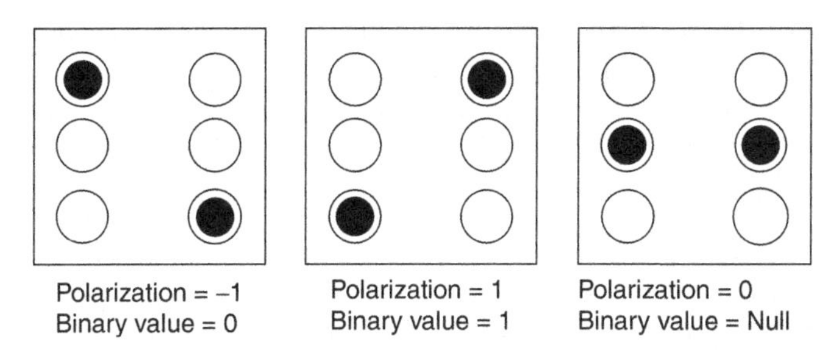

Figure 4.3. In the six-dot cell, the two center dots enable a third electronic configuration. In this third configuration, the cell is said to be in the NULL state, in which the cell does not drive adjacent cells to any particular polarization.

third low-lying electronic configuration with no net cell polarization [28, 32], as shown in Figure 4.3. To force the cell into this NULL state, the potential energy of these two sites can be controlled using coupled electrodes, and this ability provides a convenient and potentially more robust mechanism for clocking QCA circuits [22, 28, 32, 47].

Adjacent cells interact via electrostatic coupling—a quadrupole moment established from the nonuniform distribution of cell charge in each ACTIVE cell lowers the energy of one of the two ACTIVE states in neighboring cells, forcing those cells to relax preferentially to one state. The resulting cell-to-cell interaction function is highly nonlinear as shown in Figure 4.4, and the

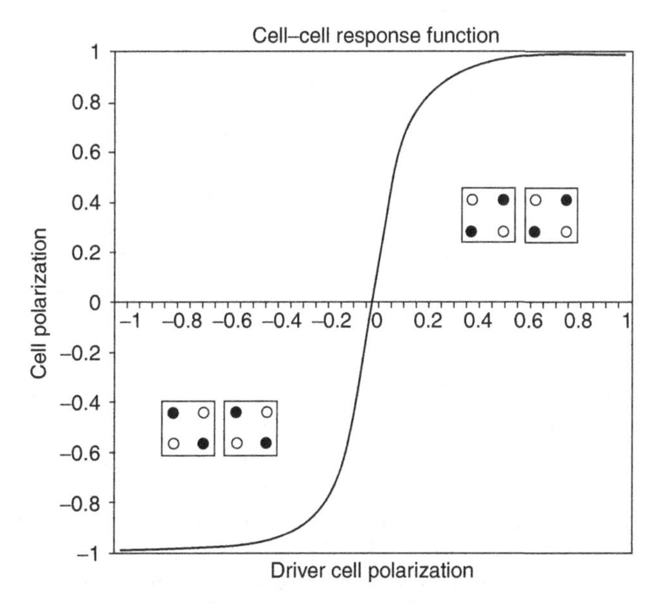

Figure 4.4. Nonlinear cell–cell response function. The output cell is almost completely polarized for a small polarization of the input cell (bold outline).

nonlinearity of this response function provides a noise margin, which is essential for signal restoration along lengthy arrays.

Because of this interaction, neighboring cells will tend to synchronize their polarization and, as a result, "wires" can be implemented from a linear array of cells.

4.3.1. Kink Energy

A full quantum mechanical model of a QCA circuit is computationally intractable. The size of the Hamiltonian matrix grows exponentially and leads to extremely large matrices for even small circuits. Assuming a two-state basis, the size of the Hamiltonian matrix is $2^N \times 2^N$, where N is the number of cells in the circuit. Various approximations can be made in order to reduce the size of the problem such that it can be solved on a standard computer. The most basic of these approximations is the Hartree approximation, in which the cells are quantum mechanically decoupled, and their interactions are modeled classically through a mean field approach. The dynamics of each individual cell are modeled quantum mechanically, but the correlations between cells are ignored and the cell is evolved in the field of all other cells, which remain stationary. When modeling the QCA cells in a two-state basis, an important parameter is the interaction coupling between two cells. The Coulombic coupling between two cells can be described by the so called *kink energy*, E_{kink}, associated with the energetic cost of two neighboring cells having opposite polarization [44]. The electrostatic interaction between charges in two four-dot cells, m and n, is

$$E_{m,n} = \frac{1}{4\pi\varepsilon} \sum_{i=1}^{4} \sum_{j=1}^{4} \frac{q_i^m q_j^n}{|r_i^m - r_j^n|},$$

where q_i^m is the charge in dot i of cell m, and r_i^m is the position of dot i in cell m. The kink energy is the difference in energy between two cells that have opposite polarization and those same two cells having the same polarization; i.e.,

$$E_{m,n}^{kink} = E_{m,n}^{\text{opposite polarization}} - E_{m,n}^{\text{same polarization}}.$$

Although not shown in Figure 4.1 and Figure 4.3, there are two positive background charges distributed throughout the cell in order to ensure charge neutrality and must also be included in the expression for kink energy. It can be shown that the kink energy decays as the fifth power of the distance between cells. This property is useful in designing circuits that do not suffer from crosstalk. Table 4.1 shows the kink energy for four different cell sizes. The reader should note that the listed dimensions represent the width and height of the cell bounding box within which the cell dots are located (in the center of the four quadrants of the box).

TABLE 4.1. Kink Energy Between Adjacent Cells

Cell Type	Cell Size	Kink Energy
Molecular QCA ($\varepsilon_r = 1$)	< 2 nm	>0.3 eV
Self-assembled ($\varepsilon_r = 12.9$)	5 nm	9.13 meV
Lithographically defined ($\varepsilon_r = 12.9$)	10 nm	4.56 meV
Lithographically defined ($\varepsilon_r = 12.9$)	20 nm	2.28 meV

4.3.2. Cell Switching

Cell switching involves the change of electronic configuration (polarization) of the cell, as illustrated in Figure 4.5. Switching is driven by the clock and the final state is determined by the neighboring cells or possibly unwanted space charges [48]. Depending on the level of kink energy, a cell can be excited into one of the other low lying states by unwanted thermal fluctuations [49, 50]. For this possibility to be avoided, the kink energy should be much larger than the thermal ambient energy, k_BT.

The attainable switching speed of a QCA circuit depends primarily on the particular choice of implementation. If the system is strongly coupled to the substrate, inelastic processes will dominate and the switching speed will be determined by the strength of the coupling between the cell and the substrate. However, if elastic processes dominate the switching, the switching speed can be determined by modeling the system using the Schrödinger equation. Tougaw et al. showed that the standard cell has a switching time as low as 2 ps [51] when operating in this coherent regime. Table 4.2 lists the order of the theoretical cell switching frequencies for different cell implementations reported in the literature.

4.4. GROUND STATE COMPUTING

QCA computing has often been referred to as ground state computation since the design of QCA circuits involves finding a layout of cells where the ground state of the layout for a particular set of boundary conditions provided by the inputs is the solution to the designed logical function. When a suitable environment is

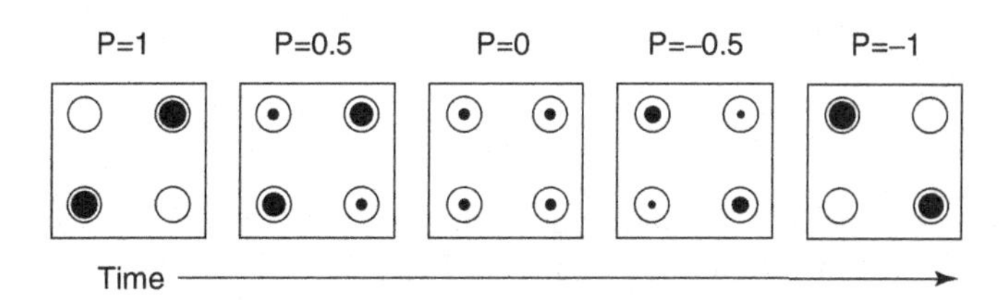

Figure 4.5. Cell switching from P=1 to P=–1.

TABLE 4.2. Theoretic Orders of QCA Cell Switching Frequencies.

Implementation	Frequency
Molecular [44]	THz
Semiconductor [51]	THz
Metallic-Island [25]	GHz
Magnetic [42]	MHz

provided, the cell will relax to the correct ground state. Changes in the boundary conditions (input values) raise the system energy and cause the system to relax to a new ground state and a new output. Original work on QCA focused on an edge-driven approach where the QCA circuit was provided inputs at the edge of a computational array, thus raising the system energy by introducing kinks at these inputs and allowing the system to relax to the correct ground state [2, 52, 53]. However, this approach limits the size of the circuit, since larger circuits will have a tendency to stick in one of many unwanted metastable states, as illustrated in Figure 4.6. The ability of the circuit to find the correct ground state can be enhanced with the application of adiabatic clocking [44].

4.5. CLOCKING

As with standard technologies, QCA clocking provides a mechanism for synchronizing information flow through the circuit. However, unlike standard

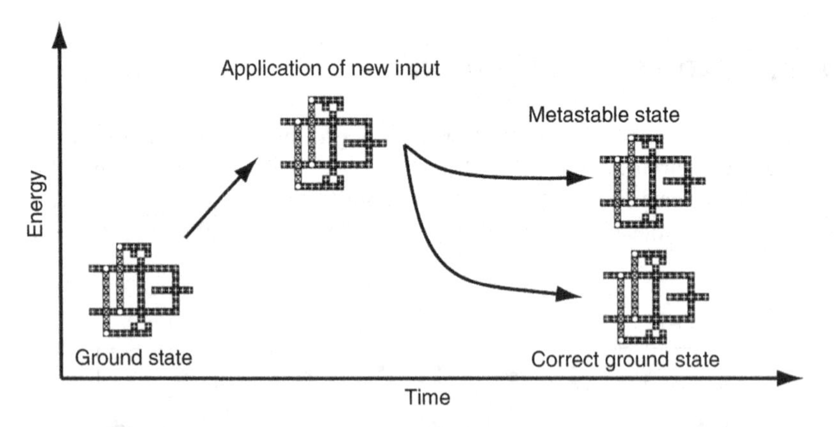

Figure 4.6. In the edge-driven computation approach, inputs are applied to one side of the circuit and the circuit is allowed to relax. Correct operation requires that the circuit settles to the correct ground state and does not get stuck in a metastable state.

technologies, which have a set input/output that provides a built-in directionality for information flow, the clock also controls the direction of information flow in a QCA circuit. In addition to this, the clock provides the power gain required for proper circuit operation [54]. By clocking the cells quasi-adiabatically, problematic metastable states can be minimized.

Clocking three-state QCA cells (six-dot cells) can be accomplished by modulating the potential of the two central dots of the cell. This applied potential modifies the potential energy of the NULL state. When the NULL state has the lowest energy, the cell will tend to relax to this state. When the clock potential is increased, the energy of the NULL state will be higher than that of the two ACTIVE states and the cell will tend to the ACTIVE state with the lowest potential energy, determined by the perturbation provided by neighboring cells.

A mechanism for clocking has also been proposed for a four-dot cell that involves a controlled ability to raise and lower the quantum mechanical tunneling barriers between the dots of the cell, thus forcing the cell into and out of a quantum mechanical superposition of the two ACTIVE states. When the tunneling barriers are high, charges in the cell become localized in one particular electronic configuration; as a result, the cell is said to be latched and the cell will not switch regardless of the influence of neighboring cells. When the tunneling barriers are low, the electronic wave function becomes delocalized ($P = 0$) and the cell is said to be in the NULL polarization state.

There are advantages to building QCA circuit design tools around this particular approach to clocking because the Hamiltonian for each cell can be reduced from three-states (3×3 matrix) to two-states (2×2 matrix). This reduction significantly lowers the computational requirements for large circuits. QCADesigner is designed around this model and is therefore capable of simulating fairly large circuits [46].

4.5.1. Zone Clocking

With zone clocking, all the cells in a design are grouped into one of four available clocking zones; that is, all the cells in a particular clocking zone are connected to one of the four available phases of the QCA clock, as shown in Figure 4.7 [55].

The successive latching and unlatching of cells connected to the different clock phases acts to pump information throughout the circuit. For example, a wire, which is clocked from left to right with increasing clocking zones, will propagate information in the same direction. As a result, QCA circuits are pipelined at the clocking zone level. This also permits more than one bit of information to be present on a particular wire. QCADesigner has been developed around this particular approach to clocking, and the example circuits presented in this chapter are based on results using this method.

Within the zone clocking scheme, each group of cells connected to a particular phase of the clock can be modeled as a D-latch [56]. As each group of cells in a particular clocking zone becomes latched, they retain their information until the clock is relaxed, independent of changes in the polarization of neighboring cells.

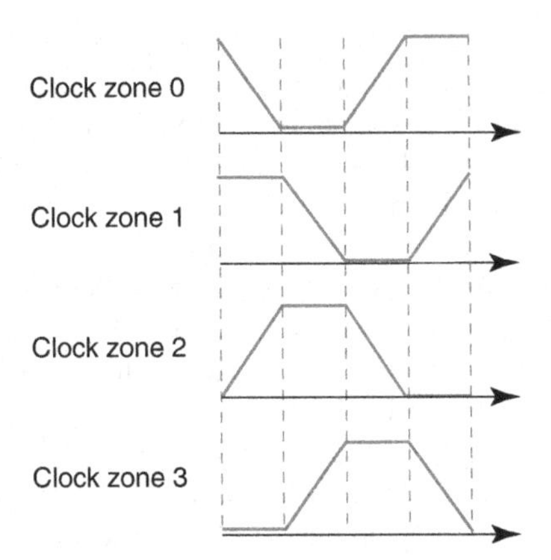

Figure 4.7. The four phases of the QCA clock used to control information flow in the QCA circuit.

A length of QCA wire can be represented schematically, as shown in Figure 4.8, where each clock zone is identified in the layout with a unique shade of gray. This deep (or fine-grained) pipelining has a major effect on the design cost function. Schematics of QCA circuits should include these D-latches to represent the clocking zones, as also shown in Figure 4.8.

The deep pipeline inherent in QCA circuits forces us to evaluate our designs differently than we would if we were designing with traditional technologies. Even in heavily pipelined transistor based logic architectures, there will normally be many gates in a combinational structure between each latch in the pipeline. In QCA, the latency is determined completely by the largest number of clocking zones between input and output, with each gate and wire connection being connected to a clocking zone. The original concept of QCA, as a quantum level implementation of classical cellular automata (CA) [2], is evident from the

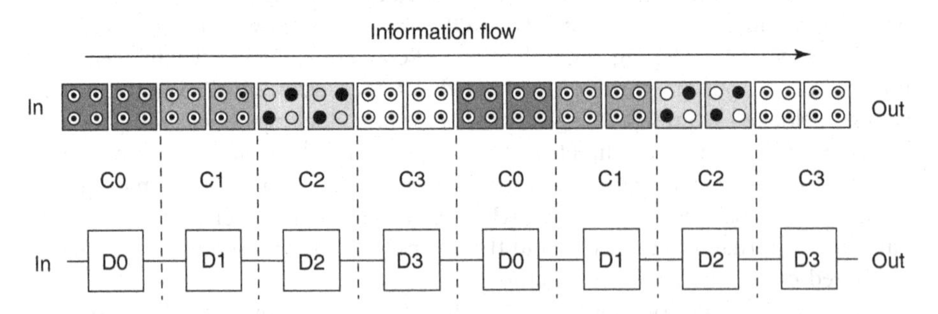

Figure 4.8. QCA wire shown with cells and schematic representation. C0, C1, C2, C3 are the four phases of the clock. Each of the clocking zones maps to a numbered D-latch in the circuit representation and is represented in the layout with a unique shade of grey. Notice that only one clocking zone is latched.

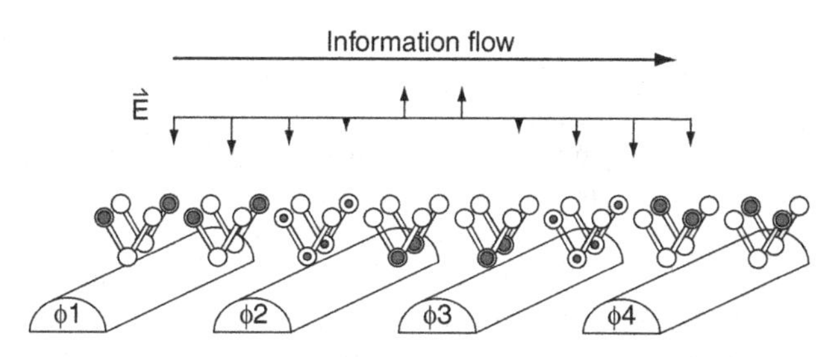

Figure 4.9. Submerged electrodes can be used to clock QCA cells by applying a forward moving electric field at the level of the cells. The figure is not shown to scale, as many cells can be placed between adjacent electrodes.

inherent clocked operation at the device level. Although CA architectures are not normally implemented in most synchronous digital systems, we can look at structures that have similar properties as being target candidates for QCA implementation. Examples are bit-serial architectures and systolic arrays. Interestingly, parallel designs, which require large fan-outs to the parallel logic components, will introduce more latency because of the very nature of QCA interconnects; thus, we will need to change our computing paradigms to match the different cost functions inherent with using future QCA technology. Significant work in these areas remains to be done.

4.5.2. Continuous Clocking

A method of clocking molecular QCA using a continuous clocking wave, realized by generating a moving electric field by a network of submerged electrodes, has been proposed in references [28, 31, 32]. Clocking networks using carbon nanotubes have been investigated in [57]. The design of three-dimensional molecular QCA cells enables a field in the vertical direction to modulate the potential energy between the NULL state and the two ACTIVE states. By applying phase-shifted sinusoidal potentials to each of the electrodes, a forward moving wave can be generated at the level of the cells. Cells are latched at the wavefront of this forward moving wave as illustrated in Figure 4.9. This enables the cells to be clocked without influencing the cell–cell interactions. This method is also valuable because there is no physical connection between the clocking network and the QCA devices. The number of cells between any two electrodes can be large, allowing the electrodes to be quite large when compared to the individual cells.

4.6. INPUT/OUTPUT INTERFACING

Interfacing to QCA circuits requires a technology that can set and detect the electronic configuration, or magnetization in the case of magnetic QCA, of single cells. I/O for molecular implementations of QCA is significantly more difficult, and there is currently no established method for I/O in these future molecular scale devices. In the case of metallic island and semiconductor QCA, the I/O mechanism is already established using single electron transistors (SETs) for reading the output and electrodes for setting the input [13, 25, 26]. The SET is a highly sensitive electrometer capable of detecting a fraction of the elementary charge present at its gate electrode. SETs operate in the Coulomb blockade regime and generally have to be cooled to extremely low temperatures. Significant research has been invested in developing techniques to realize room temperature [58] and high frequency [59] SETs since their application extends beyond sensitive electrometers.

Although magnetic QCA is still in the early stages of development, it appears that the I/O mechanism for this implementation is relatively simple to construct. At the input stage, a single wire element, through which a current is passed, provides sufficient magnetization to cause its neighboring MQCA cell to relax to one of the two polarizations [42]. At the output of the circuit a magnetization sensor, based on the giant magnetoresistance (GMR) effect, or a magnetic tunnel junction can be used to detect the magnetization of output cells.

4.7. QCA LOGIC

4.7.1. Inverter

The most common inverter design is shown in Figure 4.10. The kink energy between two cells changes sign when the cells are oriented at 45° with respect to each other. The output wire and the two legs of the inverter have this sign-change coupling and the ground state of this device has complementary input and output.

4.7.2. Majority Gate

The fundamental logic primitive available with QCA technology is the majority gate. This gate performs the following Boolean function:

$$Maj(A, B, C) = AB + AC + BC.$$

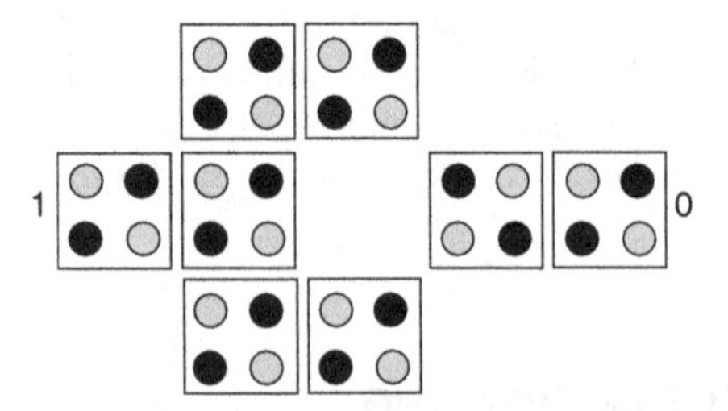

Figure 4.10. Layout of a fork inverter. The two arms of the fork reinforce the inverter function by doubling the coupling to the output cells.

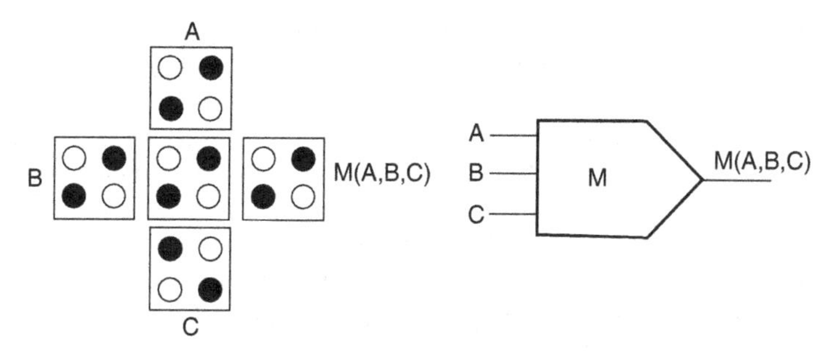

Figure 4.11. The three-input majority gate is the fundamental logic primitive available with QCA. It is created from a cross pattern of five cells. At least two of the cells must be logic '1' before the output is logic '1.'

At least two of the inputs to the gate must be of the same polarization before the output attains that polarization. This gate is a member of the higher class of threshold gates where the sum of weighted inputs must exceed the threshold before the output switches state [60]. The three-input majority gate is implemented with the layout shown in Figure 4.11.

The majority gate can be programmed to perform the standard two-input AND as well as two-input OR operations by fixing the polarization of one of the three available inputs as illustrated in Figure 4.12.

4.8. FIXED POLARIZATION CELLS

Fixed polarization cells are required to implement AND and OR functions, as discussed in the previous section. There are two possible methods for realizing

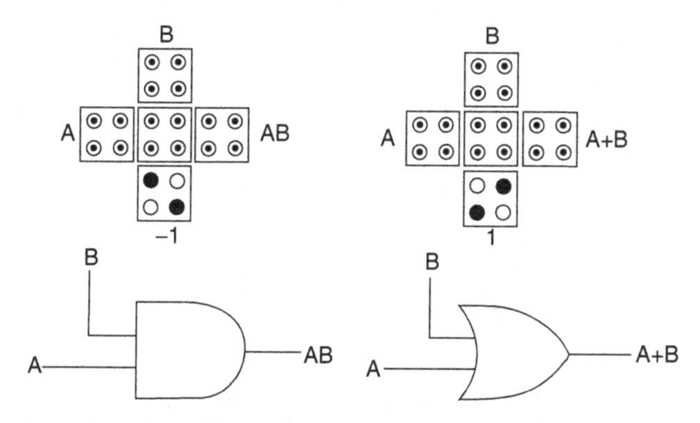

Figure 4.12. The majority gate as a programmable AND and OR gate. Fixing one of the inputs to logic '0' creates a two-input AND gate. Alternatively, if we fix one of the inputs to logic '1' we obtain a two-input OR gate.

these cells. The first involves a QCA network that distributes a constant polarization to all the AND and OR gates, inverting where necessary. In this case, the fixed polarization is provided as a separate input to the circuit. Only one such input is required. The second approach involves a dot-level manipulation of cells within the circuit or the controlled placement of fixed charges. It is theoretically possible to implement a fixed polarization cell by simply removing two quantum dots from the cell, leaving only the two dots associated with the desired diagonal. The use of fixed polarization rather than a distribution network results in circuits that consume far less area and are not prone to errors associated with the large network required by the first approach. Significant work remains to be done in this area.

4.9. WIRE CROSSING

The majority gate, inverter, and wire are insufficient to design and build a complex QCA circuit. Two other building blocks are required: the fanout and a method of wire crossing. The fanout can be implemented by joining two QCA wires to the same node. The wire crossing represents a far more difficult problem and is still being investigated. In standard technologies, wire crossings are implemented with several layers of metallic interconnects. The situation is quite different for QCA circuits, where such metal layers are will probably not be available and converting between the electronic configuration of QCA cells and voltages in wire interconnects is not practical.

4.9.1. Coplanar Crossover

In earlier research, Tougaw et al. proposed an additional type of cell to create a coplanar crossover [61]. The cell is the same size as the standard cell except the dots are rotated by 45° as shown in Figure 4.13, where we also show the information encoding in these rotated cells.

As a result of the geometry of these cells, each cell in a wire will relax to the opposite polarization of its nearest neighbors. When placed directly adjacent to one another, rotated and nonrotated cells do not interact with each other, as shown in Figure 4.14a. This lack of interaction is a result of the symmetry between

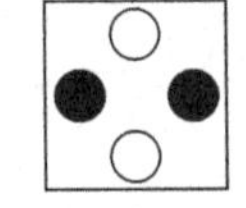

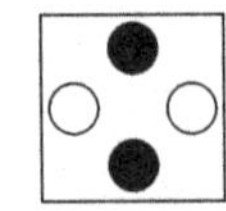

Figure 4.13. 45°-rotated QCA cells, showing the definition of polarization and binary encoding.

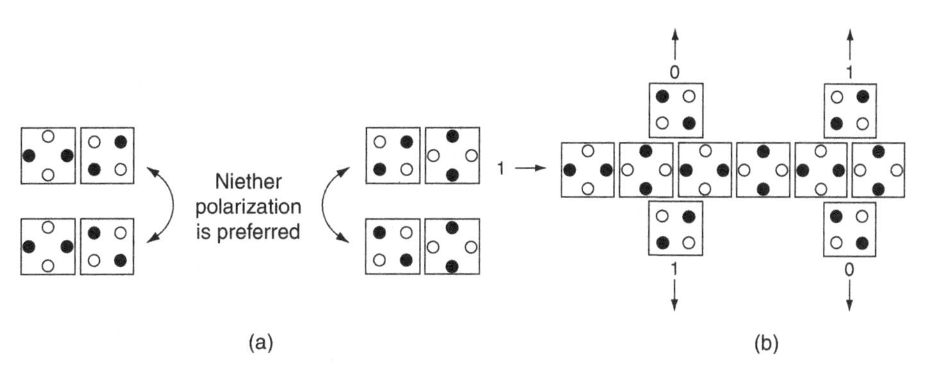

Figure 4.14. (a) Because of symmetry, when rotated and nonrotated cells are placed adjacent to one another, they do not interact. (b) Recovering a digital signal from the inverting chain is done by placing regular cells in between two of the rotated cells. We recover either the signal or its inverse, depending on where we tap off the inversion chain.

the two cells; the effect of one cell on the other does not break the energy degeneracy between the polarization of the two cells. To convert a signal from rotated to nonrotated wires the position of nonrotated cells has to be offset by a half-cell, as illustrated in Figure 4.14b.

This effect can be exploited to implement wire crossovers. Consider two standard, nonrotated cells placed at a separation of one cell. It is reasonable to assume that the polarization effect of one cell is able to traverse this gap and effectively break the energy degeneracy of the second cell. If we place a wire of rotated cells in between these two cells, this interaction will not be disrupted because of the geometric considerations discussed above. This property allows the coplanar crossover to transmit information independently along the two different cell wires: one wire comprised of regular cells and the other comprised of rotated cells, as illustrated in Figure 4.15.

If there is no crosstalk between the two interconnects, a signal is able to propagate across the gap. However, this building block is very sensitive to fabrication variations [62, 63]. Variations in the position of cells or dots in cells of this building block introduce crosstalk between the two interconnects, which quickly dominate its operation. The problem stems from the cell gap between input and output in the non-rotated cell wire. The coupling between the two wires is 1/32 that of adjacent cells and the output wire is essentially floating.

4.9.2. Coplanar Crossover with Continuous Clocking

Blair et al. proposed a new approach to the coplanar crossover that does not involve the use of rotated cells [30]. By using an eight-phase clocking scheme and generating ACTIVE domains that were shorter than the NULL domains, Blair

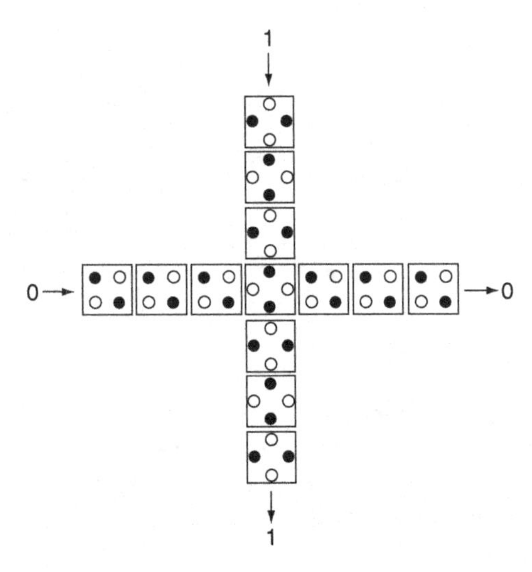

Figure 4.15. Wires created with rotated and nonrotated cells do not interact. We can exploit this to create coplanar crossovers, if the effect of one cell onto another of the same orientation is able to traverse the gap.

et al. were able to show that the progress of these domains through the junction could be interlaced, as shown in Figure 4.16. Although this building block resolves the problems associated with the previous coplanar crossover and represents the most promising approach to wire crossing, it comes at a significant cost associated with the additional area overhead; the requirement for four additional clocking phases; and the extra complexity associated with the clock timing.

4.9.3. Multilayer Crossover

Previous work has examined the possibility of building multilayer QCA [64]. The use of multilayer QCA is quite attractive as a means of wire crossing but it also introduces the possibility of a 3D computer since it is quite reasonable to assume that the additional layers could perform logical functions. The application of multilayer networks as a means of signal crossing was first explored in [62]. Many of the circuits presented later in this chapter use this approach to achieve signal crossing. Using these multilayer QCA cells, we can effectively cross signals over on another layer; such a technique has been simulated to show significantly less sensitivity to cell displacements [63]. Using this method, multilayer QCA circuits can potentially consume much less area than planar circuits (Figure 4.17).

This building block has area and displacement tolerance advantages. However, the physical realization of this building block is technically very challenging and it is difficult to predict whether such a multilayer QCA technology will be feasible in the future.

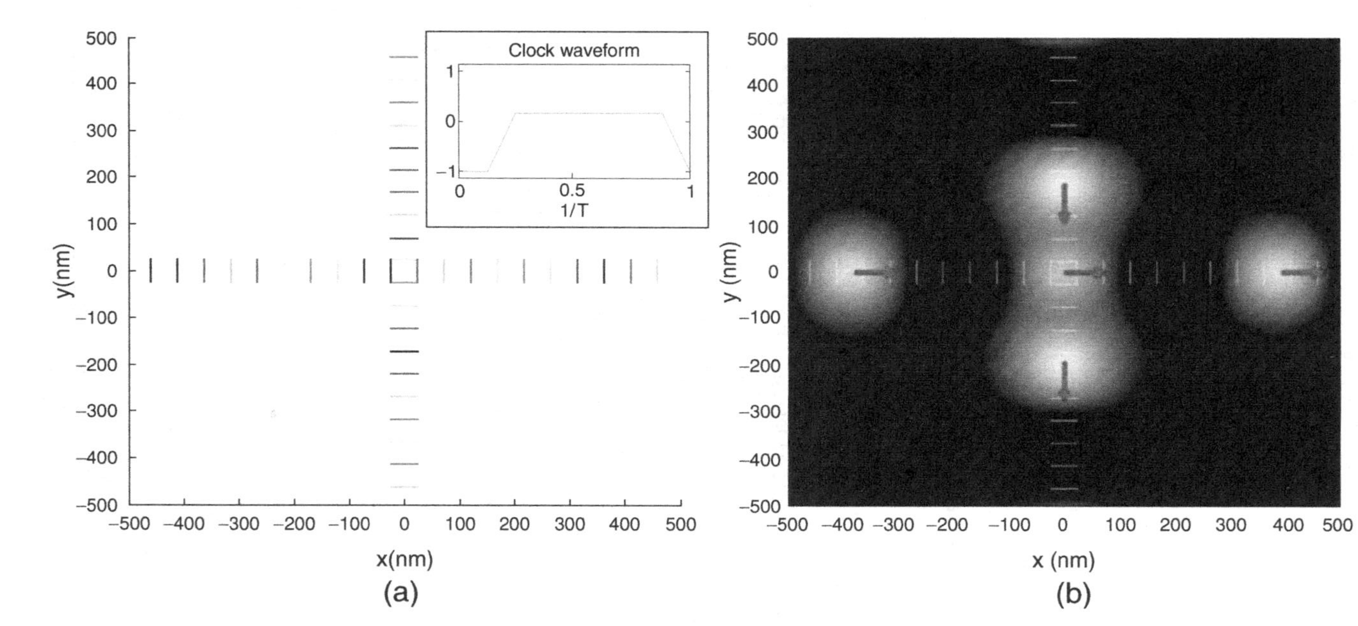

Figure 4.16. Using an eight-phase clocking network, coplanar crossover is possible by timing the information flow across a crossover point [30]. Reprinted with permission from E.P. Blair.

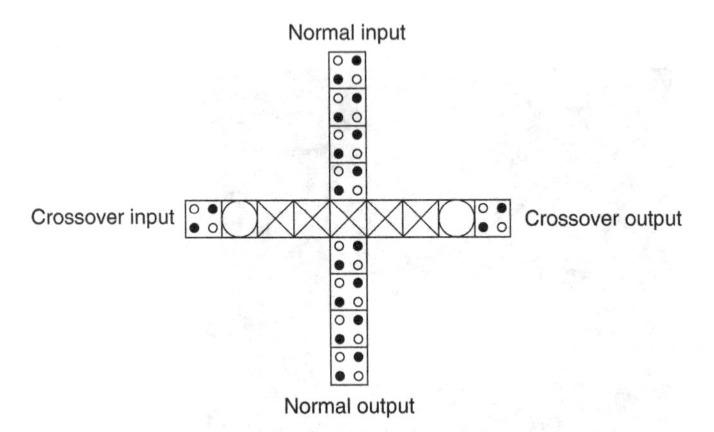

Figure 4.17. A multilayer crossover. Cells shown with an "X" are located on higher layers; cells shown with "O" are vertically stacked to produce a via.

4.10. QCADesigner: SIMULATION ENGINES

Simulation engines within QCADesigner model the static and dynamic behavior of QCA operation. Although two simulation engines are currently included with the tool, other groups are free to incorporate their own simulation engines since the source code for the tool is made available. Work on a new simulation engine tailored for continuously clocked molecular QCA has been started [65]. However, much work remains to be completed.

Models that described the static and dynamic behavior of electronic QCA have been reported in [47, 51, 54, 67, 68]. In general, quantum mechanical systems are not suitable for efficient simulation on a classical computer, and the models are approximations of the full quantum mechanical behavior of the system. Included in the current version of QCADesigner are two different simulation engines: the bistable and the coherence vector engines. The bistable engine is based on the simplest model and is very fast. There are several key assumptions made by this model. The first is the two-state approximation, which ignores higher energy configurations of the cell. The second is the Hartree approximation, which quantum mechanically decouples the cells, ignoring the intercellular correlations and reducing the size of the matrices involved. The last is that the cells remain at the ground state, no time-dependant dynamics are considered. The second simulation engine is based on the coherence vector formalism, a model that is able to model the dynamics of the cells in the presence of coupling to the substrate. This engine is also based on the two-state cell and uses the Hartree approximation to decouple the correlations between cells.

The main challenge in implementing physically accurate simulations is the lack of experimental data for the various QCA implementations. Even though several small QCA systems have been developed as proof-of-concept experiments [9, 10, 12, 20–22], these early devices do not necessarily represent scalable QCA technology. As a result, one of the main objectives of this effort is to provide motivation for further research into the implementation of such devices and development of a continued dialog between circuit designers and researchers investigating these implementations.

4.11. EXAMPLE CIRCUIT DESIGNS

The following sections describe a relatively small subset of the many circuit designs that have been already been reported in the literature. The design and layout of these example circuits has been done manually. Early work on the automatic placement and routing of QCA circuits has also been reported [69–71], as well as early work on minimizing the overall hardware requirements through majority minimization techniques [60, 72, 73].

4.11.1. QCA Addition

Of the variety of information processing tasks that computers perform, the most basic is certainly the addition of two single digit binary numbers. References [56, 74] investigate the design and layout of a QCA full adder, using QCADesigner, in more detail. The schematic and layout for the full adder are shown in Figure 4.18. It should be noted that these layouts represent a snapshot of the circuit in operation, which is why only one of the clocking phases has cells that are latched.

The apparent underutilization of space in this layout is a result of design requirements that attempt to minimize the crosstalk between different parts of the QCA circuit. The simulation results for this adder are shown in Figure 4.19. The delay between the application of the input and the appearance of the associated output is due to the inherent delay in the circuit. Figure 4.18 shows the four clocking zones (1 clock cycle) between the input and output.

A QCA ripple-carry adder is implemented by cascading n 1-bit QCA full adders. The schematic for a 4-bit adder is shown in Figure 4.20. As the inputs

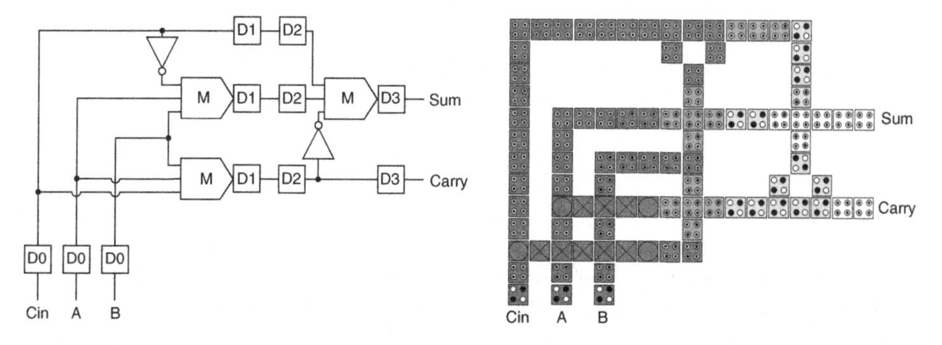

Figure 4.18. Schematic and layout of a majority-based full adder.

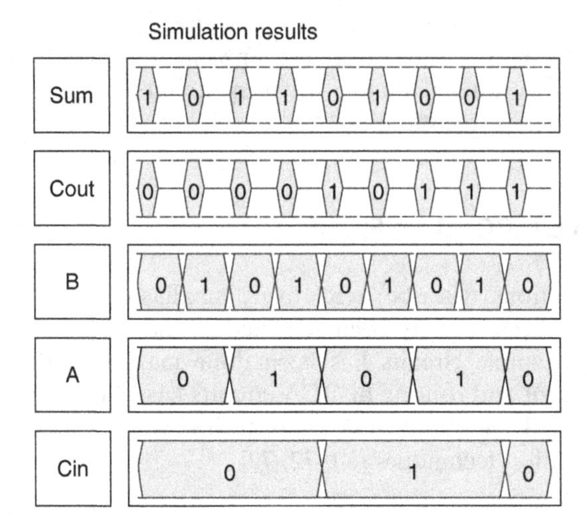

Figure 4.19. Simulation results for the full adder.

increase in significance, a larger number of clock cycles are introduced in order to synchronize the arrival of the input with the carry of the preceding adder.

The layout of the 4-bit adder is shown in Figure 4.21.

The overall latency can be obtained by considering the critical path from one of the inputs of the 1st adder to the carry output of the nth adder. Since the carry output of a QCA full adder requires 1 clock cycle to complete, the overall latency for the n-bit adder will be n clock cycles. The advantage of the QCA adder is that it is inherently pipelined. As a result, we can feed new inputs into the adder every clock cycle. When the pipeline is full, a new output will be available every clock cycle independent of the depth of the pipeline.

An n-bit QCA carry-look-ahead adder can be expressed as

$$\begin{aligned} C_i &= G_i + P_i G_{i-1} + P_i P_{i-1} G_{i-2} + \cdots + P_i \cdots P_1 C_0 \\ &= G_i + P_i (G_{i-1} + P_{i-1}(\cdots + (G_1 + P_1 C_0))) \end{aligned}$$

$$S_i = C_{i-1} \oplus P_i C_{i-1} \bar{P}_i + \bar{C}_{i-1} P_i,$$

where

$$G_i = A_i B_i,$$

$$P_i = A_i \oplus B_i = A_i \bar{B}_i + \bar{A}_i B_i.$$

Due to the limitation of QCA to two-input AND and OR gates, the critical path of a 4-bit carry-look-ahead adder consists of 10 majority gates as shown in Figure 4.22. Therefore, the best case latency of this architecture is estimated at five clock cycles.

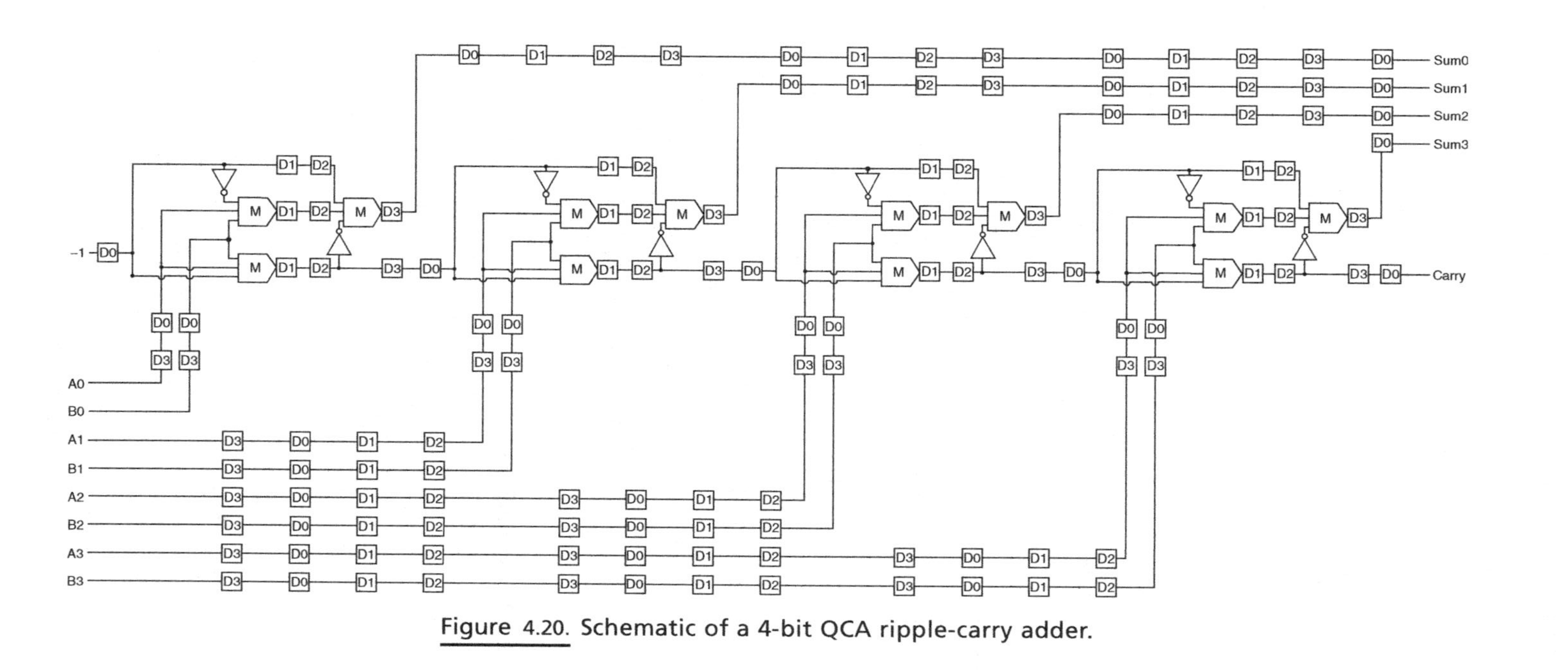

Figure 4.20. Schematic of a 4-bit QCA ripple-carry adder.

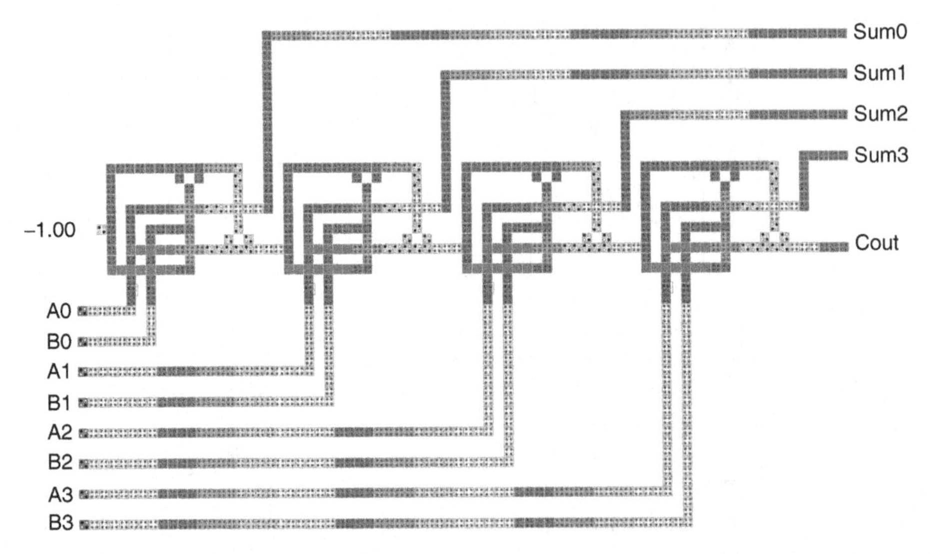

Figure 4.21. Layout of a 4-bit QCA ripple-carry adder.

In an actual layout, the latency may be larger because the long interconnects associated with the most significant carries would have to be divided into more than one clocking zone. The advantages of carry-look-ahead adders are dependent on the ability to realize AND and OR gates with more than two inputs. A ripple-carry adder is considered the slowest bit-parallel adder design in CMOS circuits. However, the QCA ripple-carry adder has better performance when compared to both the QCA carry-look-ahead and carry-select adder designs because of the inherent pipelining associated with QCA technology.

4.11.2. QCA Multiplication

A constant coefficient multiplier has been implemented using the adders described above [56]. The block schematic for a 2-bit multiplier is shown in Figure 4.23. The

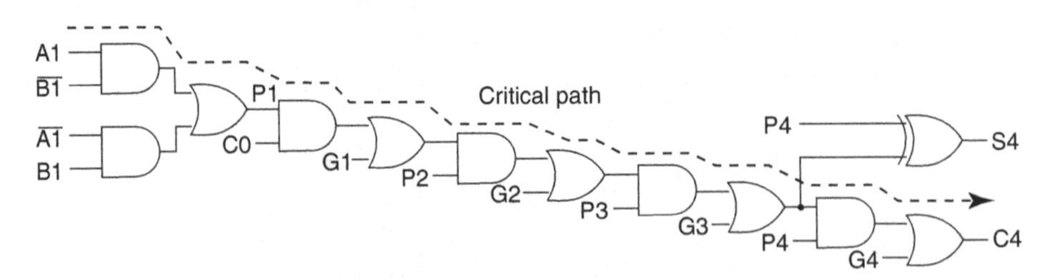

Figure 4.22. Critical path of a 4-bit QCA carry-lookahead adder from the input of the first bit to the carry output C_4 or the sum output S_4.

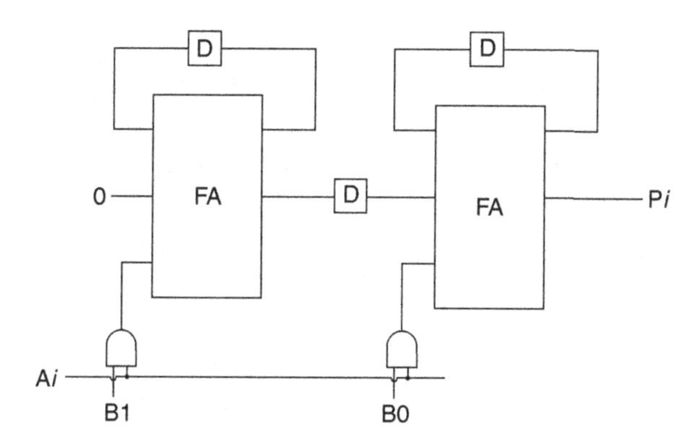

Figure 4.23. Block schematic of bit-serial multiplier.

D-latches in this schematic are required for the proper operation of the device and are not QCA zone latches. In order to map this design into a QCA circuit, we have to realize that many more D-latches are introduced from the very nature of the QCA circuit. The D-latch between the two adder blocks in the original schematic (Fig. 4.23), is implemented in QCA by adding four additional clocking zones (one clock cycle) along that path.

One of the inputs to the multiplier is serially connected across the multiplier circuit; the other is constant and implemented using fixed polarization cells. The schematic for this multiplier is shown in Figure 4.24. The overall latency for the

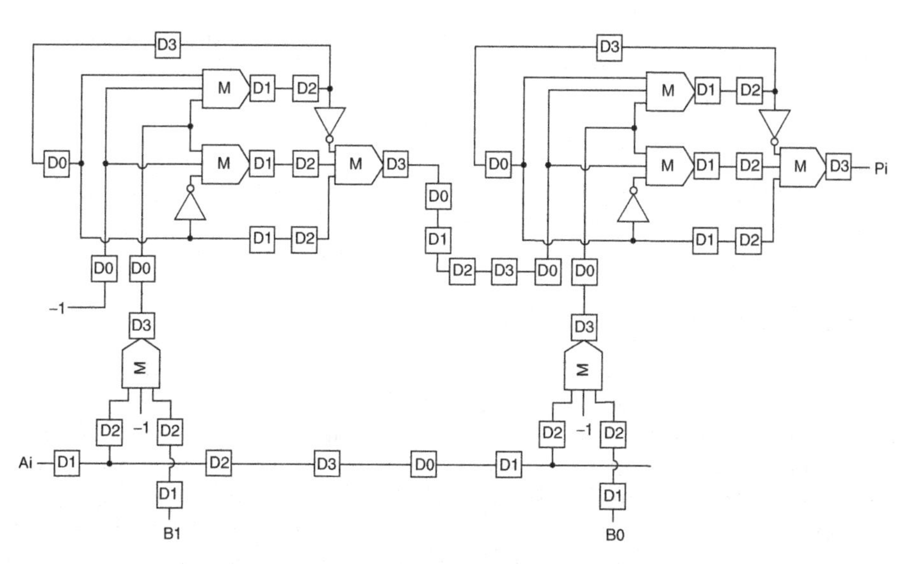

Figure 4.24. Schematic of QCA bit-serial multiplier.

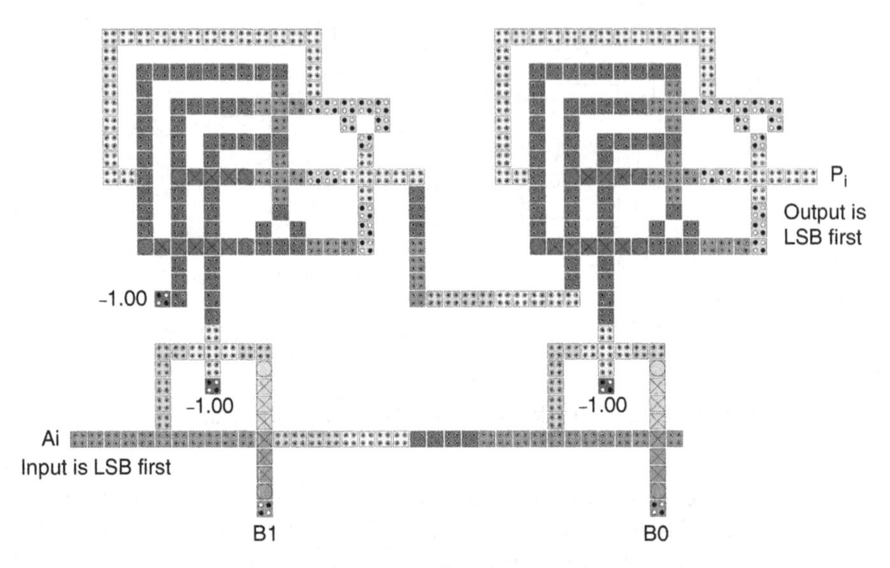

Figure 4.25. Layout of QCA bit-serial multiplier.

multiplier is three clock cycles: one clock cycle inherent to each adder and one to implement the latch between them.

The QCA layout for the multiplier is shown in Figure 4.25. The schematic is drawn to match the layout as much as possible, and the multiplier can be easily scaled by adding full adder blocks and partial product generators. We have experimented with designs as large as 32-bit using this layout. The size of the multiplier grows linearly with the number of bits, making it efficient in area. The latency also increases according to

$$L = 2n - 1,$$

where the latency, L, is measured in clock cycles, and n is the size of the multiplier in input bits.

4.11.3. Memory

Zone clocking creates a shift register with each wire of clocked cells. However, each cell is connected to a clock signal that will clear the cell contents once every clock cycle. To correct for this loss of data, small loops can be used to retain information and implement memory. The simplest memory loop consists of all four clocking zones, enabling the information to continuously circulate in the loop. A similar approach using continuous clocking has been proposed by Blair et al. [30]. Figure 4.26 shows a QCA memory loop without any mechanism for reading or writing information.

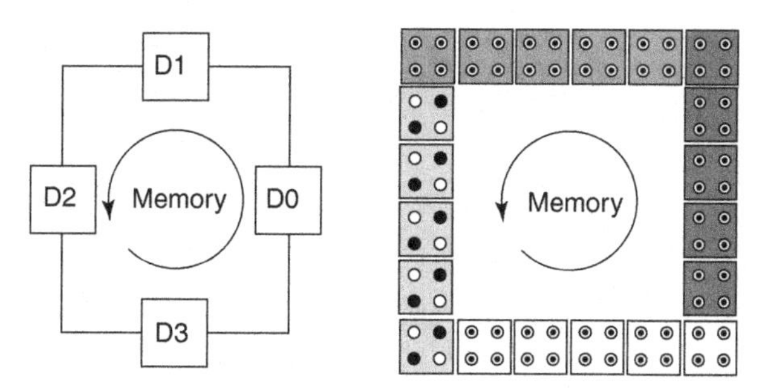

Figure 4.26. The most basic memory element in QCA is the loop. As each of the clocks latch and unlatch they circulate information around the loop.

4.11.4. Memory Cell

Extra control can be added to the memory loop to create a memory cell. The memory cell is a building block of random access memory (RAM) and implements a bit-level read/write function [75]. The QCA memory is volatile, and all stored information will be lost when the power to the circuit is disconnected. In order to control the function of this memory cell, we introduce two controls and one input. The first control wire, called the *Row Select*, serves to enable the memory cell to be used in a larger memory grid. The second control, called the *Read/Write*, selects the present operation to be performed on the memory. A schematic representation of the loop memory cell is shown in Figure 4.27.

The associated QCA circuit layout is shown in Figure 4.28. This layout attempts to maximize the application of the memory cell to larger memories, by running the *Row Select* and *Read/Write* control wires through the entire length of the cell. In this way, a row of memory can be created by simply laying out these cells in a linear array and extending the control wires to each of the cells.

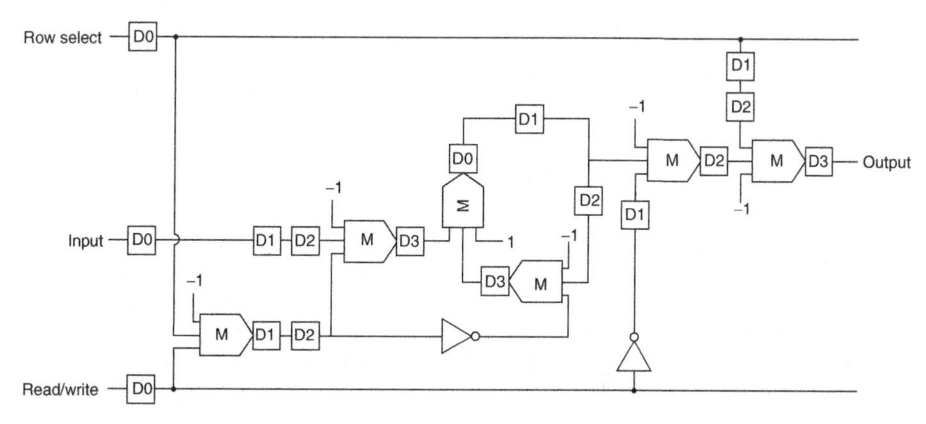

Figure 4.27. Schematic of the 1-bit memory cell.

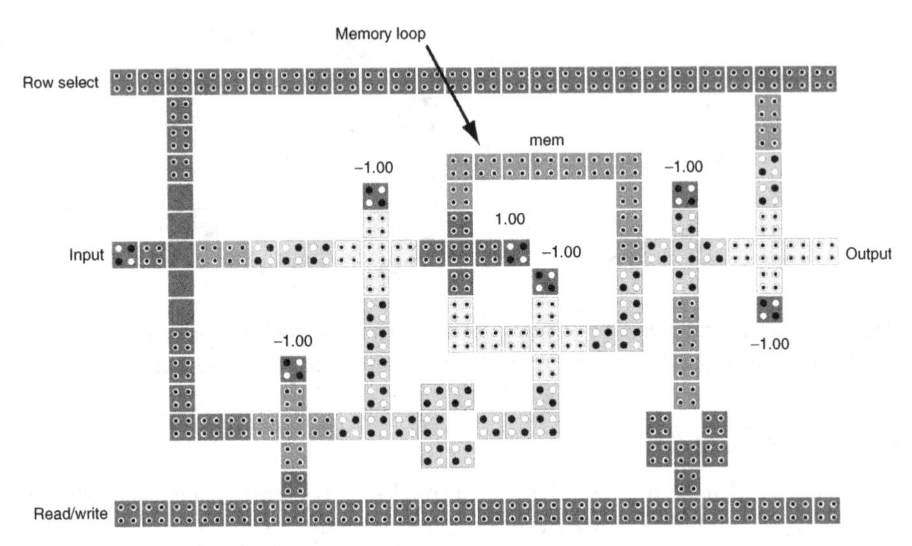

Figure 4.28. QCA layout of the memory cell.

The stored memory value is constantly circulated inside the memory loop until the *Read/Write* and *Row Select* wires are polarized to logic '1', at which time the incoming input is fed into the memory loop and circulated. If the *Row Select* is polarized to logic '1', and the 'Write/Read' is polarized to '0', the current memory value inside the loop is fed to the output. If the *Row Select* is polarized to logic '0', then the memory cell will always polarize the output cell to logic '0'. Simulation results for the memory cells are shown in Figure 4.29. Interestingly, the use of

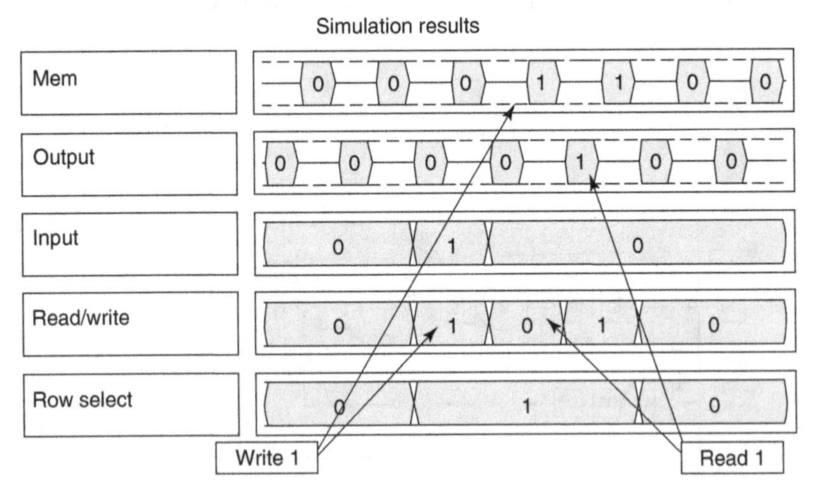

Figure 4.29. Simulation results for the memory cell.

circulating memory storage is not a new concept and can be traced back to at least the mercury delay lines that were used to implement storage in the commercial tube computers of the 1950s.

4.11.5. 4-bit Processor

A simple 4-bit processor, designed using QCADesigner, has been reported in [76]. This processor was designed mainly as a proof-of-concept to demonstrate that reasonably complex architectures are possible to build using QCA technology, as well as to create a platform for investigating the inherent zone level pipelining. This circuit is intended to demonstrate the level of complexity that can be handled by QCADesigner, not as a demonstration of the ideal architecture for computing using QCA technology. Such architectures are still being developed.

The processor is limited to operate on instructions fed into the circuit directly at the inputs; no program memory is used. The design incorporates a 4×4 random access memory (RAM) to provide temporary storage. The design is based on a simple accumulator architecture shown in Figure 4.30.

Basic arithmetic and logic operations are implemented in the arithmetic logic unit (ALU). This ALU is built by extending the 4-bit adder. The ALU layout can be seen in Figure 4.31.

An input to the ALU requires 11 clock cycles (44 consecutive clock zones) to propagate through the entire unit. Due to the natural pipelining introduced by the clocking, a new input can be applied to the ALU in each consecutive clock cycle, allowing 11 operations to be in the pipeline at any one time.

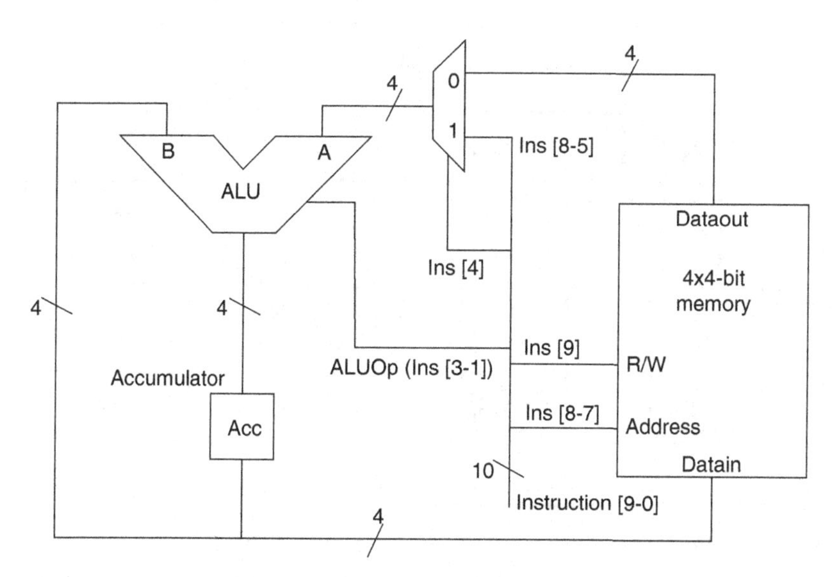

Figure 4.30. Processor architecture.

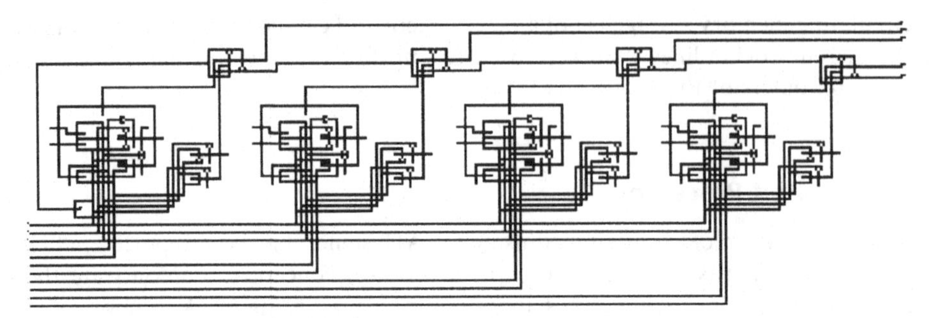

Figure 4.31. The 4-bit QCA ALU. The inputs arrive at the lower left corner, and the outputs are read at the upper right.

In order to provide the processor with data loading and storing capabilities, four 4-bit registers act as physical memory. Instructions are available to store the current accumulator value into any register, and similarly load the contents of any register into the accumulator for computations using the ALU.

Figure 4.32 shows the memory layout. It takes four clock cycles for data to be stored in a register, and five clock cycles for data to be available at the output. However, as discussed above, this reduces to one read or write operation per clock cycle in the steady state. The feed-through path, i.e., write then immediately read, consumes only six clock cycles since the read operation is not required to wait for the write to complete.

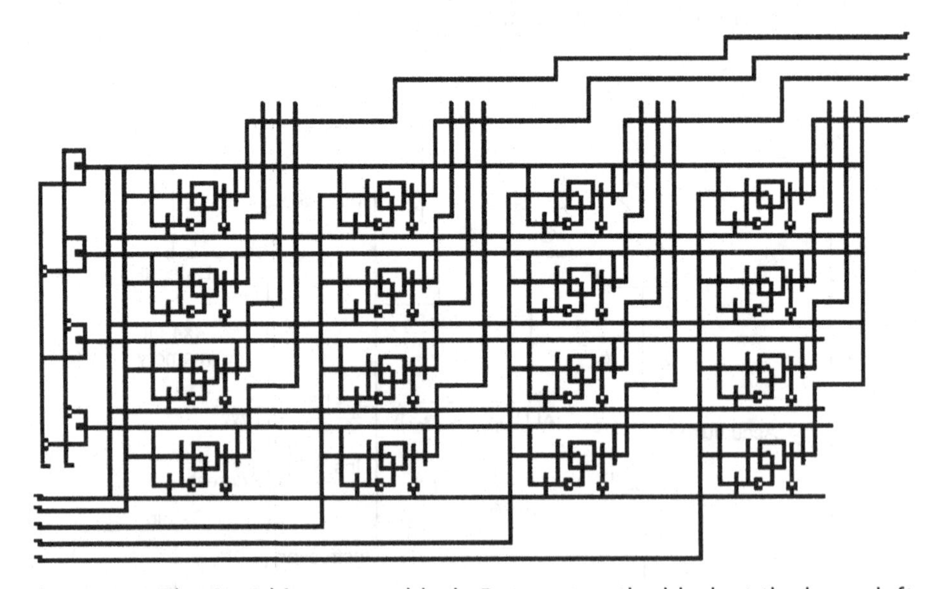

Figure 4.32. The 4×4-bit memory block. Data enters the block at the lower left, along with the register address and read/write signal. The data is then made available at the upper right.

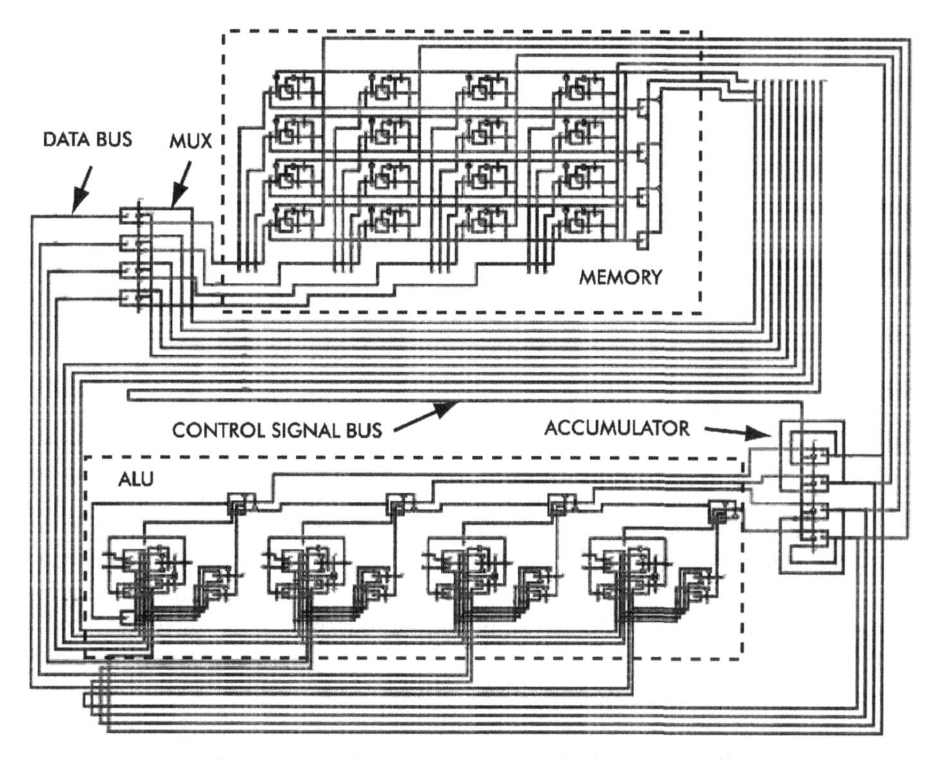

Figure 4.33. QCA layout of the 4-bit processor.

The accumulator is implemented using a 2:1 multiplexer with outputs fed back into one of the two input channels. The *Select* control enables the accumulation of the present ALU output. The propagation path in the feedback loop has a two clock cycle pipeline allowing for the accumulation of two values. These two values, however, are accessible every second clock cycle.

Figure 4.33 shows the layout of the complete QCA processor.

Each component of the processor, including the interconnect between components, has some latency associated with it, creating a systemwide pipeline. The number of pipeline stages for each component of the processor is shown in Figure 4.34.

In order to eliminate the controller logic for this simple design, we have taken advantage of this inherent pipelining. Our instructions contain control bits that determine how each of the processor components handles the arguments associated with that instruction. The control data is distributed to each of the processor components through a set of interconnects, which are clocked such that the control signals arrive synchronously, with the associated argument, to each of the processor blocks.

This design was successfully simulated using QCADesigner running on an Intel Pentium 4 computer system with 2GB of RAM. The simulation performed

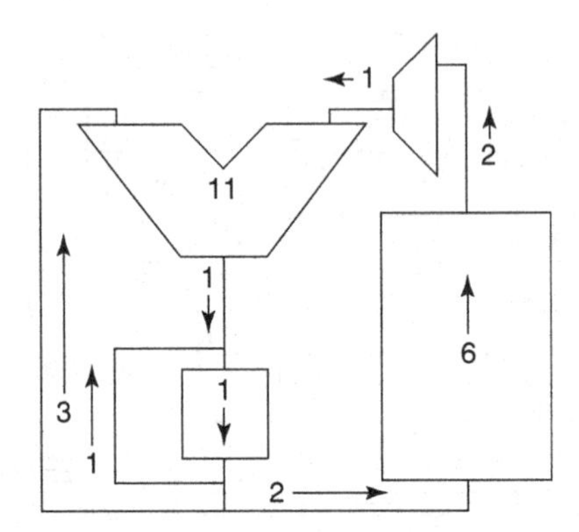

Figure 4.34. Number of clock cycles through each stage of the processor.

the addition of two pieces of data, where the result of the addition was written back to the internal memory. The total simulation time to perform this operation was approximately five hours on the Pentium computer using a time-independent simulation engine. In order to improve simulation performance, significant effort is being invested in optimizing the available simulation engines, including the use of multiprocessor computing clusters.

4.12. QCA FABRICATION

To date, QCA is a commercially unrealized technology. However, several experimental devices have been fabricated and are described in more detail in this section. The different proposed implementations attempt to realize the required bistable and locally interacting behavior required by the QCA paradigm. Of all the implementations, four distinct classes of QCA implementations have emerged and are listed below.

- Semiconductor [3–5, 77–79]
- Metal-Island [6, 7, 9–14, 16–26, 55]
- Magnetic [42, 43, 80]
- Molecular [27–35, 37–41, 81]

4.12.1. Semiconductor QCA

Room temperature QCA requires cells that are approximately 1–5 nm. Semiclassical modeling of QCA cells shows that cells made any larger would be limited to temperatures in the mK range [49, 50, 82]. Self-assembly of the quantum dots offers a promising route to fabricating room temperature QCA; however, the required homogeneity in quantum dot size and placement accuracy is not yet at a level that would support QCA technology [79]. As a result, very limited experimental results are available for this particular implementation of QCA. One of the advantages of using semiconductor QCA is that it is based on materials that are well understood and many fabrication techniques have been created to work with them. As well, there is a large economic benefit to using a material system for which significant infrastructure already exists.

Gardelis et al. demonstrated polarization transfer into and out of a QCA cell defined with metallic gates patterned on top of a 2-dimensional electron gas (2DEG) in AlGaAs/GaAs [78]. The experiment was performed in a dilution refrigerator operating at 100 mK. The use of 2DEG for implementing QCA is limited and a scalable implementation using this method is most likely not possible. In addition, the tolerance to fabrication variations appears to be too sensitive for any practical application. However, such devices can be fabricated using available techniques and are providing a valuable experimental platform to study the fundamental physics of QCA operation.

4.12.2. Magnetic QCA

Recently, significant research effort has been put into the implementation of magnetic QCA (MQCA) based on the interactions of magnetic nanoparticles [42, 43, 80]. The magnetization vector of these nanoparticles is analogous to the polarization vector in electronic QCA, and information is propagated via magnetic exchange interactions as opposed to electrostatic coupling in all other reported implementations. Although this technology is referred to as magnetic quantum-dot cellular automata, the term quantum in this case represents the quantum mechanical nature of the exchange interaction.

One of the immediate advantages of considering such a technology is that MQCA cells can operate at room temperature, even for large device features on the order of a few hundred nanometers. A potential application of such a technology would be to implement processing-in-memory where the QCA circuit can be used to implement both the information processing and storage elements, potentially eliminating the need for a read/write head, as required by conventional hard disk drives, or the interconnects required by MRAM devices. The read/write operation would be performed at the edges of a MQCA array and the information would be processed and propagated inwards by the coupled magnetic field exchange interactions. Contrary to the layout pattern of magnetic storage elements of an MRAM, where the elements are created such that there is very little interaction between them, the MQCA cells are designed to interact in order to enable information processing. Because these exchange interactions are similar to electronic QCA, the MQCA cells would also be able to perform computational tasks using majority functions. One of the drawbacks to using MQCA as a general computing platform is the low operating speed; studies report switching frequency estimates at around 100 MHz [43]. However, it is expected that the MQCA circuits will operate with a much higher tolerance to displacement and rotational defects than the electronic QCA. The QCA wires fabricated in [43] operated properly for wires of lengths between 10 and 20 cells long. It is indicated that the number of nanomagnets that can be clocked with a single phase is limited and that a multiphase clocking scheme similar to that used in electronic QCA will be required for a large-scale implementation of MQCA circuits.

4.12.3. Metallic-Island QCA

The metallic-island implementation was the first experimental realization of QCA [6, 7, 9–14, 16–26, 55]. In all these prototypical devices, the metal islands are used to replace the semiconductor quantum dots. These experiments demonstrated that single-electron tunneling between two Al metallic islands coupled with tunnel junctions would encourage single-electron tunneling in the opposite direction in two adjacent coupled metal islands, realizing this bistable behavior. This idea has been extended to majority gates and short shift registers. The advantage of this implementation is that quantized electron tunneling between metal islands could be observed for large device sizes that are relatively easily fabricated using established electron beam lithography (EBL) techniques. The schematic representation of a metal island QCA cell is shown in Figure 4.35. The cell consists of four metallic islands grouped into two sets of double-dots. The two metallic islands that make up a single double-dot are coupled to each other through tunneling junctions. The two double-dots are then coupled to each other with capacitors increasing the interactions between the two double-dots in order to enhance the bistable behavior of the cell.

The devices operate in the so called *Coulomb Blockade* regime. The island capacitance C is a parameter that determines the device operating temperature and is dominated by the capacitance of the tunnel junctions connecting the metallic island to the external electrodes and adjacent islands. Due to the relatively

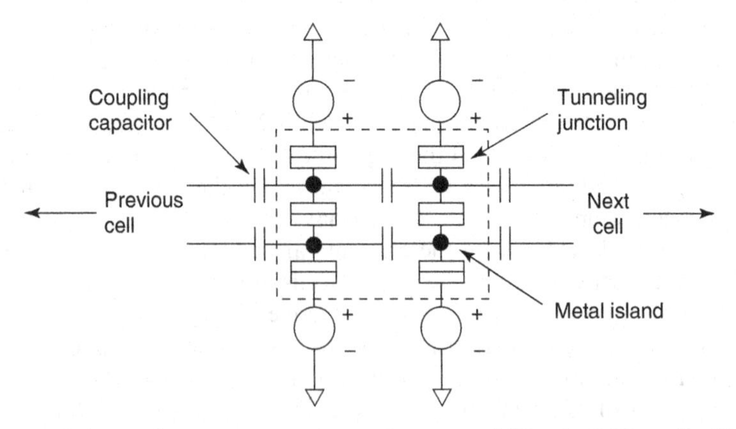

Figure 4.35. Schematic representation of a metal-island QCA cell. The two vertically stacked metal islands are coupled through tunnel junctions, which enable the quantum mechanical switching of the cell.

large area of the tunnel junctions, the devices require operating temperatures lower than 15 mK. To achieve such low temperatures in the experiments, the devices required cooling via a dilution refrigerator. At these temperatures the charging energy is much higher than the thermal ambient energy $k_B T$ facilitating the precise control over single electron charging of the metal island. Under the appropriate bias, a small change in the input can cause single electron transfer between islands, and this change in charge will cause a similar effect in the neighboring double-dot. However, due to the precise biasing requirements, each cell has several external connections and it is therefore difficult to scale this implementation of QCA to much more complex circuits.

The metal islands that make up the quantum dots are tightly coupled to the substrate or heat bath. As a result, the time evolution of the cell is not quantum mechanically coherent and can be modeled using semiclassical descriptions. Unlike the semiconductor implementation, the metal islands are filled with many conduction band electrons, and QCA operation is verified by observing the change of charge rather than the total charge at a particular site. Electron tunneling is the only nonclassical phenomenon in the metal-island implementation.

4.12.4. Molecular QCA

Developing cells using molecules represents a very promising approach to implementing QCA. Molecular QCA has the potential for room temperature operation, high device density, and high operating speed [27]. Room temperature operation is possible as a result of the high kink energies of molecular cells [53]. In theory, molecular implementations could also be fabricated with much higher uniformity than those achievable with semiconductor or metallic-island implementations [31]. Estimates using 1 nm^2 QCA cells indicate that molecular QCA could reach densities of 10^{14} devices/cm^2 [29]. However, one has to be careful when making a direct comparison of the functional capability of a QCA circuit with this density to a transistor circuit of the same density since the contribution of each QCA cell to the overall functionality is far lower than that of a transistor circuit as a majority of the cells are consumed in the interconnect network between logic gates. Although the cell density can be very high, recent research on QCA clocking is progressing towards relatively low density circuits as a result of the large area consumed in the crossovers [31]. Quantum mechanical analysis of QCA dynamics has predicted cell switching times as low as 2 ps per cell resulting in the possibility of THz operation [51]. It is not clear if the clocking network itself could operate at such high speeds and the limits to QCA operating speed could end up being determined by the clocking implementation rather than the cell dynamics.

Initial research into molecular QCA has focused on bistable mixed valence compounds similar to those first developed by Aviram [83] and Hush et al. [84]. These molecules consist of multiple redox centers, where each center can be reduced (gain an electron) or can be oxidized (lose an electron) [32]. The Aviram molecule consists of two allyl groups connected by a butyl bridging ligand. One of the allyl groups is a neutral radical, while the other is a cation. The unpaired electron forming the neutral radical can be transferred from one end group to the other without significant change to the molecular geometry. The sigma bridge separating the allyl groups provides a potential barrier and enables the measurable bistability of the molecule. The application of an electric field will cause the molecule to exhibit a change in the electronic configuration from donor-bridge-acceptor to acceptor-bridge-donor. The change in configuration can result from either electron or hole transfer in the molecule.

In order to construct a molecular implementation of a QCA cell, a molecule that exhibits bistability and charge localization is required. Lent et al. [33] analyzed the performance of the Aviram molecule (1,4-diallyl butane radical cation shown in Fig. 4.36) as a test bed for molecular QCA. *Ab initio* calculations were performed with Gaussian 98 and the unrestricted Hartree–Fock theory (UHF) and the Slater Type Orbitals simulated by the 3 Gaussians (STO-3G) basis set. Figure 4.36 shows the highest occupied molecular orbitals (HOMOs) and the isopotential surfaces for two neighboring molecules. Lent et al. also demonstrated that these molecules could be used to implement majority gates and wires. Although the molecule was found to exhibit nonlinear bistability, charge localization, and molecule–molecule interactions necessary for information transmission, other necessary features such as a mechanism for I/O and a functional group for attachment to a substrate were not designed into this molecule.

Molecular QCA was also shown to exhibit a second level of nonlinearity as a result of the coupling between electronic and nuclear degrees of freedom. This

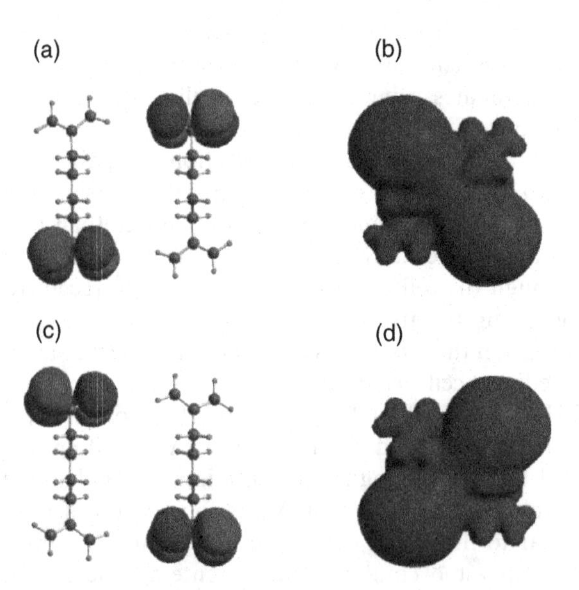

Figure 4.36. The HOMO levels (a) and (c) and isopotential sufraces (b) and (d) for the two stable states of a pair of QCA molecules studied in [33]. Reprinted with permission from [33]. (© 2003 American Chemical Society)

coupling acts to stabilize the molecule in a particular polarization and ensure that once the cell is polarized it will require an outside influence to force a change [33]. This second level of nonlinearity is not present in the other implementations. Lent et al. also point out that the kink energy in a molecular cell can be reasonably approximated with simple electrostatic calculations. For the molecular system described here, the kink energy was calculated to be 646 meV.

In order to enable the molecular QCA cells to be clocked, a third stable state is required. The molecular geometry described above was extended to include an additional allyl group and permits the molecule to have a third stable state [32]. In this molecule, alkyl bridges in a "V" shaped molecule connect 3-allyl groups (the dots). This "V" shape permits the use of a vertically aligned electric field to induce a potential difference between the upper sites (the two ends of the V) and the lower site (the vertex of the V). This potential difference can be used to clock the cell—when the potential of the lower dot is lower than the potential of the upper dots, the cell will go into the NULL state. The group is considered a molecular cation, with two neutral allyl groups and one positive allyl group. In this case it is the positive charge that is transferred between the different dots. The molecule in the three stable configurations is shown in Figure 4.37.

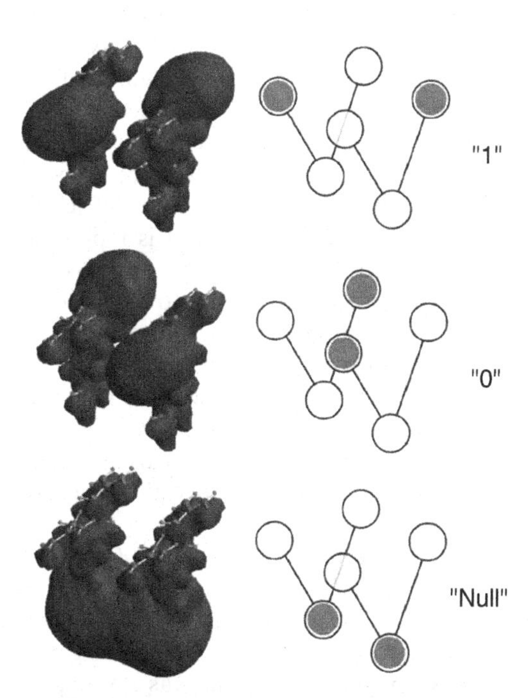

Figure 4.37. Three stable configurations of "V"-shaped QCA molecule [32]. Isopotential surfaces are shown on the left hand side and a schematic representation of the cell on the right. Reprinted with permission from [32]. (© 2003 IEEE)

The molecule requires a vertical field of 1.3 V/nm to switch from the NULL state (molecule is normally off) on one of the two active states. The state to which it is switched depends on the presence of perturbing fields (in this case, produced by a point charge a single cell distance away from the two molecules). The molecule was found to have the necessary nonlinear response function to enable QCA computing. As with the initial work, this molecule was also used as a test bed and does not have the necessary structure required for attachment to the substrate.

Significant work has also been completed with synthesized molecular QCA. Lieberman et al. have experimented with the application of the Creutz–Taube ion as a QCA cell [29]. This mixed valence compound was reported to have the necessary properties to implement a QCA cell; however, there were significant challenges with associated counterions binding to the substrate. Earlier research has shown that QCA is highly sensitive to space charges [48]; counterions on the surface close to cells can cause the cells to latch to a fixed polarization, preventing any signal propagation along that particular path.

Additional work to demonstrate a surface bound QCA cell with vertical alignment was reported by Qi et al. [34]. In this work, they successfully demonstrated a vertically aligned dinuclear Ru-Fe mixed-valence compound that has been functionalized for attachment onto Si. The change of electronic configuration from Ru^{III}-Fe^{II} to Ru^{II}-Fe^{III}, resulting from an applied vertical electric field, was measured using an AC capacitance technique. A control experiment using the inactive Ru^{II}-Fe^{II} compound showed no measurable change in the capacitance. The transition from one configuration to the other was detected as a peak in the measured AC capacitance at 1 MHz, which occurred when there was energy level alignment between the Ru and Fe centers. As these levels become aligned, the induced dipole is able to be flipped with an applied AC field; this results in an increase in AC capacitance. The flipping is a result of electron transfer between the metal centers and is representative of double-dot QCA action. The capacitance of the film was measured using a mercury probe. It is also significant that this molecule was bound to the silicon surface; this is a key step toward a final implementation of molecular QCA. To do this the mixed-valence compound was functionalized with an amine group that was subsequently bonded to the Si. The process involved termination of a Si(111) surface with chlorine. A reaction between the Si-Cl surface and the NH_2 functional group results in a strong covalent Si-N bond. The group also showed that the addition of the amine functional group did not significantly alter the properties of the mixed valence compound.

First attempts to attach and pattern molecular QCA circuits were reported by Lieberman et al. [29]. A self-assembled monolayer (SAM) of octadecyl trichlorosilane (OTS) was grown on SiO_2. When exposed to a high-energy electron beam, the OTS was shown to become more hydrophilic (contact angle changed from 110° to 80°). Using the electron beam system, the group was able to pattern lines 30 nm wide in the OTS SAM structure. Three functionally terminated SAMs were used to test the binding of their test QCA molecule onto the SAM. The first was a phosphonate-terminated surface (contact angle $<30°$), the second a hydroxyl-terminated surface (contact angle 35–52°), and the third a methyl-terminated surface (contact angle 105–110°). It was found that the test molecule formed a self-limiting monolayer on the phosphonate and hydroxyl-terminated surface, but no such layer formed on the methyl-terminated surface; this suggests that patterned changes in the hydrophobicity of a SAM could be used to create patterns of molecular QCA. One of the challenges with this approach is the resolution limits associated with using the EBL process (around 10 nm), as well as the challenges to using such a serial patterning approach for mass production of such devices. Line widths of 30 nm would create QCA wires that are many cells wide; interestingly, however, wide QCA wires have been suggested as a way of increasing the fault tolerance of QCA circuits [85].

Recent work by Hu et al. has developed the use of DNA rafts self-assembled on an e-beam patterned PMMA surface [36]. This combination of lithography and self-assembly appears to be very promising; however, much more work remains to be done.

Many exciting results have been achieved toward an implementation of molecular QCA, but many challenges to the implementation of a molecular QCA technology still remain. The placement and patterning of individual molecules is a very difficult challenge; in addition, the implementation of a mechanism to measure the electronic configuration of individual molecules, to provide a means of implementing I/O, still remains to be solved.

4.13. CONCLUSIONS

Emerging nanotechnologies that are being explored as additions to and replacements of current FET-based fabrication technologies may allow us to continue along the Moore's Law exponential path and provide the anticipated increases in circuit density that drive many advancements of our technological society. Several promising emerging technologies have already been identified and others will undoubtedly appear over the next decade or so. Whichever technologies prove to be successful replacements, and there may be several that are finally used in a mixed technology scenario, will need to be explored thoroughly in a sophisticated design and simulation environment.

This chapter has focused on the promising nanotechnology of quantum cellular automata (QCA). Our exploration has covered the basic theory of QCA including a variety of cellular architectures, logic elements, memory cells, and processor structures that have been designed and simulated using the QCADesigner tool. We have covered the proposed approaches to clocking QCA cells and circuits and provided the advantages and disadvantages of each approach. We have also provided a round up of the most promising approaches to QCA implementation including semiconductor, magnetic, metal-island, and molecular QCA.

QCA remains an uncommercialized technology; this is primarily due to the challenges with realizing a scalable implementation. However, advantages such as low power dissipation, high device density, and the lack of required connection to each device make QCA a more suitable paradigm for computing at the molecular scale. As technology limits are being reached, it may be such disruptive technologies that enable us to build the future generations of computer technology.

REFERENCES

1. ITRS. International Technology Roadmap for Semiconductors (ITRS): Emerging Research Devices. 2006.
2. C. S. Lent, P. D. Tougaw, W. Porod, and G. H. Bernstein. Quantum Cellular Automata. *Nanotechnology*, 4: pp 49–57, 1993.
3. M. Macucci, M. Gattobigio, L. Bonci, G. Iannaccone, F. E. Prins, C. Single, G. Wetekam, and D. P. Kern. A QCA cell in silicon-on-insulator technology: theory and experiment. *Superlattices and Microstructures*, 34: pp 205–211, 2003.
4. H. C. Jared, D. G. Andrew, J. W. Cameron, C L H Lloyd, and P. Steven. Quantum-dot cellular automata using buried dopants, *Physical Review B*, 71: p 115302, 2005.
5. M. Mitic, M. C. Cassidy, K. D. Petersson, R. P. Starrett, E. Gauja, R. Brenner, R. G. Clark, A. S. Dzurak, C. Yang, and D. N. Jamieson. Demonstration of a silicon-based quantum cellular automata cell. *Applied Physics Letters*, 89: p 013503, 2006.
6. A. O. Orlov, I. Amlani, G. H. Bernstein, C. S. Lent, and G. L. Snider. Realization of a functional cell for quantum-dot cellular automata. *Science*, 277: pp 928–930, 1997.
7. I. Amlani, A. O. Orlov, G. L. Snider, C. S. Lent, and G. H. Bernstein. Demonstration of a functional quantum-dot cellular automata cell. *Journal of Vacuum Science and Technology B*, 16: pp 3795–3799, 1998.
8. I. Amlani, A. O. Orlov, G. L. Snider, C. S. Lent, and G. H. Bernstein. Demonstration of a six-dot quantum cellular automata system. *Applied Physics Letters*, 72: pp 2179–2181, 1998.
9. G. L. Snider, A. O. Orlov, I. Amlani, G. H. Bernstein, C. S. Lent, J. L. Merz, and W. Porod. Experimental demonstration of quantum-dot cellular automata. *Semiconductor Science and Technology*, 13: pp A130–A134, 1998.
10. G. L. Snider, O. Orlov, I. Amlani, G. H. Bernstein, C. S. Lent, J. L. Merz, and W. Porod. A functional cell for quantum-dot cellular automata. *Solid-State Electronics*, 42: pp 1355–1359, 1998.
11. I. Amlani, A. O. Orlov, G. L. Snider, C. S. Lent, W. Porod, and G. H. Bernstein. Experimental demonstration of electron switching in a quantum-dot cellular automata (QCA) cell. *Superlattices and Microstructures*, 25: pp 273–278, 1999.
12. I. Amlani, A. O. Orlov, G. Toth, G. H. Bernstein, C. S. Lent, and G. L. Snider. Digital logic gate using quantum-dot cellular automata. *Science*, 284: pp 289–291, 1999.
13. G. H. Bernstein, I. Amlani, A. O. Orlov, C. S. Lent, and G. L. Snider. Observation of switching in a quantum-dot cellular automata cell. *Nanotechnology*, 10: pp 166–173, 1999.
14. A. O. Orlov, I. Amlani, G. Toth, C. S. Lent, G. H. Bernstein, and G. L. Snider. Experimental demonstration of a binary wire for quantum-dot cellular automata. *Applied Physics Letters*, 74: pp 2875–2877, 1999.
15. W. Porod, C. S. Lent, G. H. Bernstein, A. O. Orlov, I. Amlani, G. L. Snider, and J. L. Merz. Quantum-dot cellular automata: computing with coupled quantum dots. *International Journal of Electronics*, 86: pp 549–590, 1999.

16. G. L. Snider, A. O. Orlov, I. Amlani, G. H. Bernstein, C. S. Lent, J. L. Merz, and W. Porod. Quantum-dot cellular automata: line and majority logic gate. *Japanese Journal of Applied Physics Part 1: Regular Papers, Short Notes and Review Papers*, 38: pp 7227–7229, 1999.
17. G. L. Snider, A. O. Orlov, I. Amlani, G. H. Bernstein, C. S. Lent, J. L. Merz, and W. Porod. Quantum-dot cellular automata. *Microelectronic Engineering*, 47: pp 261–263, 1999.
18. G. L. Snider, A. O. Orlov, I. Amlani, X. Zuo, G. H. Bernstein, C. S. Lent, J. L. Merz, and W. Porod. Quantum-dot cellular automata. *Journal of Vacuum Science and Technology A: Vacuum, Surfaces, and Films*, 17: pp 1394–1398, 1999.
19. G. L. Snider, A. O. Orlov, I. Amlani, X. Zuo, G. H. Bernstein, C. S. Lent, J. L. Merz, and W. Porod. Quantum-dot cellular automata: review and recent experiments (invited). *Journal of Applied Physics*, 85: pp 4283–4285, 1999.
20. I. Amlani, A. O. Orlov, R. K. Kummamuru, G. H. Bernstein, C. S. Lent, and G. L. Snider. Experimental demonstration of a leadless quantum-dot cellular automata cell. *Applied Physics Letters*, 77: pp 738–740, 2000.
21. A. O. Orlov, I. Amlani, R. K. Kummamuru, R. Ramasubramaniam, G. Toth, C. S. Lent, G. H. Bernstein, and G. L. Snider. Experimental demonstration of clocked single-electron switching in quantum-dot cellular automata. *Applied Physics Letters*, 77: pp 295–297, 2000.
22. A. O. Orlov, R. K. Kummamuru, R. Ramasubramaniam, G. Toth, C. S. Lent, G. H. Bernstein, and G. L. Snider. Experimental demonstration of a latch in clocked quantum-dot cellular automata. *Applied Physics Letters*, 78: pp 1625–1627, 2001.
23. R. K. Kummamuru, J. Timler, G. Toth, C. S. Lent, R. Ramasubramaniam, A. O. Orlov, G. H. Bernstein, and G. L. Snider. Power gain in a quantum-dot cellular automata latch. *Applied Physics Letters*, 81: pp 1332–1334, 2002.
24. A. O. Orlov, R. Kummamuru, R. Ramasubramaniam, C. S. Lent, G. H. Bernstein, and G. L. Snider. A two-stage shift register for clocked quantum-dot cellular automata. *Journal of Nanoscience and Nanotechnology*, 2: pp 351–355, 2002.
25. R. K. Kummamuru, A. O. Orlov, R. Ramasubramaniam, C. S. Lent, G. H. Bernstein, and G. L. Snider. Operation of a quantum-dot cellular automata (QCA) shift register and analysis of errors. *IEEE Transactions on Electron Devices*, 50: pp 1906–1913, 2003.
26. A. O. Orlov, R. Kummamuru, R. Ramasubramaniam, C. S. Lent, G. H. Bernstein, and G. L. Snider. Clocked quantum-dot cellular automata shift register. *Surface Science*, 532: pp 1193–1198, 2003.
27. C. S. Lent. Molecular electronics: bypassing the transistor paradigm. *Science*, 288: pp 1598–1599, 2000.
28. K. Hennessy and C. S. Lent. Clocking of molecular quantum-dot cellular automata. *Journal of Vacuum Science and Technology B: Microelectronics and Nanometer Structures*, 19: pp 1752–1755, 2001.
29. M. Lieberman, S. Chellamma, B. Varughese, Y. L. Wang, C. Lent, G. H. Bernstein, G. Snider, and F. C. Peiris. Quantum-dot cellular automata at a molecular scale. *Molecular Electronics II*, 960: pp 225–239, 2002.
30. E. P. Blair. Tools for the design and simulation of clocked molecular quantum-dot cellular automata circuits. Master of Science Thesis, Department of Electrical Engineering, University of Notre Dame: p 92, 2003.

31. E. P. Blair and C. S. Lent. An architecture for molecular computing using quantum-dot cellular automata. In: *3rd IEEE Conference on Nanotechnology, 2*: pp 402–405, 2003.
32. C. S. Lent and B. Isaksen. Clocked molecular quantum-dot cellular automata. *IEEE Transactions on Electron Devices*, 50: pp 1890–1896, 2003.
33. C. S. Lent, B. Isaksen, and M. Lieberman. Molecular quantum-dot cellular automata. *Journal of the American Chemical Society*, 125: pp 1056–1063, 2003.
34. H. Qi, S. Sharma, Z. H. Li, G. L. Snider, A. O. Orlov, C. S. Lent, and T. P. Fehlner. Molecular quantum cellular automata cells. Electric field driven switching of a silicon surface bound array of vertically oriented two-dot molecular quantum cellular automata. *Journal of the American Chemical Society*, 125: pp 15250–15259, 2003.
35. L. Yuhui and C. S. Lent. Theoretical study of molecular quantum-dot cellular automata In: *10th International Workshop on Computational Electronics* pp 118–119, 2004.
36. W. C. Hu, K. Sarveswaran, M. Lieberman, and G. H. Bernstein. High-resolution electron beam lithography and DNA nanopatterning for molecular QCA. *IEEE Transactions on Nanotechnology*, 4: pp 312–316, 2005.
37. Y. H. Lu and C. Lent. Theoretical study of molecular quantum-dot cellular automata. *Journal of Computational Electronics*, 4: pp 115–118, 2005.
38. H. Qi, A. Gupta, B. C. Noll, G. L. Snider, Y. H. Lu, C. Lent, and T. P. Fehlner. Dependence of field switched ordered arrays of dinuclear mixed-valence complexes on the distance between the redox centers and the size of the counterions. *Journal of the American Chemical Society*, 127: pp 15218–15227, 2005.
39. K. Walus, G. Schulhof, and G. A. Jullien. Implementation of a simulation engine for clocked molecular QCA In: *Canadian Conference on Electrical and Computer Engineering*, pp 2128–2131, 2006.
40. L. Yuhui, L. Mo, and C. Lent. Molecular electronics: from structure to circuit dynamics. In: *6th IEEE Conference on Nanotechnology*: pp 62–65, 2006.
41. Y. Lu, M. Liu, and C. Lent. Molecular quantum-dot cellular automata: from molecular structure to circuit dynamics. *Journal of Applied Physics*, 102: pp 034311–034317, 2007.
42. G. Csaba, A. Imre, G. H. Bernstein, W. Porod, and V. Metlushko. Nanocomputing by field-coupled nanomagnets. *IEEE Transactions on Nanotechnology*, 1: pp 209–213, 2002.
43. A. Imre, G. Csaba, L. Ji, A. Orlov, G. H. Bernstein, and W. Porod. Majority Logic Gate for Magnetic Quantum-Dot Cellular Automata. *Science*, 311: pp 205–208, 2006.
44. C. S. Lent and P. D. Tougaw. Device architecture for computing with quantum dots. *Proceedings of the IEEE*, 85: pp 541–557, 1997.
45. J. G. Kemeny and J. Vonneumann. Theory of self-reproducing automata. *Science*, 157: p 180, 1967.
46. K. Walus, T. J. Dysart, G. A. Jullien, and R. A. Budiman. QCADesigner: a rapid design and simulation tool for quantum-dot cellular automata. *IEEE Transactions on Nanotechnology*, 3: pp 26–31, 2004.
47. J. Timler and C. S. Lent. Maxwell's demon and quantum-dot cellular automata. *Journal of Applied Physics*, 94: pp 1050–1060, 2003.
48. P. D. Tougaw and C. S. Lent. Effect of stray charge on quantum cellular-automata. *Japanese Journal of Applied Physics Part 1: Regular Papers, Short Notes, and Review Papers*, 34: pp 4373–4375, 1995.

49. C. Ungarelli, S. Francaviglia, M. Macucci, and G. Iannaccone. Thermal behavior of quantum cellular automaton wires. *Journal of Applied Physics*, 87: pp 7320–7325, 2000.

50. M. Macucci, G. Iannaccone, S. Francaviglia, and B. Pellegrini. Semiclassical simulation of quantum cellular automaton circuits. *International Journal of Circuit Theory and Applications*, 29: pp 37–47, 2001.

51. P. D. Tougaw and S. L. Craig. Dynamic behavior of quantum cellular automata. *Journal of Applied Physics*, 80: pp 4722–4736, 1996.

52. P. D. Tougaw and C. S. Lent. Quantum cellular-automata: computing with quantum-dot molecules. *Compound Semiconductors*, 1994: pp 781–786, 1995.

53. C. S. Lent, P. D. Tougaw, and W. Porod. Quantum cellular automata: the physics of computing with arrays of quantum dot molecules. In: *Workshop on Physics and Computation*: pp 5–13, 1994.

54. J. Timler and C. S. Lent. Power gain and dissipation in quantum-dot cellular automata. *Journal of Applied Physics*, 91: pp 823–831, 2002.

55. G. Toth and C. S. Lent. Quasiadiabatic switching for metal-island quantum-dot cellular automata. *Journal of Applied Physics*, 85: pp 2977–2984, 1999.

56. K. Walus, G. A. Jullien, and V. S. Dimitrov. Computer arithmetic structures for quantum cellular automata. In: *Conference Record of the Thirty-Seventh Asilomar Conference on Signals, Systems and Computers, Volume 2*: pp 1435–1439, 2003.

57. S. E. Frost, T. J. Dysart, P. M. Kogge, and C. S. Lent. Carbon nanotubes for quantum-dot cellular automata clocking. In: *4th IEEE Conference on Nanotechnology*: pp 171–173, 2004.

58. W. Yue-Min, H. Kuo-Dong, S. F. Hu, C. L. Sung, and Y. C. Chou. Coulomb blockade oscillations in ultrathin gate oxide silicon single-electron transistors. *Journal of Applied Physics*, 97: p 116106, 2005.

59. T. M. Buehler, D. J. Reilly, R. P. Starrett, D. G. Andrew, A. R. Hamilton, A. S. Dzurak, and R. G. Clark. Single-shot readout with the radio-frequency single-electron transistor in the presence of charge noise. *Applied Physics Letters*, 86: p 143117, 2005.

60. R. M. Zhang, K. Walus, W. Wang, and G. A. Jullien. A method of majority logic reduction for quantum cellular automata. *IEEE Transactions on Nanotechnology*, 3: pp 443–450, 2004.

61. P. D. Tougaw and S. L. Craig. Logical devices implemented using quantum cellular automata. *Journal of Applied Physics*, 75: pp 1818–1825, 1994.

62. K. Walus, G. Schulhof, and G. A. Jullien. High level exploration of quantum-dot cellular automata (QCA). In: *Conference Record of the Thirty-Eighth Asilomar Conference on Signals, Systems and Computers, Volume 1*: pp 30–33, 2004.

63. G. Schulhof, K. Walus, and G. A. Jullien. Simulation of random cell displacements in QCA. *Journal on Emerging Technologies in Computing Systems*, 3: p 2, 2007.

64. A. Gin, P. D. Tougaw, and S. Williams. An alternative geometry for quantum-dot cellular automata. *Journal of Applied Physics*, 85: pp 8281–8286, 1999.

65. K. Walus, G. Schulhof, and G. A. Jullien. Implementation of a simulation engine for clocked molecular QCA. In: *Canadian Conference on Electrical and Computer Engineering*: pp 2128–2131, 2006.

66. G. Snider, A. Orlov, C. Lent, G. Bernstein, M. Lieberman, and T. Fehlner. Implementations of quantum-dot cellular automata. In: *International Conference on Nanoscience and Nanotechnology*: 2006.

67. G. Toth and C. S. Lent. Role of correlation in the operation of quantum-dot cellular automata. *Journal of Applied Physics*, 89: pp 7943–7953, 2001.

68. G. Toth. Correlation and coherence in quantum-dot cellular automata. In: Department of Electrical Engineering, Ph.D. Notre Dame: University of Notre Dame, 2000, p 205.

69. R. Ravichandran, S. K. Lim, and M. Niemier. Automatic cell placement for quantum-dot cellular automata. *Integration: the VLSI Journal*, 38: pp 541–548, 2005.

70. L. Sung Kyu, R. Ramprasad, and N. Mike. Partitioning and placement for buildable QCA circuits. *Journal on Emerging Technologies in Computing Systems*, 1: pp 50–72, 2005.

71. W. J. Chung, B. Smith, and S. K. Lim. Node duplication and routing algorithms for quantum-dot cellular automata circuits. *IEEE Proceedings: Circuits Devices and Systems*, 153: pp 497–505, 2006.

72. K. Walus, G. Schulhof, G. A. Jullien, R. Zhang, and W. Wang. Circuit design based on majority gates for applications with quantum-dot cellular automata. In: *Conference Record of the Thirty-Eighth Asilomar Conference on Signals, Systems and Computers, Volume 2*: pp 1354–1357, 2004.

73. R. Zhang, P. Gupta, L. Zhong, and N. K. Jha. Threshold network synthesis and optimization and its application to nanotechnologies. *IEEE Transactions on Computer-Aided Design of Integrated Circuits and Systems*, 24: pp 107–118, 2005.

74. W. Wei, K. Walus, and G. A. Jullien. Quantum-dot cellular automata adders. In: *3rd IEEE Conference on Nanotechnology, Volume 2*: pp 461–464, 2003.

75. A. Vetteth, K. Walus, and G. A. Jullien. RAM design using quantum-dot cellular automata. In: *Nanotechnology Conference and Tradeshow*: pp 160–163, 2003.

76. K. Walus, M. Mazur, G. Schulhof, and G. A. Jullien. Simple 4-bit processor based on quantum-dot cellular automata (QCA). In: *16th IEEE International Conference on Application-Specific Systems, Architecture Processors*: pp 288–293, 2005.

77. G. Bazan, A. O.Orlov, G. L. Snider, and G. H. Bernstein. Charge detector realization for AlGaAs/GaAs quantum-dot cellular automata. In: *The 40th international conference on electron, ion, and photon beam technology and nanofabrication: Atlanta, Georgia*, pp 4046–4050, 1996.

78. S. Gardelis, C. G. Smith, J. Cooper, D. A. Ritchie, E. H. Linfield, and Y. Jin. Evidence for transfer of polarization in a quantum dot cellular automata cell consisting of semiconductor quantum dots. *Physical Review B*, 67: p 033302, 2003.

79. T. E. Vandervelde, R. M. Kalas, P. Kumar, T. Kobayashi, T. L. Pernell, and J. C. Bean. Conditions for self-assembly of quantum fortresses and analysis of their possible use as quantum cellular automata. *Journal of Applied Physics*, 97: p 043513, 2005.

80. R. P. Cowburn and M. E. Welland. Room temperature magnetic quantum cellular automata. *Science*, 287: pp 1466–1468, 2000.

81. M. Manimaran, G. L. Snider, C. S. Lent, V. Sarveswaran, M. Lieberman, Z. H. Li, and T. P. Fehlner. Scanning tunneling microscopy and spectroscopy investigations of QCA molecules. *Ultramicroscopy*, 97: pp 55–63, 2003.

82. K. Walus, R. A. Budiman, and G. A. Jullien. Impurity charging in semiconductor quantum-dot cellular automata. *Nanotechnology*, 16: pp 2525–2529, 2005.

83. A. Aviram. Molecules for memory, logic, and amplification. *Journal of the American Chemical Society*, 110: pp 5687–5692, 1988.
84. N. S. Hush, A. T. Wong, G. B. Bacskay, and J. R. Reimers. Electron and energy-transfer through bridged systems. 6. Molecular switches: the critical-field in electric-field activated bistable molecules. *Journal of the American Chemical Society*, 112: pp 4192–4197, 1990.
85. A. Fijany and B. N. Toomarian. New design for quantum dot cellular automata to obtain fault tolerant logic gates. *Journal of Nanoparticle Research*, 3: pp 27–37, 2001.

5

DIELECTROPHORETIC ARCHITECTURES

Alexander D. Wissner-Gross

Electric programmability has been the basis for decades of advances in integrated computer component performance. However, once fabricated or assembled, such components—whether blade servers in a data center or microprocessors on a circuit board—are typically stuck in place and require human intervention for reconfiguration, removal, or replacement. For continued advances at the architectural level, mechanical programmability of components may also be needed. One generally promising approach for electromechanical manipulation at the nanoscale and microscale is *dielectrophoresis*, or the net force experienced by a neutral dielectric object in a nonuniform electric field. In this chapter, we review recent advances in dielectrophoretic architectures for computation, focusing particularly on the experimental demonstration of fully reconfigurable nanowire interconnects.

Programmability in electronic systems originates from the ability to form and reform nonvolatile connections. Devices in modern programmable architectures typically derive this ability from controlled internal changes in material composition or charge distribution [1]. However, for "bottom-up" nanoelectronic systems it may be advantageous to derive programmability not only from internal state, but also from the mechanical manipulation of mobile components. Proposed applications that require component mobility include neuromorphic networks of nanostructure-based artificial synapses [2], breadboards for rapid prototyping of nanodevice circuits [3, 4], and fault-tolerant logic in which broken subsystems are replaced automatically from a reservoir [5]. In this chapter, we review recent advances in dielectrophoretic architectures for enabling such computational component mobility, focusing particularly on the experimental demonstration of fully reconfigurable nanowire interconnects—the simplest nanoelectronic components.

Bio-Inspired and Nanoscale Integrated Computing. Edited by Mary Mehrnoosh Eshaghian-Wilner

5.1. INTRODUCTION TO DIELECTROPHORESIS

Let us first begin with a discussion of the phenomenon of dielectrophoresis itself. Following Pohl's seminal work on the subject [6], we first calculate the strength of the dielectrophoretic force on a particle and then consider the consequences of dielectrophoresis in media with different electromagnetic responses.

5.1.1. The Dielectrophoretic Effect

Dielectrophoresis is the net force experienced by a dielectric object in a nonuniform electric field. This effect exists because the dipole charges $\pm q$ induced at characteristic locations $r_\pm$ are subject to different values of the electric field $E_e(r)$, as shown in Figure 5.1.

There is no net force in a homogeneous electric field, so we can expand,

$$\vec{E}_e(r_+) = \vec{E}_e(r_-) + (\vec{L} \cdot \vec{\nabla})\vec{E}_e(r_-) + O(L^2).$$

For particles that are small relative to the characteristic length of a field gradient, the net force is given by

$$\vec{F} = \vec{F}_- + \vec{F}_+ = q(\vec{E}_e(\vec{r}_+) - \vec{E}_e(\vec{r}_-)) = (q\vec{L} \cdot \vec{\nabla})\vec{E}_e = (\vec{p} \cdot \vec{\nabla})\vec{E}_e.$$

Under vector transformation,

$$\vec{F} = (\vec{p} \cdot \vec{\nabla})\vec{E}_e = \vec{\nabla}(\vec{p} \cdot \vec{E}_e) - (\vec{E}_e \cdot \vec{\nabla})\vec{p} - \vec{p} \times (\vec{\nabla} \times \vec{E}_e) - \vec{E}_e \times (\vec{\nabla} \times \vec{p}).$$

Since $\vec{\nabla} \times \vec{E}_e = 0$, and for a dielectric that is isotropically, linearly, and homogeneously polarizable, $\vec{p} = v\hat{\alpha}\vec{E}_e$ for body volume v and tensor polarizability $\hat{\alpha}$, the force reduces to

$$\vec{F} = \vec{\nabla}(\vec{p} \cdot \vec{E}_e) - (\vec{E} \cdot \vec{\nabla})\vec{p} = v\hat{\alpha}(\vec{E}_e \cdot \vec{\nabla})\vec{E}_e = \frac{1}{2} v\hat{\alpha}\vec{\nabla}\left|\vec{E}_e\right|^2.$$

Notably, the above dielectrophoretic force is invariant under field inversion.

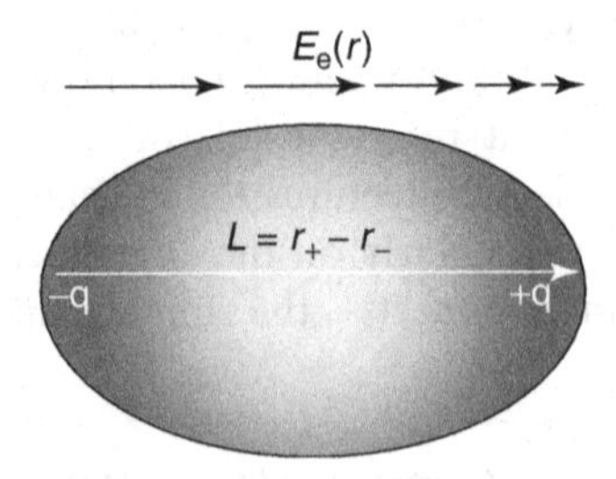

Figure 5.1. Schematic of dielectric particle in inhomogeneous field.

5.1.2. Dielectrophoresis in Media

For a particle with complex dielectric permittivity ε_p^* in a medium ε_m^*, the internal electric displacement is the sum of the particle polarization and the external displacement:

$$\vec{D}_i = \varepsilon_p^* \vec{E}_i = \vec{P} + \varepsilon_m^* \vec{E}_i.$$

Therefore, the polarization scales with the internal electric field as

$$\vec{P} = \left(\varepsilon_p^* - \varepsilon_m^*\right) \vec{E}_i.$$

This polarization generates an additional surface charge dQ on each surface element dF:

$$dQ = \sigma_{pol} \cdot dF = \vec{P} \cdot \vec{dF}.$$

So, applying Coulomb's Law and combining this result with the unperturbed field, we find that

$$\begin{aligned}\vec{E}_i(\vec{r}_1) =& \vec{E}_e(\vec{r}_1) - \oiint \frac{\vec{r}_{12}}{4\pi\varepsilon_m r_{12}^3} dQ_2 = \vec{E}_e(\vec{r}_1) - \oiint \frac{\vec{r}_{12}}{4\pi\varepsilon_m r_{12}^3} \left[\vec{P}(\vec{r}_2) \cdot \vec{dF}_2\right] \\ =& \vec{E}_e(\vec{r}_1) - \left(\varepsilon_p^* - \varepsilon_m^*\right) \oiint \frac{\vec{r}_{12}}{4\pi\varepsilon_m r_{12}^3} \left[\vec{E}_i(\vec{r}_2) \cdot \vec{dF}_2\right].\end{aligned}$$

Since we assumed that the particle is linearly polarizable and, for approximation purposes, considering only the polarization parallel to the electric field, it follows that:

$$\vec{P} = \alpha \vec{E}_e = \left(\varepsilon_p^* - \varepsilon_m^*\right) \vec{E}_i$$

$$\rightarrow \frac{\alpha}{\left(\varepsilon_p^* - \varepsilon_m^*\right)} \vec{E}_e(\vec{r}_1) \cong \vec{E}_e(\vec{r}_1) - \alpha \cdot \mathrm{Proj}_{\vec{\mathrm{E}}_e(\vec{r}_1)} \left\{ \oiint \frac{\vec{r}_{12}}{4\pi\varepsilon_m^* r_{12}^3} \left[\vec{E}_e(\vec{r}_2) \cdot \vec{dF}_2\right] \right\}$$

$$\rightarrow \frac{\alpha}{\left(\varepsilon_p^* - \varepsilon_m^*\right)} \vec{E}_e(\vec{r}_1) \cong \vec{E}_e(\vec{r}_1) - \frac{\alpha}{4\pi\varepsilon_m^*} \cdot \vec{\mathrm{E}}_e(\vec{r}_1) \oiint \frac{\vec{r}_{12} \cdot \vec{\mathrm{E}}_e(\vec{r}_1)}{\vec{\mathrm{E}}_e^2(\vec{r}_1)} \left[\frac{\vec{E}_e(\vec{r}_2) \cdot \vec{dF}_2}{r_{12}^3}\right]$$

$$\rightarrow \alpha \cong \varepsilon_m^* \cdot \frac{\left(\varepsilon_p^* - \varepsilon_m^*\right)}{\varepsilon_m^* + \frac{\varepsilon_p^* - \varepsilon_m^*}{4\pi} \cdot \oiint \frac{\vec{r}_{12} \cdot \vec{\mathrm{E}}_e(\vec{r}_1)}{\vec{\mathrm{E}}_e^2(\vec{r}_1)} \left[\frac{\vec{E}_e(\vec{r}_2) \cdot \vec{dF}_2}{r_{12}^3}\right]}$$

The above expression for α/ε_m^* is termed the Clausius–Mossotti factor and is denoted by the frequency-dependent function $K(\omega)$.

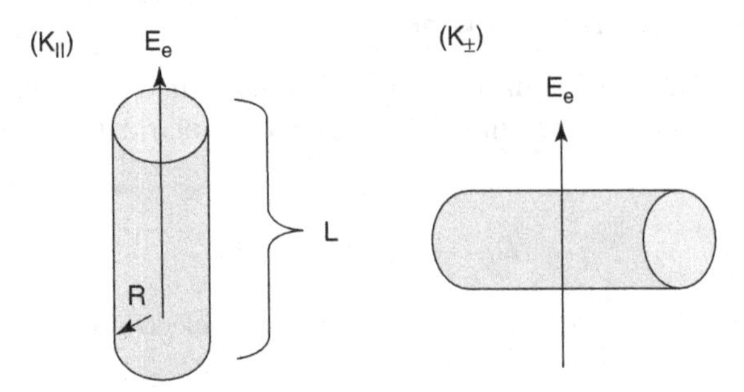

Figure 5.2. Clausius–Mossotti definitions for orientations of a cylindrical particle.

This factor, in both sign and magnitude, is essential to characterizing the dielectrophoretic behavior of a particle in a medium. When the factor is positive, the particle moves toward stronger electric fields; when the factor is negative, the particle moves toward weaker electric fields. For rotationally asymmetric particles, the magnitude can depend strongly on particle orientation relative to the field direction and, in particular, roughly corresponds to the elongation of the particle along the field axis. The nanodevices we focus on in this chapter are cylindrical, so we now present approximate Clausius-Mossotti expressions for cylindrical particles parallel and perpendicular to a field, as shown in Figure 5.2:

$$K_{\|}(f) \cong \frac{\varepsilon_p^* - \varepsilon_m^*}{\varepsilon_m^* + (\varepsilon_p^* - \varepsilon_m^*)(1 - (1 + R^2/L^2)^{-1/2})}$$

and

$$K_{\perp}(f) = \frac{\varepsilon_p^* - \varepsilon_m^*}{\varepsilon_m^*\left(1 - \frac{\pi}{8}\right) + \varepsilon_p^*\left(\frac{\pi}{8}\right)} \sim 3\,\frac{\varepsilon_p^* - \varepsilon_m^*}{\varepsilon_p^* + 2\varepsilon_m^*}.$$

5.2. DIELECTROPHORETIC ASSEMBLY AND TRANSPORT OF NANODEVICES

Having completed our discussion of the theory of dielectrophoresis, we are now ready to explore its practical applications to nanocomputation. Let us first consider the problem of assembling individual nanodevices. Various approaches for manipulating electronic nanostructures have been developed, including mechanical [7], optical [8, 9], electrostatic [10, 11], and dielectrophoretic [6] methods. Dielectrophoresis is particularly attractive for inexpensive and massively parallel manipulation [12] of neutral microscale and nanoscale objects using only standard semiconductor fabrication technologies. It has been used to trap a variety of structures from suspensions, including NiSi nanowires [13], CdS nanowires [14], GaN nanowires [15], carbon nanotubes [16], silicon microblocks [17], ZnO

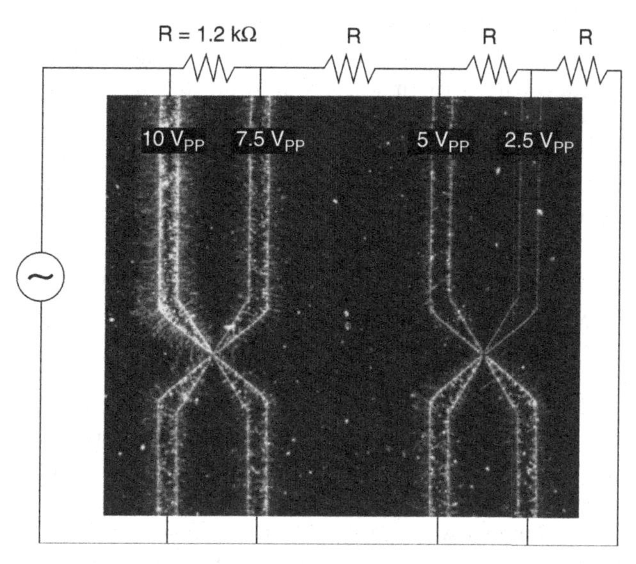

Figure 5.3. Dark-field microscope image demonstrating the trapping of large numbers of nanostructures, which are visible as lines perpendicular to the electrode faces, at relatively low voltages.

nanorods and nanobelts [18, 19], and gold nanowires [20]. An example of the ability of dielectrophoresis to trap nanostructures with relatively low voltages is shown in Figure 5.3. Previous electrical transport measurements of dielectrophoretically trapped structures have been performed either after immobilizing the structures through drying [15–19] or chemical binding [20], or performed over large films of parallel interconnects [16]. In this work we demonstrate for the first time that dielectrophoresis may also be used to reconfigure and disassemble nanowire interconnects and that our process is compatible with the assembly and electronic characterization of individual nanowire devices.

5.3. DIELECTROPHORETIC RECONFIGURATION OF NANODEVICES

Assembly of nanodevices is itself worthy of scientific interest, but the beauty of dielectrophoretic architectures lies in their versatility, since they enable not just the assembly but the reconfiguration of nanodevices as well. We now proceed to discussion of the recent experimental demonstration of reconfigurable nanowire interconnects [21], which exemplifies that versatility.

5.3.1. Fabrication of a Nanowire Trapping Architecture

For interconnects, p-type silicon nanowires were grown by the vapor-liquid-solid method [22], using 20-nm Au nanocluster catalysts (Ted Pella), and SiH_4 reactant

(99.7%) and B_2H_6 dopant (0.3%) in He carrier gas (100 ppm) at 450 torr and 450 °C. Growth was performed for 10–60 min to achieve nanowire lengths of 10–60 μm. The nanowire growth wafer was sonicated lightly in isopropanol for 1 min. The suspension was vacuum filtered using a 12-μm mesh (Millipore Isopore) in order to remove unnucleated Au catalyst particles and short nanowires. The filter mesh was sonicated in isopropanol, and the suspension was again filtered. The second filter mesh was sonicated in benzyl alcohol for 2 min and the suspension was used for trapping experiments. Benzyl alcohol was selected as a viscous, low-vapor-pressure solvent [23] for reconfiguration in order to damp Brownian motion, minimize toxicity [24], and allow ambient operation. Additionally, its static relative permittivity is slightly smaller than that of bulk Si (11.9 versus 12.1, respectively) [23], reducing van der Waals interactions at low frequencies and favouring dielectrophoretic trapping of conductive structures [6]. Heavily doped silicon nanowires were selected as interconnects to demonstrate potential compatibility of our technique with the assembly of more complex semiconducting nanostructures, such as axial heterostructures [25].

Trapping experiments were performed with 100-nm Au electrodes (5 nm Cr wetting layer) to avoid oxidative damage, on a Si wafer with a 200-nm oxide to prevent shorts. Thicker electrodes, with reduced fringing fields, were found to better allow nanowires to migrate along their edges toward the trapping region. Thinner electrodes tended to permanently pin nanowires to the top electrode faces wherever they were first trapped. The electrodes were defined by e-beam lithography with a 10° taper angle and a 1-μm tip radius of curvature.

The nanowire suspension was pipetted onto the electrode chip to form a 250-μm-thick reservoir, as shown schematically in Figure 5.4a. For trapping, electrode pairs were biased at 10 kHz to minimize both solvent electrolysis and parasitic capacitance. The bias was modulated into 10-ms bursts at 110 V_{RMS} with a period of 100 ms, which allowed migration of nanowires toward the trapping region in controlled steps. The time between bursts was manually increased to 1000 ms as nanowires approached the inter-electrode region, and the bursts were halted when the desired number of nanowires had been trapped. Movies of nanowire motion were recorded at 4 fps.

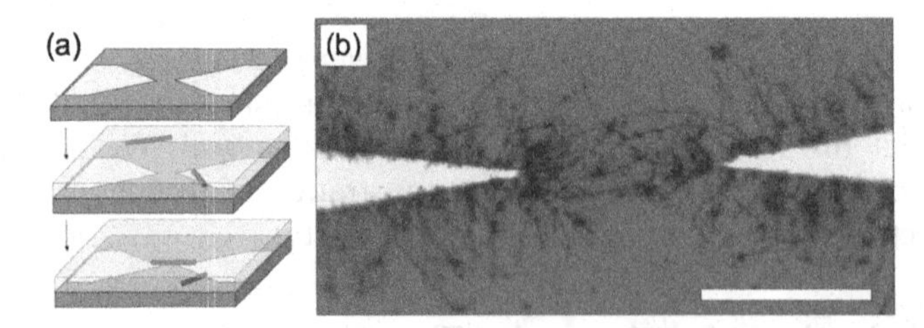

Figure 5.4. Dielectrophoretically trapped nanowires. (a) Schematic illustration of nanowire trapping process. (b) Light microscope image of multiple nanowires stably trapped between electrodes separated by 40 μm. Scale bar is 40 μm.

Electrical characterization was performed by manually switching the electrodes from the trapping voltage source to a measurement apparatus (Agilent 4156C). A 10-V sawtooth bias at 10 Hz was used and the measured currents were binned by voltage and averaged to remove hysteresis. Nanowire movement in the plane of the chip during transport measurement was minimal, since transport voltages were an order of magnitude smaller than trapping voltages.

5.3.2. Trapping Nanowire Interconnects

Nanowires up to 55 μm in length were stably trapped, as shown in Figure 5.4b, with noticeable bending due to the trapping field inhomogeneity. The longest previously reported one-dimensional nanostructures to be trapped by dielectrophoresis at both ends were less than 15 μm long and showed minimal bending [14].

For larger electrode gaps, nanowires at either tip appeared not to experience significant bending toward the opposite electrode, while shorter gaps caused significant accumulations of wires, as shown in Figure 5.5.

Confirmation that the trapping voltage was not substantially attenuated was provided by observation of a single scattering fringe around the electrodes in dark-field mode, as shown in Figure 5.6, which was consistent with the DC Kerr effect [6] at 155-V peak bias across a 10-μm gap.

The burst method was also delicate enough to enable trapping of individual nanowires when the reservoir was diluted below 2×10^{-13} M, as shown in Figure 5.7a. Before nanowire interconnects were assembled, DC transport between electrode pairs was nearly ohmic, as shown in Figure 5.7b. The measured current of 0.3 μA at 1-V bias and 40-V/s sweep rate was consistent with electrooxidation of benzyl alcohol [26, 27] at 1-μm-radius electrode tips.

However, after a 50-μm-long nanowire was trapped, the conductance became nonlinear and showed a 50% enhancement at 10-V bias, as shown in Figure 5.7c. Treating the device as an electrooxidative resistance in parallel with a series nanowire resistance and Schottky contacts [28], a nanowire transport curve was calculated, as shown in Figure 5.7d. The nanowire exhibited a calculated linear

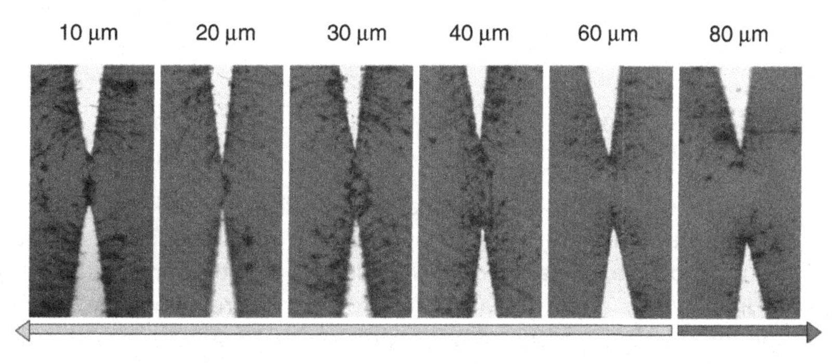

Figure 5.5. Nanowires trapped by electrodes separated by various distances.

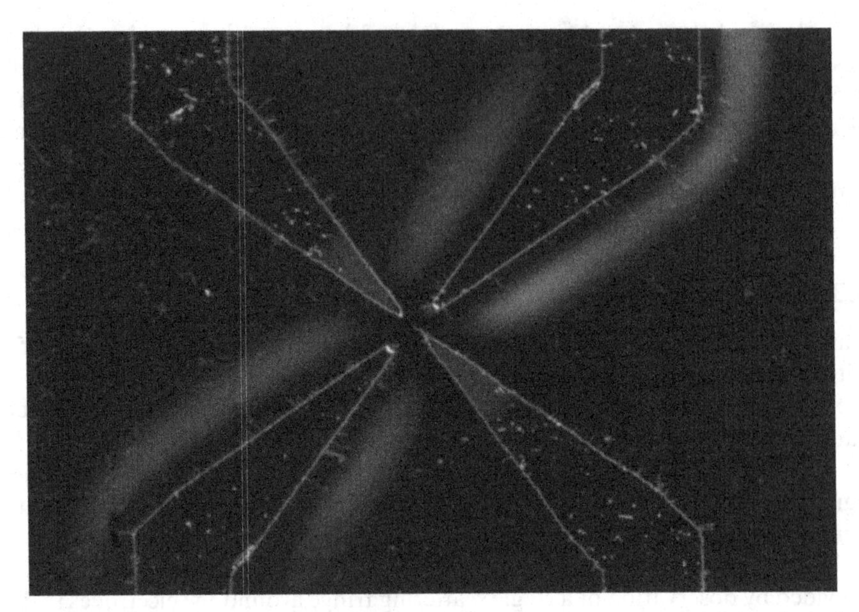

Figure 5.6. Scattering fringes about active electrodes as viewed in dark-field mode, attributed to the DC Kerr effect.

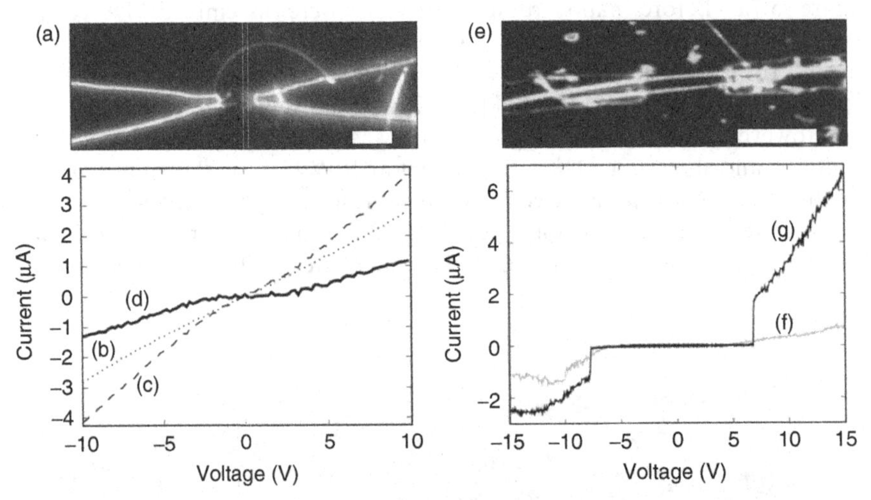

Figure 5.7. Electrical transport in trapped nanowires. (a) Dark-field microscope image of a single nanowire trapped by an electrode pair under solvent with electrical transport (b) measured with solvent, (c) measured with trapped nanowire in solvent, and (d) calculated for nanowire alone. (e) Dark-field microscope image of two trapped nanowires on dried substrate with electrical transport measured (f) immediately after drying and (g) after several voltage sweeps. Scale bars are 10 μm.

response resistivity of $3.8 \times 10^{-3}\,\Omega \cdot$ cm (6.0-MΩ resistance), consistent with a B doping ratio of 2000:1, and an estimated barrier potential of 2.0 V. The barrier potential was higher than the 0.34 V measured in evaporated Au/p-Si junctions [29], and is attributed to incomplete contact of the nanowire with the electrodes.

Further confirmation that dielectrophoretically trapped nanowires acted as interconnects was provided by substituting ethanol as a solvent and permitting the substrate to dry after trapping. A pair of nanowires thus trapped, as shown in Figure 5.7e, appeared to rest on both electrode faces but initial voltage sweeps yielded a 30-MΩ resistance, as shown in Figure 5.7f. After several sweeps, however, a sharp current turn-on was observed at 6.8–7.8 V, as shown in Figure 5.7g, which may indicate electrostatically induced bending of the nanowires to better contact the electrodes. Above the turn-on bias, the nanowires exhibited a 1.7-MΩ combined resistance, which is compatible with the solvent-based result.

5.3.3. Reconfiguring Nanowire Interconnects

The reported method for trapping nanowire interconnects furthermore enabled reconfiguration, since nanowires were maximally polarized when aligned between a pair of electrode tips. Reconfiguration of a nanowire bundle was achieved using "source" and "drain" electrodes with opposite phase and a "latch" electrode with variable phase. Several nanowires were independently trapped between the source and latch electrodes with 100-ms-period bursts, as shown in Figure 5.8a, and then bundled with 250-ms-period bursts, as shown in Figure 5.8b. The phase of the latch electrode was then inverted, with the same burst period, causing the nanowire bundle to experience a dielectrophoretic force toward the drain electrode. Because the electrode tips were arranged in an isosceles right triangle formation, the torque about the latch electrode was higher than that about the source electrode, and the nanowire remained in contact with the latch electrode during the motion, as shown in Figure 5.8c. The reconfiguration was completed 0.25–0.75 s after the phase inversion, as shown in Figure 5.8d, and could be reversed by restoring the original phase of the latch electrode, as shown in Figure 5.8e.

Similarly, parallel reconfiguration of a pair of nanowire interconnects was achieved with four electrodes in a 10-μm square, as shown in Figure 5.9. To accomplish this, the relative phases between diagonally opposite electrodes were fixed and the relative phase of the diagonal pairs was inverted. Dielectric breakdown of some nanowires occurred and short fragments were visible around the connecting nanowires.

For theoretical comparison, we now calculate how rapidly a nanowire might be dielectrophoretically reconfigured. The dielectrophoretic force per unit length on a cylindrical nanowire with length l_{wire} and diameter d_{wire} is given by

$$F_{dep} = \frac{1}{8}\varepsilon_{solv}\pi d_{wire}^{2}\mathrm{Re}\left\{\vec{K}(f)\cdot\vec{\nabla}(\vec{E}^{2})\right\}\ \hat{\nabla}(\vec{E}^{2})$$

where ε_{solv} is the solvent permittivity and $\vec{K}(f)$ is the frequency-dependent Clausius–Mossotti factor [6].

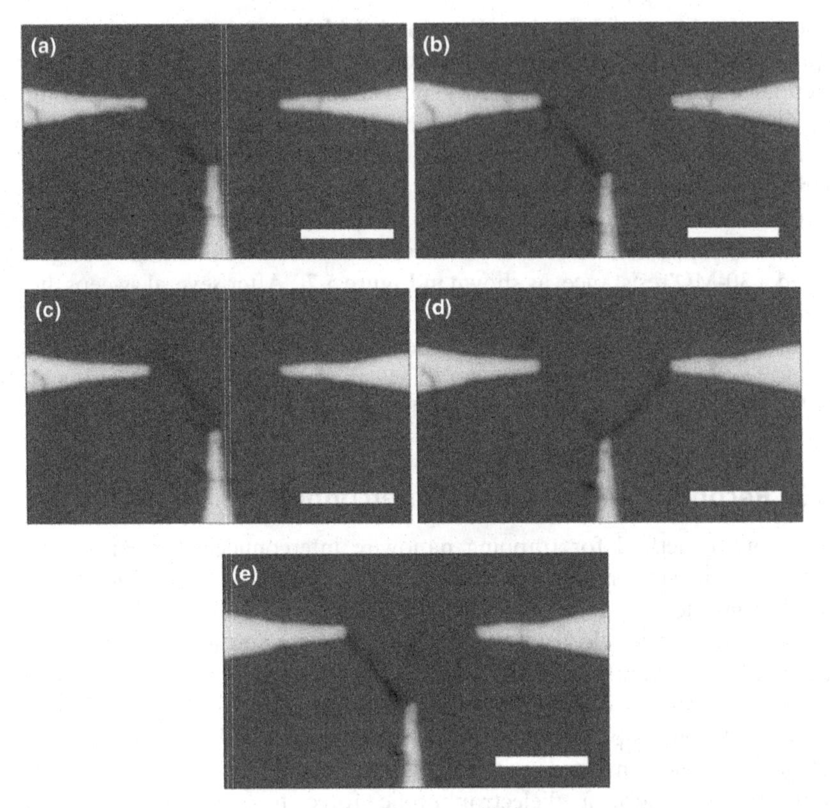

Figure 5.8. Microscope images of 3-electrode serial reconfiguration of nanowires. The relative phase between the left (source) and middle (latch) electrodes is modulated from (a,b) 180° to (c,d) 0° to (e) 180°. Scale bars are 15 μm.

For $l_{wire} >> d_{wire}$, the Clausius–Mossotti component perpendicular to the nanowire axis is approximately

$$K_{\perp} = \frac{\tilde{\varepsilon}_{wire} - \tilde{\varepsilon}_{solv}}{\tilde{\varepsilon}_{solv}\left(1 - \frac{\pi}{8}\right) + \tilde{\varepsilon}_{wire}\left(\frac{\pi}{8}\right)},$$

where $\tilde{\varepsilon}_X \equiv \varepsilon_X - i\sigma_X/(2\pi f)$ are the complex permittivities of nanowire and solvent (the solvent is assumed to be nonconductive at trapping frequencies). Approximating the gradient of the field energy density in the inter-electrode space as uniform, $\nabla(\vec{E}^2)\tilde{V}_{sd}^2/L^3$, where L is the distance between electrode tips.

The dielectrophoretic force is opposed by Stokes drag. The drag coefficient for an infinitely long, cylindrical nanowire [30] is given by

$$C_D \equiv \frac{F_{drag}}{\frac{1}{2}\rho_{wire}u^2 d} \approx \frac{8\pi}{\mathrm{Re}(2.002 - \ln \mathrm{Re})},$$

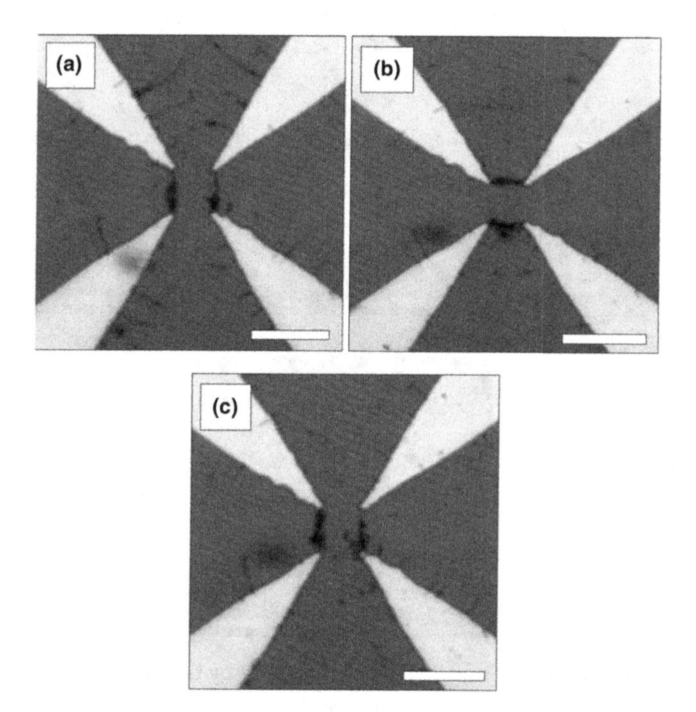

Figure 5.9. Microscope images of 4-electrode parallel reconfiguration of nanowires. Each pair of diagonally opposite electrodes is held at a constant relative phase of 180°, while the relative phase between the upper-left and upper-right electrodes is modulated from (a) 0° to (b) 180° to (c) 0°. Scale bars are 20 μm.

where F_{drag} is the drag force per unit length, ρ_{wire} is the nanowire density, u is the nanowire velocity, and $Re \sim d\rho_{solv}u/\mu$ is the Reynolds number for solvent density ρ_{solv} and dynamic viscosity μ. Assuming bulk mechanical properties of silicon and benzyl alcohol [23] and matching drag and dielectrophoretic forces, the terminal velocity during switching perpendicular to the nanowire axis is calculated to be $u_{\perp} \approx 30\,\mu m/s$, implying a 0.3-s reconfiguration time. This time is consistent with the serial reconfigurations observed, validating our model.

5.3.4. Disassembling Nanowire Interconnects

Finally, it should be mentioned that electrically driven nanowire disassembly was also found to occur. In an example of this effect, a single nanowire that was initially trapped on one electrode, as shown in Figure 5.10a, was pulled into the inter-electrode region and a gas bubble immediately formed there, as shown in Figure 5.10b. After the bubble dispersed, only short nanowire fragments remained, as shown in Figure 5.10c. This effect was observed for about 1/3 of

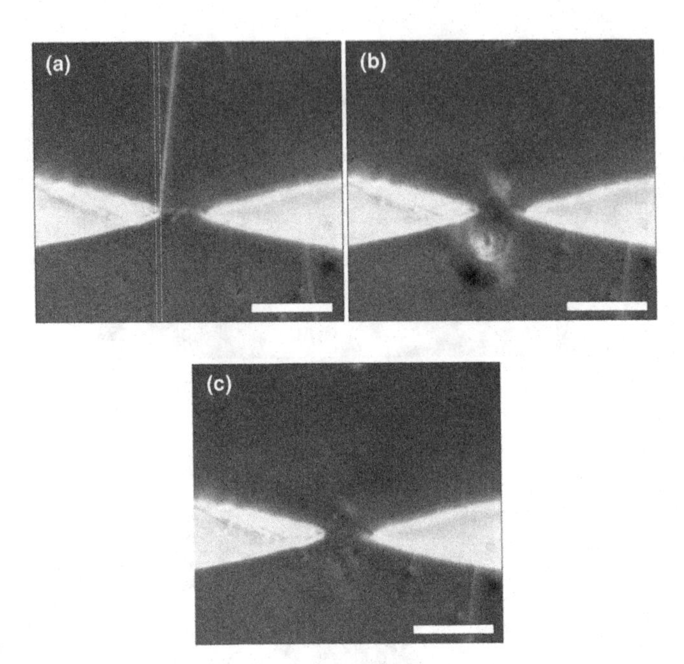

Figure 5.10. Disassembly of nanowire interconnects by thermal detonation. (a) Stably trapped nanowire before detonating voltage pulse. (b) Vapor bubble resulting from thermal detonation. (c) Only submicron fragments remain. Scale bars are 20 μm.

the nanowires trapped in inter-electrode regions and might be explained by variation in the nanowire conductivities and/or the formation of exceptionally good contacts, which could lead to current densities as high as $5 \times 10^{10}\,\mathrm{A \cdot m^{-2}}$ and thermal detonation. After the original nanowires were destroyed, the inter-electrode regions were typically able to trap new interconnects, so this effect might prove useful in fault-tolerant applications for severing connections to non-functioning components.

5.3.5. Summary of Implementation

The first assembly, reconfiguration, and disassembly of nanowire interconnects through dielectrophoresis has been presented. Silicon nanowires up to 55 μm long were trapped, and solvent-based transport studies show a 50% conductivity enhancement in the presence of the nanowires. Once assembled, these nanowire interconnects could then be reconfigured and disassembled using periodic voltage bursts. These results open up the possibility of colloidal, nanostructured connection architectures for computation.

5.4. LARGER-SCALE ARCHITECTURES

With our discussion of individual component reconfigurability concluded, we now advance to an exploration of the feasibility of larger-scale dielectrophoretic architectures. For dielectrophoresis to find real-world applications to nanocomputation, logic—at least at the gate level—must be demonstrated.

5.4.1. 1 × 1 Crossed Interconnects

In Section 5.3 we experimentally demonstrated the reconfiguration of interconnects geometrically occupying single sides or pairs of sides of a square. As a next step toward larger-scale architectures, we should also consider interconnects that do not lie on the circumference of the trapping region, but instead span it. In particular, we may estimate the value of the term ∇E^2, which is crucial to the dielectrophoretic strength, for either type of configuration, as shown schematically in Figure 5.11.

Recall that we may approximate the trapping force by assuming a uniform electric field gradient near the energy minimum. Therefore, for the geometry shown in Figure 5.11, for the noncircumferential interconnect (I),

$$\nabla E^2 \sim (V)^2/(d\sqrt{2})^3 = \frac{1}{2^{3/2}} \cdot \frac{V^2}{d^3},$$

and for the circumferential interconnect (II),

$$\nabla E^2 \sim (V/2)^2/(d)^3 = \frac{1}{2^2} \cdot \frac{V^2}{d^3}.$$

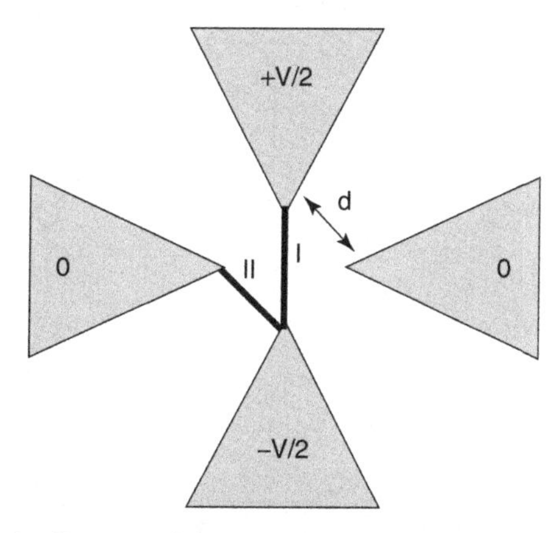

Figure 5.11. Schematic diagram for circumferential (II) versus noncircumferential (I) interconnects in a 1 × 1 crossbar.

Since the value is larger for the noncircumferential case, stable trapping of diagonal interconnects should be possible. (There is suggestive experimental data that stable diagonal interconnects can indeed be achieved [14], although the substrate in that case was dried before imaging.) Moreover, since $K_\perp \ll K_\parallel$, a short field burst perpendicular to an existing wire should not disrupt it.

5.4.2. 2 × 2 Crossbar

We can perform a similar, albeit more entailed, analysis to determine the feasibility of stable noncircumferential interconnects in larger crossbars, such as the 2 × 2 geometry we now consider.

For the specific octagonal electrode tip geometry and voltages shown in Figure 5.12, we can calculate all force values ($F \equiv \nabla E^2$):

$$\frac{F_2}{F_1} = \frac{\left(\frac{1}{2} - a\right)^2}{\left[\frac{1}{\sqrt{2}}(1-p)\right]^3}, \quad \frac{F_3}{F_1} = \frac{(2a)^2}{p^3}, \quad \frac{F_4}{F_1} = \frac{\left(\frac{1}{2} + a\right)^2}{\left[\sqrt{\frac{1}{2} + \frac{p^2}{2}}\right]^3},$$

$$\frac{F_4}{F_1} = \frac{\left(\frac{1}{2} + a\right)^2}{\left[\sqrt{\frac{1}{2} + \frac{p^2}{2}}\right]^3}, \quad \frac{F_5}{F_1} = \frac{(2a)^2}{\left[\sqrt{p^2+1}\right]^3},$$

In order to estimate the proper conditions for trapping of non-circumferential interconnects, particularly the interconnect at F_1, we are most interested in the

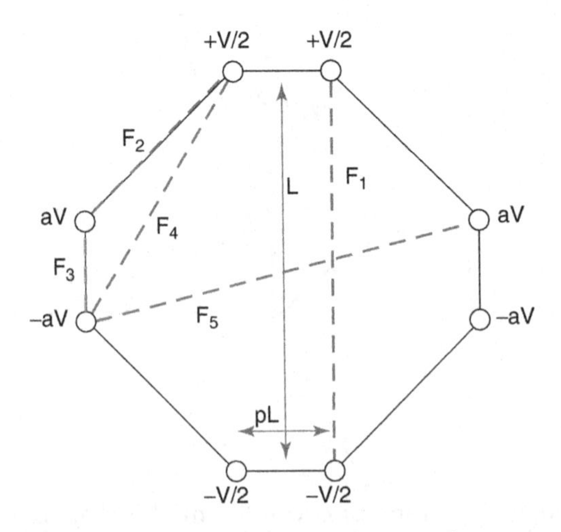

Figure 5.12. Schematic diagram for dielectrophoretic force constants for various interconnects in a 2 × 2 crossbar geometry. Circles indicate electrode tips.

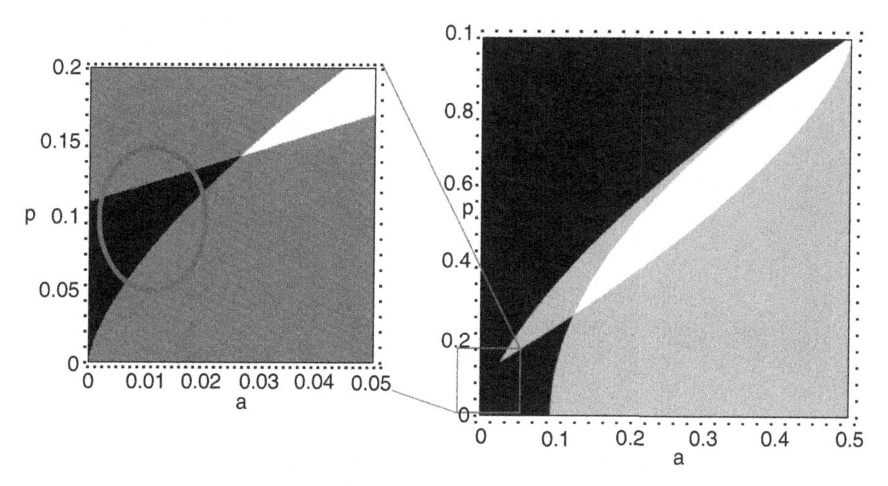

Figure 5.13. Phase diagram of dielectrophoretic forces $(F_2, F_3, F_4, F_5)/F_1$ in the space of dimensionless electrode voltages (a) versus electrode spacings (p) in a 2×2 crossbar geometry. Darker regions indicate that more of F_2, F_3, F_4, F_5 are less than F_1.

parameters $0 < a < 1/2$, $0 < p < 1$ that cause all of these ratios to fall below 1. We survey this phase space in Figure 5.13, and find that—at least by this prediction method—long crossbar-style interconnects should be possible as long as the parallel electrode pitch is 10–20 times smaller than the interconnect length ($0.05 < p < 0.1$) and electrodes uninvolved in the trapping are kept grounded ($a = 0$).

5.4.3. Field-Programmable Gates

In addition to interconnect architectures, it should also be possible to more directly build logic based on dielectrophoretic reconfiguration of nanodevices. In particular, with the proper arrangement of pairs of the latching electrodes shown in Figure 5.8, digital logic gates can be reproduced. In these gates, as depicted in Figure 5.14, the relative phase of two electrodes determines the position of interconnects which can then carry or block a current path specific to the logical operation required.

5.4.4. Packaging

In order to enable the vision of dielectrophoretic architectures for self-repairing computer components, development of stable colloidal reservoirs of components will be crucial. Fortunately, many dielectric solvents, such as the benzyl alcohol used in the reconfigurable nanowire work discussed above, have low vapor pressures. This property may enable colloidal reservoirs of replacement components to be stably sealed into a chip carrier package, as shown in Figure 5.15.

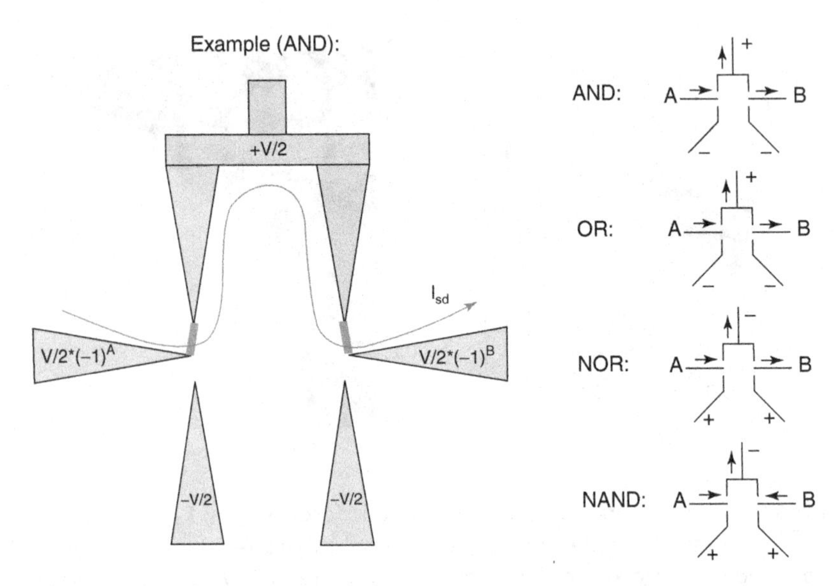

Figure 5.14. Field programmable gates in which short voltage pulses on two electrodes serve as input (phase indicated by sign), and current indicated by arrows serves as output.

Figure 5.15. Packaged nanowire reservoir.

In preliminary tests, sealed reservoirs of benzyl alcohol for dielectrophoretic manipulation remained stable, in the sense that nanowires remained suspended and there was no visible leakage of solvent, for more than a week under refrigeration.

5.5. CONCLUSIONS

We have reviewed recent progress toward the development of dielectrophoretic architectures for integrated nanoelectronic computation. Focusing on the demonstration of reconfigurable nanowire interconnects, we also derived general strategies for scaling, integrating, and packaging such dielectrophoretic systems.

ACKNOWLEDGEMENTS

The author is especially grateful to C. M. Lieber for his generous guidance and support, and for use of his laboratory, and to F. Patolsky for very helpful discussions. The author also thanks the Fannie and John Hertz Foundation for doctoral funding, and T. M. Sullivan for his editing. This work made use of Harvard Center for Nanoscale Systems and NSF/NNIN facilities. This material was supported in part by the United States Air Force and DARPA under Contract No. FA8750-05-C-0011. Any opinions expressed are those of the author and do not necessarily reflect the views of the United States Air Force or DARPA.

REFERENCES

1. J. Rose, A. el Gamal, and A. Sangiovanni-Vincentelli. Architecture of field-programmable gate arrays. *Proceedings of the IEEE*, 81(7): p 1013–1029, 1993.
2. Ö Türel, J. H. Lee, X. Ma, and K. K. Likharev. Neuromorphic architectures for nanoelectronic circuits. *International Journal of Circuit Theory and Applications*, 32: p 277–302, 2004.
3. J. C. Ellenbogen, and J. C. Love. Architectures for molecular electronic computers: logic structures and an adder designed from molecular electronic diodes. *Proceedings of the IEEE*, 88(3): p 386–426, 2000.
4. J. Moser, R. Panepucci, Z. P. Huang, W. Z. Li, Z. F. Ren, A. Usheva, and M. J. Naughton. Individual free-standing carbon nanofibers addressable on the 50 nm scale. *Journal of Vacuum Science and Technology B*, 21(3): p 1004–1007, 2003.
5. P. K. Lala. *Self-Checking and Fault-Tolerant Digital Design*. San Diego: Academic Press, p 172.
6. H. A. Pohl. *Dielectrophoresis*. Cambridge, UK: Cambridge University Press, 1978, pp 1–50.

7. Y. Huang, X. Duan, Q. Wei, and C. M. Lieber. Directed assembly of one-dimensional nanostructures into functional networks. *Science*, 291: p 630–633, 2001.
8. R. Agarwal, K. Ladavac, Y. Roichman, G. Yu, C. M. Lieber, and D. G. Grier. Manipulation and assembly of nanowires with holographic optical traps. *Optics Express*, 13(22): p 8906–8912, 2005.
9. P. J. Pauzauskie, A. Radenovic, E. Trepagnier, H. Shroff, P. Yang, and J. Liphardt. Optical trapping and integration of semiconductor nanowire assemblies in water. *Nature Materials*, 5: pp 97–101, 2006.
10. J. E. Jang, S. N. Cha, Y. Choi, G. A. J. Amaratunga, D. J. Kang, D. G. Hasko, J. E. Jung, and J. M. Kim. Nanoelectromechanical switches with vertically aligned carbon nanotubes. *Applied Physics Letters*, 87: p 163114, 2005.
11. T. Rueckes, K. Kim, E. Joselevich, G. Y. Tseng, C. L. Cheung, and C. M. Lieber. Carbon nanotube-based nonvolatile random access memory for molecular computing. *Science*, 289(5476): p 94–97, 2000.
12. P. Y. Chiou, A. T. Ohta, and M. C. Wu. Massively parallel manipulation of single cells and microparticles using optical images. *Nature*, 436: p 370–372, 2005.
13. L. F. Dong, J. Bush, V. Chirayos, R. Solanki, J. Jiao, Y. Ono, J. F. Conley Jr., and B. R. Ulrich. Dielectrophoretic controlled fabrication of single crystal nickel silicide nanowire interconnects and the investigation of their formation mechanism. *Nano Letters*, 5(10): p 2112–2115, 2005.
14. X. Duan, Y. Huang, Y. Cui, J. Wang, and C. M. Lieber. Indium phosphide nanowires as building blocks for nanoscale electronic and optoelectronic devices. *Nature*, 409: p 66–69, 2001.
15. T. H. Kim, S. Y. Lee, N. K. Cho, H. K. Seong, H. J. Choi, S. W. Jung, and S. K. Lee. Dielectrophoretic alignment of gallium nitride nanowires (GaN NWs) for use in device applications. *Nanotechnology*, 17(14): p 3394–3399, 2006.
16. Z. Chen, Y. Yang, F. Chen, Q. Qing, Z. Wu, and Z. Liu. Controllable interconnection of single-walled carbon nanotubes under AC electric field. *Journal of Physical Chemistry B*, 109(23): p 11420–11423, 2005.
17. S. W. Lee and R. Bashir. Dielectrophoresis and electrohydrodynamics-mediated fluidic assembly of silicon resistors. *Applied Physics Letters*, 83(18): p 3833–3835, 2003.
18. O. Harnack, C. Pacholski, H. Weller, A. Yasuda, and J. M. Wessels. Rectifying behavior of electrically aligned ZnO nanorods. *Nano Letters*, 3(8): p 1097–1101, 2003.
19. C. S. Lao, J. Liu, P. Gao, L. Zhang, D. Davidovic, R. Tummala, and Z. L. Wang. ZnO nanobelt/nanowire Schottky diodes formed by dielectrophoresis alignment across Au electrodes. *Nano Letters*, 6(2): p 263–266, 2006.
20. L. Shang, T. L. Clare, M. A. Eriksson, M. S. Marcus, K. M. Metz, and R. J. Hamers. Electrical characterization of nanowire bridges incorporating biomolecular recognition elements. *Nanotechnology*, 16: p 2846–2851, 2005.
21. A. D. Wissner-Gross. Dielectrophoretic reconfiguration of nanowire interconnects. *Nanotechnology*, 17: p 4986–4990, 2006.
22. Y. Cui, L. J. Lauhon, M. S. Gudiksen, J. Wang, and C. M. Lieber. Diameter-controlled synthesis of single-crystal silicon nanowires. *Applied Physics Letters*, 78(15): p 2214–2216, 2001.
23. D. R. Lide, editor. *CRC Handbook of Chemistry and Physics*, 82nd ed., New York: CRC Press, pp 3–52, 6–163; 12–59, 2001.

24. B. Nair. Final report on the safety assessment of benzyl alcohol, benzoic acid, and sodium benzoate. *International Journal of Toxicology*, 20(S3): pp 23–50, 2001.
25. M. S. Gudiksen, L. J. Lauhon, J. Wang, D. Smith, and C. M. Lieber. Growth of nanowire superlattice structures for nanoscale photonics and electronics. *Nature*, 415: pp 617–620, 2002.
26. E. A. Mayeda, L. L. Miller, and J. F. Wolf. Electrooxidation of benzylic ethers, esters, alcohols, and phenyl epoxides. *Journal of the American Chemical Society*, 94(19): p 6812, 1972.
27. S. Y. Kishioka, M. Umeda, and A. Yamada. Electrooxidation of benzylalcohol derivatives in aqueous solution. 203 rd Meeting of The Electrochemical Society, Paris, 2003, AC1, Abstract 2474.
28. M. S. Fuhrer, A. K. L. Lim, L. Shih, U. Varadarajan, A. Zettl, and P. L. McEuen. Transport through crossed anotubes. *Physica E*, 6: pp 868–871, 2000.
29. B. L. Smith and E. H. Rhoderic. Schottky barriers on p-type silicon. *Solid-State Electronics*, 14: pp 71–75, 1971.
30. D. J. Tritton. *Physical Fluid Dynamics*. Oxford: Oxford University Press, 1988, p 32.

6

MULTILEVEL AND THREE-DIMENSIONAL NANOMAGNETIC RECORDING

S. Khizroev, R. Chomko, I. Dumer, and D. Litvinov

This chapter outlines a study of multilevel (ML) and three-dimensional (3D) magnetic recording—a nanomagnetic recording technology suitable for information storage densities above 10 Tbit/in^2. To comply with the multilevel signal configuration, ML magnetic recording exploits a 3D head/media system powered with next-generation data coding methods. It is believed, when that combined with novel information processing techniques, relatively cost effective ML systems could be scaled down to a single-grain spin level, thus enabling memory with effective areal densities above 100 Tbit/in^2.

6.1. INTRODUCTION

6.1.1. Information Industries in Need of Alternative Storage Technologies

Demand for larger areal density of storage and memory devices is growing exponentially with the emergence of the Internet, explosive growth of broadband communication, increasingly complex multimedia mobile devices, and the rapid expansion of on-demand databases serving multinational businesses. Magnetic recording has been the mainstream data storage technology for more than four decades [1, 2]. After aggressive attempts to extend the life of longitudinal magnetic recording, the data storage industry is finally coming to terms with reality [3, 4]. Researchers, indeed, witness that the recorded data become highly unstable as the areal density increases beyond approximately 200 Gbit/in^2. A number of technologies have been proposed by research teams all over the world to address the fundamental (superparamagnetic) limit. The closest alternatives are believed to be

Bio-Inspired and Nanoscale Integrated Computing. Edited by Mary Mehrnoosh Eshaghian-Wilner

perpendicular magnetic recording (already commercialized) [5, 6], patterned medium [7–9], and heat-assisted magnetic recording (HAMR) [10–13]. It is expected that these technologies can defer the superparamagnetic limit somewhat above 1 TbitT/in^2. However, there are other alternatives which the industry might not seriously consider due to the lack of similarity to the conventional technology. This somewhat conservative approach (to avoid "high risk" projects) is often dictated by economical reasons. Despite the fact that the high risk, unconventional technologies often have strong potential to revolutionize the industry, they might not be considered cost effective according to the current standards of the industry. The university environment, on the other hand, may favor high risk technologies because of their scientific values and far-future potential. In this chapter, multilevel (ML) three-dimensional (3D) magnetic recording will be discussed as one of such high risk data storage systems. To comply with the multilevel signal configuration, ML magnetic recording exploits a 3D head/media system powered with next-generation data coding methods. It is believed, when that combined with novel information processing techniques, cost effective ML systems could be scaled down to a single-grain spin level, thus enabling memory with effective areal densities above 100 Tbit/in^2.

In summary, the purpose of this chapter is to diversify from the mainstream magnetic technologies and explore a novel nanoscale system suitable for unprecedented data storage densities and rates.

6.2. MULTILEVEL MAGNETIC RECORDING

6.2.1. Introduction

In this section, an unconventional approach to increase the effective areal densities is discussed. The feasibility of using more than two signal levels to code recorded information is explored. This is in contrast to conventional recording schemes in which binary coding is used methods, meaning that the signal recorded into or read back from magnetic media has only two states: presence or absence of the magnetization reversal in a bit transition. For example, Figure 6.1a and b illustrate so called frequency modulation (FM) encoding form, probably the most trivial encoding scheme, as it could be used in longitudinal and perpendicular recording, respectively. The only difference between the two recording modes is in the orientation of the magnetization, along or perpendicular to the plane of the disk, respectively. In both cases, encoding has a simple one-to-one correspondence between the bit to be encoded and the magnetization reversal pattern.

ML magnetic recording refers to the use of multiple signal values to encode data onto a magnetic disk. By using more than two levels, more information could be put in the minimum feature size. Figure 6.1c illustrates how a multilevel code could be used in a system with a 3D media with a perpendicular orientation of the magnetization. This is a simplified case; in general, the magnetization could be oriented along or at some arbitrary angle to the plane of the disk. As described

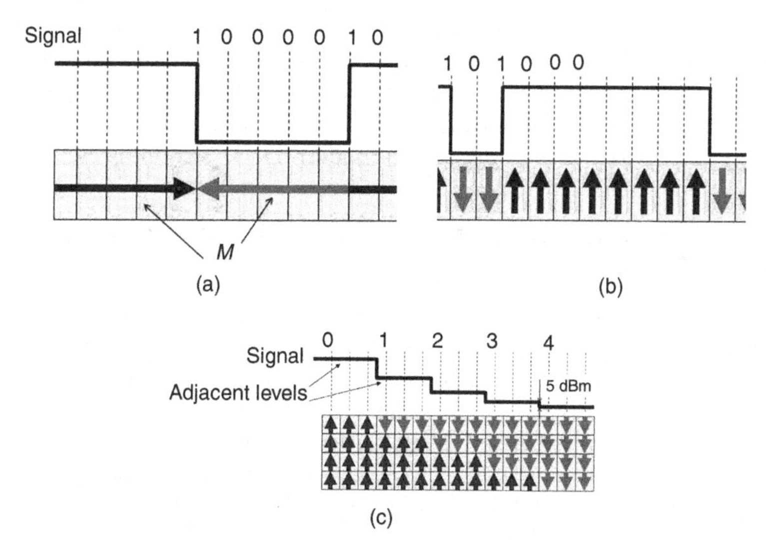

Figure 6.1. A binary signal representation with a trivial FM encoding in (a) longitudinal and (b) perpendicular recording modes. (c) A trivial example of a system with multilevel data encoding.

below in more detail, the main concern is to engineer a magnetic recording system which could maintain signal-to-noise ratio (SNR) between any two adjacent levels sufficient for a data encoding channel to maintain a certain data bit error rate. The novel multilevel data coding methods discussed in this chapter are designed to deal with SNR of approximately 5 dB. Therefore, it is clear that with the goal to develop a practical ML recording system with as many levels as possible, it is necessary to engineer a magnetic recording system with as many signal levels as possible and at least 5 dB SNR between any two adjacent levels. It is expected that the magnetic system discussed in this chapter could use at least twenty "distinguishable" signal levels. Finally, it should be mentioned that the ML approach could be applied ideally to any of the above mentioned alternative technologies to further increase the data capacity.

6.2.2. History of Multilevel Optical Recording

Before exploring the feasibility of ML magnetic recording, it may be helpful to learn from the experience of other industries with regard to ML data coding. For example, in 2004, LSI Logic, one of the electronics giants, acquired Calimetrics, the first company to focus on the development of a multilevel optical recording system for CD or DVD technologies. Though Calimetrics ceased to exist mostly because of the sudden increase in popularity of writable DVD technology, it was successful in introducing a multilevel signal standard to the world of information related technologies, especially overseas. Today, it is becoming normal for new technologies to start to prosper first overseas, as it is often easier to launch new standards because of the lack of a well established traditional technology infrastructure. It could be noted that though the ML technology by Calimetrics offered only eight signal levels and was not rewritable, it still appeared successful, especially for its tremendous impact on modern DVD technologies.

6.2.3. Apparent Differences Between ML Magnetic Versus ML Optical Technology

First, the ML technology developed by Calimetrics competed in the market of DVD/CD technologies, while the analyzed ML magnetic technology, at least at this early stage, targets the market of hard drives. Second, the Calimetrics technology was optics-based, while the currently proposed technology is magnetics-based. Consequently, the magnetic technology would be rewritable, is in contrast with the read-only Calimetrics technology. Finally, the optical technology used only eight signal levels while the ML magnetic technology has the potential for many more levels.

6.2.4. Magnetics of Multilevel 3D Recording

The concept of three-dimensional (3D) magnetic recording was proposed over 10 years ago [14, 15].

6.2.4.1. Background. Traditionally, scaling laws have been followed to advance magnetic data storage industry to a next level/generation. Scaling implies that with each next generation, all dimensions of a recording system have been respectively reduced, as schematically illustrated in Figure 6.2.

Such a straightforward approach (of scaling) has been applied since the inception of the data storage industry more than five decades ago.

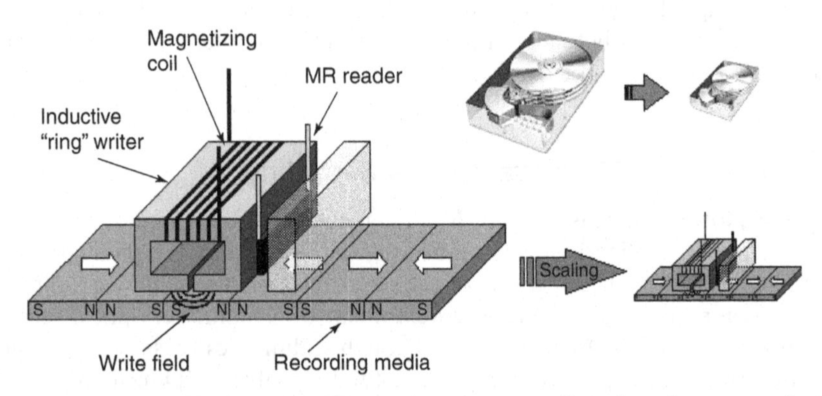

Figure 6.2. Schematic diagram illustrating how scaling has been applied to advance a technology to a next generation for the last five decades.

6.2.4.2. Superparamagnetic Limit. For the first time since the inception of the data storage, the traditional technology cannot be improved further. The reason is the superparamagnetic limit [16]. This fundamental limit is caused by thermal instabilities in the recording media when physical bit dimensions are reduced below certain fundamental values. Each bit in the recording media consists of over 100 grains. Though the grains are magnetostatically coupled, the separation between adjacent grains is large enough (~ 1 nm) to ensure the grains are "exchange" decoupled. The "exchange" coupling is a quantum mechanical coupling that exists at a separation of less than approximately a few nanometers [17]. In this case and, to some degree of approximation, the law of large number would prevail if the number of grains is maintained approximately greater than 100 [18, 19]. In other words, to maintain a certain value for the signal-to-noise ratio (SNR), e.g., 20 dB, in a recording system, it is important to have the number of grains be greater than approximately 100. Therefore, when with each next generation bit dimensions are reduced, each grain dimension also must be reduced to maintain the number of grains and thus SNR value required for adequate signal processing. A transmission electron microscopy (TEM) image of a patch of a longitudinal recording media suitable for areal densities above 100 Gbit/in^2 is shown in Figure 6.3a [20, 21]. The characteristic diameter of an average grain is approximately 6 nm. Figure 6.3b shows a schematic illustrating how a transition between two adjacent bits would be written. Now, according to the scaling law, to further increase areal density by a factor of four, i.e., up to 400 Gbit/in^2, the characteristic diameter of an average grain should be reduced by a factor of two, i.e., it should be reduced to 3 nm. Unfortunately, here also lies a problem.

The problem is the superparamagnetic limit. As the grain size is reduced (with each generation), the anisotropy energy, equal to the product of the grain volume

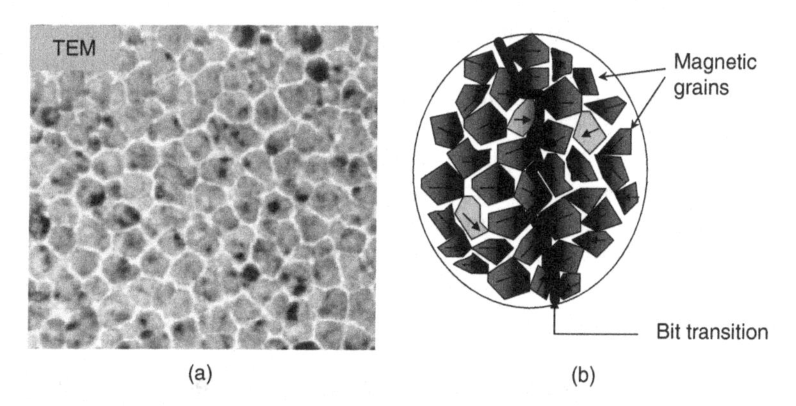

Figure 6.3. (a) A transmission electron microscopy (TEM) image of a micropatch of a typical CoCr-based longitudinal magnetic recording media. (b) A schematic illustrating how a transition between two adjacent bits of information is defined by the granular structure of the recording media.

and the anisotropy energy density, becomes comparable to the energy of thermal activations, i.e., kT (which is approximately 25 meV at room temperature). Professor Stanley Charap and his graduate students at Carnegie Mellon University were the first to use Monte Carlo-based numerical simulations to estimate that the superparamagnetic limit becomes noticeable as the ratio of the anisotropy energy and the thermal energy is reduced (approximately) below the value of 40 [22]. For example, it was found that for the ratio value to increase from 40 to 60, the relaxation time changes from 72 seconds to over 3.6×10^9 years, respectively. In other words, to avoid the superparamagnetic limit, the following condition should be satisfied:

$$\frac{K_u V}{k_B T} > 40 - 60 \tag{6.1}$$

where K_u and V are the anisotropy energy density and the volume of an average grain, respectively. For this condition to be satisfied for typical CoCr-based longitudinal media, the characteristic diameter of an average grain should be greater than approximately 3 nm. Of course, one way to increase the ratio without reducing the average grain size could be to increase the anisotropy energy density. In fact, there are ultra-high anisotropy magnetic materials, such as L_{10} and others, that could be used to achieve the goal [23]. However, there is another fundamental limit that makes it difficult to use these ultra-high anisotropy materials [24, 25]. A recording head is made of a soft magnetic material, and the maximum recording field generated by the head is limited by the saturation magnetization of the material. Unfortunately, the saturation magnetization in practical soft materials, namely the 3D transition metals (Fe, Co, and Ni), is fundamentally limited by approximately 26 kemu/cc, as described by the so called generalized Slater–Pauling curve [26]. The curve shows that the magnetic moment per atom in ferromagnetic alloys of the 3D transition metals could generally be predicated as $m = 9.6 - N_d$, where N_d is the average number of d electrons per atom. This curve works because it is energetically favorable to keep the majority-spin d-band full, independent of alloy composition. Therefore, to further extend the areal density, it is necessary to further reduce the grain size and thus further reduce the ratio in Expression 1; this implies stepping in the superparamagnetic limit. The modern laboratory demonstrations indeed indicate that the information becomes highly unstable as the areal density is increased above approximately 200 Gbit/in^2 [27]. That is why the multibillion data storage industry is currently searching for an alternative technology that could extend scaling and thus areal data density much beyond the current limits.

6.2.4.3. Temporary Solution/Patch in Layman's Terms. So called perpendicular recording promises to defer the fundamental superparamagnetic limit for several more years [28–32]. Ideally, this could bring the areal densities in demonstration data storage systems to one Tbit-per-square-inch. Figure 6.4 illustrates the difference between the conventional technology (longitudinal

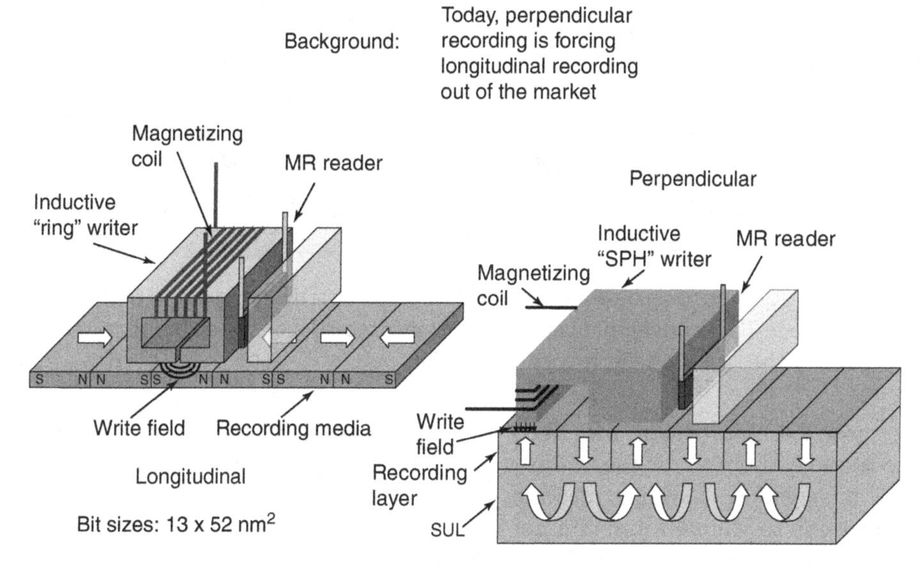

Figure 6.4. A schematic diagram to compare conventional longitudinal magnetic recording and perpendicular magnetic recording systems. In these two cases, magnetization in the recording media are directed along the plane and perpendicular to the plane of the disk, respectively. (Image from *Perpendicular Magnetic Recording* by S. Khizroev and D. Litvinov © 2004 Kluwer Academic Publishers.)

magnetic recording) and perpendicular recording. In the case of perpendicular recording, the magnetization in the disk is polarized perpendicular to the disk. This is in contrast to the conventional system in which the magnetization is polarized along the disk. According to a model in layman's terms, each bit of information is presented as a permanent magnet. Depending on the direction of the magnetic moment inside the magnet, the recorded information is "0" or "1." According to this model, one could pack more permanent magnets per unit surface area in the configuration when the magnet's axis is directed perpendicular to the disk rather than in the plane of the disk.

Unlike many other alternatives to the conventional recording (such as heat-assisted magnetic recording (HAMR) and patterned media), perpendicular recording has been quickly adopted by the industry because of its similarity to the conventional technology [33–36]. In fact, this adoption is happening today. During the last two years, most leading companies in the industry have transitioned to perpendicular recording. Such a drastic change is quite unprecedented for the industry that has been following one technology for over five decades now!

However, as mentioned earlier, perpendicular recording in its current form is only a temporary solution/patch. The solution it offers is more quantitative than qualitative and thus the expected progress is quite incremental (for only the next five years or so).

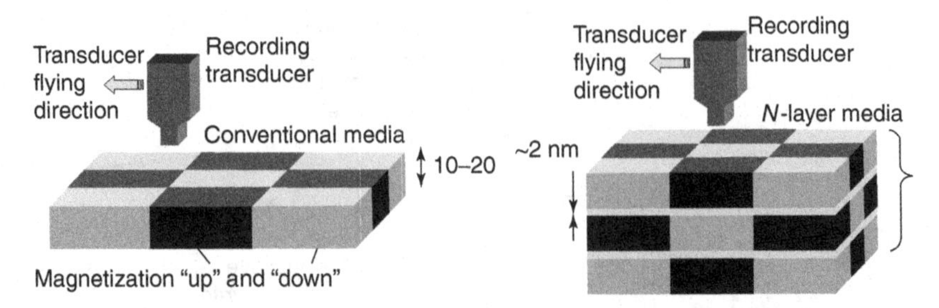

Figure 6.5. Comparison between a conventional recording system with recording on a surface (left) and a proposed 3D or/and multilevel system with recording across the thickness of the recording media.

6.2.4.4. Multilevel 3D Recording in Layman's Terms. The proposed technology differs from the traditional approach because it deals with one surface (of a magnetic recording media) only. It is proposed to use a third vertical dimension to store information not only on the surface but also in the volume of the recording media. In other words, the information is packed across the thickness, as shown in Figure 6.5. As a result, more than one signal level would be exploited to record, store, and retrieve information from the same unit surface area of the recording media. This opens a new window of opportunities to develop not only the most superior near-future data storage device (multilevel 3D (m3d) drive) but also for a long future to exploit groundbreaking advantages of an entirely new dimension.

6.2.4.5. Advantages

a. Simplicity

One of the advantages of this technology, at least at this early stage of development, is its relative simplicity. This is due to the fact that the proposed multilevel 3D magnetic recording has similarity with perpendicular recording. Similar to perpendicular recording, the information, at least at the initial stage, will be recorded perpendicular to the disk. However, other media configurations should not be disregarded. For example, longitudinally oriented media or media with the magnetization tilted at some arbitrary angle may have some advantages perpendicular media do not have. As the complexity of multilevel 3D recording system increases with each next generation, it might become beneficial to switch to one of the nonperpendicular modes.

b. Technological Superiority

As mentioned above, this is the only magnetic technology that relies on the use of a third dimension to record, retrieve, and store information in a device. In other words, the information will be recorded not only on a surface but also across the thickness or in the volume of the recording media. As a result, substantially more information could be recorded per unit area and thus the fundamental data density limit could be extended substantially further. This means that this technology has strong potential to revive the current, relatively slow progress in the multibillion-dollar data storage industry.

c. Long-term Potential

For the reason mentioned above (exploiting not only a surface but the volume of a device), the technology has potential to be extended much further into the future compared to any other alternative data storage technology.

d. Multilevel Signal

The exploitation of multilevel signals will substantially loosen fabrication requirement. In other words, unlike conventional magnetic and silicon technologies, this effort will not be limited by the many problems that arise during fabrication of sub-100-nm devices. In this context, multilevel implies the ability to record more information per unit surface area.

As mentioned above, previous research has identified the multilevel mode as a likely form of 3D recording to impact the future world of memory and storage applications [37–39]. In this mode, a varying recording field was used to sequentially record across the media thickness, as illustrated in Figure 6.6a. The research effort has triggered industry-wide interest in the new topic [40]. The current study is aimed at bringing the research to the next level. The chapter studies the feasibility of multilevel magnetic recording and attempts to consider the physical limitations of the technology designed to function at densities above 10 Tbit/in^2.

Previously, the authors developed 3D media similar to the popular perpendicular media configuration (with the magnetization perpendicular to the plane of the disk) [41]. A recording system with a single pole head and a recording media with a soft underlayer (SUL) (Fig. 6.6b [42]) was used to generate adequate perpendicular field. SUL was used to force the magnetic flux to flow in the perpendicular direction. Typical FIB-modified write head with a 80-nm trackwidth and a magnetic force microscopy (MFM) nanoprobe used to write and read information from 3D media, respectively, are shown in Figure 6.6c and d. The write head was in the form of a single pole [43], while the MFM nanoprobe was designed to read a certain component of the magnetization [44, 45].

A schematic illustrating the concept of 3D recording is shown in Figure 6.7a. For simplicity, only two magnetic layers across the thickness are discussed. The layers can be deposited via regular sputtering systems. Continuous and patterned versions of 3D media could be used. For illustration purposes and prototype development, a focused ion beam (FIB) was used to further pattern 3D media within the plane of the disk, as illustrated in Figure 6.7a. A straightforward MFM experiment can be performed to demonstrate the presence of more than one signal level. In this example, a CoCrPt-based 2-layer 3D media was sputter-deposited and then patterned via FIB into a square periodic arrays with a linear period of 80 nm [46]. Then, the media was demagnetized using an alternating external magnetic field. AFM (left) and MFM images of the surface of the fabricated

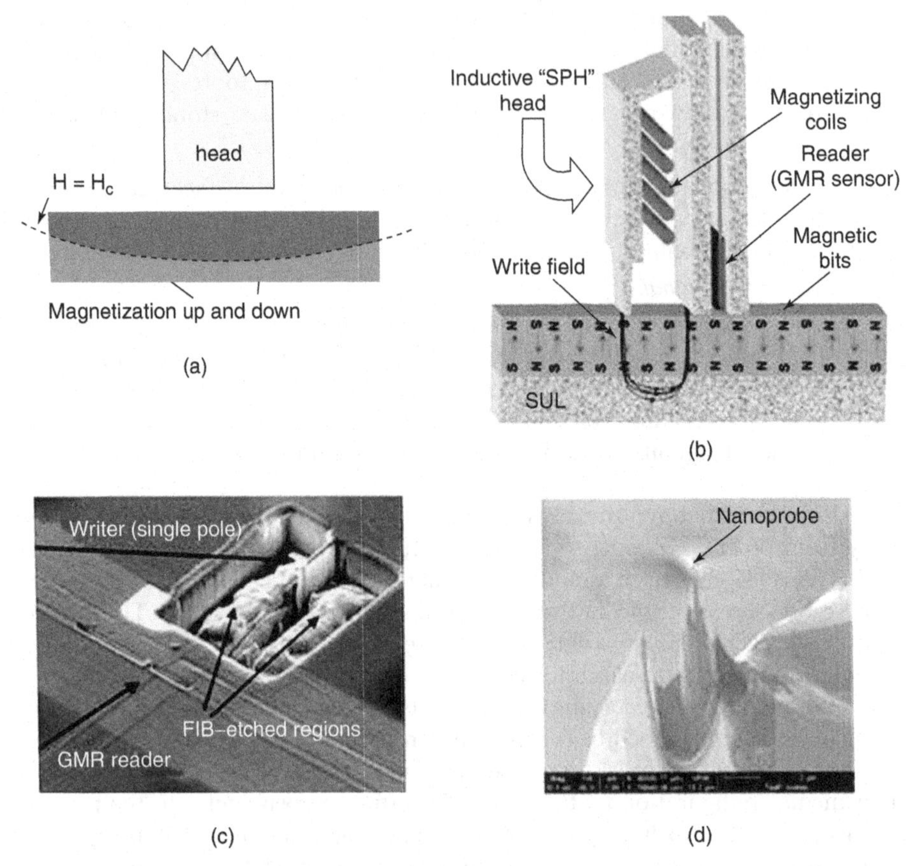

Figure 6.6. (a) Schematic illustrating recording across the thickness via continuous variation of the recording field. The boundary curve is the field profile when $H = H_c$, where H_c is the coercivity. (b) A diagram of a perpendicular recording system. FIB images of FIB-fabricated (a) single pole writer with a 80-nm trackwidth and (b) a MFM nanoprobe for reading a component of the stray magnetic field.

media are shown in Figure 6.7b. A digitized sectional profile of the signal along the highlighted line going through a track of bits is shown above the AFM and MFM images. Clearly, over three different signal levels could be detected, as shown in Figure 6.7b (top).

This experiment illustrates that the 3D media could be partially polarized across the thickness, resulting in a multilevel signal configuration. On the contrary, it is known that in conventional recording the magnetization remains in one of the two saturated states and no partial polarization is feasible. To further perfect the concept of generating a multilevel signal, various types of media have to be developed and explored. However, media development is not the main subject of this chapter. Instead, the focus is placed on the system integration.

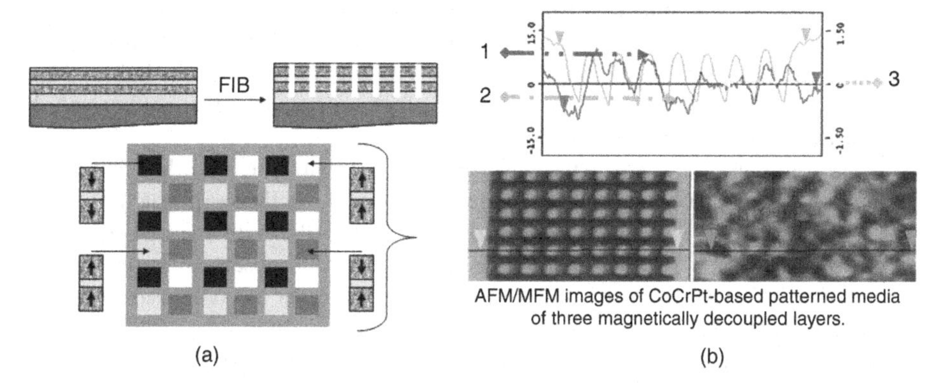

Figure 6.7. (a) A schematic illustrating a two-layer 3D media. (b) A MFM image of a patch on the surface of FIB-fabricated CoCrPt-based two-layer 3D media. A digitized sectional profile of the signal along the highlighted line is shown above the AFM/MFM images.

6.2.5. Design of a New Multilevel Magnetic System

The general task is maximizing the number of signal levels while maintaining SNR between the signal values in each pair of adjacent levels above approximately 5 dB. The 5 dB value is dictated by the requirements of the novel data encoding methods, as described below.

The magnetic system probably would be further modified and optimized with the goal to maximize the number of "distinguishable" signal levels.

6.2.5.1. Modification of a Magnetic System. As mentioned earlier, it is assumed that the material science underlying the new media type has been studied in the prior related work. To formulate the approach in one expression, it is necessary to develop media in which each magnetic grain could be recorded independently (without a chain clustering effect). In fact, a conventional CoCr-based composition (as used in conventional longitudinal recording media) could be used for this purpose, and the media could be made even thicker to fully maximize the effective number of individual Co grains participating in the process. The focus of this chapter is directed towards engineering a new system to address the above described requirements on multilevel recording. In other words, the chapter is aimed at designing a system with substantially increased recording and sensitivity fields (used for writing and reading information, respectively), thus enabling adequate SNR to accommodate as many signal levels as possible, with 5 dB difference between any two adjacent levels. The data encoding methods proposed in the chapter are aimed at providing sufficient bit error rates for maintaining SNR of the order of 5 dB.

6.2.5.2. Patterned Soft Underlayers and Interlayers. The main motivation underlying the system modification described below is to substantially increase the maximum recording field, H_{max}, so that the range of the field, $-H_{max}$ to $+H_{max}$, controlled via the variation of the drive current could also be substantially increased.

Traditionally, SUL has a flat surface boundary with the recording layer. The image model could be used to describe the effect of SUL on the recording field within the recording layer (Fig. 6.8a) [47, 48]. Ideally a factor-of-two increase in the field is expected, although from the same image model one could observe that the image is located further away from the center of the recording layer as compared to the real head. The difference between the separations of the real and image heads from the center of the recording layer (the spacing loss) is ~ equal to the thickness of the recording layer. The closer to the recording media the head is placed, the stronger and more localized the recording field would be in the media. Because the image head is located further away compared to the real head, the net effect will result in a deteriorated signal compared to the signal, due to two equally separated heads. The reciprocity principle states that the detrimental effect exists for both write and read processes [49]. The following concept is proposed to resolve the spacing loss issue.

One could recall that magnetic imaging is very much like regular mirror imaging [50]. Therefore, exactly as in the case of a convex mirror, one could use a SUL with a convex boundary to move the image closer to the "mirror's" (SUL's) boundary with the recording layer [51]. To illustrate this effect, a diagram comparing the positions of the images in the two cases is shown in Figure 6.8b. Using a convex SUL instead of a flat SUL could make the image head "be closer" to the recording layer and thus increase the areal density. Based on the above described concept, it is proposed to use a patterned SUL with each patterned island having a convex shape. Patterning is necessary to implement the novel concept on the physical scale of one bit. The cross-sectional dimensions of each patterned island correspond to the cross-sectional dimensions of the targeted bit cell. For example, for an areal density of 1 $Tbit/in^2$, assuming a square symmetry in the plane, each of the (x and y) period values T should be approximately 26 nm. Via numerical simulations, it has been discovered that this favorable effect exists not only for patterns with convex islands but also for patterns with rectangular islands (Fig. 2.8c) [52]. This is attributed to the combined effect of patterning and periodicity.

Both continuous (conventional) and patterned recording layers could be integrated with a patterned SUL, as shown in Figure 6.9a and b, respectively. Patterning in the latter case combines the described below advantages of patterned SULs with the well known advantages of patterned recording layers [53].

The key practical advantages due to the use of a convex patterned SUL are the following and described below in detail:

- SNR drastically increases, as demonstrated below.
- Patterning of SUL increases both the recording and sensitivity field gradients. The latter is critical for maximizing the areal density during writing and reading, respectively.

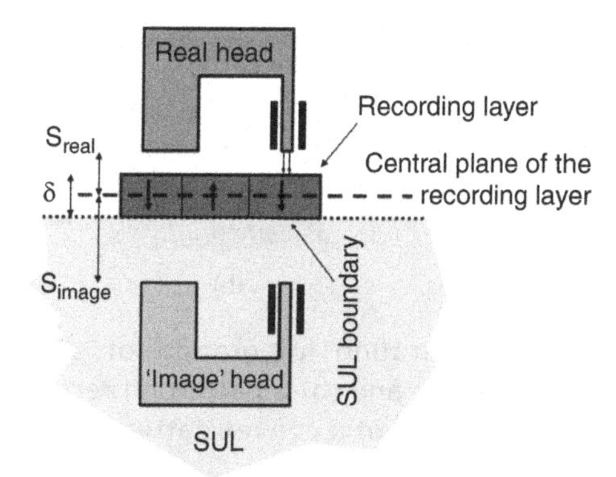

(a)

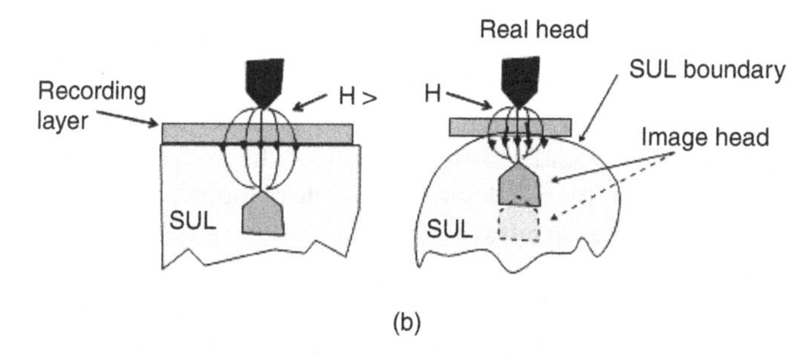

(b)

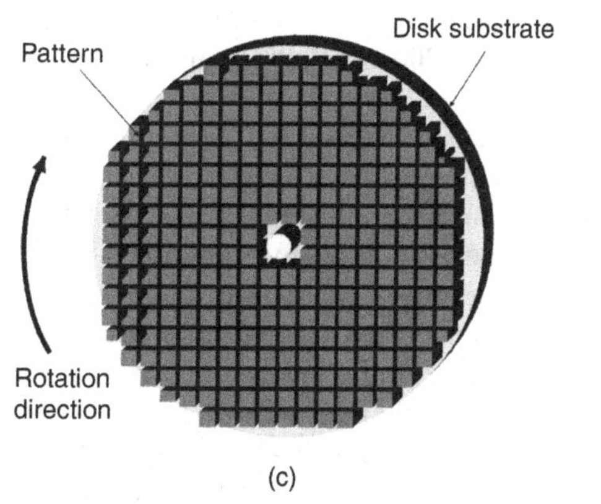

(c)

Figure 6.8. (a) A diagram of an image model to illustrate the spacing loss due to the offset in the separation between the real and image heads with respect to the central plane of the recording layer. (b) A schematic illustrating the "move" of the image closer to the recording layer if a convex SUL boundary is used. (c) Patterned SUL with rectangular islands.

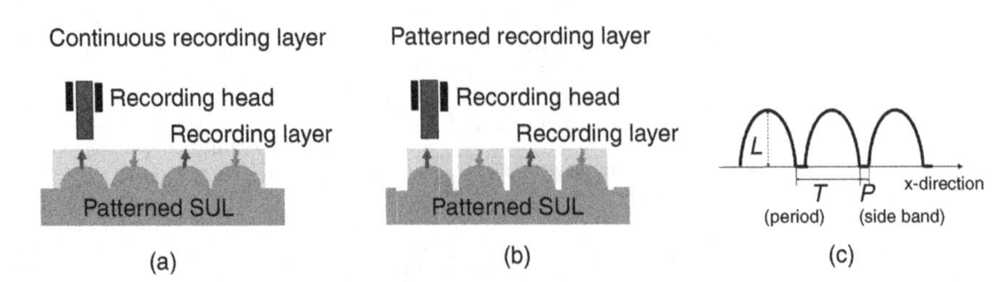

Figure 6.9. Drawings illustrating integration of a patterned SUL with (a) a continuous recording layer and (b) a patterned recording layer. (c) A schematic diagram of the cross-section of a convex patterned SUL in X-direction.

- The recording and sensitivity fields remain well localized across the entire thickness to generate a multitude of signal levels.

The period of the grid T in each direction corresponds to the bit cell dimension in this direction and thus defines the effective areal density D as $1/T^2$ (bit/in^2) (Fig. 6.9c). Below, it is shown that at a given grid period, the thickness of the island, L, could relatively and sensitively control the advantageous effects of the patterned SUL. To simplify the description, below the ratio of the grid period to the island thickness L/T is arbitrarily defined as the pattern "curvature."

Increased SNR: Figure 6.10a illustrates the simulated perpendicular recording field versus the distance along the line from the top surface of the patterned SUL to the ABS of the head (from point s to point e in Fig. 6.10b) at saturation for three "curvature" values of the patterned SUL. The three sets of three parameters, T, P, and L, are 1) 98, 2, and 200 nm, 2) 98, 2, and 100 nm, and 3) 98, 2, and 25 nm, for curvature values of 1, 2, and 4, respectively. Curvature values 1 and 4 correspond to the sharpest and flattest surfaces of the island, respectively. In all these cases, the head is centered with respect to an island in the patterned SUL, as shown in Figure 6.10b. For comparison, the solid black line in the figure shows an equivalent field line for the case of a conventional flat-surface continuous SUL. A pronounced increase of the recording field near the top SUL boundary could be observed, especially for a curvature of 1. While in the case of a conventionally flat SUL, the field reaches its maximum at the air bearing surface of the head, in the case of a patterned SUL, the field reaches its maximum in the vicinity of SUL. The recording field at SUL exceeds 30000 Oe, while at the air bearing surface of the head it barely reaches 15000 Oe. For the conventionally flat SUL, the field is less than approximately 12500 Oe when it reaches the surface of SUL. Such an increase (by a more than a factor of two) of the recording field in the region of the recording layer means that a recording media with substantially higher anisotropy could be used and thus substantially higher recording densities could be achieved. The fact that the field reaches its maximum at the side of the recording layer farthest away from the ABS is especially favorable for 3D recording. As described above, the most trivial form of recording across the thickness of 3D media could

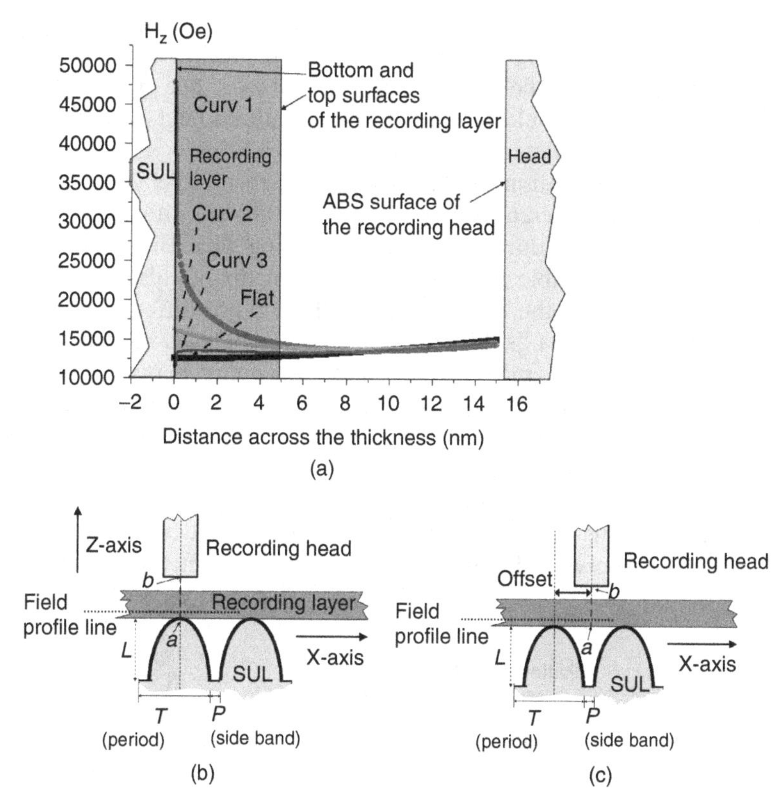

Figure 6.10. (a) The perpendicular recording field versus the distance along the line from the top surface of the SUL to the air bearing surface of the head for three different values of the curvature of the semispherical island in the SUL type under study. For comparison, an equivalent dependence for a regular flat SUL is also shown. Diagrams showing the location of the recording head with respect to an island in the convex patterned SUL (b) with the head and the island perfectly centered and (c) with a half-period offset line between the head and the island.

be achieved via sequential recording from the bottom up via continuous variation of the current in the drive coil. Patterning of SUL provides another knob to control the recording field. As seen in Figure 6.10a, at a given value of the current, the field maximum could be shifted from the bottom to the top side of the recording layer via variation of the softness of SUL. Moreover, the investigators propose to use soft interlayers (SIL) to separate magnetic layers across the thickness. As illustrated below, the use of SILs is expected to facilitate recording across the thickness substantially, thus further increasing the number of distinguishable signal levels.

Profiles of the simulated recording field along a track line in the recording layer 2.5 nm away from the top surface of SUL (Figure 6.10b, c) for a patterned

SUL and a conventional (flat) SUL are shown in Figure 6.11a. Although for the patterned SUL the recording field drops with the distance away from SUL, it is still larger compared to the field in the conventional case. The graphs also indicate that patterning of SUL would result in a more localized field. The latter effect is discussed below in more detail.

Study of the Localization of Recording and Sensitivity Field: Patterning of SUL is fundamentally different from patterning of the recording layer with respect to addressing information during writing and reading. Here, it could be reminded that SUL is an indispensable part not only of the media but also of the head [54]. The latter is not true for the recording layer. This major difference is due to the fact that SUL is made of a "soft" magnetic material. This is in contrast with the recording layer, which is made of a "hard" magnetic material. As a result, patterning of SUL effectively also "patterns" the field generated by the head. This field is the recording magnetic field or the sensitivity field depending on the writing or reading processes, respectively.

To illustrate "patterning" of the recording field because of patterning of SUL, two different locations of the recording head with respect to the patterned SUL are considered, as shown in Figure 6.10b and 6.10c, respectively. In the first case,

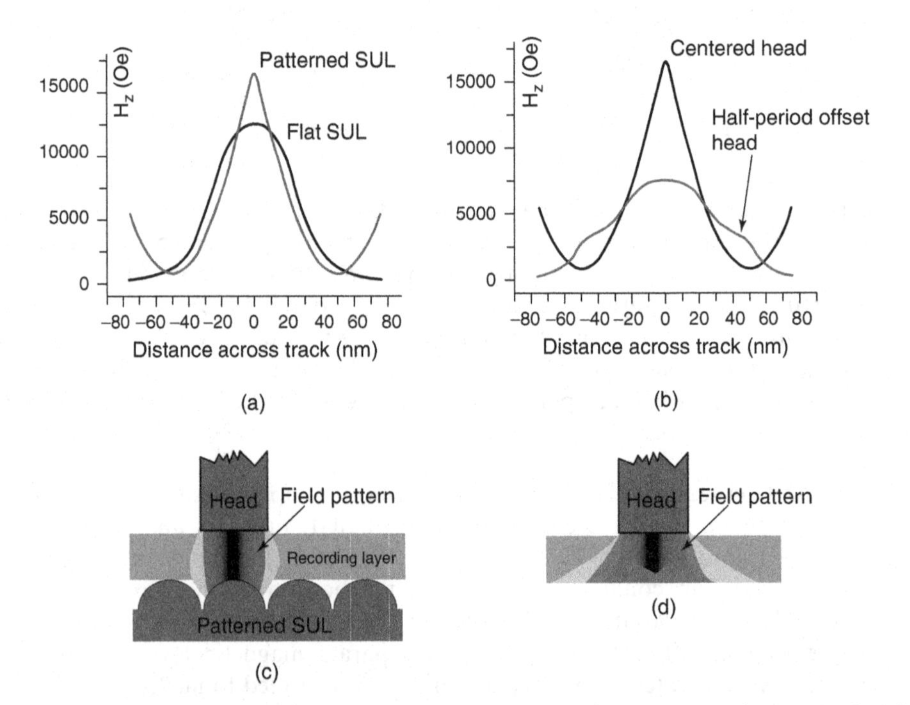

Figure 6.11. (a) Field profiles inside the recording layer for the cases of patterned and conventional SUL. (b) The field profiles along same line, 5 nm away from the surface of SUL, along a patterned direction along or across the track. The recording field profiles for two systems, (c) with a patterned SUL and (d) without a SUL, respectively.

the recording head is centered with respect to an island of the patterned SUL. In the second case, the recording head is centered with respect to a groove between two adjacent islands. In other words, in the second case, the head is a half-period offset with respect to an island. The respective field profiles along same line, 5 nm away from the surface of SUL, along a patterned direction are shown in Figure 6.11b. The fact that the field profiles are so different from each other clearly indicates the above mentioned process of "patterning" the recording field. One can observe the field in the centered case is larger (by more than a factor of two) and substantially more localized compared to the field in the off-centered case. This means that the head itself is going to substantially improve its recording quality when it is recording information in the favorite positions (compared to any off-centered positions). This is in contrast to the conventional recording process in which the recording field profile does not depend on the location of the head with respect to the media, regardless of whether the media being patterned or not.

Improved Localization of Recording and Sensitivity Field across the Thickness: Furthermore, the effective field patterning, as described above, takes place across the entire thickness of the recording media. In other words, the recording and sensitivity fields could be strongly localized not only at the air bearing surface of the head but also across the entire thickness of the recording layer. Figure 6.11c and d illustrate the recording field profiles for the two cases of a recording system with a patterned soft underlayer and without a SUL, respectively. In the latter case, one could notice the great divergence of the field away from the air bearing surface of the head. As described below, this concept can be extended to further increase the effective number of active layers by introducing so called soft interlayers (SIL) between adjacent layers.

How Patterned SUL Would Help Read Across the Media Thickness: As a simulation input, two types of information were prerecorded into the top and bottom layers of 3D recording media with a net thickness of 50 nm (Figure 6.12a). The "softness" of a patterned SUL was controlled via continous variation of SUL's bias current, as defined above. The respective signal profiles, read back at biasing current values of 5.85 and 1.56 A-turn, respectively, (Figure 6.12b) indeed

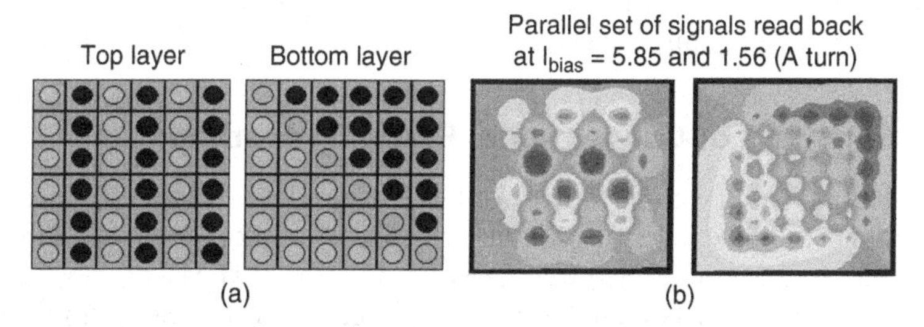

Figure 6.12. **(a)** Prerecorded magnetization in the top and bottom layers of a 20-layer recording media. **(b)** Sets of signals read back at a SUL's biasing current of 5.85 and 1.56 A-turn.

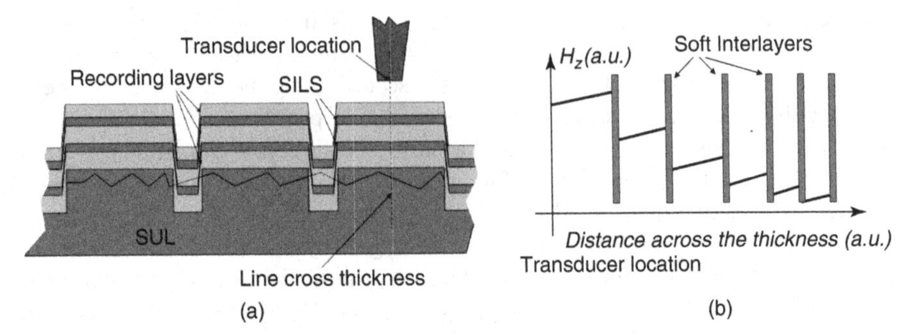

Figure 6.13. (a) Diagram and (b) simulations to illustrate how the field could be "discontinued" in the region of SILs. (c) Diagrams of various bit configurations to illustrate the effect of the stray field on the intersymbol interference.

strongly remind the prerecorded patterns. One notices here that various deconvolution based signal encoding techniques could be implemented to further optimize SNR of a multilevel system [55].

How Soft Interlayers Could Be Used to Further Separate Individual Layers across Thickness: The purpose of SILs is to create field "discontinuities" to facilitate information access not only in the top and bottom layers but also in the intermediate layers. Each SIL is patterned with a pattern reflecting the pattern of SUL (Figure 6.13a). Numerical simulations showed that a thickness (of SIL) of smaller than 2 nm is sufficient to create a detectable "discontinuity"; at the same time it is not too thick to totally shield the field from penetrating across the thickness. A typical field profile across the thickness of the media with adjacent magnetic layers separated from each other with a 2-nm thick SIL is shown in Figure 6.13b. The thickness of the magnetic layers and SILs could be optimized to maximize the effective number of layers.

6.3. MULTILEVEL DATA ENCODING

Before going into details of the new multilevel coding method, it may help to recall how conventional coding works.

6.3.1. Summary of Conventional Encoding Methods

FM encoding has a simple one-to-one correspondence between the bit to be encoded and the flux reversal pattern. One needs only to know the value of the current bit. So called modified frequency modulation (MFM) improves encoding efficiency over FM by controlling more intelligently where clock transitions are added into the data stream; this is enabled by considering not just the current bit but also the one before it. This "looking backwards" allows improved efficiency by letting the controller consider more data in deciding when to add clock reversals. Run limited length (RLL) encoding takes this technique one step further. It considers groups of several bits instead of encoding one bit at a time. The idea is to mix clock and data flux reversals to allow for even denser packing of encoded data and thus improve efficiency. The two parameters that define RLL are the *run length* and the *run limit* (and hence the name). The word "run" here refers to a sequence of spaces in the output data stream without flux reversals. The run length is the *minimum* spacing between flux reversals, and the run limit is the *maximum* spacing between them. As mentioned before, the amount of time between reversals cannot be too large or the read head can get out of sync and lose track of which bit is where. Finally, the so called partial-response maximum likelihood (PRML) sequence detection is even more advanced and is today's most common method. Partial-response means controlled inter-symbol interference (ISI). That is, the data represented by the received waveform are packed so closely together that they overlap (interfere). The "controlled" part means that there is some identifiable structure to the overlapping. This structure is "taught" to a sophisticated sequence detector that looks only for the possible controlled patterns of ISI in the received waveform. This provides a much more robust detection method in the presence of noise. This type of detection is often used with trellis coding to further improve detection. Trellis coding encodes the data such that the possible received sequences that are most similar are not used. This provides extra capability to distinguish between received waveforms in the presence of noise and distortion.

6.3.2. New Error-Correcting Techniques for Multilevel Magnetic Recording

In the previous sections, it is shown that multilevel recording can substantially increase the capacity of storage systems versus its conventional two-level counterpart. On the other hand, multilevel recording reduces the separation between different amplitude levels (given the same power constraints), which can make the entire system more error-prone as more data becomes incorrectly recorded or retrieved in the "read–write" channel. To address this problem, in this chapter, we intend to employ powerful error-correcting codes. In general, efficient error correction can reduce the overall error rates by a few orders of magnitude, and this can also lead to higher recording densities in the entire storage system. Thus, error correction becomes an efficient tool that can greatly reduce the errors caused by multilevel recording and higher densities.

The main overhead caused by many error-correcting algorithms is their excessive complexity or—equivalently—prohibitively slow data processing. Therefore, our main goal in this part of this chapter is to design the error-correcting techniques that combine powerful error-correcting performance with fast, feasible processing. Namely, we intend to design the algorithms that approach optimum, maximum likelihood (ML) decoding without significant degradation of the data transmission rate in the overall system.

The existing methods of code design can be divided into three different categories. In the first group of graph-based techniques, a linear code is associated with some bipartite graph, with code symbols on the one side and their parity checks on the other. Iterative decoding is then used similar to belief propagation in neural networks. This method combines relatively simple processing and powerful performance; however, it becomes efficient only on the relatively long blocks of thousand bits. The lack of analytical tools required to estimate low output error rates (of 10^{-12} and less) is another drawback of iterative techniques.

The second group employs algebraic methods. Here a code consists of a set of polynomials of some limited degree taken over a Galois field; a string of all values of a specific polynomial is called a codeword. This technique is especially efficient for large (nonbinary) alphabets and leads to the optimum Reed–Solomon (RS) codes, which are currently designed into most storage systems. However, RS codes have relatively high complexity in correcting multiple errors. This excessive complexity combined with long operations in Galois fields can slow down the overall performance, especially in multilevel signaling, which requires multiple error correction.

Finally, the third group consists of various concatenated techniques, which combine good short codes into the longer ones. Invented by Forney over 40 years ago, this technique still manifests itself as one of the most efficient, thanks to simple decoding procedures of multiple error correction. To further reduce decoding complexity, we intend to design new concatenated techniques using simple constituent codes.

6.3.3. Reed–Muller Codes and Their Recursive Decoding

The technique that will be used in the project has some similarities with general concatenated design. The main new feature will employ constituent codes that can be considered as "self-contained" recursive constructions. Such constructions—already known in coding theory—preserve all the properties of the shorter, building codes. These recursive techniques result in a simple, low complexity decoding, which repeatedly employs the same code properties multiple times on increasing lengths.

Reed–Muller (RM) codes represent the most notable example of recursive constructions and have been explored for five decades. These codes also yield a simple recursive decoding, which has been recently designed and analyzed in papers [56–59]. Namely, any RM code (m, r) of length $n=2^m$ and distance $d=2^{m-r}$ can be obtained from the two RM codes $(m-1, r-1)$ and $(m-1, r)$ 1 of length $n/2$. The decoding process can be relegated further using four RM codes of length $n/4$. After a few steps, the decoder finally arrives at the trivial end codes RM$(m-r, 0)$ or RM(r, r), which are either fully redundant (repetition codes) or nonredundant (full codes). For both types, it is easy to use powerful ML decoding. On the other hand, the received channel symbols are only recalculated in all intermediate recursive steps without any decoding. Due to this, decoding

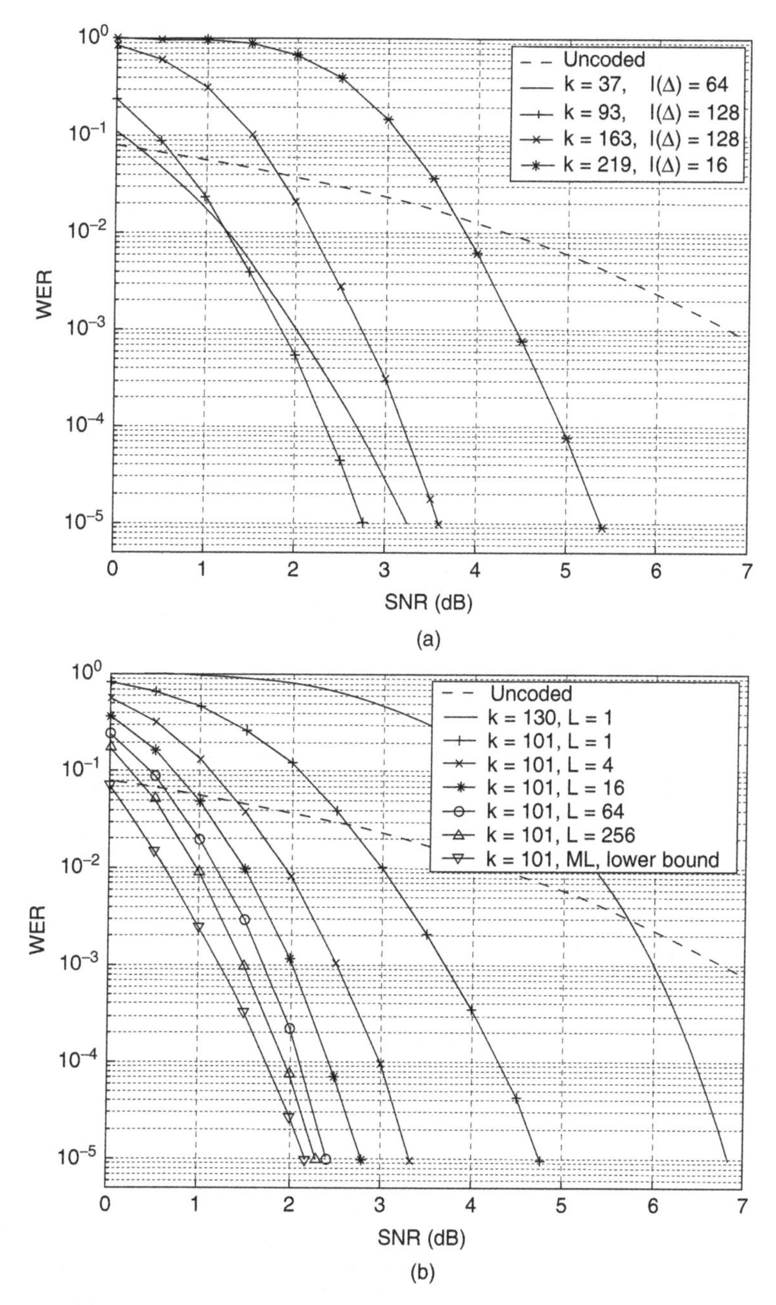

Figure 6.14. (a) Exact bounds on word error rates (WER) for ML decoding of RM codes of orders 2 to 5 on the length 256. The legend includes the list sizes $l(\Delta)$, for which recursive permutation algorithm performs within $\Delta = 0.25$ dB from ML decoding. (b) Word error rates (WER) for the (512, 101)-subcode of the (512, 130) RM code of order 3. Different curves represent WER for recursive list decoding for varying lists L.

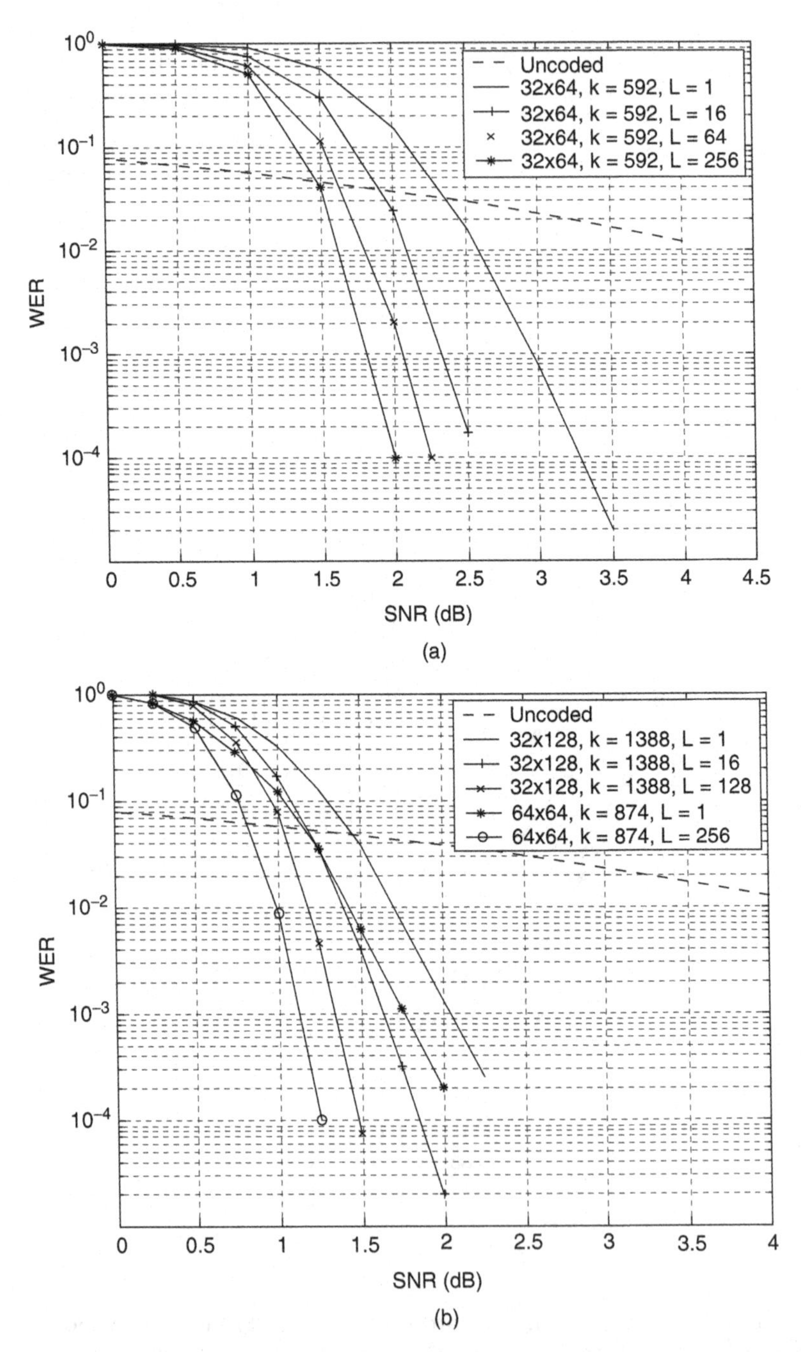

Figure 6.15. (a) Concatenated (n, k) code of length $n = n1 \times n2 = 2048$, with variable lists of size L. (b) Concatenated (n, k) codes of length $n = n1 \times n2 = 4096$, with variable lists of size L.

complexity has a low order of $n \log n$ or even a linear order n if incomplete hard decision decoding is concerned.

This combination of powerful end decoding and fast intermediate processing outperforms all other nonexponential algorithms known for RM codes. In particular, it turns out that the algorithms of [56–58] substantially outperform the two conventional techniques of bounded distance decoding and majority decoding. Namely, new recursive techniques

- increase $\ln d$ times the number of errors $d/2$ corrected by bounded-distance decoding (for most error patterns);
- double the number of errors $(d \ln d)/4$ corrected (with high probability) by conventional majority decoding; for the moderate lengths of 128 to 512, these techniques were enhanced in [4] with list decoding, which
- gives near-optimum performance (within 0.25 dB of ML decoding) with feasible complexity;
- outperforms all other decoding algorithms currently known for all existing codes.

Some results are given below in Figure 6.14a and b for different RM codes of lengths 256 and 512, respectively. An important conclusion is that RM codes yield near-optimum performance and fast processing with complexity of order n log n. Then in Figure 6.15a and b, also the results for longer blocks obtained by concatenating different RM codes are plotted. In summary, these data illustrate that concatenated RM codes combine fast feasible processing and very low SNR of 2 to 3 dB on the Gaussian channels for various block lengths of 256 to 4000 bits. These algorithms could be further enhanced in the future to correct specific (non-Gaussian) error models that pertain to magnetic recording channels.

6.4. CONCLUSIONS

This chapter outlines multidisciplinary research to develop a multilevel mode of next-generation magnetic recording, a promising and challenging solution to increasing the capacity of electronic and computer devices. The study is conducted from the system perspective. It is anticipated that through marrying the expertise in magnetic recording and state-of-the-art data encoding, the proposed multilevel magnetic recording approach could be applied to any alternative magnetic data storage technology to further increase its data capacity by several factors. The key to the multilevel system is exploitation of the recently proposed concept of three-dimensional media optionally with patterned soft magnetic underlayer and interlayers. Among the unique properties are the substantially increased (compared with conventional technologies) recording and sensitivity fields used for writing and reading information, respectively. This field increase results in adequate signal to noise ratio (SNR) to accommodate over 40 signal levels with 5 dB difference between any two adjacent levels. The proposed data encoding methods are aimed at providing sufficient bit error rates for a system with a SNR of the order of 5 dB.

ACKNOWLEDGMENTS

The authors would like to aknowledge the insighful discussions with Dr. Michael Alex of Hitachi Research and the continous support by the research groups led by Song Xue of Seagate Technology, Steve Lambert of Hitachi Global Storage Technologies, and Robert Haddon of Center for Nanoscience Innovation for Defence (CNID), University of California, Riverside. This work is partially supported by four grants from the National Science Foundation (NSF): ECS-0401297, ECS-0508218, ECS-0404308, and IIP-0712445, respectively; two grants from the U.S. Air Force Office of Scientific Research (AFOSR): FA9550-04-1-0446 and FA9550-05-1-0232, respectively; and one grant co-sponsored by Defence MicroElectronics Activity (DMEA) and Defence Advanced Research Project Agency (DARPA): DOD/DMEA-H94003-06-2-0608.

REFERENCES

1. D. A. Thompson and J. S. Best. The Extendibility of Magnetic Recording for Data Storage. IBM executive briefing: Jan 1998.
2. M. H. Kryder. Data storage technologies for advanced computing. *Scientific American*, 117 p 72, 1987.
3. S. H. Charap, P. L. Lu, and Y. He. Thermal stability of recorded information at high densities. *IEEE Transactions on Magnetics*, 33(1): pp 978–983, 1997.
4. N. H. Bertram and M. Williams. SNR and density limit estimates: a comparison of longitudinal and perpendicular recording. *IEEE Transactions on Magnetics*, 36(1): p 4, 1999.
5. S. Iwasaki and Y. Nakamura. An analysis for the magnetization mode for high density magnetic recording. *IEEE Transactions on Magnetics*, 13 p 1272, 1977.
6. S. Khizroev, M. H. Kryder, Y. Ikeda, K. Rubin, P. Arnett, M. Best, and D. A. Thompson. Recording heads with trackwidths suitable for 100 Gbit/in2 density. *IEEE Transactions on Magnetics*, 35(5): pp 2544–2546, 1999.
7. M. Albrecht, C. T. Rettner, A. Moser, M. E. Best, and B. D. Terris. Recording performance of high-density patterned perpendicular media. *Applied Physics Letters*, 81(15): pp 2875–2877, 2002.
8. C. Chappert, H. Bernas, J. Ferré, V. Kottler, J. P. Jamet, Y. Chen, E. Cambril, T. Devolder, F. Rousseaux, V. Mathet, and H. Launois. *Science*, 280 p 1919, 1998.
9. D. Weller and A. Moser, *IEEE Transactions on Magnetics*, 35 p 4423, 1999.
10. H. F. Hamann, Y. C. Martin, and H. K. Wickramasighe. Thermally assisted recording beyond traditional limits. *Applied Physics Letters*, 84(5): p 810, 2004.
11. P. C. Arnett, D. C. Cheng, S. Khizroev, and D. A. Thompson, inventors; IBM Almaden Research Center, assignee. Track width control of readback elements via ion

implantation in a bonding region of tip portion to selectively deactivate magnetic sensitivity thereof's. US patent 6,483,672. Nov 19, 2002.

12. S. Khizroev, B. W. Crue, and D. Litvinov, inventors. Seagate Technology, assignee. Method for forming a perpendicular recording read/write head. U.S. patent 6,513,228. Feb 4 2003.
13. S. Khizroev, W. Crue, and D. Litvinov, inventors. Seagate Technology, assignee. Perpendicular magnetic recording head with write pole which reduces flux antenna effect. US patent 6,646,827. Nov 11, 2003.
14. S. Khizroev, Y. Hijazi, N. Amos, R. Chomko, and D. Litvinov. Considerations in the design of three-dimensional and multi-level magnetic recording. *Journal of Applied Physics*, 100: p 63907, 2006.
15. S. Khizroev, inventor. Three-dimensional magnetic memory and/or recording device. US patent application 20060028766, filed Feb 9, 2006; 11/197,377, filed Aug 4, 2005, with provisional patent; 60/598,645, filed Aug 4, 2004, disclosure filed Feb 21, 2003.
16. S. H. Charap, P. L. Lu, and Y. He. Thermal stability of recorded information at high densities. *IEEE Transactions on Magnetics*, 33(1): pp 978–983, 1997.
17. Robert M. White. *Quantum Theory of Magnetism*, 3rd ed. Springer, 2006.
18. J. Bernoulli. *Ars Conjectandi: Usum et Applicationem Praecedentis Doctrinae in Civilibus, Moralibus et Oeconomicis*. 1713, Chapter 4. (Translated into English by Oscar Sheynin.)
19. G. R. Grimmett and D. R. Stirzaker. *Probability and Random Processes*, 2nd ed, Oxford: Clarendon Press, 1992.
20. G. D Danilatos and G. C Lewis. Integrated electron optical/differential pumping/ imaging signal detection system for an environmental scanning electron microscope. US 4,823,006 (PDF version) (1989-4-18).
21. G. D. Danilatos. Foundations of environmental scanning electron microscopy. *Advances in Electronics and Electron Physics*, 71 pp 109–250, 1988.
22. P. L. Lu and S. H. Charap. High density magnetic recording media design and identification: susceptibility to thermal decay. *IEEE Transactions on Magnetics*, 31 p 2767, 1995.
23. Y. K. Liu and M. H. Kryder. Thermally stable, soft FeXN thin films. *Applied Physics Letters*, 77 p 426, 2000.
24. L. Pauling. The nature of metals. *Pure and Applied Chemistry*, 61(12): pp 2171–2174, 1989.
25. S. X. Wang, N. X. Sun, M. Yamaguchi, and S. Yabukami. Properties of a new soft magnetic material. *Nature*, 407 pp 150–151, 2000.
26. M. Sostarich. Generalized Slatter–Pauling curve and the role of metalloids in Fe-based amorphous alloys. *Journal of Applied Physics*, 67(9): pp 5793–5795, May 1, 1990.
27. F. Liu, K. Stoev, P. Luo, Y. Liu, Y. Chen, J. Chen, J. Wang, F. G. Shan, K. T. Kung, M. Lederman, M. Krounbi, M. Re, A. Otsuki, and S. Hong. Perpendicular recording heads for extremely high-density recording. *IEEE Tranactions on Magnetics*, 39(4): pp 1942–1948, 2003.
28. S. Iwasaki, Y. Nakamura, and K. Ouchi. Perpendicular magnetic recording with a composite anisotropy film. *IEEE Tranactions on Magnetics*, 15(6): p 1456, 1979.
29. M. Mallary, A. Torabi, and M. Benakli. One terabit per square inch perpendicular recording conceptual design. *IEEE Transactions on Magnetics*, 38(40): pp 1719–1724, 2002.

30. D. A. Thompson. The role of perpendicular recording in the future of hard disk storage. *Journal of the Magnetics Society of Japan*, 21(S2): p 9, 1997.
31. F. Liu, K. Stoev, P. Luo, Y. Liu, Y. Chen, J. Chen, J. Wang, F. G. Shan, K. T. Kung, M. Lederman, M. Krounbi, M. Re, A. Otsuki, and S. Hong. Perpendicular recording heads for extremely high-density recording. *IEEE Transactions on Magnetics*, 39(4): pp 1942–1948, 2003.
32. S. Khizroev and D. Litvinov. *Perpendicular Magnetic Recording*. Amserterdam: Kluwer, 2004.
33. T. McDaniel and W. Challener. Issues in heat-assisted perpendicular recording. *IEEE Transactions on Magnetics*, 39(4): pp 1972–1979, 2003.
34. F. Chen, A. Itagi, J. A. Bain, D. D. Stancil, and T. E. Schlesinger. Imaging of optical field confinement in ridge waveguides fabricated on very-small-aperture laser. *Applied Physics Letters*, 83(16): p 3245, 2003.
35. J. Wong, A. Scherer, M. Todorovic, and S. Schultz. Farbrication and characterization of high aspect ratio perpendicular patterned information storage media in an Al2O3/GaAs substrates. *Journal of Applied Physics*, 85 p 5489, 1999.
36. E. D. Boerner, H. N. Bertram, and G. F. Hughes. Writing on perpendicular patterned media at high density and data rate. *Journal of Applied Physics*, 85 p 5318, 1999.
37. S. Khizroev, Y. Hijazi, N. Amos, D. Doria, A. Lavrenov, R. Chomko, T. M. Lu, and D. Litvinov. Three-dimensional magnetic recording: an emerging nanoelectronic technology. *Journal of Nanoelectronics and Optoelectronics*, 1 pp 1–18, 2006.
38. S. Khizroev, R. Chomko, Y. Hijazi, and N. Amos. Three-dimensional magnetic recording devices. Pentagon report: Report # A752434, 2005.
39. M. Albrecht, G. Hu, A. Moser, O. Hellwig, and B. D. Terris. Magnetic dot arrays with multiple storage layers. *Journal of Applied Physics*, 97 p 103910, 2005.
40. M. Albrecht, G. Hu, A. Moser, O. Hellwig, and B. D. Terris. Magnetic dot arrays with multiple storage layers. *Journal of Applied Physics*, 97 p 103910, 2005.
41. S. Khizroev and D. Litvinov. *Perpendicular Magnetic Recording*. Amsterdam: Kluwer, 2005.
42. D. Litvinov, M. Kryder, and S. Khizroev. Recording physics of perpendicular media: soft underlayers. *Journal of Magnetism and Magnetic Materials*, 232(1–2): pp 84–90, 2001.
43. S. Khizroev, M. H. Kryder, Y. Ikeda, K. Rubin, P. Arnett, M. Best, and D. A. Thompson. Recording heads with trackwidths suitable for 100 Gbit/in2 density. *IEEE Transactions on Magnetics*, 35(5): pp 2544–2546, 1999.
44. D. Litvinov and S. Khizroev. Orientation-sensitive magnetic force microscopy in future probe storage applications. *Applied Physics Letters*, 81(10): p 1878, 2002.
45. F. Candocia, E. Svedberg, D. Litvinov, and S. Khizroev. Deconvolution processing for increasing the resolution of magnetic force microscopy measurements. *Nanotechnology*, 15: p S575–584, 2004.
46. D. Litvinov, M. Kryder, and S. Khizroev. Recording physics of perpendicular media: recording layers. *Journal of Magnetism and Magnetic Materials*, 241(2–3): pp 453–465, 2002.
47. B. D. Cullity. *Introduction to Magnetic Materials*. Reading, MA: Addison-Wesley, 1972.
48. Yu. B. Sidorov, M. B. Fedoryuk, and M. I. Shabunin. *Lectures on Theory of Functions of Complex Variables*. 1989. Need publisher.

49. D. Litvinov and S. Khizroev. Perpendicular recording: playback. *Applied Physics Reviews*: Focused Review, JAP 97: 071101, 2005.
50. S. Khizroev, R. W. Gustafson, J. K. Howard, M. H. Kryder, and D. Litvinov. Multiple magnetic image reflection in perpendicular recording. *IEEE Transactions on Magnetics*, 38(5): pp 2066–2068, 2002.
51. Y. Hijazi, R. Ikkawi, N. Amos, A. Lavrenov, N. Joshi, D. Doria, R. Chomko, D. Litvinov, and S. Khizroev. Patterned soft underlayers for perpendicular media. *IEEE Transactions on Magnetics*, 42(10): pp 2375–2377, 2006.
52. Y. Hijazi, R. Ikkawi, N. Amos, A. Lavrenov, N. Joshi, D. Doria, R. Chomko, D. Litvinov, and S. Khizroev. Patterned soft underlayers for perpendicular media. *IEEE Transactions on Magnetics*, 42(10): p 2375, 2006.
53. E. Chunsheng, D. Smith, J. Wolfe, D. Weller, D. Litvinov, and S. Khizroev. Physics of patterned magnetic medium recording: design considerations. *Journal of Applied Physics*, 98 p 024505, 2005.
54. S. Khizroev, M. H. Kryder, Y. Ikeda, K. Rubin, P. Arnett, M. Best, and D. A. Thompson. Recording heads with trackwidths suitable for 100 Gbit/in2 density. *IEEE Transactions on Magnetics*, 35(5): pp 2544–2546, 1999.
55. F. Candocia, E. Svedberg, D. Litvinov, and S. Khizroev. Deconvolution processing for increasing the resolution of magnetic force microscopy measurements. *Nanotechnology*, 15 p S575–584, 2004.
56. I. Dumer. Recursive decoding and its performance for low-rate Reed–Muller codes. *IEEE Transactions on Information Theory*, 50 pp 811–823, 2004.
57. I. Dumer and K. Shabunov. Recursive error correction for general Reed–Muller codes. *Discrete Applied Mathematics*, 154 p 253–269, 2006.
58. I. Dumer. Soft decision decoding of Reed–Muller codes: a simplified algorithm. *IEEE Transactions on Information Theory*, 52(3): 2006.
59. I. Dumer and K. Shabunov. Soft decision decoding of Reed–Muller codes: recursive lists. *IEEE Transactions on Information Theory*, 52(3): 2006.

7

SPIN-WAVE ARCHITECTURES

Mary Mehrnoosh Eshaghian-Wilner, Alex Khitun, Shiva Navab, and Kang L. Wang

In this chapter, we present three nanoscale architectures: the reconfigurable mesh, crossbar, and fully interconnected cluster. In these architectures, both communication and computation are performed using spin waves. These architectures are capable of simultaneously transmitting multiple waves over different frequencies. In addition, they can use the superposition property of the waves and allow for concurrent writes. In these architectures, computation can be done in both analog and digital modes. On the other hand, one shortcoming of the spin-wave architectures is the limitation on the attenuation length of spin waves, about 50 microns. To address this issue, a hierarchical multiscale design is proposed that is an integration of the spin-wave modules within a MEMS architecture.

7.1. INTRODUCTION

Spinwaves have been attracting scientific interest for a long time, but only recently several experimental types of projects have been devoted to the problem of spin-wave detection using the inductive voltage measurement technique [1, 2]. The results indicate that spin-wave packets propagating in a 100 nm thick ferromagnetic films may produce an inductive voltage in the order of several mV, suitable for experimental detection [1]. The inductive voltage measurement technique appears to be one of the most convenient physical methods for spin-wave detection and integration in a semiconductor platform.

The authors of this chapter are listed alphabetically.

Bio-Inspired and Nanoscale Integrated Computing. Edited by Mary Mehrnoosh Eshaghian-Wilner

The use of spin waves for computation is an entirely new idea. The first computational architecture utilizing spin waves was for massive entanglement of distant spin-based qubits in a quantum computer as described in [3, 4]. In this research, spin waves are used for both information transmission and information processing. Moreover, the classical type of computing is employed as opposed to quantum, and the architectures are operable at room temperature.

In the following section, we present three spin-wave architectures: crossbar, reconfigurable mesh, and fully interconnected cluster. Afterward, we explain our experimental results.

7.2. SPIN-WAVE ARCHITECTURES

In the following section, after a brief introduction to spin waves, we present three nanoscale spin-wave architectures: nanoscale crossbar, spin-wave reconfigurable mesh, and spin-wave fully interconnected cluster. Then, to address the size limitation of spin-wave architectures, we propose a multiscale hierarchical architecture.

Spin wave is a collective oscillation of spins in an ordered spin lattice around the direction of magnetization. The phenomenon is similar to the lattice vibration, where atoms perform oscillation around its equilibrium position. In other words, a collection of electron spin precessions about the magnetic field is a spin wave [5], as shown in Figure 7.1. The magnitude of the spin wave is determined by precession angle. A propagating spin wave changes the local polarization of spins in the ferromagnetic material. In turn, the change in magnetic field results in an inductive voltage. Recently published experimental results indicate that an inductive voltage signal of the order of mV produced by spin waves propagating through a nanometer thin ferromagnetic film are detectable at the distances of up to 50 microns at room temperature. Our idea is to use the phase of the spin wave for both information exchange and information processing [6].

Typically, the speed of the spin wave is 10^4 m/s. In the current experimental results, the switching frequency is in the order of GHz, but up to the order of THz is possible. As mentioned above, the attenuation of spin waves is around 50 microns, which makes it a suitable candidate for nanoscale communication.

In the following section, we show how spin waves can be used in implementing nanoscale architectures.

7.3. SPIN-WAVE CROSSBAR

In this section, a nanoscale crossbar architecture that is interconnected with ferromagnetic spin-wave buses is introduced [7, 8]. The power consumption of

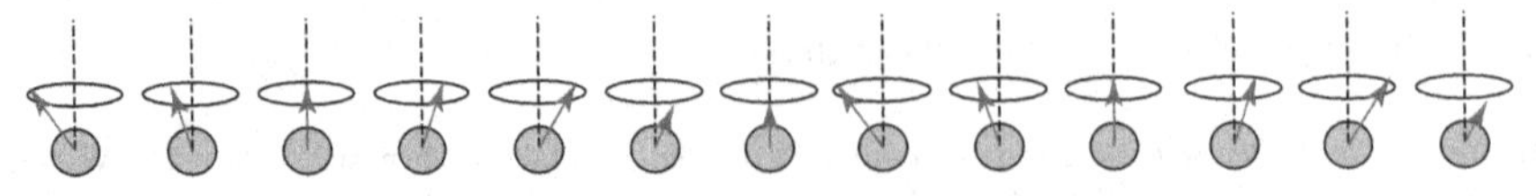

Figure 7.1. Spin wave: a collective oscillation of spins.

these proposed modules is relatively low, and the classical type of computing, as opposed to quantum, is employed. By using phase logic, simple operations such as AND/OR/NOT can be performed efficiently on the transmitted waves. Signal detection is accomplished by the time-resolved inductive voltage measurement technique. In this structure, the sphere of interactions is limited by the spin-wave attenuation length; therefore, it is suitable to be used as nanoscale computing modules that can be ported or integrated onto higher-scale chips or devices.

The architecture described here, while requiring the same number of switches as a standard crossbar, is capable of simultaneously transmitting N waves on each of the spin-wave paths. As compared to the known molecular crossbars, this design is fault tolerant: In case there is a failure in one of the N channels, any of the other channels can be reconfigured to transmit the data. This is possible since all the channels are accessible by all the ports and each channel can handle multiple data.

Crossbars are attractive architectures because they can realize any permutations of N inputs to N outputs. However, their main shortcoming is due to the fact that N^2 switches are used to transmit only N pairs of data. As mentioned earlier, each spin-wave bus is capable of carrying multiple waves at any point of time. Therefore, each of the N inputs can broadcast its data to all of the N outputs in parallel. Using this type of architecture, it would be possible to efficiently realize highly connected types of computations such as the Hopfield Neural Network model, as shown in [9]. An example of the proposed spin-wave crossbar architecture is shown in Figure 7.2. A set of column spin-wave buses on the bottom and a set of row spin-wave buses on the top are connected via the vertical spin wave switches.

An important element required for the spin-wave crossbar is the spin-wave switch. We define a spin-wave switch as a device that has an externally controllable magnetic phase. In the "On" state, the switch transmits spin waves;

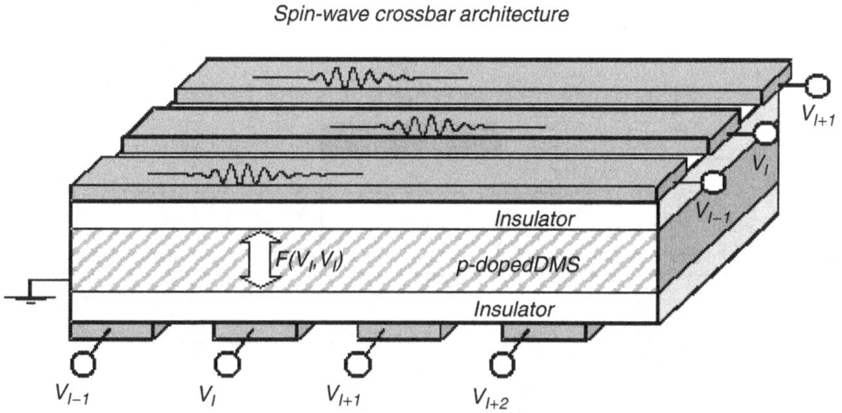

Figure 7.2. Spin-wave crossbar architecture.

in the "Off" state, it reflects any incoming spin wave. A possible structure of the spin-wave switch is shown in Figure 7.3.

A ferromagnetic film is divided by a region of diluted magnetic semiconductor (DMS) used as a magnetic channel; the magnetic phase is controlled by the applied electric field via the effect of hole-mediated ferromagnetism [10]. Figure 7.3a shows the cross-section of a metal-insulator-diluted magnetic semiconductor structure that is used for experimental study of the effect of hole-mediated ferromagnetism in [10]. A negative gate bias increases the hole concentration in the DMS region, resulting in the paramagnetic-to-ferromagnetic transition, whereas a positive bias has an opposite effect (as shown in Fig. 7.3b). The cross-section of the spin-wave switch is shown in Figure 7.3c. The ferromagnetic film is divided by the DMS cell. Spin waves can propagate through the DMS cell only if it is in the ferromagnetic phase. While in the paramagnetic phase, the DMS cell reflects all incoming spin waves.

As it is shown, there is a common p-doped DMS layer for the whole structure. The magnetic phase in every region is controlled by the hole concentration. Using conducting ferromagnetic films (NiFe, for example), it is possible to control the hole distribution along the structure by the applied electric field. The hole concentration is high when both the top and the bottom ferromagnetic films are

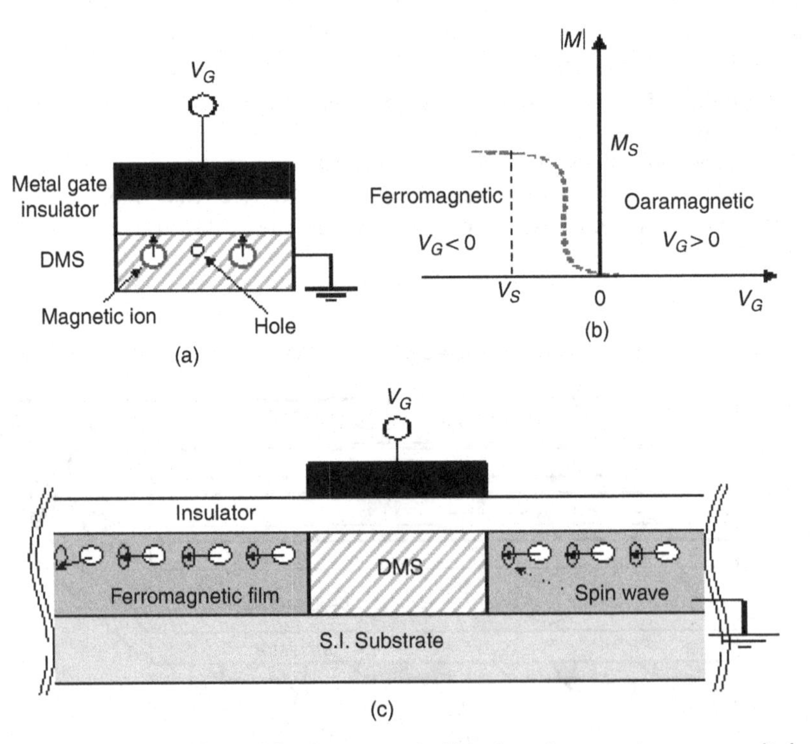

Figure 7.3. DMS cells and ferromagnetic film forming a spin-wave switch.

biased by a negative voltage. There is a self-alignment mechanism that makes the proposed structure tolerant to structure imperfections.

The detailed analysis of the spin wave switch can be done on the base of the Heisenberg model, treating the ferromagnetic film and the DMS cell as two media having different exchange integrals J. The Hamiltonian for the system ferromagnetic film plus DMS cell can be expressed as follows:

$$\mathrm{H} = -\sum_{i,j} J_{ij}[\vec{S}_i \cdot \vec{S}_j],$$

where J is the exchange integral for the nearest spins in the lattice. The propagation/reflection of spin waves through the interface between the DMS cell and the ferromagnetic film depends on the ratio J_{DMS}/J_{film}, where J_{DMS} and J_{film} are the exchange integrals for the DMS cell and the ferromagnetic film, respectively. The exchange integral in the ferromagnetic film is constant, while the exchange integral in the DMS cell is a function of the gate voltage $J_{DMS} \equiv J_{DMS}(V_G)$. In the two ultimate limits $J_{DMS}/J_{film} = 1$ and $J_{DMS}/J_{film} = 0$, we have complete transmission and complete reflection, respectively.

The same structure of the spin-wave switch based on the effect of hole-mediated ferromagnetism may be used for the vertical integration of the spin-wave buses. In Figure 7.4 we have schematically shown two vertically separated ferromagnetic films with a spin-wave switch in between. The switch serves as a connector between two spin-wave busses in this structure.

At the negative applied bias (ferromagnetic phase), the switch allows spin wave propagate from one film to another. At the positive bias (paramagnetic film), spin waves can not propagate through. The use of the spin-wave switches for vertical integration makes possible array-based architectures similar to the one proposed for the array of nanowires [11]. The advantage of the proposed architecture is that signal information can be encoded in the phase of the propagating spin wave rather than in the number of electrons to be transmitted via conducting wires.

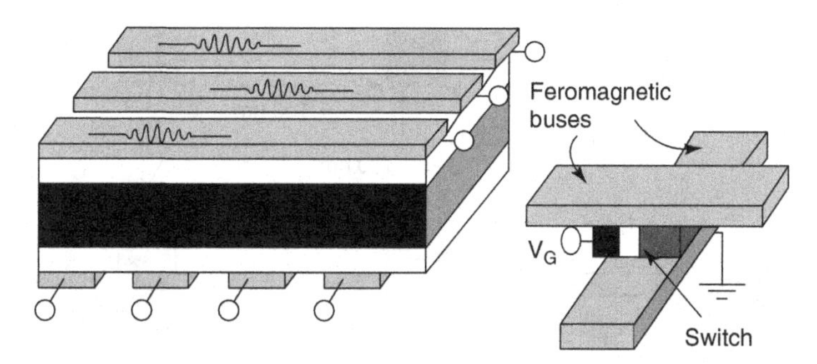

Figure 7.4. Spin-wave crossbar with spin-wave switches.

7.4. SPIN-WAVE RECONFIGURABLE MESH

A nanoscale spin-wave reconfigurable mesh of size N^2 consists of an $N \times N$ array of processors connected to a reconfigurable spin-wave bus grid, where each processor has a locally controllable bus switch [12], as shown in Figure 7.5.

In the proposed spin-wave reconfigurable mesh architecture with spin-wave buses, a set of column spin-wave buses on the bottom and a set of row spin-wave buses on the top are connected via the spin-wave switches. Each switch is placed at the grid point of the mesh, as shown in Figure 7.6. Basically, the nanoscale spin-weave reconfigurable mesh is similar to the standard reconfigurable mesh, except for the spin-wave buses and switches.

It is worth noting that, similar to the standard reconfigurable mesh, the nanoscale reconfigurable mesh of size N occupies $N \times N$ area, under the assumption that processors, switches, and the link between adjacent switches occupy unit area. However, the main difference in terms of area here is that the unit of area is at nanoscale level as opposed to the standard reconfigurable meshes that are at microscale level of integration.

Each of the spin-based devices serves as a one-bit input/output port. One spin-wave packet excited by one port can be superposed with many packets generated by the nearby ports. The length of interaction among the spin-based devices is limited by the spin-wave attenuation caused mainly by the scattering on phonons. The length of interaction is defined by the material properties of the ferromagnetic film, film size, and the operation temperature. According to our experimental data it can be as high as tens of microns.

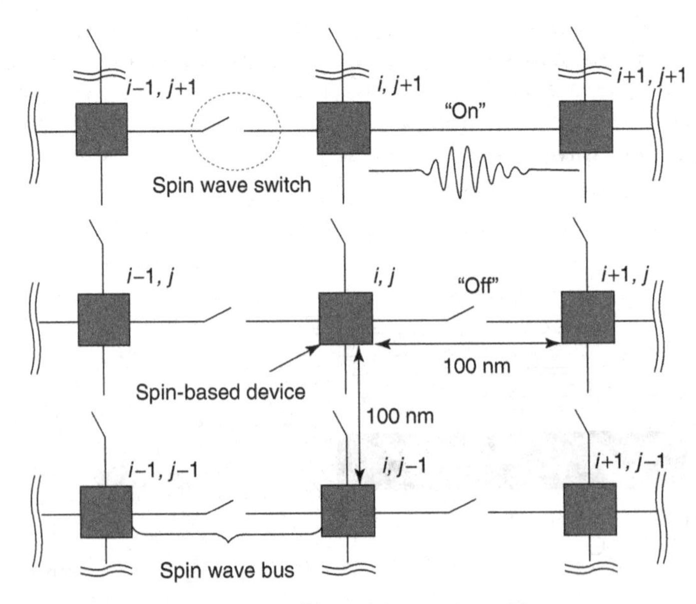

Figure 7.5. Reconfigurable mesh architecture.

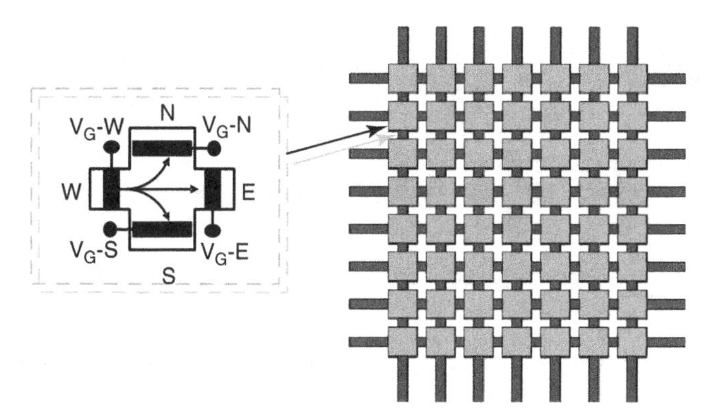

Figure 7.6. Reconfigurable spin-wave switches at the grid points of the mesh.

In Figure 7.7, the cross-section of the reconfigurable spin-wave switch is shown. A ferromagnetic film is divided by regions of diluted magnetic semiconductor (DMS).

The reconfigurable switches work based on the same mechanism as a regular spin-wave switch used in a crossbar architecture. The difference is that the reconfigurable switches are controlled by four different control signals to form 15 possible interconnections, as shown in Figure 7.8.

7.5. SPIN-WAVE FULLY INTERCONNECTED CLUSTER

The architecture presented in this section is a nanoscale fully interconnected architecture [13]. In this architecture, each node can broadcast using spin waves to all other nodes simultaneously, and similarly, each node via spin waves can receive and process multiple data in parallel. The significance of this design is that the communication between the nodes can be done in constant time. This is

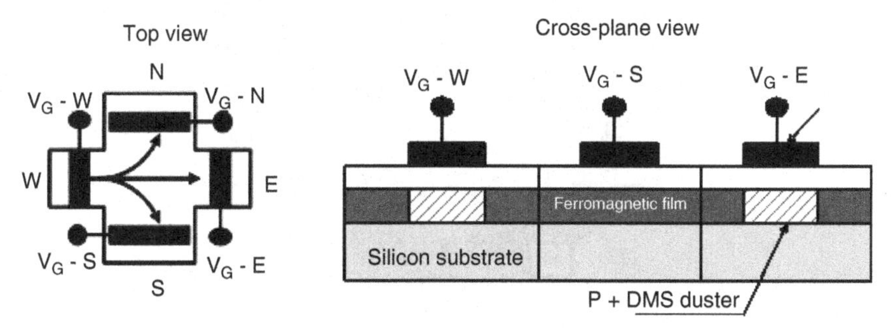

Figure 7.7. Reconfigurable spin-wave switches.

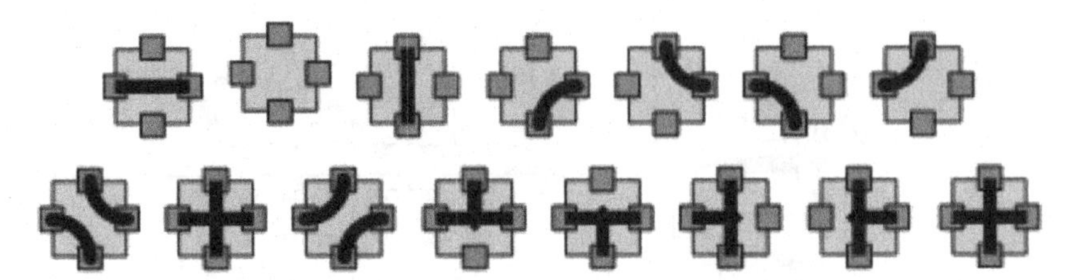

Figure 7.8. Reconfigurable spin-wave switch possible connections.

a significant improvement considering the $\Omega(\log N)$ lower-bound on the time delay for implementing such networks in VLSI using traditional electrical interconnects. Another key advantage is that the spin-wave fully interconnected network can be laid out in $O(N^2)$ area as opposed to $O(N^4)$. Of course, the unit of area is in order of nanometer as opposed to the standard micron technology. In this architecture, information is encoded into the phase of spin waves, and no charge is transmitted. As a result of this, power consumption can be significantly reduced in this architecture (compared to other nanoscale architectures).

Figure 7.9 shows the top view of the architecture in which the N computing nodes are placed around a circle on a magnetic film. Each node is an asymmetric coplanar strip (ACPS) line (described later in Section 7.7), which can be used as a sender or receiver at each point of time. Figure 7.10 shows the cross view of the layout of this architecture on a semiconductor chip. The area requirement of this architecture is $O(N^2)$, as opposed to the $O(N^4)$ area requirement if electrical interconnects were to be used. We should also note that in this architecture all the distances are in nanoscale. Also unlike the network with electrical interconnections,

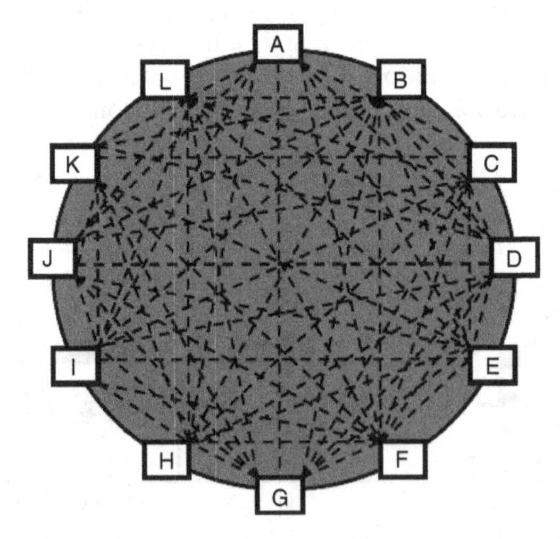

Figure 7.9. The top view of the architecture with full spin-wave interconnectivity.

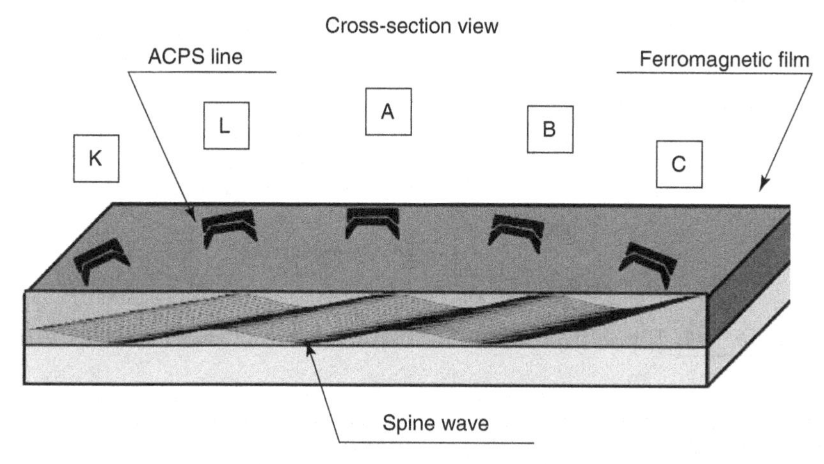

Figure 7.10. Cross-section view of spin-wave fully interconnected cluster.

in which only one transmission can be done at a time, in this architecture, multiple simultaneous permutations are possible. This can be done by transmitting the spin waves over different frequencies. The information is coded into the phase of the spin waves in the sender and is detected by the receivers. In addition, within each frequency, data can be sent to one or more other nodes from each node. In the following we present a brief discussion on the placement of nodes and their intercommunication mechanism.

Normally, in architectures where the phases of the waves are the means of information transmission, the exact location of the nodes with respect to the size of topology is an important design issue. The distance between the sender and receiver has to be a multiple of the wavelength; otherwise, the receiver might receive the wave at a different phase. For instance, if the wave is received with a π radian phase-shift, a "0" is received instead of a "1" or vice versa. In our design, however, this is not an issue since the wavelength of spin waves is considerably larger than the distance between the nodes. The speed of spin waves is around 10^5m/s. Assuming the input frequency range of 1–10 GHz, the wavelength will be in the order of 10^{-4} to 10^{-5} m, while the distances are at the nanoscale, or 10^{-9} m. In other words, the wavelengths of the spin waves are orders of magnitude greater than the distances between the nodes. Therefore, all the nodes receive the same phase regardless of their location, and there is no need to place the nodes at specific distance relative to the other ones.

7.6. MULTISCALE HIERARCHICAL ARCHITECTURE

As mentioned earlier, the attenuation of a spin wave is about 50 microns, which dictates the maximum size of spin-wave architectures. In order to overcome this

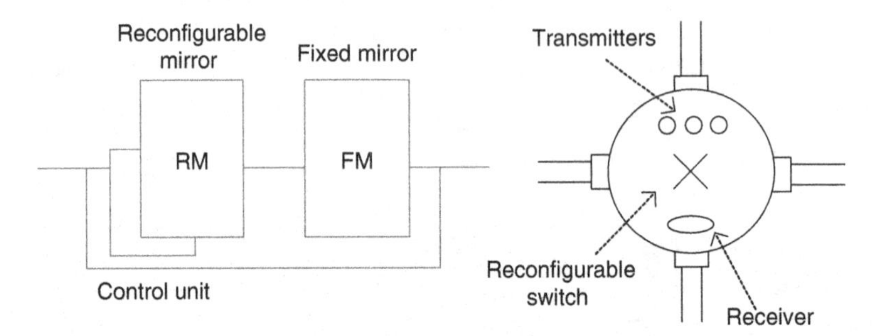

Figure 7.11. The deflection unit in hierarchical multiscale architecture.

size limitation, we propose using a hierarchical multiscale architecture, which will embody the spin-wave modules described above.

The proposed multiscale architecture is based on the optical reconfigurable mesh (ORM) architecture [14], which consists of two layers: the deflection layer and the processing layer. In this architecture, similar to ORM, the deflection layer consists of N^2 deflecting units, while the processing layer consists of N^2 processing units.

In this architecture, in addition to the standard processing layer of ORM at microscale level, there is another processing layer at nanoscale level which consists of a set of spin-wave modules. These modules can be spin-wave crossbars, reconfigurable meshes, or fully interconnected clusters. Each deflecting unit consists of a fixed mirror, a reconfigurable mirror, and a unit to control the direction of the reconfigurable mirror, as shown in Figure 7.11. Each processing unit at the microscale has an optical receiver, optical transmitters, and a regular reconfigurable electrical switch. These processing units are interconnected as a standard reconfigurable mesh and can also intercommunicate optically using the deflection layer.

Data at the processing module level can be routed in three different ways: electrical routing through electrical buses, optical routing via free-space optical interconnections, and electro-optical routing that uses electrical and optical free-space connections to allow a complete connection among N processors [14]. Each of these routing mechanisms is described below, assuming we index the processing unit in the ith row and the jth column of the mesh as $P(i,j)$ and the mirror above it as $M(i,j)$, in which $1 \leq i, j \leq N$.

7.6.1. Electrical Routing

The electrical routing in ORM is similar to the routing in the standard reconfigurable meshes. This type of routing uses the electrical buses in the reconfigurable mesh for one-to-one routing or broadcasting. This type of communication is suitable for providing arbitrary configuration of the buses in the processing layer.

7.6.2. Optical Routing

The optical routing in ORM is the routing through optical free-space interconnections. The data transfer does not use any electrical bus in the system. All N^2 processors can communicate in unit time delay as long as there is only one read or write from or to each location. In the following, we describe how such an optical connection is established between two processors through the RM.

A connection phase consists of two cycles. In the first cycle, each processor sends the address of its desired destination processor to the arithmetic control unit of its associated mirror using its dedicated laser. The arithmetic control unit of the mirror computes a rotation degree such that both the source and the destination processors have equal angle with the line perpendicular to the surface of the mirror in the plane formed by the mirror, the source processor, and the destination processor. Once the angle is computed, the mirror is rotated to point to the desired destination. In the second cycle, the connection is established by the laser beam, carrying the data from the source to the mirror and then from the reflected mirror towards the destination. The read operation has two phases. In the first phase, the read requirement and the reader's address are sent to the processor which stores the desired data. In the second phase, the data is sent back to the reader, depending on the reader's address. Both phases use the two-cycle write routing method.

7.6.3. Electro-Optical Routing

This communication mechanism establishes an efficient full connectivity among only the N processors situated diagonally in the processing layer on the N^2 processors in the ORM (i.e., for processors $P(j,j)$ where $1 \leq j \leq N$). This routing technique uses electrical buses on the processing layer and fixed mirrors on the deflection layer.

The connection for electro-optical routing is implemented as follows. Each processor $P(j,j)$ is associated with the jth row of the deflection unit, where the row contains N fixed mirrors. The ith fixed mirror in that row for $1 \leq i \leq N$ is directed to the processing unit $P(i, i)$. In an exclusive read exclusive write (EREW) scenario, any processing element $P(i,i)$ sends data to $P(k,k)$ in the following way: First, $P(i, i)$ sends the data to $P(i,k)$ through the electrical row bus; then, $P(i,k)$ sends data to $P(k,k)$ through the transmitter and its deflector $M(i,k)$. The variety of techniques available in this architecture makes ORM a very powerful computing model [15].

In our proposed hierarchical multiscale architectures, the processing elements are interconnected using spin waves at the nanoscale level, and via ORM's electrical, optical, or electro-optical routing at the microlevel [16]. Figure 7.12 shows this organization.

As an example, an electro-optical routing is shown in Figure 7.12. In this routing, as mentioned previously no reconfiguration of mirrors is necessary, and only fixed mirrors are used. The data communication in each row is through

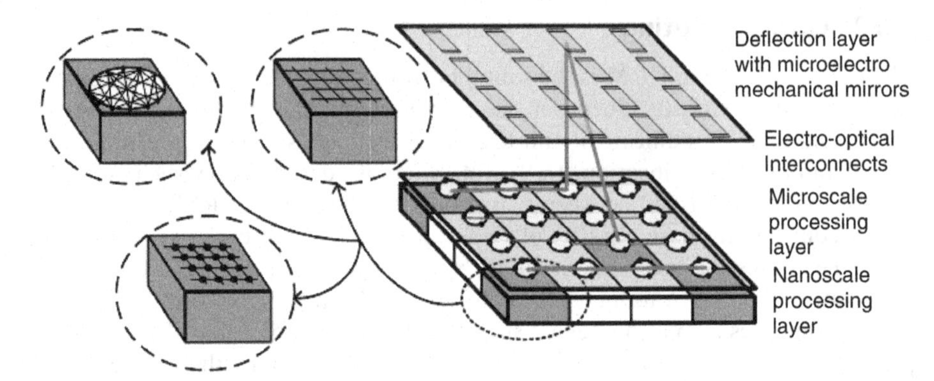

Figure 7.12. The hierarchical multiscale architecture.

electrical interconnections, while the fixed mirrors provide vertical paths among processors.

Since only electrical switches and fixed mirrors are used, this architecture has a switching time of nanoseconds. In the example shown in Figure 7.12, the electrical connection between P(4,1) and P(4,4) has been highlighted, and the electro-optical routing from processor P(1,1) to processor P(3,3) is shown. P(1,1) makes an electrical connection to P(1,3) while P(1,3) is connected to P(3,3) using M(1,3) fixed mirror.

7.7. PRELIMINARY IMPLEMENTATIONS

While the architectures presented in the previous section have not yet been fully implemented, we have some preliminary experimental results based on a simple prototype that is built and tested by our collaborators. The significant point about these experimental results is that the architectures are operable at room temperature and currently have GHz switching speed. In the following, we first show some numerical simulations that illustrate the functionality of a simple logic device. Next, our prototype and experimental results are presented, along with a theoretical discussion explaining the operation of the prototype.

7.7.1. Spin-Wave-Based Logic Devices

Spin waves can be used to provide an "LC" coupling of devices, without dissipative resistance. With the spin-wave concept, the spin rotates as a propagating wave and there is no particle (electron/hole) transport. In Figure 7.13 we have schematically shown a prototype structure for spin-wave bus demonstration. The core of the structure consists of a ferromagnetic film (NiFe, for example) grown on a semi-insulating substrate. The film is polarized, along with the x axis. There are three asymmetric co-planar strip (ACPS) transmission lines on the top of the

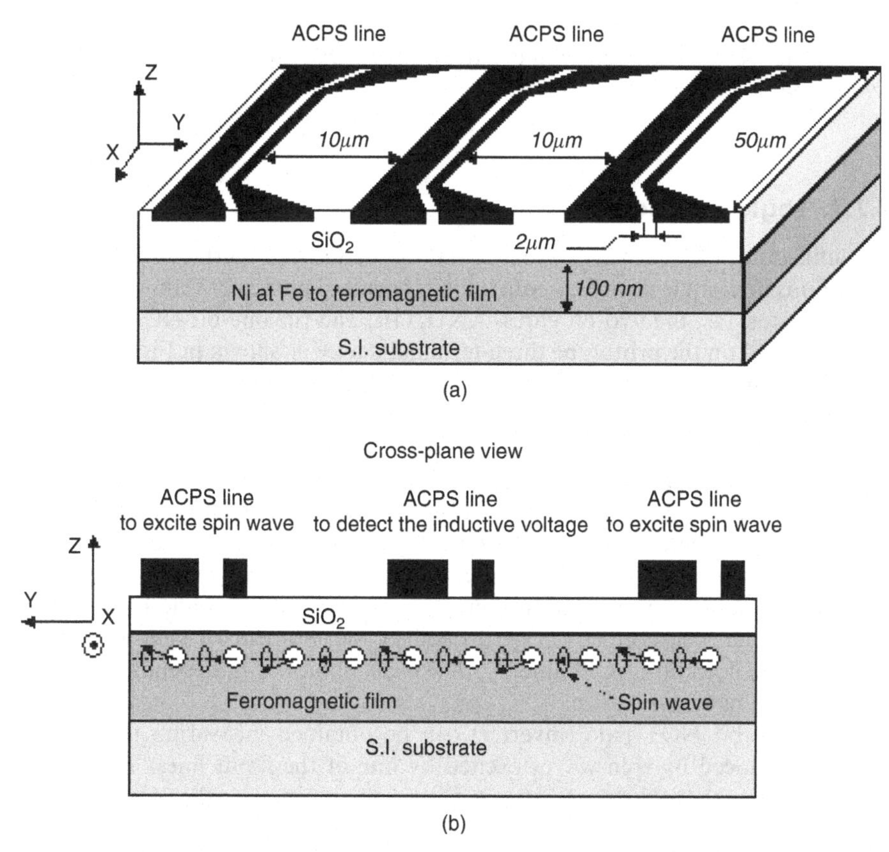

Figure 7.13. Prototype of a spin-wave logic device.

structure. The lines and the ferromagnetic layer are isolated by the silicon oxide layer.

The thickness of the ferromagnetic layer can be as thin as tens of nanometers, and the thickness of the oxide layer is several hundreds of nanometers. The dimensions of the ACPS lines are defined by the frequency of the transmitting signal. Each of the ACPS lines can be used for spin-wave excitation and detection. A voltage pulse applied to the ACPS line produces magnetic field perpendicular to the polarization of the ferromagnetic film, and, thus, generates a spin wave (spin-wave packet). Being excited, the spin wave propagates through the ferromagnetic film. As it reaches the nearest ACPS line, the amplitude and the phase of the spin wave is detected by the inductive voltage measurements. For example, the edge ACPS lines can be considered as the input ports, and the ACPS in the middle as the output port. The middle ACPS line detects the inductive voltage produced by the *superposition* of two waves. Depending on the relative phase of the spin waves, the amplitude of the inductive voltage may be enhanced (when two waves are in phase) or decreased (when two waves are out of phase) in comparison to the inductive voltage produced by a single spin wave. As we will show below, it is possible to realize different logic gates AND, OR, and NOT controlling the *relative phase* of the spin waves.

7.7.2. Logic Functionality

The utilization of spin waves provides an opportunity to perform different logic operations in a single device by controlling the initial phases of spin waves. The set of logic gates, i.e., the two-bit gates, AND, OR, and the one-bit NOT gate can be demonstrated on the prototype three-terminal device as shown in Figure 7.13. The edge ACPS lines are considered as the input terminals, and the ACPS line shown in the middle is the output port. The input information is coded into the polarity of the voltage pulse applied to the edge ACPS lines (for example, $V_{input} = +10\,V$ corresponds to the logic state 1>, and $V_{input} = -10\,V$ corresponds to the logic state 0>). We assume the input terminals are placed equidistantly from the output one. In order to detect the output signal V_{ind} we use the time-resolved inductive voltage measurement. To recognize the output logic state we introduce a reference voltage V_{ref}. We assign the output logic state 1> if $V_{ind} > V_{ref}$, and logic state 0> otherwise. The measurement is performed at the moment of spin-wave packet arrival to the detecting ACPS line, t_m. The exact choice of t_m depends on the logic function one needs to realize.

The one-bit NOT gate (inverter) can be obtained measuring the inductive voltage produced by spin waves excited by one of the input lines. Taking $t_m = \pi g / v_{ph}$ (where g is the distance between the contacts and v_{ph} is the spin-wave phase velocity), and $V_{ref} = 0\,V$, we achieve the required logic correlation. Next, the two-bit AND gate can be realized when $t_m = 2\pi g / v_{ph}$, and $V_{ref} = 0.5\,mV$. Two spin-wave packets coming in phase enhance the amplitude of the produced inductive voltage, and cancel each other when coming out of phase. The nonzero reference voltage is needed to avoid the effect caused by the finite size of the detecting ACPS line. The OR gate can be realized by analogy, taking $t_m = 2\pi g / v_{ph}$, and $V_{ref} = -0.5\,mV$. In Table 7.1, we have summarized the input/output correlations into the truth tables for different logic gates.

TABLE 7.1. Truth Tables for AND and OR Gates

AND						OR					
$V_{ref} = 0.5\,mV$						$V_{ref} = -0.5\,mV$					
Input voltage		Logic state		Output voltage	Logic state	Input voltage		Logic state		Output voltage	Logic state
+1 V	+1 V	1	1	+27 mV	1	+1 V	+1 V	1	1	+27 mV	1
+1 V	−1 V	1	0	0 mV	0	+1 V	−1 V	1	0	0 mV	1
−1 V	+1 V	0	1	0 mV	0	−1 V	+1 V	0	1	0 mV	1
−1 V	−1 V	0	0	−27 mV	0	−1 V	−1 V	0	0	−27 mV	0<

7.8. EXPERIMENTAL RESULTS

In Figure 7.14, we have schematically shown the prototype of the spin-wave logic device structure. The core of the structure consists of a 100 nm-thick CoFe film deposited on a silicon substrate. There are two asymmetric co-planar strip (ACPS) transmission lines on the top of the structure. The distance between the lines is 8 μm. The lines and the ferromagnetic layer are isolated by a 300 nm silicon oxide layer. The dimensions of the ACPS lines are adjusted to match 50 Ω of the external coaxial cable. One of the ACPS lines (shown on the right) is used for spin-wave excitation. Hereafter, we will refer to this ACPS line as the "excitation" line. A voltage pulse applied to the excitation line produces a magnetic field and excites a spin wave (spin-wave packet) in the ferromagnetic layer. Being excited, the spin-wave propagates through the ferromagnetic film. The other ACPS line (shown on the left) is used to detect the inductive voltage signal produced by the propagating spin wave. Hereafter, we will refer to this line as the "detection" line.

Figure 7.15 shows the experiment setup, where an input voltage pulse with an amplitude of 24.5 V, rising time of 1.2 ns, and pulse length of 20 ns is applied to the

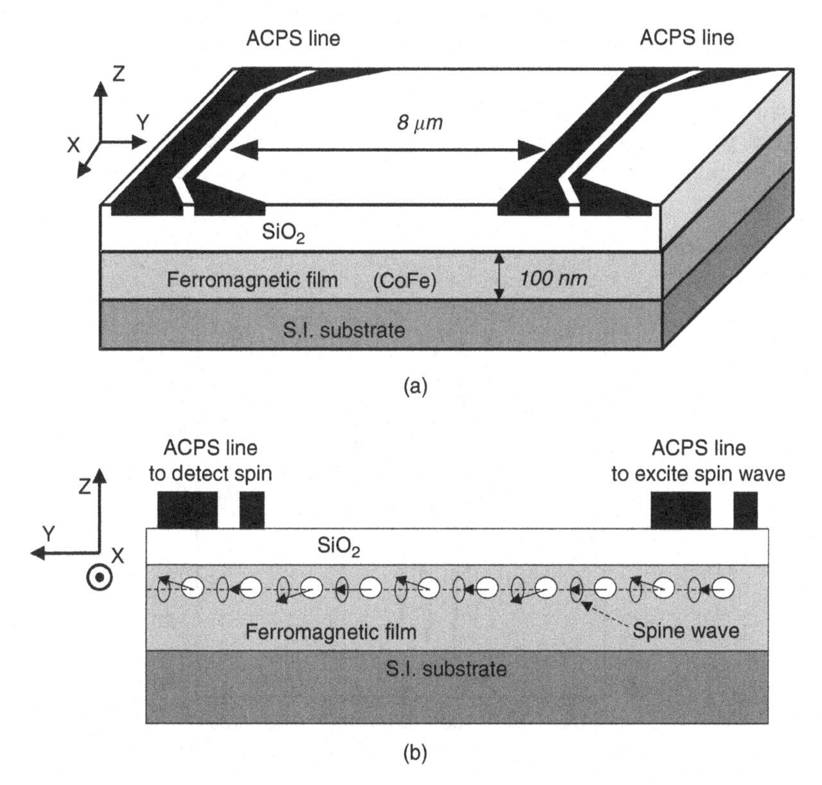

Figure 7.14. Prototype of spin-wave device used in the experiments.

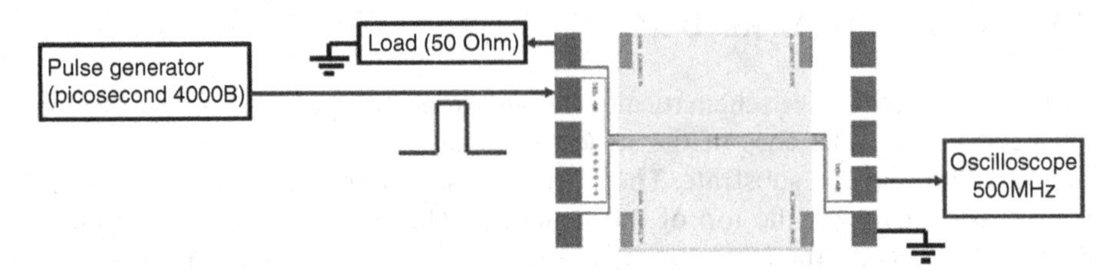

Figure 7.15. Spin-wave voltage detection set-up.

input port. In Figure 7.16, we present our experimental results on spin-wave detection by the time-resolved inductive voltage measurement technique [1]. The dashed line depicts the voltage pulse applied to the excitation line. One can see the inductive voltage oscillation at the detection line, which is caused by the inductive coupling via the spin waves. The output voltage signal has maximum pulse amplitude of 26 mV, and the period of oscillation is 9 ns.

These experimental data illustrate the possibility of signal transmission by spin waves over micrometer range distances. The attenuation time is about 20 ns, and the signal-to-noise ratio is satisfactory (at least for 8 μm propagation distance). We would like to stress that the utilization of spin waves is prominent for short-range on-chip communication. There are two important issues that require additional consideration: (i) power dissipation in spin-wave bus and (ii) signal gain. In principle, the energy per spin wave can be scaled down to several kT, just above the thermal noise level. On the other hand, in order to

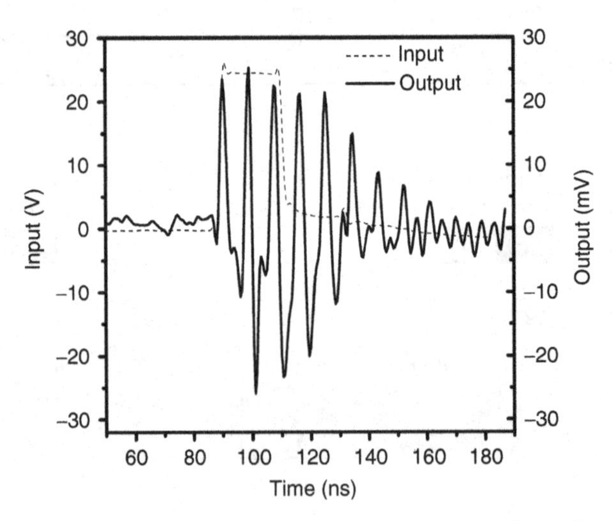

Figure 7.16. Experimental data on spin-wave detection.

compensate the damping of spin waves, one needs to include an amplification mechanism (gain). One of the possible gain mechanisms may be parametric spin-wave amplification, which has been experimentally demonstrated [17].

7.9. THE PRINCIPLE OF OPERATION

In order to illustrate the principle of operation, we consider the propagation of a spin wave excited by an ACPS transmission line. The spin dynamics can be described using the Landau–Lifshitz's equation as follows:

$$\frac{d\vec{m}}{dt} = -\frac{\gamma}{1+\alpha^2}\vec{m} \times \left[\vec{H}_{eff} + \alpha\vec{m} \times \vec{H}_{eff}\right], \tag{7.1}$$

where $\vec{m} = \vec{M}/M_s$ is the unit magnetization vector, M_s is the saturation magnetization, γ is the gyro-magnetic ratio, and α is the phenomenological Gilbert damping coefficient. The first term of Equation 7.1 describes the precession of magnetization about the effective field and the second term describes its dissipation. $\vec{H}_{eff}$ is the effective field that consists of the superposition of the external field and contributions from anisotropy, and exchange fields.

$$\vec{H}_{eff} = -\nabla^2\Phi + \frac{2A}{M_s}\nabla^2\vec{m} + \frac{2K}{M_s}(\vec{m}\cdot\vec{e})\vec{e} + \vec{H}_{pulse}, \tag{7.2}$$

where$\nabla^2\Phi = 4\pi M_s \nabla \cdot \vec{m}$, A is the exchange constant, K is the uniaxial anisotropy constant, and $\vec{e}$ is the unit vector along with the uniaxial direction, $\vec{H}_{pulse}$ is the pulse field produced by the source-drain current. The first terms in Equation 7.2 are defined by the material properties of the ferromagnetic sample, while the last term, the excitation pulse field, can be artificially controlled. Theoretically, by adjusting the form of the external field pulse, it is possible to excite a spin wave of any desired frequency and amplitude. To be realistic, we use experimental data and an analytical model developed in [2], where spin wave excitation in NiFe film by applying current pulses has been investigated in detail. It was found that a short current pulse (<100 ps) through a conducting strip placed close to the NiFe film (0.54 μm oxide thickness) excites a spin-wave packet, which is a linear superposition of spin waves. An analytical solution to Equation 7.1 was found, and it describes the propagation of the wave packet through the ferromagnetic film. The results of the numerical simulations demonstrated a good agreement with experimental data [2].

In Figure 7.14, we have show two ACPS lines, one as input and one output. If we add another ACPS line, it can be used as a second input port. Each of the input devices (ASPC lines) excites a spin-wave packet consisting of a Gaussian distribution of wave vectors that is $2/\delta$ in width and centered about k0. The wave packet propagates along with the y direction and can be described with one magnetization component My as follows:

$$M_y = \frac{C\exp(-t/\tau)}{\delta^4 + \beta^2 t^2}\exp\left[\frac{-\delta^2(y - vt)^2}{4(\delta^4 + \beta^2 t^2)}\right] \times \cos(k_0 y - \omega t + \phi), \tag{7.3}$$

where C is a constant proportional to the amplitude, τ is the decay time, ϕ is the initial phase, $v = \partial\omega/\partial k$ ($k = k_0$) and $\beta = (1/2)\partial^2\omega/\partial k^2$ ($k = k_0$) are the coefficients of the first and second order terms, respectively, in the Taylor expansion of the nonlinear dispersion, $\omega(k)$. The dispersion relation for spin waves propagating orthogonally to the magnetization is given by

$$\omega = \gamma\left\{8\pi K + (2\pi M_s)^2[1 - \exp(-2kd)]\right\}^{1/2}, \tag{7.4}$$

where d is the thickness of the film. In numerical simulations, we used NiFe material characteristics: $A = 1.6 \times 10^{-6}$ erg.cm^{-1}, $4\pi M_s = 10$ kG, $2K/M_s = 4$ Oe, $\gamma = 19.91 \times 10^6$ rad/s Oe, $\alpha = 0.0097$, known from the literature [1, 18]. Taking the fitting parameters for the wave packet obtained in [2], we use $\tau = 0.6$ ns, $k_0 = 0.25\ \mu\text{m}^{-1}$ and $\delta = 5.7\ \mu$m for $d = 27$ nm. In Figure 7.17, we have show the results of numerical simulations illustrating spin-wave packet propagation.

The distance between the excitation point and the point of observation is 50 μm. The spin waves produce perturbation in spin orientation perpendicular to the direction of magnetization, whose amplitude is much less than the saturation magnetization $|M_y|/M_s \ll 1$.

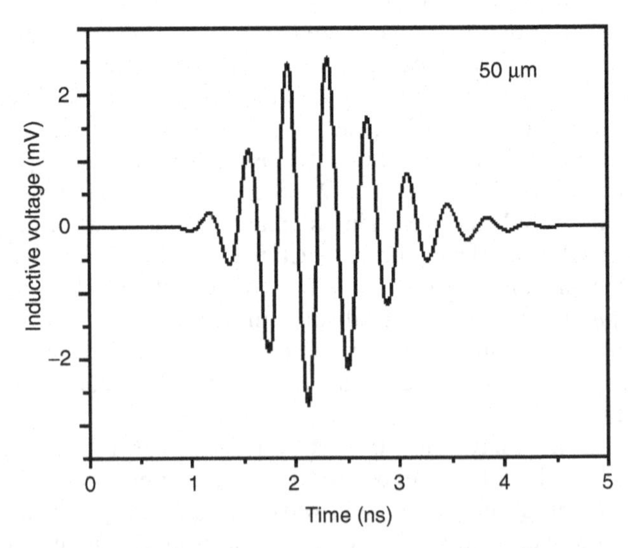

Figure 7.17. Numerical simulations results: propagation of a spin-wave packet.

Next, we consider the combined effect of spin-wave packets produced by two input devices. We assume that each of the input devices generates a spin-wave packet, which is described by Equation 7.3. The amplitudes of the input signals are the same, while the relative phase between the signals can be controlled, for example, by the polarity of the applied current pulses. The current pulses having the same polarity produce local magnetic fields oriented in the same direction, so the generated spin-wave packets have the same initial phase ($\phi_1 = \phi_2$). In the other case, when the current pulses have different polarity, the produced spin-wave packets have relative phase difference. ($\phi_1 = \phi_2 = \pi$)

In order to find the magnetization change caused by two spin-wave packets, we calculated the resultant magnetization as a superposition of waves of the same frequency from each packet.

$$\bar{M}_y = \frac{1}{w} \int_{-w/2}^{w/2} \int_{\omega} \left[M_{1y}^2(\omega) + M_{2y}^2(\omega) + 2M_{1y}(\omega)M_{2y}(\omega)\cos(\phi_1 - \phi_2) \right] d\omega dy \quad (7.5)$$

In above equation, w is the width of the detecting device along the Y axis (the gap between the strips), and the subscripts depict the magnetization components of the first and the second packets, respectively. We made integration over the finite length ($w = 200$ nm) to take into account the effect of dephasing. Then, we calculated the inductive voltage according to [1]:

$$V_{ind} = \left(\frac{\mu_0 l d f(z,w)}{4} \right) \left(\frac{Z}{Z + 0.5 R_{dc}} \right) \frac{d\bar{M}_y}{dt}, \quad (7.6)$$

where μ_0 is the magnetic constant, l is the length of the sample, $f(z,w)$ is the spacing loss function, Z is the strip line resistance, and R_{dc} is the total ACPS line

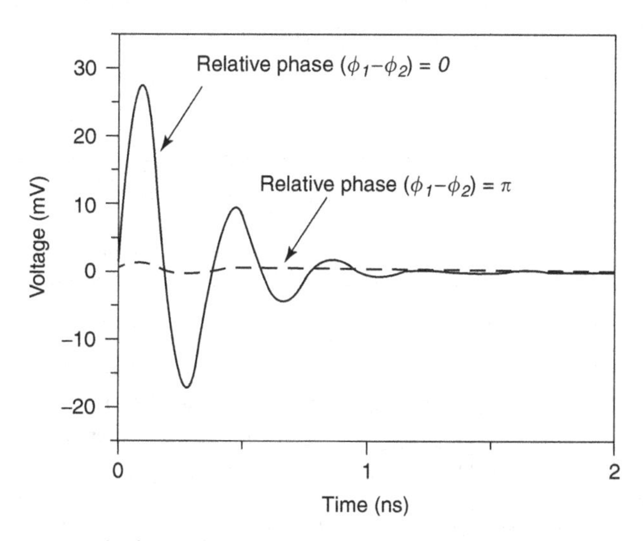

Figure 7.18. Numerical simulations results: output of two spin-wave packets.

dc resistance. In our calculation, we neglected the additional losses due to the ACPS line resistance. In Figure 7.18, we present the results of numerical simulations showing the inductive voltage as a function of time obtained in two cases: Spin-wave packets are excited in phase (the solid line) and out of phase (the dashed line).

The results of simulations illustrate the output of the prototype logic device described above. As shown in Figure 7.18, the amplitude of the inductive voltage is maximum when the spin-wave packets are coming in phase. In the other case, the waves from two packets compensate each other and produce inductive voltage of much less amplitude.

7.10. CONCLUSIONS

In this chapter, three nanoscale architectures were presented: the reconfigurable mesh, crossbar, and fully interconnected cluster. In these architectures, both communication and computation are performed using spin waves. Furthermore, computation is done in both analog and digital modes. A hierarchical multiscale architecture was also presented that is an efficient integration of the spin-wave modules within a MEMS architecture.

REFERENCES

1. T. J. Silva et al. Inductive measurement of ultrafast magnetization dynamics in thin-film permalloy. *Journal of Applied Physics*, 85(11): p 7849–7862, 1999.
2. M. Covington, T. M. Crawford, and G. J. Parker. Time-resolved measurement of propagating spin-waves in ferromagnetic thin films. *Physical Review Letters*, 89(23): p 237202-1-4, 2002.
3. A. Khitun, R. Ostroumov, and K. L. Wang. Spin-wave utilization in a quantum computer. *Physical Review A*, 64(6): p 062304/1–5, 2001.
4. A. Khitun, R. Ostroumov, and K. L. Wang. Feasibility study of the spin wave quantum network. 10th International Symposium on Nanostructures: Physics and Technology. *Proceedings of the SPIE*, 5023 pp 449–451, 2003.
5. M. Wu, C. E. Patton, and B. A. Kalinikos. Generation of spin wave envelope solitons through modulational instability, integrable systems, and applications, International Conference on Nonlinear Waves, June (2005).
6. A. Khitun and K. L. Wang. Nanoscale computational architectures with spin-wave bus. *Superlattices and Microstructures*, 38(3): pp 184–200, Sep 2005.
7. M. M. Eshaghian-Wilner, A. Khitun, S. Navab, and K. L. Wang. A Nanoscale Crossbar with Spin Waves. Proceeding of the 6th IEEE Conference on Nanotechnology: pp 326–329, July 2006.
8. M. M. Eshaghian-Wilner, A. Khitun, S. Navab, and K. L. Wang. Nanoscale modules with full spin-wave interconnectivity. *Journal of Experimental Nanoscience*, 2(1–2): pp 73–86, Mar 2007.

9. M. M. Eshaghian-Wilner, A. Friesz, A. Khitun, A. C. Parker, S. Navab, and K. L. Wang. Emulation of neural networks on a nanoscale architecture. *Journal of Physics: Conference Series*, pp 288–292, 2007.
10. Y. Ohno *et al.* Electrical spin injection in a ferromagnetic semiconductor heterostructure. *Nature*, 402(6763): pp 790–792, 1999.
11. A. Dehon. Array-based architecture for FET-based, nanoscale electronics. *IEEE Transactions on Nanotechnology*, 2(1): pp 23–32, 2003.
12. M. M. Eshaghian-Wilner, A. Khitun, S. Navab, and K. L. Wang. A nanoscale reconfigurable mesh with spin waves. Proceeding of the ACM International Conference on Computing Frontiers, Italy: pp 65–70, May 2006.
13. M. M. Eshaghian-Wilner, A. Khitun, S. Navab, and K. L. Wang. Nanoscale modules with spin-wave inter-communications for integrated circuits. Proceedings of NSTI Nanotech 2006, Boston: pp 320–323, 2006.
14. M. M. Eshaghian-Wilner and L. Hai. Application-specific design of the optical communication topology in ORM. Proceedings of Supercomputing Applications in the Transportation Industries, International Symposium on Automotive Technology and Automation, Florence, Italy: June 1997.
15. L. Hai. Efficient parallel computing with optical interconnects. Ph.D. Thesis, New Jersey Institute of Technology: 1997.
16. M. M. Eshaghian-Wilner, A. Khitun, S. Navab, and K. Wang. Hierarchical multiscale architectures with spin waves. Proceedings of the 2006 International Conference on Computer Design CDES 2006, Las Vegas, Nevada: pp 220–226, June 2006.
17. B. A. Kalinikos, et al. Parametric frequency conversion with amplification of a weak spin-wave in a ferrite film. *IEEE Transactions on Magnetics*, 34(4, pt.1): pp 1393–1395.
18. W. K. Hiebert, A. Stankiewicz, and M. R. Freeman. Direct observation of magnetic relaxation in a small permalloy disk by time-resolved scanning Kerr microscopy. *Physical Review Letters*, 79(6): pp 1134–1137, 1997.

8

PARALLEL COMPUTING WITH SPIN WAVES

Mary Mehrnoosh Eshaghian-Wilner and Shiva Navab

In this chapter, we study the algorithm design aspects of three newly developed spin-wave architectures. The three nanoscale architectures, the spin-wave reconfigurable mesh, spin-wave crossbars, and spin-wave fully interconnected clusters, are capable of simultaneously transmitting multiple signals using different frequencies, and allow for concurrent read/write operations. Exploiting such parallel features, we introduce a set of generic parallel processing techniques that can be used for routing and design of fast algorithms on these spin-wave architectures. We also present a set of application examples to illustrate the operation of the proposed generic parallel techniques on these new spin-wave models.

8.1. INTRODUCTION

When mapping parallel algorithms onto parallel architectures, the performance of the algorithms highly depends on the underlying interconnectivity. In the past decades, various architectures with different types of interconnectivity were introduced; their aim was to produce fast algorithms. The major shortcoming in most of bus-based designs has been due to the fact that the use of bus was restricted to one pair of processors at any given time. Here, we deal with architectures with spin-wave buses that allow multiple pairs of processors to intercommunicate over each bus at any given time.

In the following chapter, we study the algorithm design aspect of the three spin-wave architectures presented in Chapter 7. Exploiting the highly parallel features of these architectures, we introduce a set of generic parallel processing

The authors of this chapter are listed alphabetically.

Bio-Inspired and Nanoscale Integrated Computing. Edited by Mary Mehrnoosh Eshaghian-Wilner

techniques that can be used for routing and design of fast algorithms on these spin-wave architectures. We also present a set of application examples to illustrate the operation of the proposed generic parallel techniques on these new spin-wave models.

8.2. PARALLEL ALGORITHM DESIGN TECHNIQUES

In this section, we present three generic parallel processing techniques that are mainly designed for spin-wave reconfigurable meshes. The implementation of some of these techniques using spin-wave crossbar and fully interconnected clusters is also briefly discussed. When using these architectures, one should take into consideration the fact that two different types of data detections are possible at the nodes. Once the spin waves are detected by the receiver, the transmitted data can either be digitized or left as analog.

In analog detection mode, the receiver detects a voltage that is the *superposition* of multiple waves. For example, if ten waves are sending a "1," then their analog sum through their cumulative amplitude is computed instantly as 10. Also, this property can be used to compute logical functions as described previously. In digital detection mode, this value is digitized to just a "1," and then the computations are continued digitally. In the following section, the generic parallel techniques are presented [1, 2].

8.2.1. Find the Maximum/Minimum

To find the maximum or minimum of $O(N^2)$ inputs (with a maximum value of 2^N) on a reconfigurable mesh, we first assign each value to a processing node on the grid points of the reconfigurable mesh. Next, each node checks its most significant bit (MSB). If MSB is 1, the node broadcasts a "1" to all the other processing nodes. If the MSB is 0, on the other hand, the processing node listens to the channel and gets disabled if it detects a 1 on the bus. At the next step, the nodes check their next most significant bit. This procedure is repeated for all the bits. Since the maximum value of the inputs is N, this procedure takes $O(\log N)$ time [3]. If the number of inputs is at most N, the minimum and maximum numbers can be found in $O(1)$ as presented in [4].

Finding the minimum is similar to finding the maximum except that in the case of MSB = 1, the node broadcasts a "1" and at the same time listens to the channel. If there is a 1 on the bus, that node becomes disabled.

The same implementation technique applies when using one single column of a spin-wave crossbar. In that case, the max/min of at most N inputs can be found using the N processing units. One of the columns would be used as the shared medium, where nodes broadcast their MSB. All the switches on that column must be turned on to connect the processing units to the shared bus.

This algorithm can also be implemented on a fully interconnected cluster in a similar fashion, except that in the fully interconnected cluster, there are no switches to be set. The nodes are all connected to the shared medium, where they can broadcast their MSB or receive signals from other nodes.

Note that on all of the spin-wave architectures, this whole routine can be performed simultaneously on disjoint sets of input, using different frequencies for each set. This cannot be done on a standard VLSI reconfigurable mesh due to the conflicts on the buses. Finding the maximum/minimum routine is used in several applications. For instance, it is used in the labeling algorithm and finding the nearest neighbor, explained in Chapter 19.

8.2.2. Find the First/Last in the List

In many applications, we need to find the first or last element of a list. In geometric algorithms, for instance, this operation is useful where we find the extreme points of a figure. The convex hull algorithm presented in Chapter 19 and the graph formation for multiple sequence alignment algorithms, presented in Chapter 14, are application examples that use this routine.

To find the first or last element of at most N inputs, the first step is to store the inputs in the processing elements in a column. We find the local topmost nodes on that column by making each node send a signal downward. This procedure begins by turning off all the switches above each node (disconnect the channel) to force the signal sent by each node downwards. Then all nodes simultaneously send out a signal downward and turn on all switches (reconnecting the channel), thereby lettings all signals pass through the bus. Each node that receives a signal will become disabled. As the result, the only active nodes are the local topmost nodes that have not received a signal. This procedure is performed in $O(1)$ time. An example is shown in Figure 8.1, where the first A in the list is found by disabling the rest of the As.

If different sets exist among the inputs, it is possible to find the first element of each set using frequency division multiplexing. For instance, in the above example, there are two sets (As and Bs). As shown in Figure 8.2, we find the first A and B in the list simultaneously by performing the routine on two different frequencies for A and B.

Unlike a reconfigurable switch that can be controlled in four different directions, a crossbar switch has just two possible statuses: "on" or "off." Therefore, it is not possible to force the wave in only one direction. As a result, the first/last routine cannot be implemented on a crossbar in a similar fashion as a reconfigurable mesh.

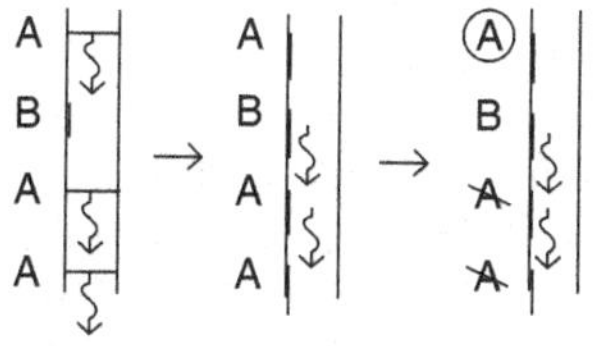

A B A A → A B A A → Ⓐ B A A

Figure 8.1. Finding the first element of a list.

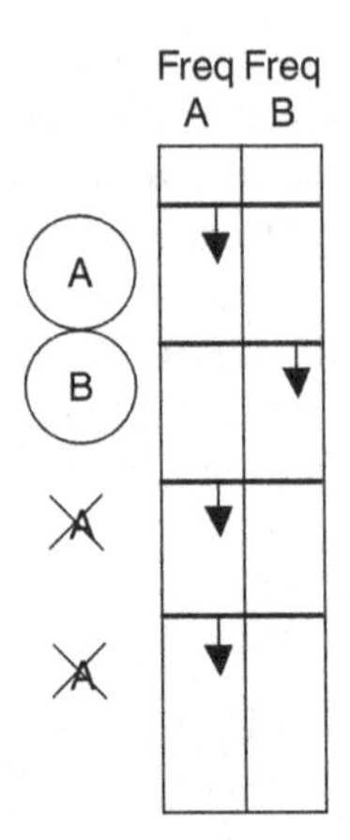

Figure 8.2. Finding the first element of different sets.

This implementation does not apply to a fully interconnected network either, since in that symmetric architecture, there is no notion of first or last.

8.2.3. Find the All Prefix Sum

To find the all prefix, one should compute the sum of the first j elements in an array and store the sum in the jth element. Using this algorithm, the sum of all the input values will be computed and stored in the last element. Here we show how this operation is performed in $O(1)$ time on spin-wave architectures.

To perform the all prefix sum routine for most N inputs, first we map the inputs to the processing nodes on one column of the spin-wave reconfigurable mesh. At the beginning, all of the switches are turned on. Then, each node sends a signal with an amplitude equal to the value it has stored and immediately turns off the switch above it, directing the generated wave downwards. Note that for synchronization purpose, each node generates its spin-wave signal after a short period of time to compensate the spin-wave propagation delay in the ferromagnetic channel. This short period of time has to be greater than or equal to the distance between adjacent nodes divided by the speed of spin waves, which is in the order of $10^{-9}/10^{4}=10^{-13}$. Considering the fact that the frequency is in the order of GHz or even THz, this whole process is considered as constant time. The sum of previous values can be found at each element by using the superposition property of spin-waves. The main idea behind superposition is that when nodes transmit signals on the same frequency, the signals will superpose in amplitude when they meet. The amplitude of the superposed wave received by each node will be the prefix sum on that node. This routine is used in the graph formation algorithm for partial-order multiple sequence alignment, presented in Chapter 14.

A special case of finding all prefix sum routine is counting the number of elements preceding a certain element in the list. To solve this problem, all the nodes transmit a signal downwards with an amplitude of 1. In each node, the total amplitude of all the superposed signals sent by the nodes above shows the number of its preceding elements in the list.

Another special case of finding the all prefix sum routine would be the summation routine, where the summation of all nodes is found and stored in constant time. This is equivalent of all the prefix sum routine when performed at the very last node in the list. All the processing elements send downwards a signal with an amplitude equal to the value they have stored. Utilizing the superposition property of spin-waves, the last node receives the "sum" of the input values in constant time. This routine has been used in implementing the graph formation of the POMSA algorithm [5] described in Chapter 14. Note that the all prefix sum routine can be performed on at most N inputs, on a single column of the reconfigurable mesh. However, the summation routine can have as many as N^2 inputs, mapped to all the nodes of the reconfigurable mesh. In that case, all the nodes send out their signals and the superposed signal or the "sum" received by all the nodes in $O(1)$ time.

Since it is not possible to send signals downwards in a crossbar or in a fully interconnected cluster, the all prefix sum routine can not be implemented on these two architectures. However, the special case of finding the summation of all inputs, for at most N inputs, can be implemented on one single column of a crossbar and on a fully interconnected cluster.

As another application example, consider the problem of creating a histogram of input values where all such values are in the range of 1. N and each value represents a type. We show that this problem can be solved in $O(1)$ time on a spin-wave reconfigurable mesh. Assuming that the number of inputs is at most N^2, the inputs can be mapped to processing nodes. There are two approaches to store the result of the histogram. The first one is to choose one node of each type to be the representatives of that type. We can find the first of each type in the list to represent its type, using the routine explained in the previous section. The second approach, which we choose, is to let each node keep track of the number of nodes from its type.

The next step is to count the number of times each value has been repeated in the list and store the sum at each (or the representative) node. As explained above, this problem is a subset of finding the all prefix sum routine. In this case all the nodes send a value of "1" and the result would be the number of inputs in that category.

Note that this routine should be performed on each type of input; however, using different frequencies, these separate routines can be done simultaneously. In other words, each type's sender and receiver are tuned on a distinct frequency. Due to superposition characteristic of the waves, each node receives the sum of 1 s sent on its tuned frequency, which is the number of nodes of that type.

8.2.4. Shifting the Elements of a List

In several applications, the elements of a list needed to be shifted. The shifting operation can be implemented on a row or a column of a spin-wave reconfigurable mesh. We explain two methods for implementing the shifting operation on one row of the spin-wave reconfigurable mesh. In both methods, first the elements of the list are mapped to the processing nodes.

In the first method, the shift operation is performed by using the spin-wave switches. As explained in the previous section, spin waves can be forced in one direction by using the spin-wave switches. To implement shifting to the right, all the nodes first turn off their left switch and send out a spin-wave with an amplitude equal to their data while their right switch is open. This forces the waves to be directed to the right. As soon as the nodes send their waves, they immediately open their left switch and close their right switch. This allows each node to receive the spin wave from its left neighbor, while blocking the channel and preventing this received wave to go through to the next node on the right hand side. Similarly, shifting to the left can be implemented by forcing the waves to the left and not letting the wave pass through the immediate neighboring node. This method can not be used in a crossbar or fully interconnected cluster, since there are no switches between two adjacent nodes.

No spin-wave switches are used as the second method for implementing the shift operation. The nodes are required to dynamically tune their frequency. In this method a distinct frequency is assigned to each node as its receiving frequency. In addition, each node's sending frequency is tuned on its right neighbor's assigned frequency. For instance, the sending frequency of node x_i is tuned on f_{i-1}. Since these data transmissions between adjacent nodes are done at different frequencies, all can be performed simultaneously. This method can be used in implementing the shift operation on the reconfigurable mesh, as well as the crossbar and fully interconnected cluster.

One of the applications of the shifting operation is in implementing a spin-wave finite impulse response (FIR) filter module. An FIR filter is one of the primary types of filters used in digital signal processing [6]. An FIR filter works by multiplying an array of the most recent n data samples by an array of constants (called the tap coefficients), and summing up the elements of the resulting array. The filter then inputs another sample of data, which causes the oldest piece of data to be discarded, and repeats the process [7].

FIR filters are implemented using a finite number M delay taps on a delay line and M computation coefficients to compute the algorithm (filter) function. Figure 8.3 illustrates the block diagram of an FIR filter with M taps [8]. In the FIR filter, the output at time n, $y(n)$, is computed as the summation of inputs multiplied by the tap coefficient as illustrated by the following equation:

$$y(n) = b_0 x(n) + b_1 x(n-1) + \cdots + b_M x(n-M).$$

To implement a spin-wave FIR filter, we use the superposition property of the waves to realize the summation function and we employ spin-wave switches to implement the shifting operation. First, each coefficient is stored in the nodes with respect to the node's index. In the next step, each processing node computes the multiplication of its fixed coefficient and the input mapped to it. All the processing

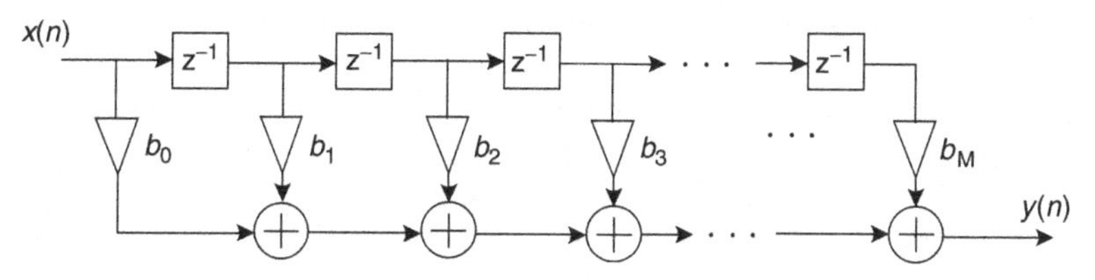

Figure 8.3. Block diagram of an FIR filter.

nodes then broadcast a spin wave on the bus with an amplitude equal to their data. These values, which are sent at the same frequency, are superposed and form a wave with amplitude equal to the summation of all. The output node's receiving frequency is tuned on the sending frequency of all the input nodes; therefore, it receives the result of this summation. At this stage, all the previous inputs are shifted right by one position, the new data is inputted to the leftmost node, and the summation process is repeated. The structure of a spin-wave FIR filter module is shown in Figure 8.4. This structure includes $M+1$ input node, where M is the number of taps.

In this implementation, the shift operation is performed by using the spin-wave switches as explained previously. (These switches are all on in the summation step.)

As mentioned, it is possible to use another method for implementing the shift operation in which distinct frequencies are assigned to different nodes. Note that in implementing the FIR filter, after each shifting operation, all the nodes' sending and receiving frequencies must be tuned on a common frequency to perform the summation. Therefore, in each step the nodes are required to dynamically tune their frequencies.

Having explained some of the parallel techniques for algorithm design on spin-wave architectures, in the next section we describe some of their parallel routing schemes.

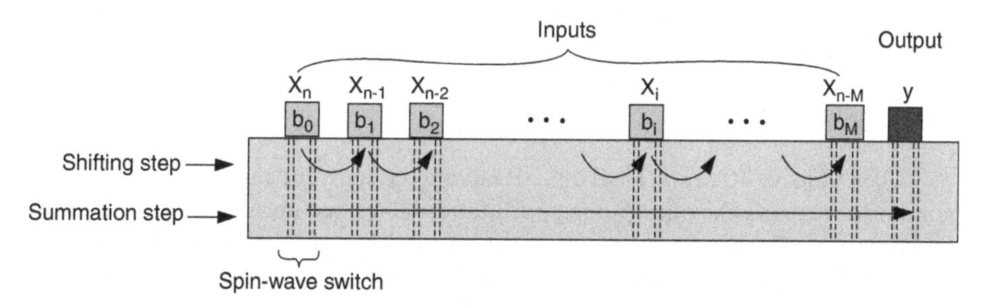

Figure 8.4. Structure of a nanoscale spin-wave FIR filter.

8.3. PARALLEL ROUTING AND BROADCASTING

In the following section, we illustrate the routing features of our three spin-wave architectures. We focus primarily on the routing on a spin-wave crossbar, and we explain how the routing on the other two architectures differs from the routing on the crossbar. We also discuss the enhanced multiple multicasting feature on the fully interconnected cluster.

8.3.1. On Spin-Wave Crossbar

Our spin-wave crossbar has several parallel and fault tolerant routing features [9]. We concentrate the routing features of this architecture in three different scenarios. These techniques are then compared to those for the reconfigurable spin-wave architecture and the fully connected spin-wave architecture.

It is well known that all crossbars are capable of realizing any arbitrary one-to-one permutation. In a standard VLSI crossbar, however, unless there are broadcasting buses on each row, at any single point in time, only one switch is turned on in each row and each column. Spin-wave crossbars, on the other hand support additional features such as broadcasting and concurrent receiving as described below.

ARBITRARY PERMUTATIONS. Similar to any standard crossbar, a spin-wave crossbar realizes arbitrary permutations. As described in the previous section, in the crossbar architecture the signals are directed in each row and each column through spin-wave buses. As an example of a one-to-one permutation realization, assume that input 3 needs to send a message to output 6. In that case, the switch in row 3 and column 6, represented as $s(3,6)$ should be set to on. In addition, the receiver's frequency of node 6 should be tuned to sender's frequency of node 3. The switches can be set to on according to the following mechanism: A fixed frequency is assigned to each column, and on top of each switch there is a receiver that is tuned to the frequency assigned to its column. As soon as the switch receives a signal on its frequency, it is activated and can route the data. For instance, switch $s(3,6)$ is tuned onto the frequency assigned to column 6, f_6. Input node 3 sends a signal on frequency f_6 on row 3, which turns on $s(3,6)$. Now, the third row is connected to the sixth column, and permutation (3,6) is realized. Figure 8.5 shows this communication on a crossbar of size 6. Note that there is a switch located on each of the grid points, but here we are just showing the activated one.

CONCURRENT RECEIVE FEATURE. Realizing concurrent receive feature is very similar to realizing the one-to-one permutation described above. A fixed frequency is assigned to each column (each receiver), and the senders tune their sending frequency to that frequency. As explained in the previous section, one of the important features of a spin-wave crossbar is allowing concurrent write. For instance node 2, 3, and 4 can all send a message to node 5, as shown in Figure 8.6.

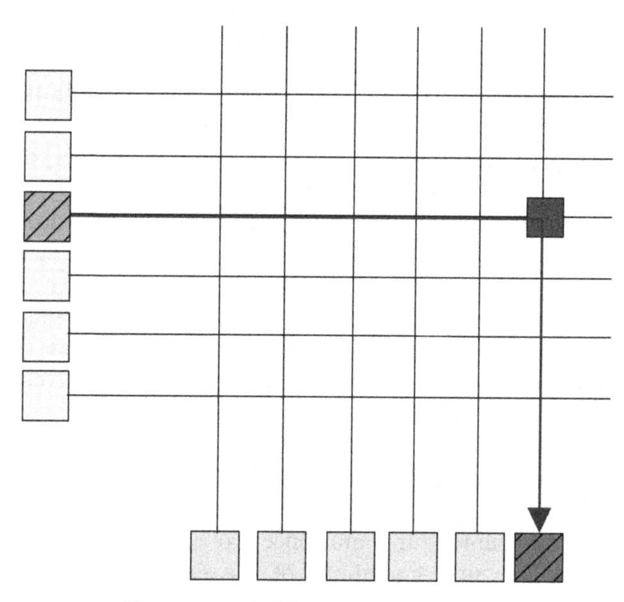

Figure 8.5. Arbitrary permutation.

Due to the superposition property of waves, output 5 receives a signal which is the sum of these three waves. In a standard VLSI crossbar, it is not possible to perform these three communications simultaneously because such a situation will cause a conflict on column 5.

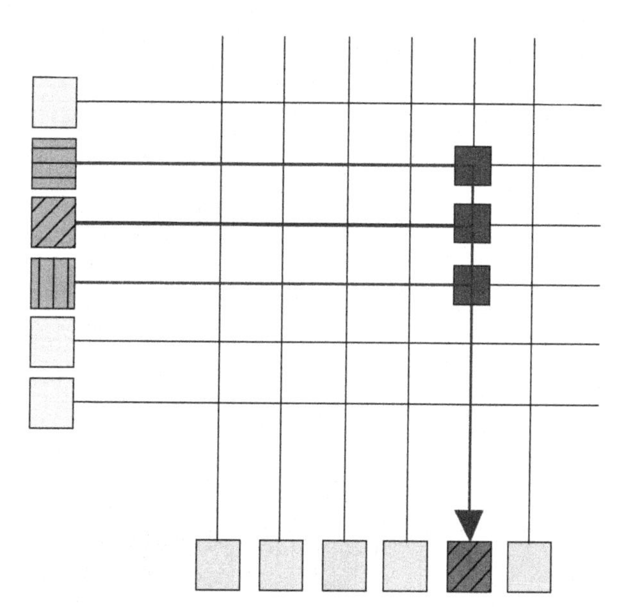

Figure 8.6. Concurrent receive feature.

BROADCASTING FEATURE. Broadcasting at a node happens when that node sends a single message to multiple receivers. Realizing broadcasting in a spin-wave crossbar is slightly different from realizing concurrent receive. In this case, a fixed frequency is assigned to each row (each sender), and the receivers tune their receiving frequency to that fixed frequency. As explained earlier, one of the most important advantages of a spin-wave crossbar is that one input can broadcast to multiple outputs simultaneously. For instance, node 3 can broadcast a message to output 2, 4, 5, and 6 at the same time, as shown in Figure 8.7. The only constraint is that the receivers should be tuned on the sender's frequency.

Note that different senders can broadcast to different sets of inputs on different frequencies. However, the sets must be disjoint since the receivers in different sets need to be tuned on different frequencies respectively.

FAULT-TOLERANT ROUTING. Fault tolerance is one of the most important requirements for nanoscale devices and architecture because these devices suffer from dramatically increased permanent and transient failure rates. These failures are mainly due to the quantum nature (and hence probabilistic behavior) of the devices as well as the fundamental limitations of the fabrication processes [10]. The achievable degree of fault-tolerance to the defects present in the nanoscale devices will be the main concern in the adoption of new approaches in nanotechnology.

One well known approach for developing reliable architectures to address both types of manufacturing and transient defects in nanoscale devices is to incorporate spatial and/or temporal redundancy [11]. In recent years, different tools have been developed to evaluate certain design trade offs in the nanoscale architectures. The

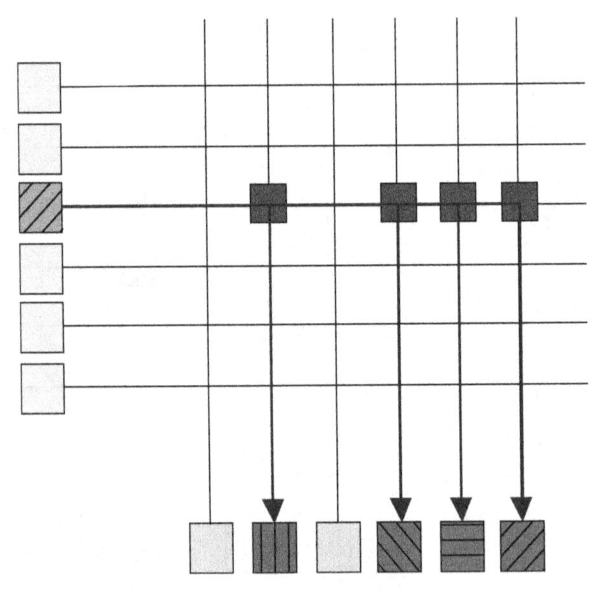

Figure 8.7. Broadcasting feature.

most important trade offs are between granularity and reliability [12] and between redundancy and reliability [13]. It is shown that to make nanoscale systems reliable, different degrees of redundancy need to be applied at different granularity levels (i.e., gate and reconfigurable logic block levels) [13].

Several designs and techniques have been studied to increase the tolerance of nanoscale architectures to both transient and fabrication defects [14–31]. For instance, a number of redundancy schemes, including von Neumann's multiplexing logic, N-tuple modular redundancy, and interwoven redundant logic, have been presented in [20]. In addition, a new fault-tolerant design approach based on coding theory has recently been proposed at HP Labs [21]. In this approach, by using a crossbar architecture and adding 50% more wires, nanoelectronic circuits with nearly perfect yields can be fabricated even though the probability of broken components is high. To implement fault-tolerant quantum computers, quantum error correcting codes are being developed and elementary quantum gates are being constructed to form the basic building block of these computers [22]. Furthermore, three logic mapping algorithms with defect avoidance have been presented in [23] to circumvent clustered defective crosspoints in nanowire reconfigurable crossbar architectures.

Fault tolerance is achieved in all these architectures by adding some level of redundancy. While redundancy is needed for reliable computation, economic constraints also need to be considered in choosing the redundancy factor [11]. The advantage of the fault-tolerant scheme presented in this chapter as compared to the other methods mentioned previously is in its smaller degree of required redundancy. In this section, we concentrate on the fault-tolerance features of a spin-wave nanoscale crossbar architecture. We show that by employing the parallel features of this architecture the amount of the spatial redundancy required for a fault-tolerant design is significantly reduced [9]. In the following section, we briefly talk about fault diagnosis for our spin-wave crossbar and then present a simple fault recovery scheme.

Fault Diagnosis. There are many different ways in literature to detect a defective switch [24–28]. Here we choose a very simple method, in which an acknowledgment is sent from the receiver back to the sender for each transmission. If the sender does not receive an acknowledgment from the receiver after a fixed amount of time, existence of a fault has been determined. The sender then tries to resend the message through another route.

Fault Recovery. In the example presented earlier where input 3 was sending to output 6, assume that the switch $s(3,6)$ is defective. So now after node 3 sends a message to node 6, it does not receive the acknowledgement. Therefore, sender 3 will attempt to resend the message to node 6 through a new path. There are several schemes to reroute the message. The method we employ to reroute the path is simple. It is basically performed by adding an extra column of switches to the crossbar. In case of a fault on any of the switches on row i, the input, using the extra column, connects with the spin-wave bus path on row $i+1$ (or row $i-1$ if in

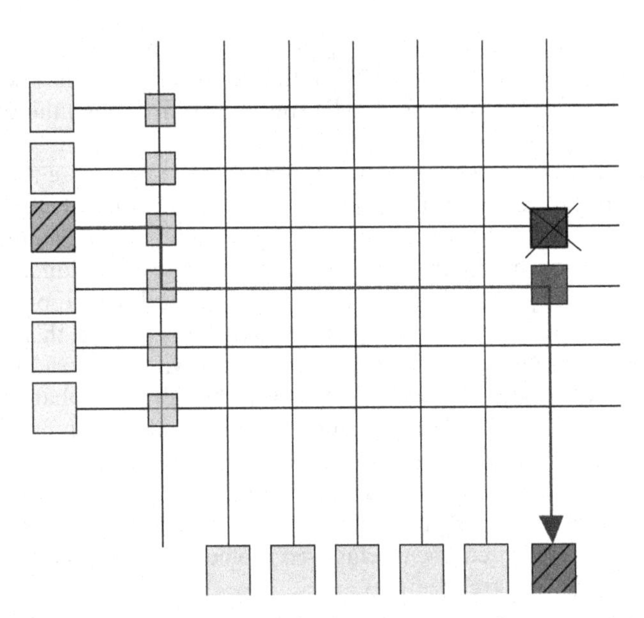

Figure 8.8. Fault recovery example.

the last row) and then from there goes to its destination, as shown in the example below. We refer to the switches in the extra routing column as switches in column 0. In our example, node 3 connects to row 4 via this switch and reroutes the path through $s(4,6)$, as shown in Figure 8.8.

The significant advantage of a spin-wave crossbar over a standard VLSI crossbar is that this simple scheme for rerouting does not collide with other intercommunications along the same row or column. For instance, as shown in Figure 8.9, in the same example, node 4 can still send a message to node 5 while 3 sends a message to 6 via $s(4,6)$ in row 4. Although the rerouted path passes through row 4, these communications can be done in parallel with no conflict. This is due to the fact that nodes 3 and 4 use different frequencies, so their signal waves pass through each other without interference.

Note that in this example, the two signals from input 3 and 4 go through row 4, as well as both columns 5 and 6. So the two input messages reach both outputs 5 and 6; however, output 5 detects the message from input 4 on its tuned frequency, while output 6 receives the signal from input 3.

8.3.2. On Spin-Wave Reconfigurable Mesh

The routing on a reconfigurable mesh is similar to a crossbar. However, this routing can be from any of the N^2 processing elements to any other one, so there can exist up to $N^2 \times N^2$ different routing schemes.

The routing mechanism in a spin-wave reconfigurable mesh is as follows: To send information from $P_{i,j}$ to $P_{k,l}$, the sender, $P_{i,j}$ sends the signal to switch $s(i,l)$ to be routed to $P_{k,l}$ as shown in Figure 8.10.

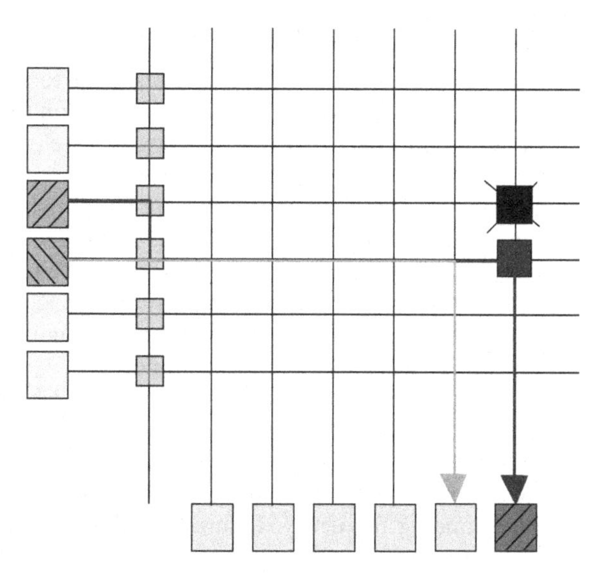

Figure 8.9. Parallel communications on spin-wave bus.

As mentioned in the previous section, the significance of spin-wave architectures is that multiple waves on different frequencies can pass through the same bus without any conflict. For instance $P_{3,2}$ can send a signal to $P_{6,5}$, while $P_{3,3}$ is sending a signal, on the same row and column, to $P_{5,5}$.

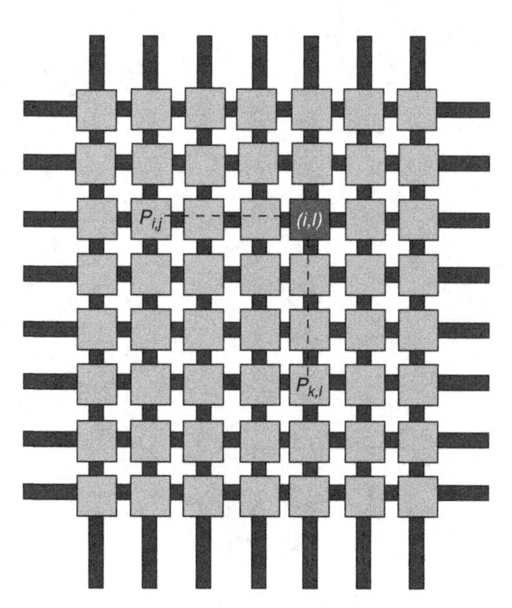

Figure 8.10. Routing in reconfigurable mesh.

8.3.3. On Spin-Wave Fully Interconnected Cluster

The routing on a fully interconnected cluster is quite similar to the routing on a spin-wave crossbar, except the fact that there are no switches on this architecture. In addition, the fully interconnected cluster has an extra feature (which we explain later in this section).

Similar to a crossbar, the concurrent receive feature applies here as well. At a given frequency, a node can listen to multiple waves simultaneously. Using the superposition property of waves, it receives the sum of all waves destined to it. Consider the scenario where multiple senders send data to a node, for instance *G*, at the same time. *G* receives the sum of those signals, providing that all the nodes transmit on the frequency at which *G*'s receiver is tuned.

Multiple broadcasting is possible here too. To distinguish the data being transmitted to different nodes, transmissions are done at distinct frequencies, using frequency division multiplexing. In a way, this is similar to having various radio stations, each broadcasting at a different frequency. To listen to a specific station, one tunes to the corresponding frequency. Figure 8.11 shows an example, where node *A* is sending to a set of nodes, while *C* is sending to another set.

Note that since different senders broadcast to different sets on different frequencies, the sets must be disjoint. However, as pointed out earlier, the fully interconnected network has an additional feature that is different than the other two

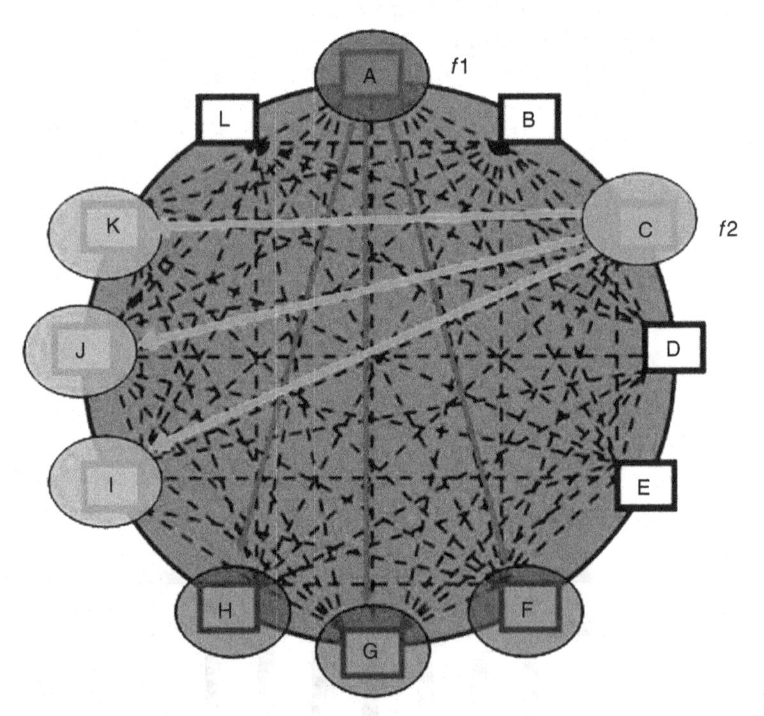

Figure 8.11. Multiple broadcasting on disjoint sets of receivers.

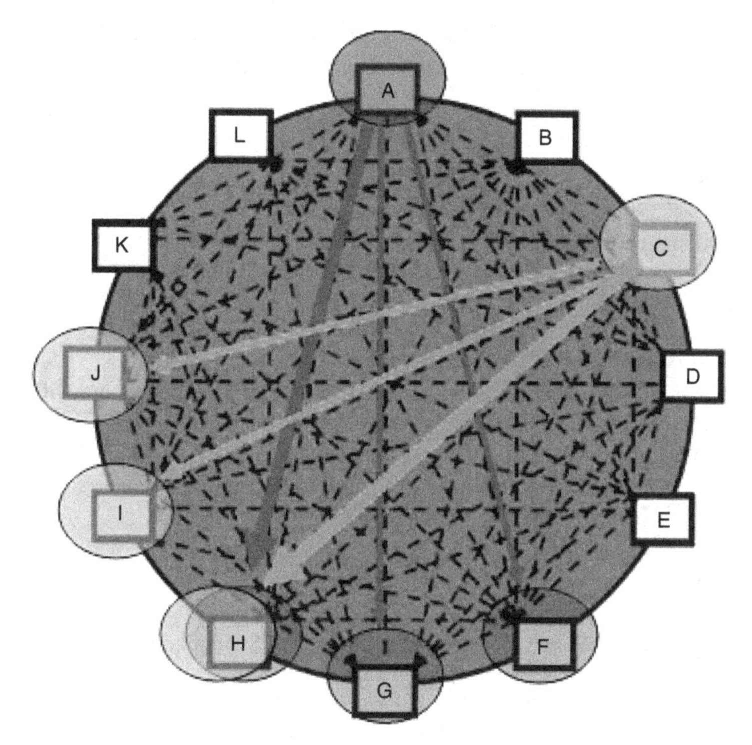

Figure 8.12. Multiple broadcasting on disjoint sets of receivers.

architectures. This feature allows multiple braodcasting to sets which are not disjoint. This is basically the combination of concurrent receive and multiple broadcasting.

Assume the scenario shown in the Figure 8.12 where one of the *A*'s and *C*'s destinations is the same (node *H*). This requires the sending frequency of A and C to be the same as the receiving frequency of H. However, if the sending frequency of these two nodes are the same, the receiving frequency of *K*, *J*, *G*, and *F* has to be the same too, which causes each of these nodes to receive the superposition of the signals sent by *A* and *C*.

One approach to solve this problem, is using phased array techniques explained in [32] to direct the waves to specific locations. It is also possible to combine the phased array technique with multiple frequencies. As a result, for each frequency, some of the waves are only transmitted to desirable directions and are received by the intended sources.

8.4. CONCLUSIONS

In this chapter, the algorithm design aspects of the spin-wave reconfigurable mesh, spin-wave crossbars, and spin-wave fully interconnected were studied. The architectures simultaneously transmit multiple signals using different frequencies, and allow for concurrent read/write operations. Exploiting such parallel features, a set of generic and fast parallel processing techniques were introduced.

REFERENCES

1. M. M. Eshaghian-Wilner and S. Navab. Generic parallel processing techniques for nanoscale spin-wave architectures. 2007 World Congress in Computer Science, Computer Engineering, and Applied Computing WORLDCOMP'07. In: Proceedings of the 2007 International Conference on Parallel and Distributed Processing Techniques and Applications (PDPTA'07): pp 221–227.
2. M. M. Eshaghian-Wilner and S. Navab. Efficient parallel processing with spin-wave nanoarchitectures. *Journal of Supercomputing*; October 2008.
3. M. M. Eshaghian-Wilner, S. Navab, A. Khitun, and K. L. Wang. The Spin-wave Reconfigurable Mesh and Labeling Problem. *ACM Journal on Emerging Technologies in Computing Systems*, 3(2, no. 5): July 2007.
4. M. M. Eshaghian-Wilner, A. Khitun, S. Navab, and K. L. Wang. A nanoscale architecture for constant time image processing. *Physica Status Solidi A*, 204(6): pp 1931–1936, June 2007.
5. M. M. Eshaghian-Wilner, L. Lau, S. Navab, and D. Shen. Parallel graph formations of partial-order multiple-sequence alignments using nano-, micro-, and multi-scale reconfigurable meshes. Submitted to IEEE *Transactions on NanoBioScience*: 2008.
6. A. V. Oppenheim, R. W. Schafer, and J. R. Buck. *Discrete-Time Signal Processing*. Upper Saddle River, NJ: Prentice Hall, 1999.
7. http://www.bores.com/courses/intro/filters/4_fir.htm.
8. http://www.freqdev.com/guide/DgtlFltrDsgnGd.pdf.
9. M. M. Eshaghian-Wilner and S. Navab. Parallel and fault-tolerant routing in the nanoscale spin-wave architectures. 2007 World Congress in Computer Science, Computer Engineering, and Applied Computing WORLDCOMP'07. In: Proceedings of the 2007 International Conference on Computer Design (CDES'07): pp 3–9.
10. C. Constantinescu. Trends and challenges in VLSI circuit reliability. *IEEE Microwaves*, 23: pp. 14–19, Jul–Aug 2003.
11. S. Roy and V. Beiu. Majority multiplexing—Economical redundant fault-tolerant design for nano architectures. *IEEE Transactions on Nanotechnology*, 4(4): pp 441–451, Jul 2005.
12. D. Bhaduri, S. K. Shukla, and Nanoprism. A tool for evaluating granularity vs. reliability trade-offs in nano architectures. GLSVLSI, ACM, Boston, Apr 2004.
13. S. K. Shukla, G. Norman, D. Parker, and M. Kwiatkowska. Evaluating the reliability of defect-tolerant architectures for nanotechnology with probabilistic model checking. Proceedings of the International Conference on VLSI Design 2004.
14. P. T. Gauehan, B. V. Dao, S. Yalamanchili, and D. E. Schimmet. Distributed, deadlock-free routing in faulty, pipelined direct interconnection networks. *IEEE Transactions on Computers*, 6: pp 651–665, 1996.
15. B. Almohammand and B. Bose. Fault-tolerant communication algorithms in toroidal networks. IEEE Transactions on Parallel and Distributed Systems, 10: pp 976–983, 1999.

16. V. P. Roychowdhury, D. B. Janes, S. Bandyopadhyay, and X. Wang. Collective computational activity in self-assembled arrays of quantum dots: a novel neuromorphic architecture for nanoelectronics. *IEEE Transactions on Electron Devices*, 43: 1996.
17. R. M. P. Rad and M. Tehranipoor. A reconfiguration-based defect tolerance method for nanoscale devices. 21st IEEE International Symposium on Defect and Fault-Tolerance in VLSI Systems (DFT'06): pp 107–118, 2006.
18. K. Nikolic, A. Sadek, and M. Forshaw. Fault-tolerant techniques for nanocomputers. *Nanotechnology*, 13(3), pp 357–362(6), 2002.
19. J. Byunghyun, Y. Kim, and F. Lombardi. Error tolerance of DNA self-assembly by monomer concentration control. 21st IEEE International Symposium on Defect and Fault-Tolerance in VLSI Systems (DFT'06): pp 89–97, 2006.
20. J. Han, J. Gao, Y. Qi, P. Jonker, and J. A. B. Fortes. Toward hardware-redundant, fault-tolerant logic for nanoelectronics. *IEEE Design and Test of Computers*, 22(4): pp 328–339, July 2005.
21. http://www.extremetech.com/article2/0,1558,1826021,00.asp.
22. D. P. Vasudevan, P. K. Lala, and J. P. Parkerson. Fault tolerant quantum computation with new reversible gate. Proceedings of the NSTI Nanotechnology Conference: pp 744–747, 2005.
23. Y. Yellambalase, M. Choi, and Y. Kim. Inherited redundancy and configurability utilization for repairing nanowire crossbars with clustered defects. Proceedings of the 21st IEEE International Symposium on Defect and Fault-Tolerance in VLSI Systems (DFT'06): pp. 98–106, 2006.
24. B. Almohammand and B. Bose. Fault-tolerant communication algorithms in toroidal networks. IEEE Transactions on Parallel and Distributed System, 10: pp 976–983, 1999.
25. J. Zhou and F. Lau. Adaptive fault-tolerant wormhole routing with two virtual channels in 2D meshes. International Symposium on Parallel Architectures, Algorithms and Networks: p 142, 2004.
26. G. Wang and J. Chen. A new fault-tolerant routing scheme for 2-dimensional mesh networks. Proceedings of the Fourth International Conference on Parallel and Distributed Computing: 2003.
27. X. Fan, W. Moore, C. Hora, M. Konijnenburg, G. Gronthoud. A gate-level method for transistor-level bridging fault diagnosis. 24th IEEE Proceedings of VLSI Test Symposium: 2006.
28. J. Zhou and F. Lau. Adaptive fault-tolerant wormhole routing with two virtual channels in 2D meshes. International Symposium on Parallel Architectures, Algorithms and Networks: p 142, 2004.
29. G. Wang and J. Chen. A new fault-tolerant routing scheme for 2-dimensional mesh networks. Proceedings of the Fourth International Conference on Parallel and Distributed Computing: 2003.
30. X. Fan, W. Moore, C. Hora, M. Konijnenburg, and G. Gronthoud. A gate-level method for transistor-level bridging fault diagnosis. 24th IEEE Proceedings of VLSI Test Symposium: 2006.
31. M. B. Tahoori and S. Mitra. Fault detection and diagnosis techniques for molecular computing. NanoTech Conference: 2004.
32. R. C. Hansen editor. *Significant Phased Array Papers*. Norwood, MA: Artech House, 1973.

9

NANOSCALE STANDARD DIGITAL MODULES

Shiva Navab

In this chapter on employing the computing features of spin waves, we present several nanoscale standard combinational modules. We show that a number of these widely used arithmetic and logic modules, including full adders, multipliers, decoders, encoders, multiplexers, and demultiplexers, can be implemented by employing the concurrent write feature and superposition property of the spin waves. In addition, we demonstrate how to implement more complex modules, such as priority encoders and shifters, by using spin-wave switches. The universality of the presented digital architectures ensures the possibility of realizing any logic switching functions. Moreover, these nanoscale modules can be integrated by adding latches to implement sequential systems.

9.1. INTRODUCTION

The significance of efficiently implementing standard combinational arithmetic and logic modules is due to the fact that they are the basic blocks of any digital system. All the sequential systems can be implemented by just adding latches to the combinational circuit. The universality of the combinational modules ensures the possibility of realizing any logic switching function. A set of combinational modules are said to be universal if any combinational system can be implemented using the elements from just that set. In addition to logic gates such as {NAND} and {NOR}, standard combinational modules can generate universal sets as well. For instance, a network of multiplexers can implement any switching function according to Shannon's decomposition [1]; therefore the set {MUX} is universal. {Decoder, OR} is another example of a universal set. We show how each of these modules can be implemented using spin waves.

Bio-Inspired and Nanoscale Integrated Computing. Edited by Mary Mehrnoosh Eshaghian-Wilner

The architectures presented here use spin waves for information transmission and data processing. Spin wave is a collective oscillation of spins in an ordered spin lattice around the direction of magnetization [2]. A propagating spin wave changes the local polarization of spins in ferromagnetic material. In turn, the change in magnetic field results in an inductive voltage. Recent published experimental results indicate that an inductive voltage signal of the order of mV produced by spin waves propagating through a nanometer thin ferromagnetic film are detectable at the distances of up to 50 microns at room temperature [3, 4].

A number of spin-wave architectures have been recently proposed, namely, logic gates [5–7], a crossbar [8, 9], a reconfigurable mesh [10, 11], and a fully interconnected cluster [12]. Similar to those architectures, in the spin-wave digital modules presented here, information is encoded into the phase of the spin waves. Moreover, all these architectures are capable of transmitting multiple waves simultaneously.

In this chapter, we present a set of widely used digital architectures and demonstrate how they can be implemented at the nanoscale [13]. A number of these standard and combinational arithmetic and logic modules, including full adders, multipliers, logic gates, decoders, encoders, multiplexers, and demultiplexers can be implemented by employing the concurrent write and superposition property of the spin waves. To implement more complex modules such as priority encoders and shifters, spin-wave switches [8] are used as well.

The rest of this chapter is organized as follows: In Section 9.2, we present our proposed spin-wave combinational arithmetic modules. We then describe the proposed spin-wave standard combinational logic modules in Section 9.3. We present more complex digital modules in Section 9.4, followed by our conclusion and future work in Section 9.5.

9.2. NANOSCALE SPIN-WAVE STANDARD ARITHMETIC MODULES

In this section, we first present a digital to analog module that is used in the design of several other arithmetic and logic modules. Afterwards, we show how to implement spin-wave full adders and multipliers.

9.2.1. Nanoscale Spin-Wave Digital to Analog Converter

As mentioned in Chapter 7, the computation using spin waves can be done in both digital and analog forms. We assume that upon receiving an analog signal, analog to digital conversion is performed in each processing node by comparing the signal value to several threshold voltages. In this section, we illustrate a simple method for converting a digital value to an analog signal to be transmitted via the spin-wave bus. The analog value is computed as the weighted sum of input bits, where the weight of the ith value is 2^i.

$$\text{Analog value} = \sum_{i=0}^{N-1} i \times 2^i$$

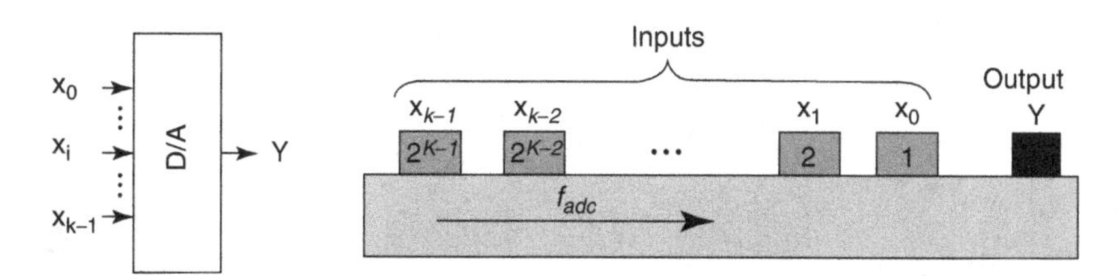

Figure 9.1. Structure of a nanoscale spin-wave digital to analog converter.

To implement an N-bit digital to analog converter (D/A), we use N nodes as inputs and one node as the output, all connected to the spin-wave bus. Assuming that the output is the rightmost node, the N-bit digital value gets mapped to the N input nodes such that the most significant bit (MSB) is mapped to the leftmost node, and the least significant node (LSB) to the rightmost node. The index of these nodes starts at zero for the rightmost node and continues to $N-1$ for the leftmost node. A fixed value equal to 2^{index} is assigned to each of these nodes.

To compute the analog value, each node checks its input value; if equal to "1," it sends a spin wave on the bus with an amplitude equal to its assigned fixed value. Note that all the senders' frequencies are the same as the frequency at which the output is tuned. The output node receives the superposition of all the waves, which is equal to the analog value represented by the N input bits. Figure 9.1 illustrates the structure of a spin-wave digital to analog converter.

The digital to analog converter module has several applications. For instance, it is used in the design of decoders, multiplexers, and demultiplexers as shown in the next sections.

9.2.2. Nanoscale Spin-Wave Full Adder

A k-input full adder is a device that adds the two k-bit input values and a carry-in value and generates a k-bit sum and a 1-bit carry-out [1]. The structure of a spin-wave k-bit full adder is shown in Figure 9.2. It consists of k nodes representing input x, k nodes representing input y, one node for c_{in}, and one node for analog value of the output, z. These nodes intercommunicate via the spin-wave bus.

All the input nodes' sending frequency is tuned on the receiving frequency of the z. The inputs x and y are converted to analog value using two digital to analog converters (D/A). The c_{in} node broadcasts its value on the spin-wave bus at the same time as all the other input nodes broadcast $2^i \times x_i$ and $2^i \times y_i$, for $i = 0 \cdots k-1$, explained in Section 9.2. All these waves get superposed and the output node z receives the analog value of this summation. Similar to a spin-wave encoder, the internal A/D converts this value to $k+1$ bit binary value. The most significant bit (MSB) of this binary value is the c_{out}, while the rest of the bits represent the *sum* value.

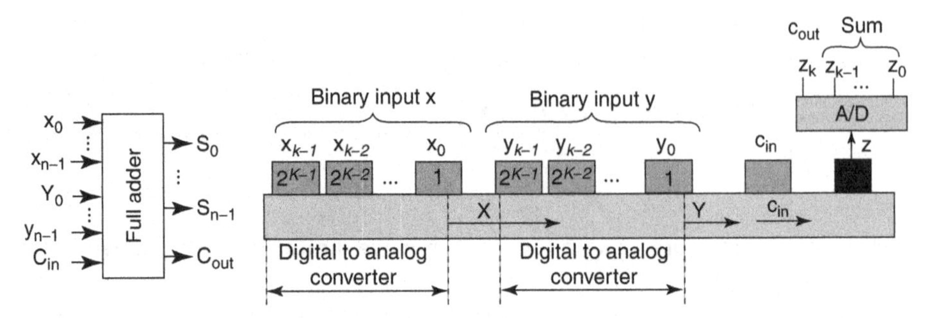

Figure 9.2. Structure of a nanoscale spin-wave full adder.

9.2.3. Nanoscale Spin-Wave Multiplier

A k-input multiplier is a device that multiplies two k-bit input values and generates a $2k$-bit product. A spin-wave multiplier consists of k nodes representing input x, k nodes representing input y, 2^i intermediate nodes, t, and one node for analog value of the output, P. These nodes intercommunicate via the spin-wave bus, as shown in Figure 9.3.

The multiplication is implemented as x number of times adding the value of y. This process is performed in three steps. Similar to a spin-wave adder, both binary input values are converted to analog using digital to analog converters:

First Step. First the receiving frequency of $t_0 \cdots t_{N-1}$ (t nodes) is tuned on the sending frequency of x nodes. Each x_i node (which has a value 1) sends a spin-wave on the bust with its assigned amplitude, 2^k, to implement the digital to analog conversion. Therefore, all the t nodes receive the analog value of x.

Second Step. In the next step, the receiving frequency of t nodes is tuned on the sending frequency of y nodes. Step 1 is repeated for y and the analog value of y is received by t nodes.

Third Step. Each t node compares its index to value of x. If its index is less than x, the t node broadcasts a spin-wave with an amplitude equal to the value of y.

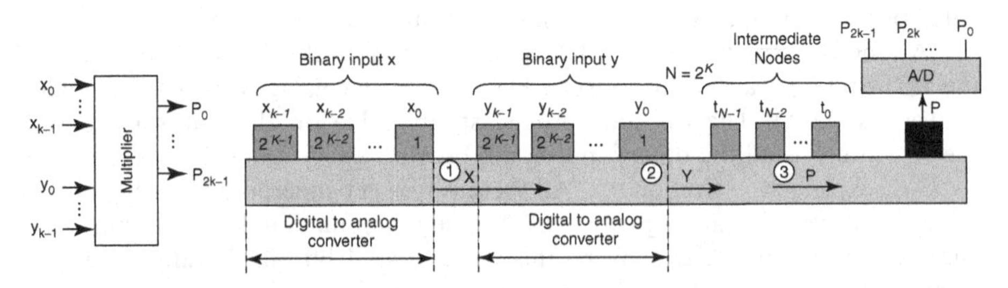

Figure 9.3. Structure of a nanoscale spin-wave multiplier.

TABLE 9.1. Frequency Assignments in the Multiplier Implementation

Node	Sending frequency	Receiving frequency
Binary x input nodes (step 1)	f_{adc}	–
Intermediate nodes t (step 1)	–	f_{adc}
Binary y input nodes (step 2)	f_{adc}	–
Intermediate nodes t (step 2)	–	f_{adc}
Intermediate nodes t (step 3)	f_{out}	–
Output node (step 3)	–	f_{out}

The output node's receiving frequency is tuned on the t nodes' sending frequency, so that it receives the superposition of these waves. In this fashion, x number of y is accumulated to generate the multiplication operation. Note that 2^k nodes are required as t since maximum value of x is 2^k.

Table 9.1 shows the frequency assignments for implementing a spin-wave multiplier.

9.3. NANOSCALE SPIN-WAVE STANDARD LOGIC

In this section we demonstrate how standard combinational logic modules can be implemented at the nanoscale. We employ parallel features of spin waves for implementing these logic modules: logic gates, decoders, encoders, multiplexers, and demultiplexers.

9.3.1. Logic Gates

The utilization of spin waves provides an opportunity to perform different logic functions in one device controlling the initial phases of spin waves and the voltage threshold to which the output voltage is compared. As explained in Chapter 7, Khitun and Wang showed how the two-bit gates, AND, OR, and the one-bit NOT can be implemented on the simple spin-wave prototype device [5, 6]. In this section we first briefly explain how they have implemented these logic gates, and next, we show how to extend their design to implement other logic functions such as NAND, NOR, XOR, and XNOR.

As explained in Chapter 7, the input information is coded into the polarity of the voltage pulse applied to the input ACPS lines and is detected at the output port [5]. The output logic state is determined by comparing the detected voltage to a reference voltage, V_{ref}. Assume that the measured inductive voltage produced by a single spin wave at the output port is V_1. The reference voltage V_{ref} can be a function of V_1 as shown below. The measurement is performed at the moment two spin wave packets arrive at the detecting ACPS line t_m. The exact choice of t_m depends on the logic function one needs to realize.

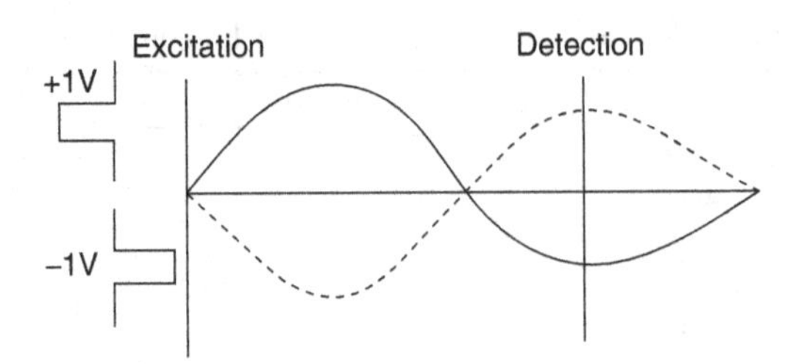

Figure 9.4. Encoding the information in the phase of the spin-wave.

The one-bit NOT gate can be obtained measuring the inductive voltage, V_{out}, at the output line produced by spin waves excited by the input line. Taking $t_m = \pi g/v_{ph}$ (where g is the distance between the contacts and v_{ph} is the spin-wave phase velocity), and $V_{ref} = 0\,V$, we achieve the required logic correlation [7]. At this t_m, the spin waves are detected with 180° phase difference comparing to the original phase as shown in Figure 9.4.

Next, the two-bit AND gate can be realized when $t_m = 2\pi g/v_{ph}$, and $V_{ref} = V_1$ (in Table 9.2, $V_1 = 10\,mV$). Two spin-wave packets coming in phase enhance the amplitude of the produced inductive voltage, and cancel each other when coming out of phase. The nonzero reference voltage is needed to avoid the effect caused by the finite size of the detecting ACPS line. The OR gate can be realized by analogy, taking $t_m = 2\pi g/v_{ph}$ and $V_{ref} = -V_1$. In Table 9.2, the input/output correlations are summarized for different NOT, AND, and OR logic gates as shown in [8].

Extending the same idea, a two-bit NAND gate can be implemented when $t_m = \pi g/v_{ph}$ and $V_{ref} = -V_1$. Similarly a two-bit NOR gate can be realized when $t_m = \pi g/v_{ph}$ and $V_{ref} = V_1$. Table 9.3 shows the input/output correlations for NAND and NOR gates.

To realize XOR and XNOR gates, we need two reference voltages, V_{ref1} and V_{ref2}. In these two gates, we define the output to be in state "1" if the detected output voltage is between V_{ref1} and V_{ref2} and in state "0" if it is greater than V_{ref1} or less than V_{ref2}. For both gates, $V_{ref1} = V_1$ and $V_{ref2} = -V_1$. For an XOR gate

TABLE 9.2. Truth Table for AND and OR Gates [6]

AND							OR						
$V_{ref} = 0.5\,mV$							$V_{ref} = -0.5\,mV$						
Input voltage		Logic state		Output voltage	Logic state		Input voltage		Logic state		Output voltage	Logic state	
+1 V	+1 V	1	1	+27 mV	1		+1 V	+1 V	1	1	+27 mV	1	
+1 V	−1 V	1	0	0 mV	0		+1 V	−1 V	1	0	0 mV	1	
−1 V	+1 V	0	1	0 mV	0		−1 V	+1 V	0	1	0 mV	1	
−1 V	−1 V	0	0	−27 mV	0		−1 V	−1 V	0	0	−27 mV	0	

TABLE 9.3. Truth Table for NAND and NOR Gates

NAND							NOR						
$V_{ref} = -0.5\,mV$							$V_{ref} = 0.5\,mV$						
Input voltage		Logic state		Output voltage	Logic state		Input voltage		Logic state		Output voltage	Logic state	
+1 V	+1 V	1	1	+27 mV	0		+1 V	+1 V	1	1	+27 mV	0	
+1 V	−1 V	1	0	0 mV	1		+1 V	−1 V	1	0	0 mV	0	
−1 V	+1 V	0	1	0 mV	1		−1 V	+1 V	0	1	0 mV	0	
−1 V	−1 V	0	0	−27 mV	1		−1 V	−1 V	0	0	−27 mV	1	

$t_m = 2\pi g/v_{ph}$, while in an XNOR gate $t_m = \pi g/v_{ph}$. Table 9.4 summarizes the input/output correlations for XOR and XNOR gates.

9.3.2. Nanoscale Spin-Wave Binary Encoder

A binary encoder is used when an occurrence of one of several disjoint events needs to be represented by an integer identifying the event [1]. A 2^k-input binary encoder is a combinational system that has 2^k binary inputs and k binary outputs. Note that at any given time only one of the encoder's inputs can be 1, and all the others have to be 0. The structure of a spin-wave encoder is shown in Figure 9.5. To implement an encoder with $N = 2^k$ input bits, we use N nodes for the inputs and one node as the output. These nodes intercommunicate via the spin-wave bus to which they are connected.

The sending frequency of all nodes is tuned at the same frequency, namely f_T. The output's receiving frequency is tuned on f_T as well. As soon as the input value of a node becomes 1, it broadcasts a spin wave with amplitude equal to its index; the output node receives this broadcasted value. In this architecture, the output node will contain the "analog" value equal to the index of the input containing a 1. As mentioned previously, we assume that this analog value can be digitized at

TABLE 9.4. Truth Table for XOR and XNOR Gates

XOR							XNOR						
$V_{ref1} = 0.5\,mV$, $V_{ref2} = -0.5\,mV$							$V_{ref1} = 0.5\,mV$, $V_{ref2} = -0.5\,mV$						
Input voltage		Logic state		Output voltage	Logic state		Input voltage		Logic state		Output voltage	Logic state	
+1 V	+1 V	1	1	+27 mV	0		+1 V	+1 V	1	1	+27 mV	1	
+1 V	−1 V	1	0	0 mV	1		+1 V	−1 V	1	0	0 mV	0	
−1 V	+1 V	0	1	0 mV	1		−1 V	+1 V	0	1	0 mV	0	
−1 V	−1 V	0	0	−27 mV	0		−1 V	−1 V	0	0	−27 mV	1	

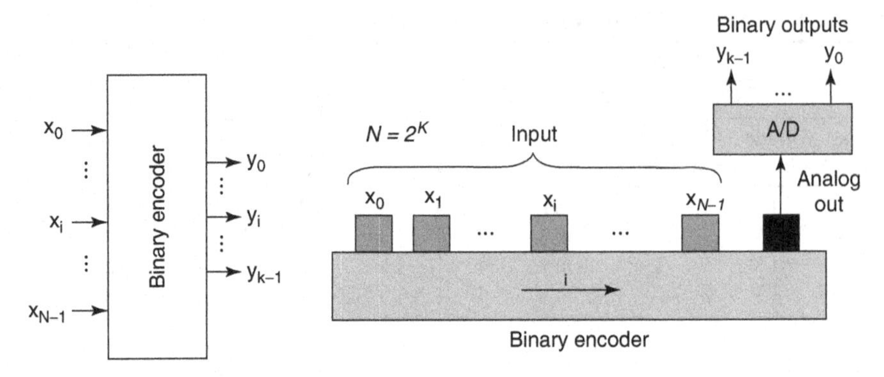

Figure 9.5. Structure of a nanoscale spin-wave binary encoder.

the output node based on several reference voltage thresholds. The digitized value will be $k = \log N$ bits which are the outputs of the binary encoder.

9.3.3. Nanoscale Spin-Wave Binary Decoder

A k-input binary decoder is a combinational system that has k binary inputs and 2^k binary outputs. The inputs represent an integer in the range of $0 \ldots 2^k - 1$ [1]. In other words, assuming $N = 2^k$, the binary decoder has N outputs and $k = \log N$ inputs. A spin-wave binary decoder consists of k input and N output nodes connected to a spin-wave bus. We show two different methods for implementing a nanoscale binary decoder using spin waves. The structure of a spin-wave decoder that has been implemented based on these two methods is shown in Figure 9.6.

In both designs, the binary input value is first converted to analog using a digital to analog converter. An additional node, which is in fact the output of the digital to analog converter, stores the analog value of the input represented by the log N bit inputs. To find this analog value, the binary input nodes send their value $(i \times 2^i)$ on frequency f_{adc}. The analog input node's receiving frequency is tuned on f_{adc} and it receives the index value. At the next step, the sending

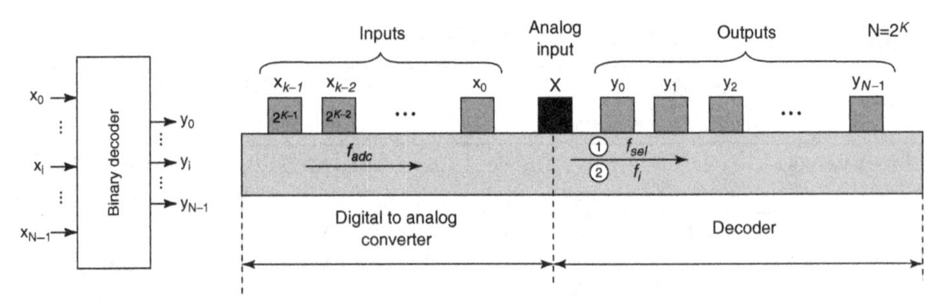

Figure 9.6. Structure of a nanoscale spin-wave binary decoder.

TABLE 9.5. Frequency Assignments in the First Method of Binary Decoder Implementation

Node	Sending frequency	Receiving frequency
Binary inputs	f_{adc}	–
Analog input	f_{sel}	f_{adc}
Outputs	–	f_{sel}

frequency of this analog input node is determined depending on the method we choose for implementing the decoder.

In the first method, the input node broadcasts the index of the output it is selecting on a common frequency, f_{sel}, at which all the output nodes are tuned. In this implementation, all the output nodes are required to know their index. They receive the broadcasted value, compare it to their index, and the one whose index has been broadcasted outputs 1. Table 9.5 shows the sending and receiving frequencies the nodes have to be tuned on.

In the second method, the output nodes do not need to have any information regarding their indices. Instead, a distinct frequency is assigned to each output node. For instance, output i is tuned on f_i. The input node, which contains the index of the output node, sends a signal on the specific frequency assigned to that output. The output node that receives this signal outputs 1. Since the output nodes are tuned on distinct frequencies, only one output node receives this signal. Table 9.6 shows the frequency assignments for the second method of implementing a spin-wave binary decoder.

Binary decoders are widely used in different logic systems. One of their most important applications is in decoding the address when accessing memory cells.

9.3.4. Nanoscale Spin-Wave Multiplexer

A multiplexer is a device that selects one of many data sources and outputs that source into a single channel. In other words, multiplexers function as multiple-input, single-output switches. A 2^k-input multiplexer is a combinational system that has 2^k binary inputs, k binary control (select), and one data output. The output of the multiplexer is equal to the data input selected by the control input [1].

TABLE 9.6. Frequency Assignments in the Second Method of Binary Decoder Implementation

Node	Sending frequency	Receiving frequency
Binary inputs	f_{adc}	–
Analog input	*Variable* (f_i)	f_{adc}
Output i	–	f_i

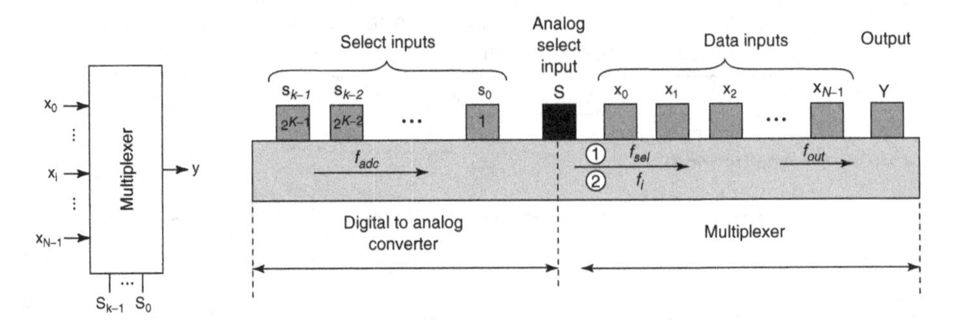

Figure 9.7. Structure of a nanoscale spin-wave multiplexer.

We present two different methods for implementing a nanoscale spin-wave multiplexer, in which input and output nodes intercommunicate via a spin-wave bus, as shown in Figure 9.7. Similar to the implementation of the binary decoder, in both designs, the binary select-inputs value is first converted to analog using a digital to analog converter.

In the first method of implementing a multiplexer, the analog-select-input node broadcasts on frequency f_{sel}, the index number of the input it is selecting. The data-input nodes, which are tuned on f_{sel}, receive the broadcasted value, compare it to their indices, and check if they have been selected. The selected data-input node sends its data on f_{out} that is the frequency the output node's receiver is tuned on. Table 9.7 shows the sending and receiving frequency assignment of the nodes.

In the second method, a distinct frequency is assigned to each data-input node. In other words, data input i is tuned on f_i. Assuming that the analog-select-input node contains the index i, it broadcasts a signal on f_i to be received by the ith data-input node. Upon receiving a signal, this node sends its data on the frequency at which the output node is tuned. Thus, the output node receives the data sent by the selected data-input node. Table 9.8 shows the frequency assignments for the second method of implementing a spin-wave multiplexer.

Multiplexers are widely used in various digital circuits. As an example, we mention one of their applications in digital signal processing. The multiplexer takes several separate digital data streams and combines them together into one data stream of a higher data rate. This allows multiple data streams to be carried from one place to another over one physical link.

TABLE 9.7. Frequency Assignments in the First Method of Multiplexer Implementation

Node	Sending frequency	Receiving frequency
Binary select inputs	f_{adc}	–
Analog select input	f_{sel}	f_{adc}
Data inputs	f_{out}	f_{sel}
Output	–	f_{out}

TABLE 9.8. Frequency Assignments in the Second Multiplexer Implementation

Node	Sending frequency	Receiving frequency
Binary select inputs	f_{adc}	–
Analog select input	*Variable* (f_i)	f_{adc}
Data inputs	f_{out}	f_i
Output	–	f_{out}

9.3.5. Nanoscale Spin-Wave Demultiplexer

A demultiplexer is a device that takes a single input which selects one of many data-output-lines and connects the single input to the selected output line. A demultiplexer is often used in the receiver with a complimentary multiplexer on the sender end. A 2^k-output demultiplexer is a combinational system that has one binary input, k binary control (select), and 2^k data output. This module routes the input data to the output selected by the control input; all other outputs are zero [1]. The structure of a spin-wave demultiplexer consists of k select-input, one data input, and $N = 2^k$ output nodes that are connected to a spin-wave bus, as shown in Figure 9.8.

The implementation of the spin-wave demultiplexer is quite similar to the spin-wave multiplexer (second method) implementation in which a distinct frequency is assigned to each of the output nodes. In the first step the binary select-input value is converted to analog value using a digital to analog converter. Note that unlike the decoder and multiplexer, the implementation of the demultiplexer requires no additional node to keep the analog select value, since this value is received by the data-input node. The data-input node, which has received the index of the output node, sends a signal with an amplitude equal to its data on the specific frequency assigned to the output node. The output nodes are tuned on distinct frequencies, therefore only the selected output node receives the

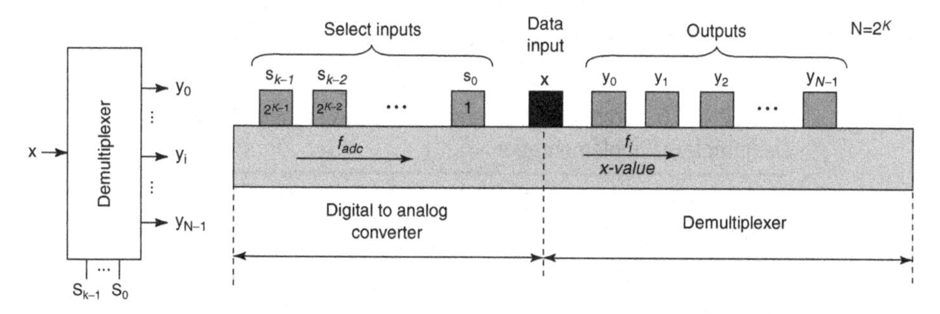

Figure 9.8. Structure of a nanoscale spin-wave demultiplexer.

TABLE 9.9. Frequency Assignments in the First Method of Demultiplexer Implementation

Node	Sending frequency	Receiving frequency
Binary select inputs	f_{adc}	–
Data input	*Variable* (f_i)	f_{adc}
Output i	–	f_i

data sent by the data-input node. Table 9.9 shows the frequency assignments for the first method of implementing a spin-wave demultiplexer.

The other possible implementation would require the demultiplexing to be performed in two steps. In the first step, all the output nodes' receiving frequencies are tuned on a common frequency f_{adc} so that all the nodes receive the analog value of select-input. All the output nodes compare the select value to their own index to see which output is selected. In the second step, the selected output node tunes its frequency on the input node's sending frequency. The data-input node broadcasts a spin-wave with an amplitude equal to its data, and the selected output receives and outputs it. Table 9.10 shows the frequency assignments for the second method of implementing a spin-wave demultiplexer.

9.4. MORE COMPLEX NANOSCALE SPIN-WAVE DIGITAL MODULES

In this section, we demonstrate how to implement more complex modules such as priority encoders and shifters. We present two implementations for a simple shifter and show how to extend one to design a spin-wave p-shifter. Next, we show how to implement a regular and also programmable priority encoder by using spin-wave switches.

9.4.1. Nanoscale Spin-Wave Shifter

In this section we first present two methods for implementing a simple shifter. In the first method, we use assigned frequencies for parallel intercommunications between adjacent nodes. In the second method, we use spin-wave switches to

TABLE 9.10. Frequency Assignments in the Second Method of Demultiplexer Implementation

Node	Sending frequency	Receiving frequency
Binary select inputs	f_{adc}	–
Outputs (step 1)	–	f_{adc}
Data input	f_{out}	–
Output i (step 2)	–	f_{out}

implement the shifting operation [14]. Afterwards, we show how to extend the first implementation if a simple shifter to design a p-shifter.

A SPIN-WAVE SIMPLE SHIFTER. A simple shifter is a combinational system having $(N+2)$-bit data-input, N-bit data output, and two one-bit control inputs: d (shifter direction) and s (shift or no shift). This system shifts the input data by one bit either to the right or to the left (depending on the value of d), or delivers the input data unchanged [1]. In implementing a simple shifter, we need $N+2$ processing nodes. Except the first and last nodes which are just input nodes, The nodes act as both inputs and outputs, except the first and last nodes that are just inputs. We present two methods for implementing a simple shifter.

In the first method, a single frequency is assigned to each of the nodes. The receiving frequencies of all nodes are tuned on these distinct frequencies. In addition, each node is aware of the frequency of its neighboring nodes. Each node determines its sending frequency based on the control signals. If control signal s shows "no shift," no node sends out any signal. On the other hand, if s commands to "shift," each node's sending frequency will be tuned on its right or left neighbor's frequency depending on the direction shown by control signal d. For instance, node x_i sends its data on frequency f_{i+1} if d indicates shift to the left and on f_{i-1} for shift to the right. Since these data transmissions are done on different frequencies, all can be performed simultaneously. The structure of this shifter (except s and d) is shown in Figure 9.9.

In the second method, all the nodes are tuned on the same frequency, and the shifting operation is performed by using spin-wave switches. The structure of this shifter is shown in Figure 9.10. If s indicated "no shift," all the nodes keep their switches off and no wave is transmitted; otherwise, the shifting operation is performed by using the spin-wave switches. Note that spin-waves propagate in both directions in the ferromagnetic bus; however, they can be forced in one direction by using the spin-wave switches. The structure of spin-wave switches were described in Chapter 7.

To implement shifting to the right, all the nodes turn off their left switch and send out a spin-wave with an amplitude equal to their data while their right switch is open. This forces the waves to be directed to the right. As soon as the nodes send their waves, they immediately open their left switch and close their right switch.

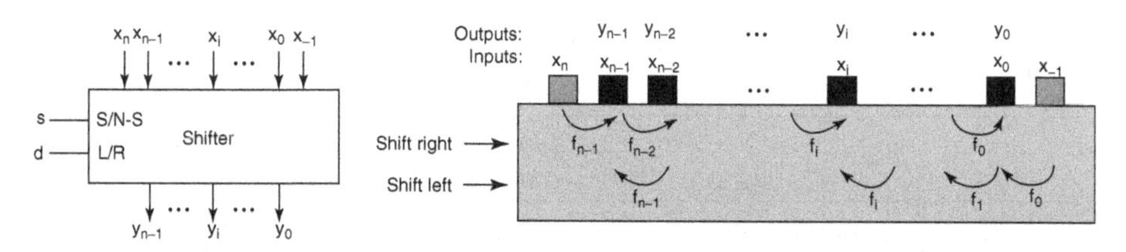

Figure 9.9. Structure of a nanoscale spin-wave shifter using multiple frequency assignments.

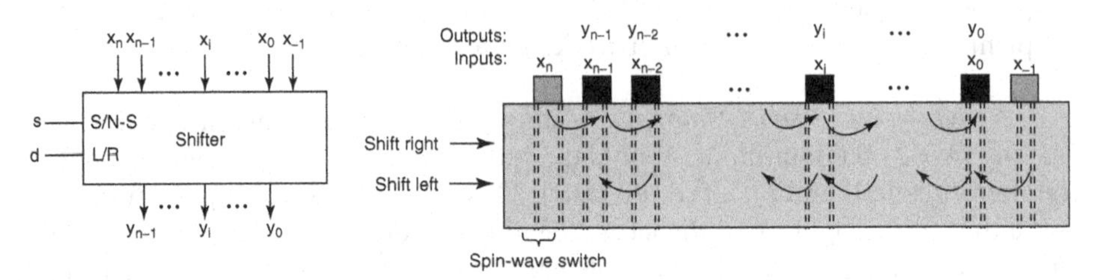

Figure 9.10. Structure of a nanoscale spin-wave shifter using spin-wave switches.

This allows each node to receive the spin wave from its left neighbor, while blocking the channel and preventing this received wave to go through to the next node on the right hand side. Similarly, shifting to the left can be implemented by forcing the waves to the left and blocking them at the immediate neighboring node.

A Spin-wave P-shifter. A p-shifter is a generalization of the simple shifter, which has a $(N+2p)$-bit data-input, N-bit data output, a one-bit control inputs d (for shifter direction), and a $(\log p)$-bit control signal s (shift distance). In this system, the data-input is shifted 0, 1,..., p positions (according to s) to either the right or left (according to d) [1]. The structure of a p-shifter is shown in Figure 9.11.

This structure contains $\log p$ nodes that represent the binary value of p, $N+2p$ input nodes, and N output nodes. All the input nodes' receiving frequency is tuned on a common frequency f_{adc}, and the output nodes' receiving frequencies on distinct frequencies. First the binary value of p is converted to analog as explained in Section 9.2.1. No additional node is required to keep this analog value, since all

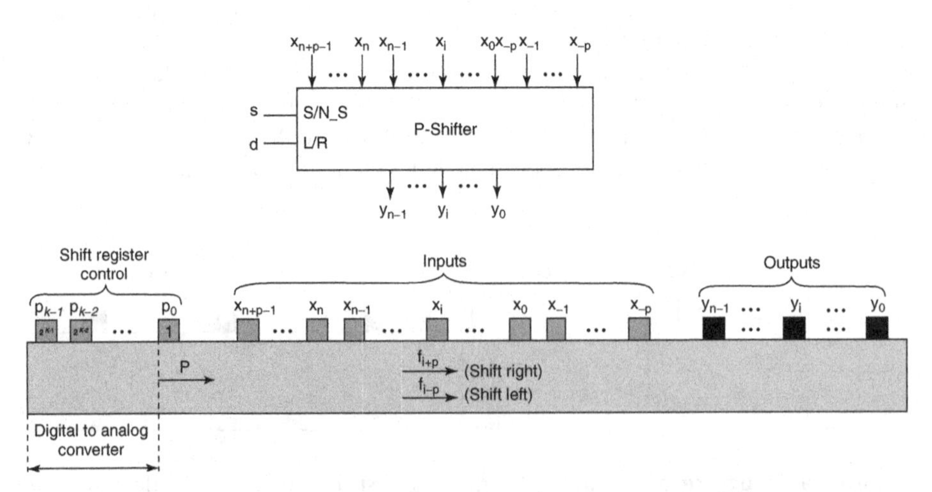

Figure 9.11. Structure of a nanoscale spin-wave p-shifter.

TABLE 9.11. Frequency Assignments in the First Method of *p*-shifter Implementation

Node	Sending frequency	Receiving frequency
Binary p inputs	f_{adc}	–
Input nodes	*Variable* (f_i)	f_{adc}
Output node i	–	f_i

the input nodes receive the analog value of p. Each input node knows its own index i and has received the value p. Depending on the shift direction d, each input node finds the index of the output node to which its data should be sent. If the shifter is in "shift to the right" status, each node x_i should send its data to the output node with index $i-p$; if it is in "shift to the left" status, the data is sent to the node with index $i+p$. Therefore node x_i sends its data on frequency f_{i-p} if d indicates shift to the right and on f_{i+p} for shift to the left. Table 9.11 shows the frequency assignments for implementing a spin-wave p-shifter.

Note that in order to use the same nodes as both input and output similar to a simple shifter, the shifting operation will have two main steps. In the first step, all the nodes' receiving frequency is tuned on a common frequency f_{adc} so that all receive the analog value of p. In the second step, the nodes' receiving frequencies are dynamically tuned on distinct frequencies to receive the data as explained in the previous method. Table 9.12 shows the frequency assignments for implementing a spin-wave p-shifter. This structure is shown in Figure 9.12.

9.4.2. Nanoscale Spin-Wave Priority Encoder

A priority encoder is used to select, according to a predefined priority, one of the several events that can occur simultaneously; the selected event is represented by an integer [1]. In this section we first explain the design of a standard priority encoder with fixed priority order, and afterwards, the implementation of a programmable priority encoder.

SPIN-WAVE PRIORITY ENCODER. A standard priority encoder has N-bit inputs and k-bit outputs, where $N = 2^k$. Unlike the binary encoder that can have one

TABLE 9.12. Frequency Assignments in the Second Method of *p*-shifter Implementation

Node	Sending frequency	Receiving frequency
Binary p inputs	f_{adc}	–
Input/output nodes (step 1)	*Variable* (f_i)	f_{adc}
Input/output node i (step 2)	–	f_i

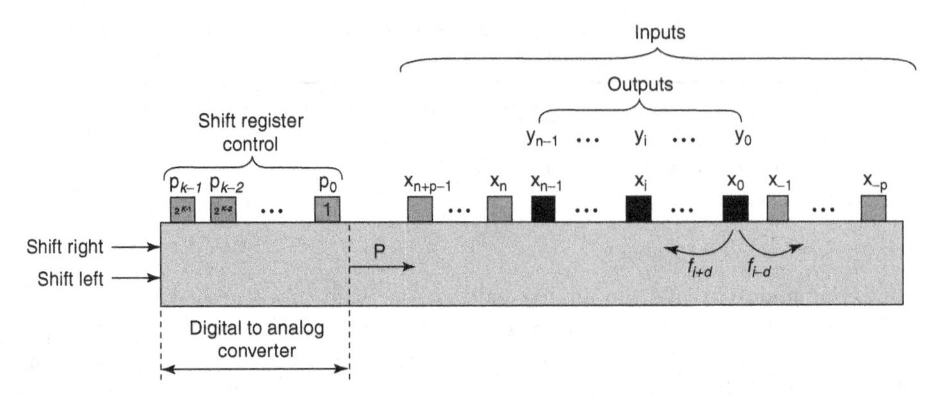

Figure 9.12. Structure of a nanoscale spin-wave p-shifter with overlapping input/output nodes.

input with value 1, the priority encoder may have several 1-valued inputs. The output represents the index of the highest priority input having the value 1. The priority is a fixed ordering implemented by the encoder; usually, x_{N-1} has the highest priority, whereas x_0 has the lowest [1].

The general way to implement a priority encoder is to have a module called "priority resolution" which keeps the highest priority 1 and sets all the other inputs to 0. These data will be the input to a standard binary encoder which gives the index of the only input with value 1. In implementing the spin-wave priority encoder, however, we do not use a "priority resolution" module to reset the lower priority 1-valued inputs to 0. Instead, we use spin-wave switches to force the waves in the desired direction and arrange the nodes in a fashion that the output node receives the index of the highest priority 1-valued input.

The structure of a spin-wave priority encoder is shown in Figure 9.13. In this design, the highest priority node, x_{N-1}, is placed closest to the output node, while the lowest priority node, x_0, the farthest. Note that in the structure shown in

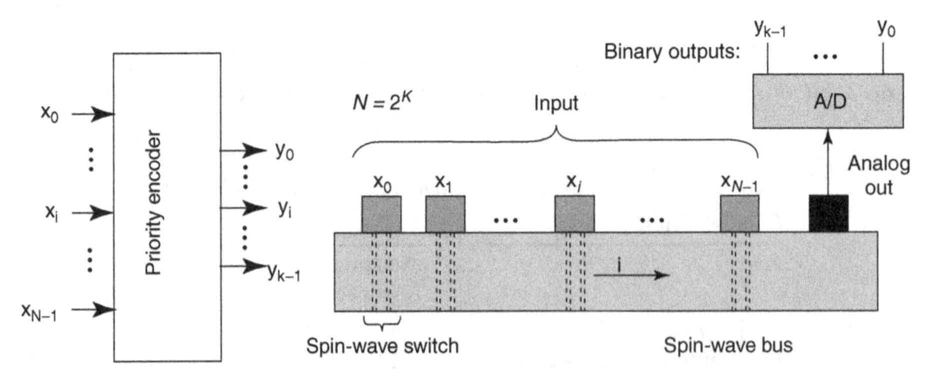

Figure 9.13. Structure of a nanoscale spin-wave priority encoder.

Figure 9.13, the output node is placed on the right hand side of the inputs, so all the nodes should direct their waves to the right. First each node that has a value 1 turns off its left switch to direct the wave to the right. Next, it sends out a spin wave with an amplitude equal to its index and immediately turns off its right switch to block the wave coming from its left hand side. As a result, the output node will just receive the spin wave sent by the highest priority 1-valued node. This analog value is converted to digital at the output node.

One application of priority encoder can the control of interrupt requests to a processor in a computer system. Since several interrupts might occur at the same time, a mechanism is required to determine which request should be handled. In such a scenario, a priority encoder can be used to identify the interrupting signal with the highest priority.

Spin-wave Programmable Priority Encoder. Similar to a priority encoder, a programmable priority encoder is used to select one of the several events that can occur simultaneously; however, in a programmable priority encoder, this selection is not according to a predefined order, but at an order defined at run-time. The output represents the index of the highest priority 1-valued input.

The structure of a spin-wave programmable priority encoder is shown in Figure 9.14. This design includes a digital to analog converter, a shifter, and a priority encoder. First, the analog value of p, which shows the highest priority node's index, is found by the digital to analog converter on frequency f_{adc}. At this step, the receiving frequency of all the nodes, including the input and the output nodes, is tuned on f_{adc}, so all receive p.

In the next step, the data is rotated p times, so that the highest priority node is mapped to the closest node to the output. For rotating the data, each input node computes the sum of p and its index and finds the frequency of the destination node to which it should sent its data. For node x_i, the index of the destination would be $(i+p)$ mod N.

After the rotation is done, the next step will be finding the index of the highest priority node that contains a 1. This step is performed as discussed in the implementation of a priority encoder. The spin-wave switches let the index of

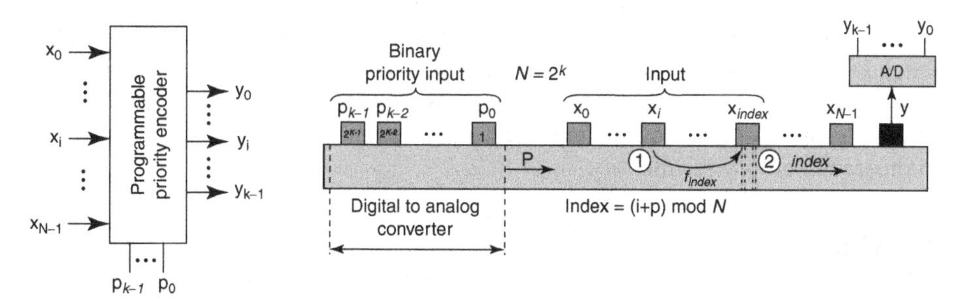

Figure 9.14. Structure of a nanoscale spin-wave programmable priority encoder.

TABLE 9.13. Frequency Assignments in the Programmable Priority Encoder Implementation

Node	Sending frequency	Receiving frequency
Binary p inputs	f_{adc}	–
Input nodes (step 1)	–	f_{adc}
Output node (step 1)	–	f_{adc}
Input nodes (step 2)	*Variable* (f_i)	–
Output node i (step 2)	–	f_i

the highest priority node reach the output node. Note that this index is the index of the highest priority node when the data has been shifted p times. Therefore, as the last step, the output node finds [(received index-p) mod N] as the final output of the programmable priority encoder. Table 9.13 shows the frequency assignments for implementing a spin-wave programmable priority encoder.

The programmable priority encoder can be used to determine which interrupt request should be handled in the case that the priority order of the interrupts is not fixed. For instance, a programmable priority encoder can be used in designing a "fair" system. In a fair system, whenever an interrupt request is handled, its priority becomes the lowest priority for the next interrupt handling cycle. By using a programmable priority encoder the winning interrupt node will determine the next priority order.

9.5. CONCLUSIONS

In this chapter, we presented a set of nanoscale, standard combinational modules by exploiting the computing features of spin waves. We showed that a number of these widely used, standard arithmetic and logic digital modules including full adders, multipliers, decoders, encoders, multiplexers, and demultiplexers can be implemented by employing the concurrent write feature and superposition property of the spin waves. In addition, we demonstrated how to implement more complex modules such as priority encoders and shifters by using spin-wave switches. The universality of these modules ensures the possibility of realizing any logic switching functions. Sequential systems can be implemented by adding latches to the combinational circuit. The latches or D flip-flops can be easily simulated by the processing nodes. One possible extension of this research could the investigation of how nanoscale latches can be implemented using the nanoscale memory technology.

REFERENCES

1. M. Ercegovac, T. Lang, and J. Moreno. *Introduction to Digital Systems*. New York: Wiley, 1999.

2. C. Kittel. *Introduction to Solid State Physics*. New York: Wiley, 1986.
3. T. J Silva, et al. Inductive measurement of ultrafast magnetization dynamics in thin-film permalloy. *Journal of Applied Physics*, 85(11): pp 7849–7862, 1999.
4. M. Covington, T. M. Crawford, and G. J. Parker. Time-resolved measurement of propagating spin-waves in ferromagnetic thin films. *Physical Review Letters*, 89(23): p 237202-1-4.l, 2002.
5. A. Khitun and K. L. Wang. Nanoscale computational architectures with spin-wave bus. *Superlattices and Microstructures*, 38(3): pp 184–200, Sep. 2005.
6. A. Khitun and K. L. Wang. Nanologic circuits with spin wave bus. In: Proceedings of the Third International Conference on Information Technology: New Generations (ITNG'06).
7. A. Khitun, M. Bao, J. Y. Lee, K. L. Wang, D. W. Lee, S. Wang, and I. Roshchin. Inductively coupled circuits with spin wave bus for information processing. Submitted for publication in *IEEE Transactions on Electron Devices*, 2006.
8. M. M. Eshaghian-Wilner, A. Khitun, S. Navab, and K. L. Wang. A nanoscale architecture with spin-wave crossbar. In: Proceeding of the 6th IEEE Conference on Nanotechnology: pp 326–329, July 2006.
9. M. M. Eshaghian-Wilner, A. Khitun, S. Navab, and K. L. Wang. Nano-scale modules with full spin-wave interconnectivity. *Journal of Experimental Nanoscience*, 2(1–2): pp 73–86, Mar 2007.
10. M. M. Eshaghian-Wilner, A. Khitun, S. Navab, and K. L. Wang. A nanoscale reconfigurable mesh with spin-waves. In: Proceeding of the ACM International Conference on Computing Frontiers: pp 65–70, May 2006.
11. M. M. Eshaghian-Wilner, A. Khitun, S. Navab, and K. L. Wang. The spin-wave reconfigurable mesh and labeling problem. *ACM Journal on Emerging Technologies in Computing Systems*, 3(2, no. 5): July 2007.
12. M. M. Eshaghian-Wilner, A. Khitun, S. Navab, and K. L. Wang. Nanoscale modules with spin-wave intercommunications for integrated circuits. In: Proceeding of the NSTI Nanotech 2006, Boston: pp 320–323, May 2006.
13. S. Navab. Nanoscale standard combinational modules. 2007 World Congress in Computer Science, Computer Engineering, and Applied Computing WORLD-COMP'07. In: Proceedings of the 2007 International Conference on Computer Design (CDES'07): pp 10–16.
14. M. M. Eshaghian-Wilner and S. Navab. Efficient parallel processing with spin-wave nanoarchitectures. *Journal of Supercomputing*, October 2008.

10

FAULT- AND DEFECT-TOLERANT ARCHITECTURES FOR NANOCOMPUTING

Sumit Ahuja, Gaurav Singh, Debayan Bhaduri, and Sandeep Shukla

In the recent past, CMOS manufacturing technology has been successfully downscaled to create feature sizes below 100 nm; it is predicted to reach the 22 nm mark very soon. But with the predicted demise of Moore's law, continued success of the electronic industry will increasingly depend on emerging nanotechnologies. These emerging nanotechnologies have influenced the rapid development of nonsilicon nanodevices such as carbon nanotubes and molecular switches. CMOS or not, affordable manufacturing of defect-free nanosystems seems unlikely. Besides manufacturing defects, various transient faults will affect these systems. Therefore, there is a need for developing computing architectures that are tolerant to defects and faults and that may be one of the ways to avoid an industrial meltdown. This chapter provides a survey of techniques and architectures for providing defect- and fault-tolerance to nanocomputing.

10.1. INTRODUCTION

10.1.1. The Micro to Nano Trend

For four decades, the rapid pace of improvement in microelectronics has been based on the ability to decrease the minimum feature sizes used to fabricate integrated circuits. The different parameters associated with integrated circuits that have improved due to such feature scaling are outlined in Table 10.1. One of the improvements that is frequently cited is the improvement in integration level; this is expressed as Moore's law—the number of components per chip doubles every 24 months.

Bio-Inspired and Nanoscale Integrated Computing. Edited by Mary Mehrnoosh Eshaghian-Wilner

TABLE 10.1. Parameters that have Improved Due to feature Scaling [1]

Parameter	Example
Integration Level	Components/chip, Moore's law
Cost	Cost per function
Speed	Microprocessor clock rate, GHz
Power	Laptop or cell phone battery life
Compactness	Small and light-weight products
Functionality	Nonvolatile memory, imager

The scaling of CMOS technology has faced many barriers, but clever engineering solutions and new device architectures have thus far broken through such barriers. However, the size of a silicon atom will be an indisputable barrier in CMOS scaling [2]. Since 2001, the ITRS Roadmap has challenged the practicality of CMOS scaling projections beyond MOSFET channel lengths of 9 nm and has addressed the need for non-CMOS technologies.

In the recent past, a number of novel non-CMOS nanotechnologies have emerged that have shown potential in enhancing the CMOS platform; they have also demonstrated promise developing fundamentally new approaches to information processing. Chemical self-assembly of molecular devices is one such nanotechnology that has been proposed for the development of reconfigurable molecular nanofabrics. Some of these nanotechnologies have led to the development of exciting memory and logic nanodevices. Engineered tunnel barrier memory, polymer memory, and molecular memory are some of the novel nanomemories under research, whereas CNTs, RTDs, molecular and spin devices are some of the non-CMOS logic devices that have matured significantly. These logic devices are predicted to have high switching speed, low power consumption, and good demonstrations of scaling potential.

These nanodevices represent charge-based logic and their scaling is limited by the minimum switching energy per binary operation, also called the thermodynamic limit [3]. Beyond this limit, the challenge is to invent and develop nanotechnologies based on something other than electronic charge. Ferromagnetic logic and spin gain devices have been identified as some of the first potential non-charge-based devices.

10.1.2. Challenges

Nanoelectronics has advanced appreciably in the recent past, and it has shown potential for large scales of integration, specifically on the order of a trillion (10^{12}) devices in a square centimeter. But at the same time, some of the characteristics of these nanodevices and their fabrication methods pose prominent limitations to such ultra-scale integration. Some of these characteristics are manufacturing defects, unreliable device performance, interconnect limitations, and thermal power dissipation [1, 4, 5]. The two major categories of defects occur because of interconnections and power dissipation.

The unreliability of nanoscale devices is a consequence of the inherent variability in fabrication processes and the physical principles that govern their operation. Self-assembly methods may have to be used at dimensions below those for which conventional lithographic-defined subtractive processing methods are used. Since variability and imprecision are inherent in such self-assembly processes, it is estimated that a significant number of devices, up to many percent, may suffer from manufacturing defects. The other source of unreliability is due to the physical principles of these devices that cause reduced noise tolerance and higher susceptibility to external influences, such as electromagnetic interference, thermal perturbations, and terrestrial radiation, resulting in in-service transient faults.

The problem of interconnects is due to a number of fundamental challenges. These are (i) the geometrical challenge of interconnecting devices at the nanoscale dimensions at high speed and bandwidth; (ii) the process of interfacing these devices with the macroscopic world; and (iii) the challenge of transforming long-distance communications to short-distance communications for nanodevices that have low drive capabilities. To tackle some of these challenges, fabrication processes for producing aligned wires and highly regular, homogeneous and locally connected parallel architectures have been proposed.

The other challenge is thermal power dissipation that comes from device switching energy and the energy needed for driving signals. Due to this, there is a trade off between clock speed and device density—clock speeds need to be decreased for high device densities, and densities need to be lowered for high clock speeds. The problem of power dissipation sets a general limit to the operational speed of any charge-based nanodevice. To solve these problems, researchers are looking at approaches like charge recovery [6], reversible computing [7], and adiabatic computing [8].

10.2. FAULT TOLERANCE THROUGH REDUNDANCY

For the management of defective and error-prone devices, several fault- and defect-tolerant nanoarchitectures have been proposed in order to aid the reliable integration of nanodevices, hence allowing the demonstration of their full potential [9]. Techniques such as redundancy and reconfiguration are used in such architectures to achieve defect and fault tolerance. Here we discuss various redundancy techniques commonly used to achieve fault tolerance.

10.2.1. Structural Redundancy

Redundancy has been widely used for fault tolerance in hardware. A structurally redundant architecture is one which mitigates the effects of faults in the devices and interconnects that make up the architecture and guarantees a given level of reliability. Various structural redundancy-based techniques such as triple modular redundancy (TMR), Cascaded triple modular redundancy (CTMR), von Neumann multiplexing and their variations have been proposed to provide fault tolerance in hardware. As opposed to structural redundancy, temporal redundancy entails repetition of the same computation if the computation fails to compute correctly. A classical temporal redundancy mechanism used in hardware is based on checkpoints and roll-back recovery [10, 11]. This chapter focuses on fault tolerance techniques based on structural redundancy.

10.2.1.1. Triple Modular Redundancy (TMR). The concept of TMR [12] is to have three functionally identical units working in parallel and comparing their outputs with a majority (MAJ) gate to produce the final output. The units could be gates, logic blocks, logic functions, or functional units. TMR provides a functionality similar to one of the three parallel units but provides a better probability of working [12]. The MAJ gate is a single failure point and three of these can be used instead of just one to improve the reliability of the system further.

The main drawback of simple, modular level TMR technique is that it does not provide robust error recovery mechanisms [13]. For example, if the triplicated units in TMR architectures have random logic with sequential elements and any one of the units are in error, there is no mechanism by which such an error can be detected until it manifests at the output of the modular unit. At that point in the execution of the system, the internal states of the erroneous and nonerroneous redundant modules will be inconsistent; Ref. 13 discusses the significance of feeding back the voted result to all the voted sequential elements to resynchronize all the redundant modules and avoid error build up. This implies that a Boolean network needs to be apportioned into combinational and sequential parts; then, redundancy may be inserted. This is shown in Figure 10.1 where triple MAJ gates are used along with feedback.

10.2.1.2. Cascaded Triple Modular Redundancy (CTMR). CTMR [12] is similar to TMR, wherein the units working in parallel are TMR units combined

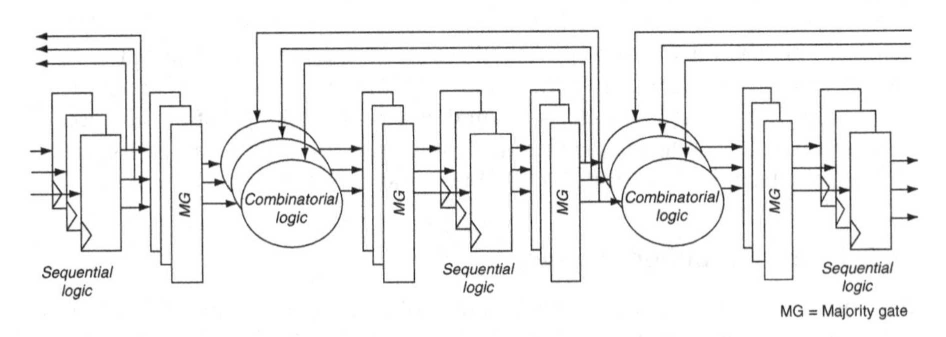

Figure 10.1. TMR for sequential and combinational logic.

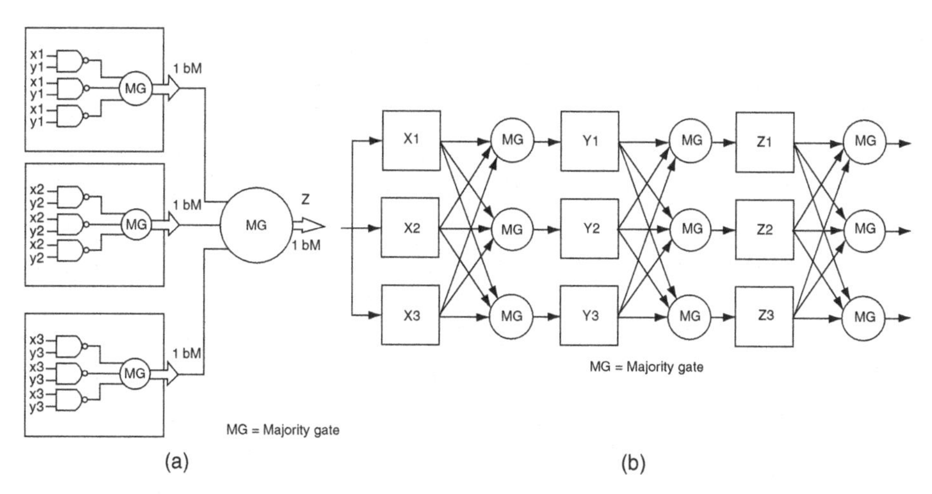

Figure 10.2. Different CTMR configurations. (a) Generic CTMR: multilayer voting; (b) CTMR with triple voters: smaller granularity.

with a MAJ gate. Figure 10.2a shows a first-order CTMR configuration where the parallel processing units in each of the three TMR units are NAND gates. Due to the area and latency overheads associated with this technique, the triplicated units in the CTMR with a multilayer voting scheme are normally functional units or logic blocks, not single gates as shown in Figure 10.2a. Since the triplicated functional units or logic blocks may consist of a large number of gates, their failure probability is more than individual gates. Hence, the multilevel CTMR with triple voters as shown in Figure 10.2b may be used to apportion the system into optimally sized functional units or logic blocks to effectively allow the architecture to withstand more errors across the triplicated units [14].

10.2.1.3. Triple Interwoven Redundancy (TIR). The idea of TIR is based on von Neumann's multiplexing techniques and interwoven redundant logic [15]. A TIR architectural configuration has three times as many gates and interconnects as compared to the nonredundant network. The interconnections are arranged in random patterns. Such inherent randomness in the interconnections makes this structural redundancy technique favorable for the integration of molecular devices, since the manufacturing method for such devices is most likely to be stochastic chemical assembly. Figure 10.3 shows a nonredundant half adder and its corresponding TIR implementation. For a particular interconnect pattern, [16] shows that the TIR actually works as a TMR configuration, implying that TMR is a specific implementation of TIR.

10.2.1.4. Multiplexing Techniques. Structural redundancy-based architectures that can circumvent transient faults can affect both the computation and communication in nanosystems. Interestingly, von Neumann addressed this issue

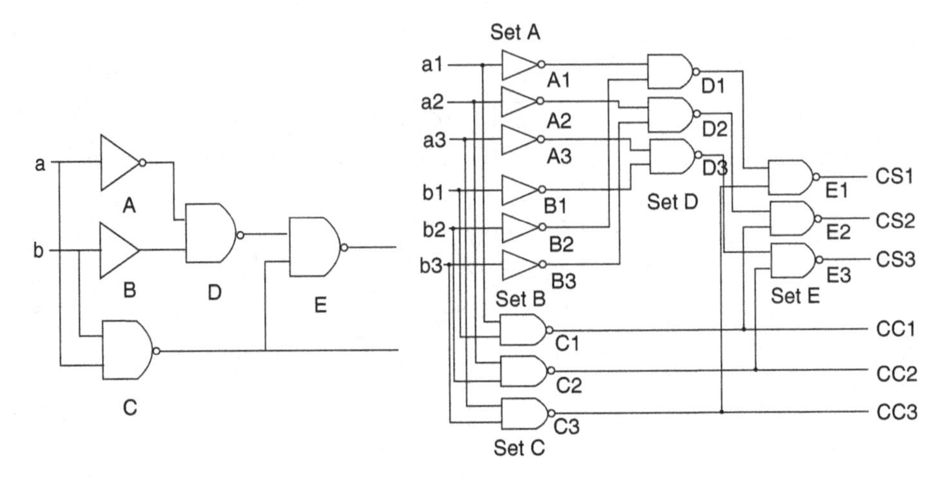

Figure 10.3. Nonredundant and TIR implementation of half adder.

in 1952 and developed a technique called multiplexing. He introduced multiplexing as a technique for constructing a system whose malfunction cannot be caused by the failure of a single device or a small set of devices. It has been identified as one of the most effective techniques for transient fault mitigation. Von Neumann proposed multiplexing architectures based on two universal logic functions—NAND and MAJ—which will be discussed in detail later in the chapter.

10.2.2. Information Redundancy

Information redundancy, i.e., providing more information, is used in applications for error detection and correction to make systems more reliable. Below, some basic information redundancy techniques are discussed.

10.2.2.1. Parity. A parity bit is a binary digit that indicates whether the number of bits with value of 1 in a given set of bits is even or odd. There are two types of parity bits: even and odd. An even parity bit is set to 1 if the number of ones in a given set of bits is odd (thus making the total number of ones even). An odd parity bit is set to 1 if the number of ones in a given set of bits is even (thus making the total number of ones odd). If an odd number of bits (including the parity bit) are changed in transmission of a set of bits then parity bit will be incorrect and will thus indicate that an error during transmission has occurred. Therefore, parity bit is used as an error-detecting code. Note that it is not an error-correcting code as there is no way to determine which particular bit is corrupted. If an error occurred during transmission, then the entire data must be discarded, and retransmitted from scratch.

10.2.2.2. Cyclic Redundancy Check (CRC). The CRC is a very powerful and easily implemented technique to obtain data reliability. CRC is a type of hash function used to produce a checksum, a small fixed number of bits against a block of data. The checksum holds redundant information about the block of data that helps the recipient detect errors. A CRC is computed and appended before transmission or storage, and verified afterwards by the recipient to confirm that no changes occurred during transmission. It is one of the most widely used techniques for error detection in data communications. The technique is popular because CRCs have extreme error detection capabilities, have little overhead and are easy to implement. Moreover, they are simple to implement in binary hardware and are easy to analyze mathematically. Even parity is actually a special case of a CRC, where the 1-bit CRC is generated by the polynomial $x+1$.

10.2.2.3. Error-Correcting Code (ECC). An error-correcting code is an algorithm for expressing a data signal such that any errors which are introduced can be detected and corrected, within certain limitations, based on the other parts of the signal. In an ECC, each data signal conforms to specific rules of construction so that departures from this construction in the received signal can generally be automatically detected and corrected. It is used in computer data storage and transmission. The simplest error-correcting codes can correct single-bit errors and detect double-bit errors. There are other codes which can detect or correct multi-bit errors. Some of the examples of ECC are Hamming code, BCH code, Reed-Muller code, Binary Golay code, and convolutional code. ECC-based computer memory provides greater data accuracy and system uptime by protecting against soft errors.

10.3. DEFECT TOLERANCE THROUGH RECONFIGURATION

Chemically self-assembled molecular nanofabrics are by nature very regular and homogeneous, hence, well-suited for reconfigurability [17–19]. It is reported that the reconfiguration is the most effective technique to cope with manufacturing defects in nanodevices. Reconfigurable fabrics are composed of programmable logic elements [like configurable logic blocks (CLBs)] and interconnects which can be configured to implement any logic circuit. It is expected that reconfigurable fabrics made from next generation fabrication processes will go through a post-fabrication defect mapping phase during which these fabrics are configured for self-diagnosis [20, 21]. Thus, defect tolerance in such fabrics can be achieved by detecting faulty components during an initial defect map phase and excluding them during actual configuration. In other words, design of reliable digital logic and architectures on unreliable nanofabrics will require defect mapping followed by defect avoidance to circumvent hard faults. Defect mapping is the process of finding defect locations in a nanofabric, and defect avoidance is the process of mapping a computing logic on a faulty nanofabric knowing its defect-maps.

While such reconfigurable architectures may aid in circumventing manufacturing defects at the nanoscale, they will not provide tolerance to natural external transient faults. The addition of structural redundancy a priori may enhance the reliability of such systems in the presence of transient faults.

There are two general classes of defect mapping and avoidance techniques: (i) techniques that use test circuits to find the location and number of defects on a reconfigurable nanofabric [19, 21], and (ii) broadcast-based methods that flood test packets through the whole nanofabric to locate nonreachable nodes [22]. The test circuits or packets placed on the fabric during the self-diagnosis phase utilize resources that are available later for normal logic mapping. Although such defect mapping techniques can be performed with massive parallelism, they have been reported to be expensive in terms of cost and time.

In [23–25], a defect-mapping and hierarchical redundancy insertion methodology is presented which is discussed below in detail. First, we present the target nanofabric and fault models for this methodology followed by the details of the methodology and a proposed automation framework.

10.3.1. Nanofabric and Fault Models

10.3.1.1. Nanofabric Model. The target nanofabric model is composed of processing elements (PEs) built from crossbars. Figure 10.4 shows a nanofabric composed of a grid of PEs connected together. This model is a combination of the nanofabric models used in [22] and [26]. The PEs are modeled to function as simple single-bit arithmetic logic units (ALUs), 8-bit adders or 8-bit combinational multipliers depending on the application being targeted. In this model, each PE has 4 transceivers that are connected to the 4 asynchronous bidirectional interconnects. Figure 10.4 also shows vias that are used to interface the nanofabric with external circuitry. The nanofabric model has three levels of hierarchy: regions, mapping units (MUs) and components (defect-mapping technique discussed in [22] uses a monolithic model).

10.3.1.2. Fault Model. The fault model considers the effects of both manufacturing defects and transient faults. Since the proposed work focuses on crossbar-based nanofabrics, the faults relevant to such nanofabrics are discussed. The manufacturing defects in the crossbar cause stuck-open and stuck-closed faults at the junctions and wires. Since the yield of the crossbar is very low in the presence of stuck-closed faults, it is possible to bias the chemical self-assembly process to decrease the probability of such faults [19]. Hence, only stuck-open faults at the junctions and wires are considered.

The methodology [23–25] models these faults as the probability of failure to program the molecular diodes or latches to the appropriate logic value. Also, the faults at the interconnects are modeled as Gaussian failure distributions that quantify the effect of signal noise. It is also assumed that the faults at the junctions and interconnects are independent, identical and uniformly distributed (i.i.u). This fault model can be changed in the proposed toolset discussed below and the

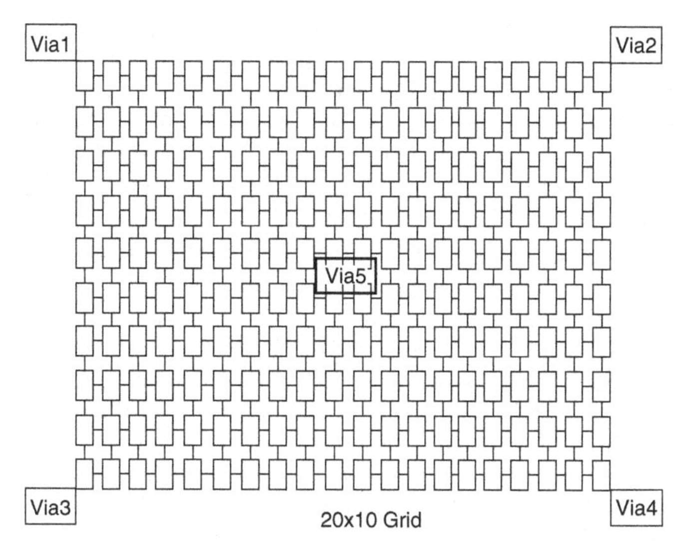

Figure 10.4. Molecular Nanofabric.

current model is used to represent the fault classes that have the highest probability of occurrence in chemical self-assembly fabrication processes.

10.3.2. Methodology

The instance-based methodology proposed in [23–25] can be used to design and analyze molecular logic systems. The methodology entails (i) developing a script to translate specifications of molecular nanofabric instances in terms of number of PEs, size of crossbar for each PE, etc. into probabilistic transition models, (ii) determining the exact failure probability of each junction by running test circuits physically on the nanofabrics and using these values on the nanofabric models to generate probabilistic defect maps, (iii) developing a script to insert behavioral redundancy in the designs and/or structural redundancy at the different structural hierarchies of the nanofabrics and (iv) developing Markovian and state space traversal techniques to analyze redundant instances of different designs that are mapped onto the nanofabrics.

Many specific design instances have to be analyzed by this methodology to predict general performance trends of general structurally redundant architectures, a drawback that is common for instance-based methodologies. This issue is addressed by hierarchical modeling and state space partitioning techniques discussed later in this section.

10.3.2.1. Test Circuits. The probabilistic defect mapping technique is based on limited on-line testing of nanofabrics. Simple test circuits are configured on the nanofabrics to get a notion of the probability of the crossbar-based PEs being defective. One of the easiest test circuits proposed for such purposes is a counter circuit [21]. Counter circuits with thresholds of t can indicate $0,1,2,\ldots,t$ or more than t defects. Since the lowest element of the structural hierarchy are regions (composed of 8 PEs), the threshold of such a test circuit may be set to 8.

Another set of simple circuits that can be used for this purpose are linear feedback shift registers (LFSRs). LFSRs are used widely in built-in self test (BIST) of RAMs. These can be configured on MUs and run autonomously with sets of inputs. parallel signal analyzers (PSAs) are also configured on the MUs. These PSAs are used as parallel-to-serial compression circuits to avoid testing large number of pseudo-random vectors generated by the LFSRs. The compression of a number of patterns using a PSA is called a signature. If there are signature mismatches, the LFSRs may be split into smaller units and configured on each region of the specific MU instance. Note that even then if there are no signature mismatches, it does not mean that the constituent nodes are defect free. This is due to a non-zero probability of aliasing. A way to minimize this aliasing problem is to use maximal-length PSAs and to frequently compare the signatures with the expected values [27].

The test circuits discussed here are just a small subset of the different test circuits that can be used to compute the probability of PEs being defective. The crossbar junction failure probability can be computed using these probabilities. Note that determining the junction failure probability is required for both the defect mapping and structural redundancy insertion methodologies, since logic functions need to be mapped to the crossbar-based PEs in both the cases.

10.3.2.2. Defect-Mapping Technique. The proposed nondeterministic defect-mapping technique [23–25] is based on a variant of the RPF broadcast scheme [22] and on the defect-mapping methodology proposed in [21]. Simple test circuits from [21] are configured physically on the nanofabrics to determine the probability of the crossbar junctions being defective. These defect rates are used in determining the probability of successfully configuring all the PEs with the same broadcasting functionality. Once this configuration probability is computed, probabilistic models representing such PE-based nanofabrics are constructed and the broadcast-based algorithm is used to determine the reachability graph of each PE from the vias.

The broadcast-based algorithm used is a modification of the deterministic algorithm proposed in [22]. The algorithm has been modified such that defect maps represented by probabilistic broadcast trees can be generated, hence reflecting the non-deterministic defects in molecular nanofabrics. In this algorithm, the probability of a PE being reached depends on the probability of the packet reaching the previous PE that forwarded the packet and the probability of the interconnect on which the packet was transmitted being defect free. The methodology also stores the via number from which a PE is reachable with highest probability, since this facilitates the physical logic configuration of the fabric at a later stage if there is more than one available via. Since the methodology applies this modified broadcasting algorithm on a nanofabric model, the dynamics of the physical broadcast are mimicked in the model. Note that the broadcast-based part of the defect-mapping technique is executed off-line, i.e., on an external system.

The methodology also partitions the nanofabric models to decrease the computational complexity of the off-line probabilistic broadcasting algorithm. A molecular nanofabric is partitioned into equal-sized smaller units such that the algorithm can be run simultaneously on all the units. This implies that in certain cases some of the PEs may be a part of more than one unit and hence can be reached from more than one via. The defect-mapping mechanism pin-points the via from which a specific PE can be reached with the highest probability. If the highest probability value of reaching a PE from any of the vias is lower than a user-specified threshold, that PE is marked defective.

In summary, the methodology combines the good features of test circuit-based and broadcast-based defect mapping techniques. The differences between this methodology [234–25] and the methodologies in [21, 22] are as follows:

- [21] uses extensive on-line testing to improve the recovery metric whereas this methodology only uses limited on-line testing.
- [22] proposes a deterministic broadcast algorithm that has been extensively modified to a non-deterministic broadcast algorithm.
- [23] generates on-line defect maps by physically broadcasting test packets in the nanofabrics whereas in the methodology discussed such broadcasts are mimicked in the nanofabric models.
- The discussed methodology uses state space partitioning techniques on the nanofabric models to improve off-line computational complexity. This was not considered in both [21, 22].

10.3.2.3. Hierarchical Redundancy Insertion Methodology. The problem of mapping logic onto ultra-dense nanofabrics has been handled well in [26]. In this methodology, the same procedure is used to map covers of different designs onto the structural hierarchy of the nanofabric models. Note that a cover is the decomposition of a monolithic data flow graph (DFG) of a design, where a DFG of a design represents the behavior of a design. Since there can be more than one way of decomposing a DFG for a system into a cover, the DFG of a system may have more than one cover. Figure 10.5 shows one of the possible covers for an auto regression (AR) filter design.

This procedure is extended so that behaviorally redundant systems can be mapped onto structurally redundant molecular nanofabrics. Redundancy can be added to the covers by replicating the nodes. Figure 10.6a shows how a non-redundant cover with a single flow changes to a cover with two flows when an operation (Node 1) is triplicated. Similarly, structural redundancy can be inserted either at the crossbar level; by adding columns or at any of the three tiers of the nanofabric hierarchy. Figure 10.6b shows TMR configurations for the different structural hierarchies. The TMR configurations are representative examples of structurally redundant fault-tolerant configurations. The discussed hierarchical

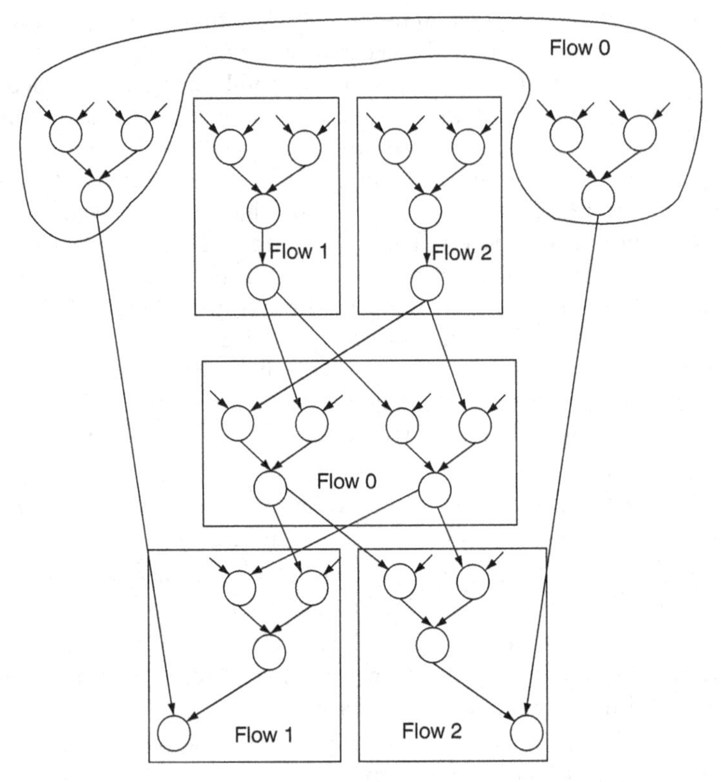

Figure 10.5. Cover for AR filter design.

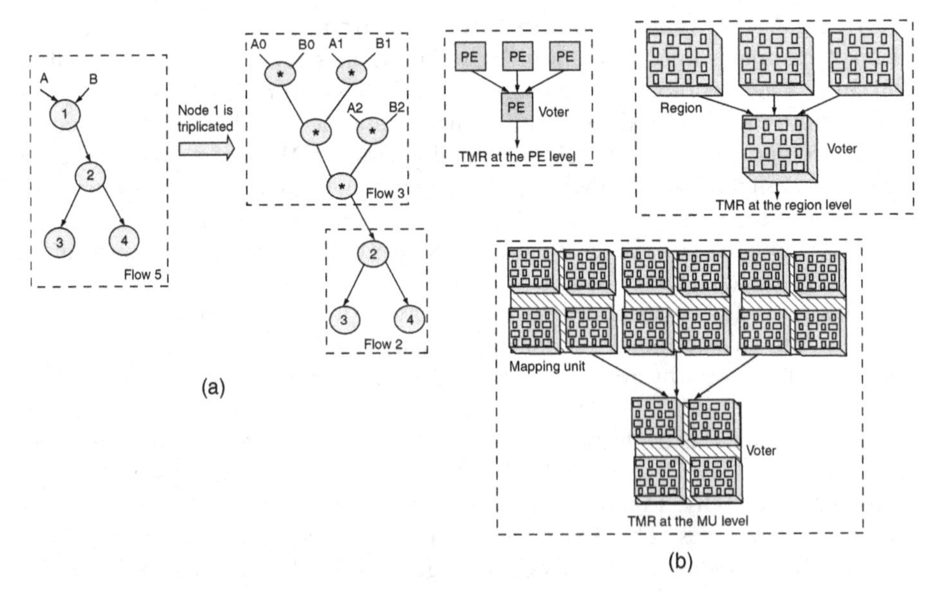

Figure 10.6. Behavioral and structural redundancy.

redundancy insertion methodology maps the non-redundant or redundant covers of the systems onto the structurally redundant nanofabrics by generating *pseudo-random* maps.

These pseudo-random maps (i) avoid defective components if and only if a defect map is available for the nanofabric, since a designer can opt not to generate a defect map, (ii) introduce redundant structural units at the different hierarchies of the nanofabrics, and (iii) attempt to reduce routing latencies while inserting structural redundancy. As part of this work, scripts are developed that use these maps to hierarchically translate each primitive flow of the covers onto the structurally redundant nanofabric models. The stress is on hierarchy, since the computational time required for steady state probability analysis of hierarchically mapped designs is less as compared to flat models [28]. Such a specific behavioral-structural mapping of a design is called a *cover-map*.

10.3.2.4. Design Flow. Figure 10.7 shows the design flow of the proposed design and analysis methodology. Below, the different steps are outlined in detail.

1. Test circuits are run physically on the nanofabric to compute the approximate junction defect rate. A model of the specific molecular

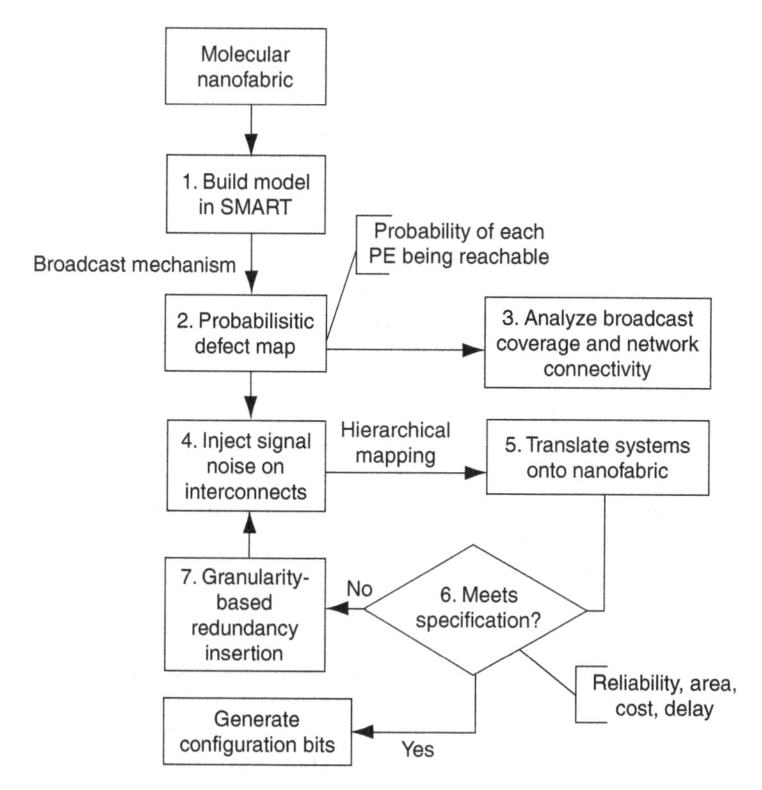

Figure 10.7. Design Flow.

nanofabric is built in SMART [29] (SMART is probabilistic model checking tool which can be used to model and analyze complex probabilistic systems) by creating an array of PEs and interconnects. Since each PE needs to have some functionalities to support the proposed broadcast-based algorithm, the nanofabric model enforces that each PE must at least have logic functionalities to support 4 transceivers, a simple single bit ALU and control circuitry for routing. The junction defect rate is used to compute the probability of the PEs being successfully configured to achieve such minimal functional capabilities.

2. The nondeterministic broadcast-based algorithm (Section 3.2.2) is used to generate a defect map, if the designer opts to generate one.
3. If a defect map is generated, the designer can analyze the performance of the defect mapping scheme in terms of the broadcast coverage and recovery.
4. A transient fault injection library is developed to model signal noise at the interconnects. This library can be used to model noise as Gaussian distributions with different means and variances.
5. The hierarchical logic mapping scheme proposed in [26] and the pseudo-random maps are used to translate a nonredundant or redundant cover (Fig. 10.6) onto the nanofabric model composed of regions, MUs, and components.
6. The mapped design is analyzed to check conformance to required specifications in terms of reliability, cost, latency and area. If the specifications are met, the system is physically configured on the nanofabric.
7. If the specifications are not met, structural redundancy is added at the different hierarchies of the nanofabric by using the redundancy insertion methodology discussed above.

This methodology can be applied for the design and analysis of mission-critical systems and systems that can tolerate certain degrees of failure as shown in Figure 10.8. If the system is mission-critical and needs a reliability guarantee that is arbitrarily close to 100%, the probabilistic defect mapping mechanism can be used with high reachability threshold values—implying that each PE needs to be accessed from one of the vias with very high probability. PEs that are not reachable are marked defective and are not considered while mapping the system onto the nanofabric. The nondefective nodes are partitioned into regions, MUs and components, and the DFG of the system is mapped to the defect mapped nanofabric model with suitable redundancy. If the system can tolerate low but nonzero failure probabilities, it may not be required to generate defect maps for the nanofabric. Instead, such a system can be mapped onto the defective nanofabric with suitable redundancy. These will allow the system to be tolerant to transient faults, achieve the required reliability guarantee and entail less computational complexity for its design. In both the system types, the value of R and architectural level for redundancy insertion is dependent on the system and

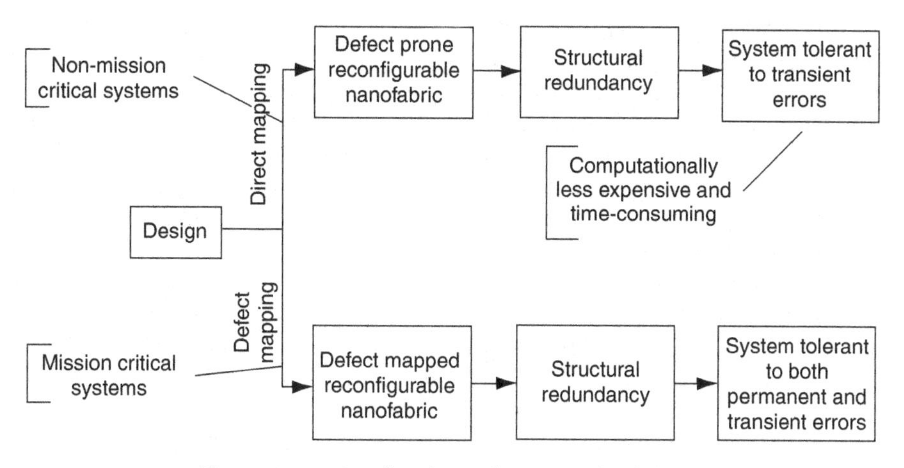

Figure 10.8. Application of our methodology.

its operational environment, hence, determining these parameters is experimental. The toolset is used to determine whether these parameters provide the necessary reliability without causing unacceptable delay, area and cost penalties.

10.4. RELIABILITY EVALUATION OF DEFECT/FAULT-TOLERANT NANOCOMPUTING

As discussed earlier, due to increased defects in nanoscale devices, designing reliable digital circuits with such devices is a major challenge. Development of accurate reliability estimation methodologies is critical in analyzing the robustness of circuits designed from unreliable nanoscale devices.

Recently, a number of methodologies have been developed to evaluate circuit reliability, methodologies that analyze the reliability independent of the technology used to fabricate nanoscale devices. Methodologies based on probabilistic model checking (PMC) have been developed to evaluate circuit architecture reliability and demonstrated its applicability at the logic gate and logic block levels. [30, 31] developed matrix-based gate-level models called probabilistic transfer matrices (PTM) that are used for circuit reliability analysis. Concurrently, but independently [32], proposed a technique to develop probabilistic models called probabilistic gate models (PGM) for unreliable logic gates and used these models to analyze the circuit reliability.

10.4.1. Probabilistic Model Checking (PMC)

PMC is an algorithmic procedure for ascertaining whether a given probabilistic system satisfies probabilistic specifications such as *the probability of logical correctness at the output of a logic network must be at least* 90% *given that each gate has a failure probability of* 0.001. The system is usually modeled as a state transition system with probability values attached to the transitions. Examples of such transition systems are discrete time Markov chains (DTMCs), continuous time Markov chains (CTMCs) and Markov decision processes (MDPs). The specifications or properties to be verified are specified typically in probabilistic extensions of temporal logic. A probabilistic model checker applies algorithmic techniques [33] to analyze the state space of the input model to verify the probabilistic properties.

The two most common temporal logics used for specifying properties of probabilistic systems are PCTL [34, 35] and CSL [36, 37], both extensions of the logic CTL. PCTL is used to specify properties for DTMCs and MDPs and CSL is used for CTMCs. One common feature of the two logics is the probabilistic Π operator. For example, the formula $\Pi_{\geq 1}[\Diamond\, terminate]$ states that the system will eventually terminate with probability 1. On the other hand, the formula $\Pi_{\geq 0.95}[\neg repair\ U^{\leq 200}\ terminate]$ asserts that the system should terminate within 200 time steps without requiring any repairs with a probability 0.95 or greater. In addition to the Π operator, CSL also provides the Σ that helps in specifying steady-state behavior. For instance, $\Sigma_{<0.01}[queue\_size = max]$ states that the probability that a queue is full is strictly less than 0.01 in the long run. Further properties can be analyzed by introducing the notion of costs (or, conversely, rewards). If each state of the probabilistic model is assigned a real-valued cost, one can compute properties such as the expected cost to reach specific states, the expected accumulated cost over some time period, or, the expected cost at a particular time instant.

Methodologies based on PMC techniques can be applied to evaluate the reliability of nanoarchitectures. PMC-based tools like PRISM [38] and SMART [29] can be used for such analysis. PRISM [38, 39] is an example of model checker that supports the analysis of three types of probabilistic models: DTMCs, CTMCs and MDPs. DTMCs can be used to model conventional digital circuits since DTMCs are suitable for such modeling [12]. The DTMC model of computation specifies the probability of transitions between states such that the probabilities of performing a transition from any given state sums up to 1. Modeling of fault-tolerant architectures can be done as templatized DTMCs with probabilistic assumptions about the occurrence of noise at the gates and interconnects and using Markov analysis techniques to evaluate different properties of the architectures.

In [40, 41] PMC is used to evaluate the reliability of small circuits assuming that each gate fails independently. Circuits are specified as DTMCs, due to their suitability for modeling digital systems [12]. The reliability of a circuit is evaluated by computing the probability of reaching specific DTMC states, states that represent the correct Boolean values at the circuit outputs for a certain probability distribution at the inputs. Below, we discuss this PMC-based reliability estimation methodology with an illustrative example.

Figure 10.9 shows the DTMC for a NAND gate with inputs $A0$ and $B0$, and output *out*. The DTMC has 15 states and 22 transitions. The state variables are s, $A0$, $B0$ and *out*, where s is used to represent the level in the DTMC. Formally, each

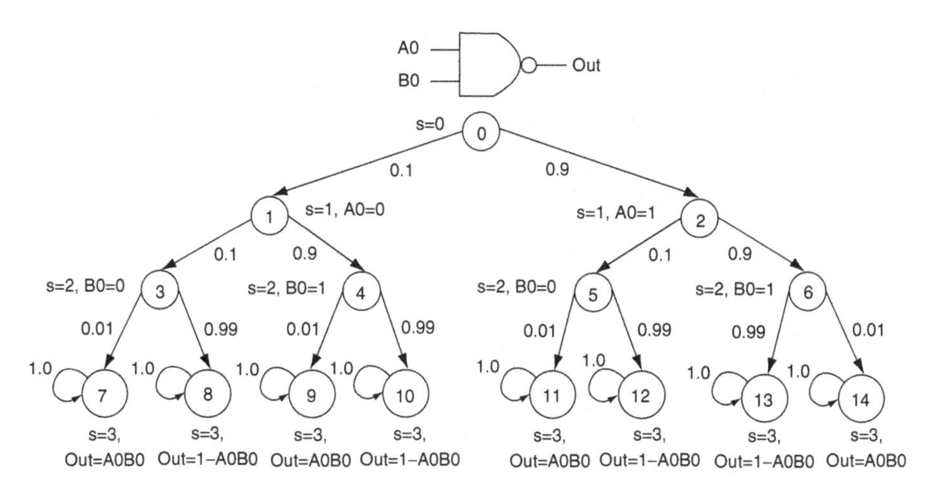

Figure 10.9. DTMC of a NAND gate.

DTMC state is encoded as a quadruple $(s, A0, B0, out)$, where $s \in \{0, 1, 2, 3\}$ and $A0$, $B0$ and out are Boolean variables. The transitions in Figure 10.9 have probability values attached to them. For instance, the probability of 0.9 for transitioning from state 0 (encoded as (0,0,0,0)) to state 2 (encoded as (1,1,0,0)) implies that input $A0$ has a 0.9 probability of being stimulated, i.e., being in logic "1" state. Similarly, a 0.01 probability of moving from state 6 (2,1,1,0) to state 14 (3,1,1,1) implies that the NAND gate can fail independently with a 0.01 probability.

Figure 10.10 is a one step transition probability matrix of the DTMC in Figure 10.9. Each row of the matrix represents the transition flow out of the corresponding state. For instance, the first row of the matrix shows that there is a probability of 0.1 for transitioning from DTMC state 0 to DTMC state 1, and a probability of 0.9 of moving from state 0 to state 2. Since the accumulative transition flow out of each state must be 1 in a DTMC, the rows of the matrix must sum to 1.

Usually these probability matrices are sparse, and can hence be represented compactly. PRISM uses multiterminal binary decision diagrams (MTBDDs) to represent these matrices. MTBDDs were introduced in [42] as a generalization of binary decision diagrams (BDD) [43]. Like BDDs, they are rooted directed acyclic graphs whose non-terminal nodes are labeled with Boolean variables from an ordered set. However, unlike BDDs, the terminal nodes are labeled with values from a finite set M, where usually $M \subset P$, not just 0 and 1. An MTBDD with n Boolean variables and terminals from M can be considered as a map $f : \{0, 1\}^n \rightarrow M$.

$$
\begin{bmatrix}
0{:}0 & 0{:}1 & 0{:}9 & 0{:}0 & 0{:}0 & 0{:}0 & 0{:}0 & 0{:}0 & 0{:}0 & 0{:}0 & 0{:}0 & 0{:}0 & 0{:}0 & 0{:}0 & 0{:}0 \\
0{:}0 & 0{:}0 & 0{:}0 & 0{:}1 & 0{:}9 & 0{:}0 & 0{:}0 & 0{:}0 & 0{:}0 & 0{:}0 & 0{:}0 & 0{:}0 & 0{:}0 & 0{:}0 & 0{:}0 \\
0{:}0 & 0{:}0 & 0{:}0 & 0{:}0 & 0{:}0 & 0{:}1 & 0{:}9 & 0{:}0 & 0{:}0 & 0{:}0 & 0{:}0 & 0{:}0 & 0{:}0 & 0{:}0 & 0{:}0 \\
0{:}0 & 0{:}0 & 0{:}0 & 0{:}0 & 0{:}0 & 0{:}0 & 0{:}0 & 0.01 & 0.99 & 0.0 & 0.0 & 0.0 & 0.0 & 0.0 & 0.0 \\
0{:}0 & 0{:}0 & 0{:}0 & 0{:}0 & 0{:}0 & 0{:}0 & 0{:}0 & 0{:}0 & 0{:}0 & 0{:}01 & 0{:}99 & 0{:}0 & 0{:}0 & 0{:}0 & 0{:}0 \\
0{:}0 & 0{:}0 & 0{:}0 & 0{:}0 & 0{:}0 & 0{:}0 & 0{:}0 & 0{:}0 & 0{:}0 & 0{:}0 & 0{:}0 & 0{:}01 & 0{:}99 & 0{:}0 & 0{:}0 \\
0{:}0 & 0{:}0 & 0{:}0 & 0{:}0 & 0{:}0 & 0{:}0 & 0{:}0 & 0{:}0 & 0{:}0 & 0{:}0 & 0{:}0 & 0{:}0 & 0{:}0 & 0{:}99 & 0{:}01 \\
0{:}0 & 0{:}0 & 0{:}0 & 0{:}0 & 0{:}0 & 0{:}0 & 0{:}0 & 1{:}0 & 0{:}0 & 0{:}0 & 0{:}0 & 0{:}0 & 0{:}0 & 0{:}0 & 0{:}0 \\
0{:}0 & 0{:}0 & 0{:}0 & 0{:}0 & 0{:}0 & 0{:}0 & 0{:}0 & 0{:}0 & 1{:}0 & 0{:}0 & 0{:}0 & 0{:}0 & 0{:}0 & 0{:}0 & 0{:}0 \\
0{:}0 & 0{:}0 & 0{:}0 & 0{:}0 & 0{:}0 & 0{:}0 & 0{:}0 & 0{:}0 & 0{:}0 & 1{:}0 & 0{:}0 & 0{:}0 & 0{:}0 & 0{:}0 & 0{:}0 \\
0{:}0 & 0{:}0 & 0{:}0 & 0{:}0 & 0{:}0 & 0{:}0 & 0{:}0 & 0{:}0 & 0{:}0 & 0{:}0 & 1{:}0 & 0{:}0 & 0{:}0 & 0{:}0 & 0{:}0 \\
0{:}0 & 0{:}0 & 0{:}0 & 0{:}0 & 0{:}0 & 0{:}0 & 0{:}0 & 0{:}0 & 0{:}0 & 0{:}0 & 0{:}0 & 1{:}0 & 0{:}0 & 0{:}0 & 0{:}0 \\
0{:}0 & 0{:}0 & 0{:}0 & 0{:}0 & 0{:}0 & 0{:}0 & 0{:}0 & 0{:}0 & 0{:}0 & 0{:}0 & 0{:}0 & 0{:}0 & 1{:}0 & 0{:}0 & 0{:}0 \\
0{:}0 & 0{:}0 & 0{:}0 & 0{:}0 & 0{:}0 & 0{:}0 & 0{:}0 & 0{:}0 & 0{:}0 & 0{:}0 & 0{:}0 & 0{:}0 & 0{:}0 & 1{:}0 & 0{:}0 \\
0{:}0 & 0{:}0 & 0{:}0 & 0{:}0 & 0{:}0 & 0{:}0 & 0{:}0 & 0{:}0 & 0{:}0 & 0{:}0 & 0{:}0 & 0{:}0 & 0{:}0 & 0{:}0 & 1{:}0
\end{bmatrix}
$$

Figure 10.10. Transition probability matrix for NAND DTMC.

[42] shows how to represent probability matrices as MTBDDs. Consider a square $2^m \times 2^m$ matrix A (non-square matrices can be padded) with elements from M. Its elements a_{ij} may be viewed as values of a function $f_A : \{1, \cdots, 2^m\} \times \{1, \cdots, 2^m\} \rightarrow M$, where $f_A(i,j) = a_{ij}$. Using the Boolean encoding $c : \{0,1\}^m \rightarrow \{1, \cdots, 2^m\}$, f_A may be interpreted as a Boolean function $f : \{0,1\}2^m \rightarrow M$, where $f(x,y) = f_A(c(x), c(y))$ for $x = (x_1, \cdots, x_m)$ and $y = (y_1, \cdots, y_m)$. The variables for the rows and columns are interleaved, implying that the MTBDD interpreted from the function $f(x_1, y_1, \cdots, x_m, y_m)$ can be used to represent the matrix A. This interleaving imposes a recursive structure from which efficient recursive algorithms for all matrix operations are derived [42].

For instance, the transition probability matrix of Figure 10.10 can be represented as a MTBDD with Boolean row variables s_0, s_1, A_0, B_0, out and column variables $s'_0, s'_1, A'_0, B'_0, out'$. Note that since $s \in \{0,1,2,3\}$, it is encoded as $\{0,1\}^2 \rightarrow \{0,1,2,3\}$. $f(s_0, s'_0, s_1, s'_1, A_0, A'_0, B_0, B'_0, out, out')$ is the interleaved function from which the MTBDD is interpreted in this case. Figure 10.9 indicates that only 52 MTBDD nodes are required to represent the 15×15 probability matrix. Thus, the PMC-based methodology entails modeling circuits as DTMCs and the PRISM tool uses compact MTBDDs to represent and analyze the DTMCs.

10.4.1.1. Evaluating Multiplexing Architectures. The effect of noisy communication on computing systems has been theoretically analyzed in terms of information theoretic entropy, and bounds have been established on the noise levels that can be tolerated during transmission [44]. Error-correcting codes have been used extensively to abate transient faults at interconnections [45, 46]. The theory to treat the effect of noise on gates was first developed by von Neumann [47], and later extended in [3, 48–54].

In his seminal work [47], von Neumann proposed a gate-level fault tolerance technique called multiplexing (MUX) for both the majority (MAJ) and NAND gates. He also showed analytically how a NAND MUX architecture can be used to enhance the reliability of a system in the presence of limited computation noise and large bundle size (N). *Bundle size* (N) is defined as the number of replicated transmission lines and *redundancy factor* (R) is defined as the ratio of the circuit sizes of the redundant and nonredundant designs for a nanosystem. Note that R is dependent on the value of N. Reliability evaluation of multiplexing (MUX) architectures is based on PMC technique discussed earlier in this chapter.

Von Neumann addressed a fundamental problem in his seminal work [47], the problem that the probability (δ) of malfunction of a system cannot be less than probability (ε) of failure of the last device in the design. He proposed multiplexing (MUX) as a technique to achieve $\delta < \varepsilon$. The main trick behind MUX is in replacing a single transmission line by a bundle of N parallel transmission lines. Since a signal bundle is used to carry messages, a single logic gate is replaced by N copies. The N devices process the signals in parallel to give N outputs. Each element of the output set will be identical and equal to the original output of the logic gate, if all the copies of the inputs and devices are reliable. However, if the inputs or devices are in error, the outputs will not be identical. To tackle such an error-prone scenario, von Neumann defined a critical level $\Delta \in (0, 0.5)$. The output of the MUX unit is considered stimulated (taking logical value true) if at least $(1 - \Delta) \cdot N$ of the outputs are stimulated and non-stimulated (taking logical value false) if no more than $\Delta \cdot N$ outputs are stimulated. Any other number of stimulated lines is interpreted as a malfunction.

The MUX unit formed by the signal bundles and logic gates is known as the executive stage since this performs the basic logic function. In Figure 10.11, the executive stage of the NAND and MAJ MUX consist of N NAND and MAJ gates, respectively. The advantage of implementing this technique at the gate level is to increase the reliability of the basic building blocks of a logic network, hence, enhancing the reliability of more complex units built from these.

In [47], von Neumann observed that for MAJ MUX whose outputs are governed by a two-to-one majority of the input bundles (two inputs are either stimulated or non-stimulated while the third one represents a different logic state), the error at the output bundle is the sum of errors in two governing input bundles.

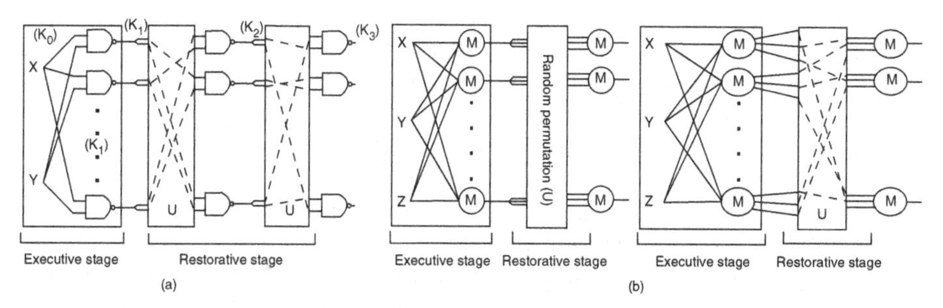

Figure 10.11. Original von Neumann MUX schemes.

Similarly, the output bundle error for a NAND MUX system is the sum of the errors at the two input bundles, if both the input bundles are prevalently stimulated. This implies that certain worst-case input configurations may produce outputs lying in the intermediate region of uncertain information. Thus, von Neumann proposed that a logical stage is required to transform an input bundle with a stimulation level close to zero or one into an output bundle with stimulation level closer to the corresponding extremes [47].

Hence, the von Neumann multiplexed systems must contain a signal restoration stage that reduces the degradation caused by the executive stage for certain input bundle configurations. The signal carried by each input bundle of an executive stage can be thought of as a probability wave that deviates from the prevalent logic values due to signal noise at the executive stage gates. The purpose of a restorative stage is to counteract this deviation and increase the probability of the output bundle being in a valid stimulated or non-stimulated state. Figure 10.11 shows the restorative stages for the NAND and MAJ MUX schemes. Correlation between the input bundles will destroy the signal restoration process, hence, the logic circuit U (see Fig. 10.11) provides a randomizing effect to provide statistical independence between the output bundle of the executive stage and the input to the parallel NAND or MAJ gates that form the restorative stage.

The number of restorative stages can be increased arbitrarily, but it has been shown in [41, 55] that the output distributions of the MUX systems stabilize very quickly as the number of restorative stages increases while causing degradation in speed and increase in the redundancy factor R. It can be observed from Figure 10.11, that R for a MUX system is computed as a function of the bundle size (number of replicated gates in each MUX stage) and the number of MUX stages. For NAND MUX, R is computed as *Bundle Size* $\times$ (1 + 2 $\times$ *Number of Restorative Stages*), whereas, for MAJ MUX, R is evaluated as *Bundle Size* $\times$ (1 + *Number of Restorative Stages*). For instance, a NAND MUX system with a bundle size of 10 and 2 restorative stages (5 MUX stages) has a R of 50, and a MAJ MUX system with a similar configuration has a R of 30 (3 MUX stages).

Although von Neumann had proposed both NAND and MAJ MUX systems in [47], most researchers extended his analysis of the NAND MUX architecture. von Neumann had analyzed the NAND MUX system for $N > 1000$ by exploiting the Gaussian nature of the output distribution. This theory was extended to $N < 1000$ in [55], and it was proved that the output distribution is binomial for smaller values of N. [56] introduces a modification of the von Neumann NAND MUX technique called parallel restitution. Parallel restitution is simply a methodology to apply the NAND MUX scheme to a large system to make it fault-tolerant to noise. [57] was probably the first theoretical attempt at analyzing the reliability of MAJ MUX. In that work, the authors analyzed the reliability of MAJ MUX architectures theoretically for small R and extended this scheme by eliminating unnecessary restorative stages.

The integration of reconfigurable architectures and NAND MUX has also been evaluated in [16] to circumvent both permanent and transient faults. In [58–60], different fault-tolerant techniques including NAND MUX were used on a chip composed of 10^{12} devices to compare their fault mitigation potential. It was shown that a NAND MUX based architecture on such a chip could be reliable 90% of the time if the devices had a failure probability ≤ 0.01, however, the R required to achieve this was $\approx 10^6$.

Reference [28] provides a comparative study of the von Neumann NAND and MAJ MUX architectures. These comparisons have been made in the presence of (i) only noisy gates and (ii) noisy gates and interconnects, for different values of R. Some of the results are discussed below. The first five results are obtained in the presence of only noisy gates while the rest of them are determined in the presence of both noisy gates and interconnects.

- For large gate failure probabilities, the performance of the MAJ MUX architecture is substantially better than the NAND MUX system.
- As the gate failure probability increases beyond a certain threshold, increasing N degrades the performance of the NAND MUX system. But for MAJ MUX, increasing N results in marginal improvement of performance.
- For small gate failure probabilities, the performance of the MAJ MUX architecture is better than NAND MUX at lower R. But both MUX architectures tend to reach similar steady state reliability values at higher R.
- For small gate failure probabilities, the increase in N results in a higher probability of malfunction. This is because the chance of error introduction due to more noisy gates may be counterproductive [57]. Hence a higher R is required to improve the performance of the architecture. This is true for both MUX schemes.
- The individual device failure probabilities that can be tolerated by incorporating a MAJ MUX scheme with $50 \leq R \leq 100$ on a chip are determined and compared to theoretical results.
- MAJ MUX has a higher reliability than NAND MUX in the presence of small and large noise spikes at the inputs and interconnects.
- In the presence of signal noise at the inputs and interconnects, a MUX system with a large bundle size N must have higher R to provide better reliability than a MUX system with small N.

10.4.2. Probabilistic Transfer Matrices (PTMs)

PTMs have been used in [61] to model defective logic gates using matrix representations, an idea that dates back to [62]. The PTM for a NAND gate with a failure probability of ε is given by

$$PTM_{NAND} = \begin{bmatrix} \varepsilon & 1-\varepsilon \\ \varepsilon & 1-\varepsilon \\ \varepsilon & 1-\varepsilon \\ 1-\varepsilon & \varepsilon \end{bmatrix}, \tag{10.1}$$

and its output probability is given by $[p_0\, p_1] = [p_{00}\, p_{01}\, p_{10}\, p_{11}] \times PTM_{NAND}$. The PTM representation encompasses all the input combinations of the NAND gate and hence is very similar to logic compatibility functions (Section 4.4) or DTMC models (Section 4.1) for NAND logic.

A PTM for a specific circuit is formulated by composition of the gate PTMs, the composition being dependent on the logic dependency of the circuit. The authors propose a framework in [31] based on PTMs that can be used for computing the output probabilities for combinational circuits. This involves the composition of gate PTMs in terms of the logic dependency of a circuit. This composition technique takes into account signal dependencies between gates by considering the underlying joint probabilities; it also considers the effects of logical masking.

The authors in [31] use algebraic decision diagrams (ADDs) to alleviate a potential memory bottleneck for PTMs representing large circuits. The ADD representations result in elimination of identical information and compression of the PTMs. PTM operations such as probability value extraction are performed on the ADDs.

There are a number of similarities between the PMC and PTM approaches. First, instead of specifying the initial probabilities as a tuple ($[p_{00}\, p_{01}\, p_{10}\, p_{11}]$), if PTM_{NAND} is modified to encode them, the modified PTM becomes the transition probability matrix shown in Figure 10.10. Second, both the PMC and PTM approaches require a matrix of the order of $O(2^{m+n})$ entries to evaluate a circuit with m inputs and n outputs. Therefore, the PMC and PTM methodologies use compact MTBDD and ADD (another name for MTBDDs [63, 64]) matrix representations, respectively. Third, even the usage of these compression techniques does not help these two reliability evaluation methodologies to scale for large circuits.

10.4.3. Probabilistic Gate Models (PGMs)

The PGM-based methodology [32] entails the formulation of a PGM for each logic gate type. For instance, the PGM for the NAND gate in Figure 10.9 with a failure probability of ε is

$$p(out) = [1 - p(A0)p(B0) \quad p(A0)p(B0)] \cdot \begin{bmatrix} 1-\varepsilon \\ \varepsilon \end{bmatrix}, \qquad (10.2)$$

where $p(A0)$ and $p(B0)$ are the probabilities of inputs $A0$ and $B0$ being stimulated. Hence, $p(out)$ is the probability of the faulty NAND output being logic high. This formulation can be applied iteratively to compute circuit reliability. This shows that the PGM approach requires computation of the output distribution, given the input probability distributions and logic function (truth table) performed by the gate. The DTMC model of a gate in the PMC methodology can be interpreted as a representation of the input probability distribution and the logic function; hence, the PMC and PGM approaches are similar. At the core, these methodologies construct probabilistic models that can be derived from one another; assume that if each gate fails independently, the scalability problem affects them equally.

10.4.4. Markov Random Fields (MRFs)

Reference [65] proposed a probabilistic approach based on Markov Random Fields (MRFs) for analyzing nanocircuits. An MRF is defined as a finite set of random variables, $\Lambda = \{\lambda_1, \lambda_2, \ldots \ldots, \lambda_k\}$. Each variable λ_i has a neighborhood N_i which has variables from $\{\Lambda - \lambda_i\}$. The probability distribution of a given variable depends only on a typically small neighborhood of other variables that is called a clique. As per the Hammersley–Clifford theorem [66],

$$P(\lambda_i | \{\Lambda - \lambda_i\}) = \frac{1}{Z} e^{\frac{-1}{KT} \sum_{c \in C} U_c(\lambda)}. \qquad (10.3)$$

The conditional probability in Equation 10.3 is the Gibbs distribution. Z is the normalizing constant and for a given node i, C is the set of cliques. U_c is the clique energy function [65] and depends only on the neighborhood of the node whose energy state probability is being calculated.

The idea of this model of computation is to use such a Gibbs distribution-based technique to characterize the logic functionality of each gate and maximize probability of being in valid energy configurations at the gate outputs. The logic functionality of each gate is represented by a logic compatibility function which is similar to a truth table. But instead of only considering the valid logic combinations, i.e., for given logic values at the inputs, the corresponding logic value at the gate output is correct; the logic compatibility function considers the invalid logic logic operation scenarios as well (output value is invalid for a given logic combination at the inputs). Such a function is used to represent the logic or clique energy for each Boolean function, and hence formulate energy-based transformation for the Boolean function.

Due to such a formulation of the logic compatibility function, this computational scheme implicitly considers structural defects during the circuit formulation and eliminates the need for defect mapping followed by defect avoidance. The reliability of Boolean networks can be evaluated by representing circuits as MRF-based formulations of logic gates, applying Belief Propagation to the output energy distributions of each logic gate, and computing the probabilities of the signals at the prime outputs. These output distributions are evaluated for specific probability distributions at the primary inputs and signal noise at the interconnects. Note that this model of computation encodes signals over a *continuous* energy distribution, unlike conventional computational models where signals are bimodal (logic low or high).

Let us take a specific NAND gate example to walk through the methodology in [65]. For a two-input NAND gate, there are three nodes in the assumed MRF: the inputs x_0 and x_1, and the output x_2. Figure 10.12 shows x_0 and its

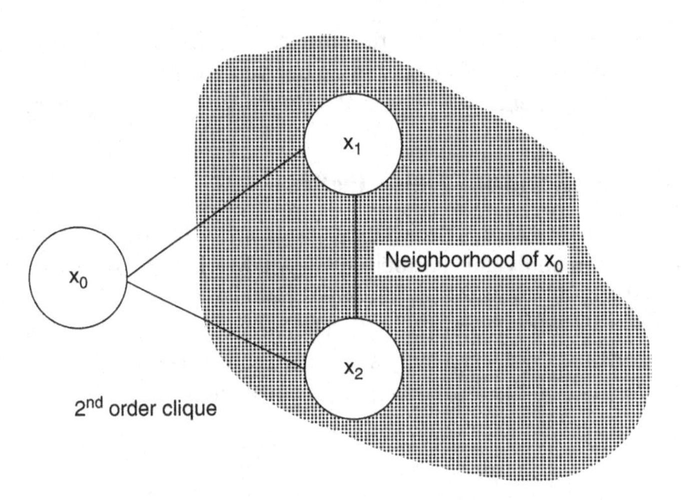

Figure 10.12. NAND gate depicted as an MRF.

neighborhood, since the energy state of this node depends on its neighboring nodes. The edges in Figure 10.12 depict the conditional probabilities with respect to the other input x_1 and the output x_2 (nodes in the same clique). The operation of the gate is designated by the logic compatibility function | (x_0,x_1,x_2) shown as a truth table in Table 10.2. $|=1$ when $(x_2=(x_0 \wedge x_1)'$ (valid logic operations). Such a function takes all valid and invalid logic functionality scenarios into account so as to represent a Gibb's energy-based transformation of the NAND logic.

Entropy has been defined in many different yet equivalent ways. In this discussion, we define entropy as a measure of the disorder of a system. It is considered to have high values when the system under consideration is very disordered. The concept of entropy originated from classical thermodynamics, but has found widespread application in dynamical systems theory, communication theory, information theory, etc. Considerable work has been done on statistical thermodynamics [67], which became the inspiration for adopting the word entropy in information theory. Let us consider a random variable X, which must take on

TABLE 10.2. Logic Compatibility Function of a NAND Gate

I	x_0	x_1	x_2	F
0	0	0	1	1
1	0	0	0	0
2	0	1	1	1
3	0	1	0	0
4	1	0	1	1
5	1	0	0	0
6	1	1	0	1
7	1	1	1	0

one of the values $x_1, x_2, \ldots, x_n$ with respective probabilities $p_1, p_2, \ldots, p_n$. Then, the expected degree of uncertainty (randomness) in the system that is dependent upon X is

$$H(X) = -\sum_i p_i \log(p_i). \tag{10.4}$$

This is information entropy [68] of the random variable X—the average amount of uncertainty associated with the random variable X.

It is also worth mentioning that Equation 10.3 relates logic and thermal energy, hence providing a means of measuring the effect of thermal perturbation. The thermal energy KT (K is the Boltzmann constant and T is the temperature in Kelvin) is normalized to the logic or clique energy. For example, $KT = 0.1$ can be interpreted as unit logic energy being 10 times the thermal energy. The logic margins of nodes in a Boolean network decrease at higher values of KT and increase at lower values. The logic margin in this case is the difference between the probabilities of occurrence of a logic low and a logic high. Higher logic margins result in decrease in entropy or uncertainty in computation, and hence better reliability of computation. If we consider high thermal perturbations, the reliability of computation is likely to be adversely affected; if we can keep our systems far from these temperature values, the reliability is likely to improve. The model of computation in [65] thus considers such thermal perturbations, along with discrete errors and continuous signal noise, as sources of errors.

10.4.5. Scalability Problem

Evaluating the reliability of large circuits is important since realistic electronic systems are significantly large, having millions to billions of devices. The aforementioned reliability analysis methodologies do not scale well and, hence, are not efficient at evaluating sufficiently large circuits. For a circuit with m inputs and n outputs, the PMC-based methodology has a worst-case space and run-time complexity of $O(2^{m+n})$. This is because the PMC-based methodology involves the complete enumeration of all the possible input and output combinations. Although multiterminal binary decision diagrams (MTBDD) [42] are used to compress the state space representation, there is a high demand on memory and run-time for analyzing large circuits. The space and time complexity of a circuit PTM is also of order $O(2^{m+n})$. The PTM method uses algebraic decision diagrams (ADD) [64], synonymous to MTBDDs [63, 64], to compress the PTM representation. Similar to the PMC-based methodology, this compression does not help in reducing the space and time complexity for many typical large circuit PTMs. The PGM-based methodology has also been shown to have run-time issues for circuits with large m [32].

Figure 10.13 shows a circuit with inputs $A0$, $B0$, $C0$ and $D0$, each having a probability of 0.9 of being logic high. The NAND and NOR gates can fail independently with a probability of 0.001. There are four intermediate outputs $o1$,

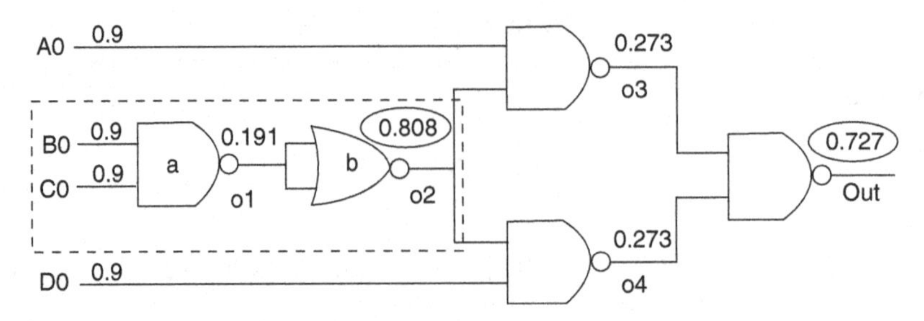

Figure 10.13. Example circuit.

*o*2, *o*3 and *o*4 and one primary output marked as *out*. The probability of each of these outputs being logic 1 are manually computed, given the defect rate of 0.001. The NAND and NOR gates marked *a* and *b*, respectively, form a subcircuit with output *o*2. The probability of *o*2 being logic high is computed to be 0.808, whereas the primary output of the circuit in Figure 10.13 has a 0.727 probability of being logic 1. These probability values indicate that the reliability of a circuit may be very different from the reliability of the smaller sub-circuits that compose the circuit. For large nanoscale circuits that may consist of millions to billions of devices, this difference will only widen. Hence, analyzing smaller sub-circuits does not accurately reflect the impact of each nanoscale process technology on the reliability of large circuits.

On a practical note, Table 10.3 indicates the exponential state space increase when the input width of a ripple carry adder is increased. It has been shown that for adders with width ≥ 4, if the inputs have nonzero probability of being both stimulated and nonstimulated at the same time, the analysis becomes intractable in terms of memory for even high-end workstations.

To alleviate this scalability problem, [69] presents a scalable technique. The basic concept in this technique is to iteratively build and analyze probabilistic gate-level models and propagate the probability values through the circuit according to the logic dependencies. This concept is dissimilar to the direct PMC or PTM approach because it does not build a monolithic probabilistic model or PTM for the circuit. Instead, it folds space into time by iteratively computing

TABLE 10.3. State Space Size for Multibit Adders

Design	DTMC states	MTBDD nodes
1 bit	847	618
2 bit	10191	1455
3 bit	84943	2580
4 bit	682959	3965
5 bit	5467087	5644
6 bit	43740111	7579
7 bit	349924303	9806
8 bit	2799397839	12307

and propagating reliability values in stages, exploiting the inherent structural decomposition of the circuit. To demonstrate how this general scalable technique can make the reliability evaluation methodologies more scalable, [69] develop an algorithm specifically for the PMC-based methodology. Since the PTM and PGM approaches are similar to the PMC approach, minor modifications to this algorithm can also scale the PTM and PGM methodologies.

10.5. CONCLUSIONS

This chapter provides a very short and cursory glimpse of the different techniques that are being used to mitigate the high percentage of defective devices per chip produced by nanoscale fabrication or self-assembly processes. Many of these techniques have high cost in terms of redundancy factor (the ratio of number of devices needed to enforce tolerance in presence of defects and faults to the number of devices needed to implement a logic in absence of faults/defects), area, and power dissipation. An argument in support of using such expensive techniques has often been that at the nanoscale, there will be an abundance of devices, many more than needed for the functionality to be implemented. However, in some cases, the cost is exponentially high, and hence cannot be paid in practice. As a result, the reconfiguration techniques to avoid defects and faults seem to be more attractive solutions, in light of the facts that self-assembled nanofabrics are likely to be regular; with such large number of devices, it is likely that regular fabrics will be the best substrate to create complex logics on, rather than the current practice of random logical layouts on chips.

In any case, we hope that this chapter provides the reader with the preliminary background needed to delve into further details on the various redundancy techniques to mitigate defects and faults in the nanoscale computing. A large body of literature exists on these techniques (pointers to some of which can be found in the bibliographic references) but a lot of the important pointers can be found within the articles referenced here. Similar comments apply to the extensive literature on reconfigurable nanocomputing. The choice of references and topics in this chapter does not indicate any valuation on our part as to which of these articles constitute the most important contributions in this field, but are rather based on our familiarity with the chosen topics and articles.

REFERENCES

1. International technology roadmap for semiconductors, 2005. http://www.public.itrs.net.
2. E. J. Nowak. Maintaining the benefits of CMOS scaling when scaling bogs down. *IBM Journal of Research and Development*, 46(2–3): pp 169–180, 2002.
3. C. H. Bennett. The thermodynamics of computation–a review. *International Journal of Theoretical Physics*, 21(905–940): 1982.

4. M. Forshaw, R. Stadler, D. Crawley, and K. Nikolic. A short review of nanoelectronic architectures, *Nanotechnology*, 15: pp S220–S223, 2004.
5. J. Meindl, Q. Chen, and J. Davis. Limits on silicon nanoelectronics for terascale integration. *Science*, 293: pp 2044–2049, 2001.
6. S. Kim, C. H. Ziesler, and M. C. Papaefthymiou. Charge-recovery computing on silicon, 54: pp 651–659, 2005.
7. R. Landauer. Uncertainty principle and minimal energy-dissipation in the computer, *International Journal of Theoretical Physics*, 21: pp 283–297, 1982.
8. W. C. Athas, L. J. Swensson, J. G. Koller, and N. Tzartzanis. Low-power digital systems based on adiabatic-switching principles, *IEEE Transactions on VLSI Systems*, 2: pp 398–407, 1994.
9. P. Beckett and A. Jennings. Towards nanocomputer architecture. In: ACS Conferences in Research and Practice in Information Technology (CRPIT), ser. Computer Systems Architecture, 6. Australian Computer Society, 2002. http://www.cellmatrix.com/entryway/ products/pub/publications.html.
10. A. Tannenbaum. *Distributed Systems: Principles and Paradigms*. Upper-Saddle River, NJ: Prentice Hall, 2002.
11. P. Jalote. *Fault Tolerance in Distributed Systems*. Upper Saddle River, NJ: Prentice Hall, 1994.
12. D. P. Siewiorek and R. S. Swarz. *Reliable Computer Systems: Design and Evaluation*, 2nd ed. Burlington, MA: Digital Press, 1992.
13. S. Habinc. Funtional triple modular redundancy (ftmr): Vhdl design methodology for redundancy in combinatorial and sequential logic. NASA office of logic design, Technical Report, 2002. http://klabs.org/richcontent/fpga_content/pages/notes/seu_hardening.htm.
14. P. Graham and M. Gokhale. Nanocomputing in the presence of defects and faults: A survey. In: *Nano, Quantum and Molecular Computing: Implications to High Level Design and Validation*. Amsterdam: Kluwer, 2004, pp 39–72.
15. W. H. Pierce. *Fault-Tolerant Computer Design*. New York: Academic Press, 1965.
16. J. Han and P. Jonker. A defect- and fault-tolerant architecture for nanocomputers. *Nanotechnology*, pp 224–230, 2003.
17. A. DeHon. Array-based architecture for fet-based nanoscale electronics. *IEEE Transactions on Nanotechnology*, 2: pp 223–232, 2003.
18. A. DeHon and M. J. Wilson. Nanowire-based sublithographic programmable logic arrays. In: International Symposium on FPGAs, 2004: pp 123–132.
19. S. C. Goldstein and M. Budiu. Nanofabrics: Spatial computing using molecular electronics. In: Annual International Symposium on Computer Architecture (ISCA), July 2001: pp 178–189.
20. J. Heath, P. Kuekes, G. Snider, and R. Williams. A defect tolerant computer architecture: Opportunities for nanotechnology. *Science*, 80: pp 1716–1721, 1998.
21. M. Mishra and S. C. Goldstein. Defect tolerance at the end of the roadmap. In: Test Conference (ITC), Charlotte, North Carolina, Sep 30–Oct 2, 2003.
22. J. Patwardhan, C. Dwyer, A. Lebeck, and D. Sorin. Evaluating the connectivity of self-assembled networks of nano-scale processing elements. In: IEEE International Workshop on Design and Test of Defect-Tolerant Nanoscale Architectures, May 2005.

23. D. Bhaduri, S. K. Shukla, P. Graham, and H. Quinn. Transient error tolerant configuration of nanofabrics using a reliability driven probabilistic approach. In: International Conference on Bio-Nano- Informatics Fusion, July 2005.

24. D. Bhaduri, S. K. Shukla, P. Graham, and M. Gokhale. Reliability analysis of fault tolerant reconfigurable architectures. In: IEEE International workshop on Design and Test of Defect-Tolerant Nanoscale Architectures (NANOARCH), May 2005. http://fermat.ece.vt.edu/Publications/pubs/techrep/techrep0415.pdf.

25. D. Bhaduri and S. Shukla. Probabilistic analysis of self-assembled molecular networks. In: Proceedings of Foundations of NANOSCIENCE (FNANO), May 2005.

26. M. Jacome, C. He, G. Veciana, and S. Bijansky. Defect tolerant probabilistic design paradigm for nanotechnologies. In DAC, June 2004: pp 596–601.

27. J. Koeter. What's an lfsr? (rev. a). Texas Instruments. Tech. Report, 1996. http://www.ti.com/sc/docs/psheets/abstract/apps/scta036a.htm.

28. D. Bhaduri, S. Shukla, P. Graham, and M. Gokhale. Comparing reliability-redundancy trade-offs for two von neumann multiplexing architectures. *IEEE Transactions on Nanotechnology*, to appear 2007. http://fermat.ece.vt.edu/Publications/online-papers/Nano/MUX_TNANO.pdf.

29. http://www.cs.wm.edu/~ciardo/SMART/.

30. K. Patel, I. L. Markov, and J. P. Hayes. Evaluating circuit reliability under probabilistic gate- fault models. In: International Workshop on Logic Synthesis (IWLS), 2003: pp 59–64.

31. S. Krishnaswamy, G. F. Viamontes, I. L. Markov, and J. P. Hayes. *Accurate Reliability Evaluation and Enhancement via Probabilistic Transfer Matrices*. New York: ACM Press, 2005, pp 282–287.

32. J. Han, E. Taylor, J. Gao, and J. Fortes. Faults, error bounds and reliability of nanoelectronic circuits. In: ASAP, 2005: pp 247–253.

33. D. Parker. Implementation of symbolic model checking for probabilistic systems. University of Birmingham. Ph.D. dissertation, 2002.

34. H. Hansson and B. Jonsson. A logic for reasoning about time and probability. *Formal Aspects of Computing*, 6(5): pp 512–535, 1994.

35. A. Bianco and L. de Alfaro. Model checking of probabilistic and nondeterministic systems. In: Foundations of Software Technology and Theoretical Computer Science (FSTTCS'95), ser. LNCS, 1026,New York: Springer, 1995: pp 499–513.

36. A. Aziz, K. Sanwal, V. Singhal, and R. Brayton. Verifying continuous time Markov chains. In: Conference on Computer-Aided Verification (CAV' 96), July 1996: pp 269–276.

37. C. Baier, J. P. Katoen, and H. Hermanns. Approximate symbolic model checking of continuous time Markov chains. In: International Conference on Concurrency Theory (CONCUR'99), Eindhoven, August 1999, pp 146–161.

38. www.cs.bham.ac.uk/~dxp/prism/.

39. M. Kwiatkowska, G. Norman, and D. Parker. *Prism: probabilistic symbolic model checker*. In: TOOLS 2002, ser. LNCS, 2324. New York: Springer April 2002: pp 200–204.

40. D. Bhaduri and S. Shukla, Nanoprism: A tool for evaluating granularity vs. reliability trade-offs in nano-architectures, In: *GLSVLSI*. Boston:, ACM, April 2004.

41. G. Norman, D. Parker, M. Kwiatkowska, and S. Shukla. Evaluating the reliability of nano multiplexing with prism. *IEEE Transactions on Computer-Aided Design of Integrated Circuits and Systems*, 24(9): to appear September 2005.
42. E. Clarke, M. Fujita, P. McGeer, J. Yang, and X. Zhao. Multi-terminal binary decision diagrams: An efficient data structure for matrix representation. In: International Workshop on Logic Synthesis (IWLS), 1993.
43. E. Clarke, E. Emerson, and A. Sistla. Automatic verification of finite state concurrent systems using temporal logic specifications. In: 10th Annual Symposium on Principles of Programming Languages, 1983.
44. C. E. Shannon. A mathematical theory of communication. *Bell Systems Technical Journal*, 27: pp 379–423–623–656, 1948.
45. R. W. Hamming. Error-detecting and error-correcting codes. *Bell Systems Technical Journal*, 29: pp 147–160, 1950.
46. R. Hamming. Coding and Information Theory, 2nd ed. Prentice Hall, 1986.
47. J. von Neumann. Probabilistic logics and synthesis of reliable organisms from unreliable components. *Automata Studies*, pp 43–98, 1956.
48. P. Gacs and A. Gal. Lower bounds for the complexity of reliable Boolean circuits with noisy gates. *IEEE Transactions on Information Theory*, 40: pp 579–583, 1994.
49. C. H. Bennett. Notes on the history of reversible computation. *IBM Journal of Research and Development*, 32(16–23): 1988.
50. R. Landauer. Irreversibility and heat generation in the computing process. *IBM Journal of Research and Development*, 5(183–191): 1961.
51. R. Landauer. Computation: a fundamental physical view. *Physical Science*, 35(88–95): 1987.
52. N. Pippenger. Reliable computation by formulas in the presence of noise. *IEEE Transactions on Information Theory*, 34(2): pp 194–197, 1988.
53. N. Pippenger. *On Networks of Noisy Gates*. 26. New York: IEEE, 1985, pp. 30–38.
54. N. Pippenger. *Developments in the Synthesis of Reliable Organisms from Unreliable Components*. 50. Providence, RI: American Mathematical Society, 1990, pp. 311–323.
55. J. Han and P. Jonker. A system architecture solution for unreliable nanoelectronic devices. *IEEE Transactions on Nanotechnology*, 1: pp 201–208, 2002.
56. A. S. Sadek, K. Nicoli'c, and M. Forshaw. Parallel information and computation with restitution for noise-tolerant nanoscale logic networks. *Nanotechnology*, 15(1): pp. 192–210, Jan 2004.
57. S. Roy and V. Beiu. Majority multiplexing-economical redundant fault-tolerant designs for nanoarchitectures. *IEEE Transactions on Nanotechnology*, 4(4): pp 441–451, 2005.
58. K. Nikolic, A. Sadek, and M. Forshaw. Architectures for reliable computing with unreliable nanodevices. In: Proceeding of the IEEE-NANO'01. IEEE, 2001, pp. 254–259.
59. K. Nikolic, A. Sadek, and M. Forshaw. Fault-tolerant techniques for nano-computers. In: *Nanotechnology*(13): pp 357–362, 2002.
60. M. Forshaw, K. Nikolic, and A. Sadek. Ec answers project (melari 28667). Tech. Rep. http://ipga.phys.ucl.ac.uk/research/answers.

61. K. N. Patel, J. P. Hayes, and I. L. Markov. Evaluating circuit reliability under probabilistic gate level fault models. In: IWLS: pp 59–64, May 2003.

62. V. L. Levin. Probability analysis of combination systems and their reliability, *Engineering Cybernetics*, 6, pp 78–84, Nov–Dec 1964.

63. L. de Alfaro, M. Kwiatkowska, G. Norman, D. Parker, and R. Segala. Symbolic model checking of concurrent probabilistic processes using mtbdds and the kronecker representation. In: TACAS, 2000.

64. I. Bahar, E. Frohm, C. Gaona, G. Hachtel, E. Macii, A. Pardo, and F. Somenzi. Algebraic decision diagrams and their applications. *Journal of Formal Methods in Systems Design*, 10(2–3): pp 171–206, 1997.

65. R. I. Bahar, J. Mundy, and J. Chen. A probability-based design methodology for nanoscale computation. In: International Conference on Computer-Aided Design. San Jose, CA, Nov 2003.

66. J. Besag. Spatial interaction and the statistical analysis of lattice systems. *Journal of the Royal Statistical Society, Series B*, 36(3): pp 192–236, 1994.

67. A. Sommerfeld. *Thermodynamics and Statistical Mechanics: Lectures on Theoretical Physics*. New York: Academic Press, 1964.

68. http://www.pha.jhu.edu/~xerver/seminar2.

69. D. Bhaduri, S. Shukla, P. Graham, and M. Gokhale. Scalability techniques and tools for reliability analysis of large circuits. International Conference on VLSI Design: pp 705–710, Jan 2007.

11

MOLECULAR COMPUTING: INTEGRATION OF MOLECULES FOR NANOCOMPUTING

James M. Tour and Lin Zhong

Molecular computing seeks to implement nanoscale computation systems using molecules or molecular bundles. Enormous progress has been made in constructing nanoscale switching devices from molecules. Devices based on individual molecules and molecular bundles have been demonstrated for logic gates and memory units. However, the construction of a practical molecular computer will require the molecular switches and their related interconnect technologies to behave as large-scale heterogeneous logic, with input/output wires scaled to molecular dimensions so that we can take advantage of the ultrasmall size of the molecules. While numerous innovations have been made in fabrication and design methodologies, no implementations have yet been reported to be at any scale comparable to what can be achieved by the conventional silicon MOSFET technology. In this chapter, we will describe the switching and memory effects of molecular bundles and address practical circuit and architecture solutions for large-scale integration of molecular switches for computing.

We further show that molecules can be employed to enhance silicon-based computing. The drain current I_D and threshold voltage V_T in pseudo-MOSFETs can be controllably modulated by grafting a monolayer of molecules atop oxide-free H-passivated silicon surfaces, and this effect should be extenuated in nanoscale devices where the surface-to-volume ration increases. The technology provides a new mechanism to control process variations and leakage power consumption in conventional MOSFET circuits. Moreover, intrinsic silicon nanowires with different surface states exhibited a p-type doping behavior with clear differences in resistivity and hole mobility. By controlling the surface state density, acceptable electrical operation can be achieved directly on intrinsic Si nanowires without bulk doping. The technology provides a potential alternative to bulk impurity doping for nanoscale silicon structures.

Bio-Inspired and Nanoscale Integrated Computing. Edited by Mary Mehrnoosh Eshaghian-Wilner

11.1. INTRODUCTION

Silicon metal-oxide-semiconductor field-effect transistors (MOSFETs), the building elements of modern computing, have already entered the nanometer era. Nonstandard MOSFET and non-MOSFET nanometer devices have received intense research and are predicted to dominate the landscape of computing in 10 years. Molecular computing seeks to build computational systems wherein individual or small collections of molecules serve as discrete device components or play a significant role in them. Physicists and chemists have bravely ventured into measuring and modeling electronic properties of molecules as well as synthesis and assembly of molecules of desirable electronic properties. Typical molecules used in these molecular switches are orders of magnitudes smaller than the state-of-the-art silicon MOSFET. Such an advantage in size can potentially lead to computing systems with much higher density and performance yet lower power and lower cost than what promised by silicon MOS technologies. Inspired by the vision of molecular computing, computer engineering researchers have explored its circuit, logic, architecture, and even system-level design.

While molecules are approximately one million times smaller than their present day solid-state counterparts, this small size brings with it a new set of challenges. First, although molecules can be synthesized in large quantities relatively easily, they can be difficult to arrange on a surface or in a three-dimensional array such that each molecule is addressable. It is equally difficult to ensure that every molecule stays in place. Second, although individual molecules have been shown to be switching, it is very difficult, if possible at all, to interconnect them or selectively interface them with microscale[1] input/output. This is particularly true for three terminal molecules because it is far more difficult to bring three probes into close proximity than to bring two probes into near contact.

These challenges invite a rethinking of computing design and implementation. Present day computing is based on a top-down process in which predesigned patterns are fabricated exactly as specified and small features, such as transistors, are precisely etched into silicon using resists (chemicals) and light (photolithography). Such a what-you-design-is-what-you-get is no longer true in the nanometer domain. Uncertainty in conventional silicon MOSFET fabrication has already made it impossible to fabricate precisely as specified, leading to great variations in performance and power characteristics in circuits of the same design and fabrication process. Thus various design methodologies have been introduced to keep the top-down process alive in silicon MOSFET-based computing. Unfortunately, the top-down process alone is unlikely to work for molecular computing because it is extremely difficult to position and interconnect molecules, as highlighted above. The fabrication of molecular switches is essentially a bottom-up process, i.e., self-assembly of molecules into higher ordered structural units. Self-assembled structures form the basis of the first generation test

[1] We use microscale to refer to what can be patterned using the state-of-the-art photolithography. We use nanoscale to refer to things that are too small for the state-of-the-art photolithography.

molecular devices that have been made to date. However, the bottom-up process alone is unlikely to produce large-scale, integrated, heterogeneous structures for computing because self-assembly usually leads to homogeneous structures. Therefore, a hybrid or meet-in-between paradigm that combines the strength of both bottom-up and top-down processes is more likely to address the design and fabrication of molecular computing. In such a paradigm, nanoscale elements, including molecules, are fabricated through a bottom-up process into microscale units, which are patterned and programmed to achieve large-scale integration through a more conventional top-down process.

In addition to building computing circuits directly, molecules can be employed to change the properties of semiconductor devices for computing. It has been demonstrated that a monolayer of molecules can dramatically change the threshold voltage, thus leakage current and delay, of silicon MOSFETs, providing a new mechanism to counter process variations and control leakage power consumption, which have haunted deep-submicron silicon MOSFET circuits.

In this chapter we will review recent progress in molecular computing, in particular, our own endeavors in addressing its challenges. While it is impossible to survey such a rapidly developing research field, we seek to address research in building and characterizing molecular devices based on bundles of molecules, circuit and architecture solutions for molecular computing, and molecular grafting for silicon-based computing, with a focus on the integration of molecules into large-scale computing systems. While many assumptions and speculations can be made in molecular computing research, we strive to achieve a balance between fabrication and theoretical design. We refer readers to monographs on molecular computing [1, 2] for more comprehensive treatments. In particular, the chapter does not cover the large body of literature on defect tolerance in molecular computing [3–8] or devices and characterization based on a single molecule [9], notably work in single-molecule (including carbon nanotube) transistors [10–12] and molecular wires [13].

11.2. SWITCHING AND MEMORY IN MOLECULAR BUNDLES

While single-molecule switching and memory devices are desirable, positioning and interconnecting such devices still pose a great research challenge. Therefore, bundles of hundreds or thousands of molecules, or *molecular bundles*, have been employed as switches. Many works that claim single-molecule devices are indeed based on measurement on molecular bundles.

11.2.1. Self-Assembly and Molecular Ordering

The assembly of massive numbers of molecules onto a predefined surface underpins most molecular switches fabricated and characterized so far [14–17]. It is also critical to molecular grafting of semiconductor surface, a complementary method, in some cases of nanodevices, and to impurity doping as will be addressed later [18]. Numerous molecules with sophisticated looks are now accessible to researchers wherein function is built into the molecule, 10^{23} units at a time. Significant research progress has been made to orient 10^{14} molecules/cm^2 in a predefined addressable array on a chip. The atomic order of a surface coupled with molecular packing requirements can give rise to thermodynamically driven self-assembly over large surfaces. The gold-thiolate (R-S-Au) system is the most commonly studied self-assembled monolayer (SAM) [19]. The assembly takes place within seconds or minutes. Sometimes 24-hour periods are used for the assembly to permit dense packing of otherwise slow-to-assemble molecules. Crystalline ordering of the system can occur in domains ranging to hundreds of square nanometers. The gold-thiolate bond is quite robust relative to typical metal-molecule bonds that can be formed at ambient temperature. A SAM also slows the inherent surface reconstruction, a phenomenon that often causes rapid randomization of surface atoms. Figure 11.1 illustrates the self-assembly process. Molecules are constructed that bear some functional component, such as the nitroaromatic as a memory storage unit. These are randomized in solution in which a gold substrate is dipped. In a matter of a few seconds to minutes, the molecules faithfully arrange on the substrate surface. The aromatic thiolates have a tilt angle of approximately 20° relative to the surface normal, while alkanethiolate are tilted approximately 30° to the normal. Few molecules stand perpendicular to the surface; the tilt is a function of the hybridization state of the sulfur and the intermolecular packing requirements.

Besides gold, there are numerous other surfaces that have been employed in SAM construction, including other conductors such as copper, silver, and palladium; semiconductors such as silicon, gallium arsenide, and cadmium sulfide; and insulators such as silicon oxide. Other alligator clips that have been studied

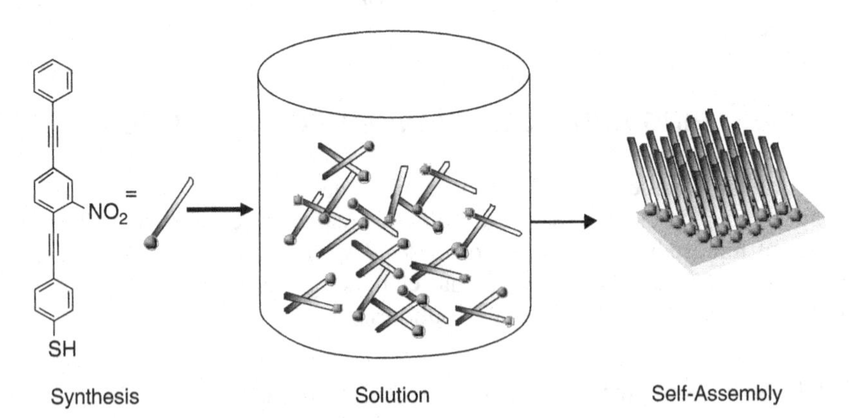

Figure 11.1. The self-assembly process wherein molecules that have programmed function built into their structure then self-order on a surface upon dipping the surface into a solution of the molecules. Once the surface is removed and rinsed, it will bear a monolayer of molecules that possesses long-range order.

include selenols (–SeH), phosphines (R_3P), carboxylates (–CO_2^-) and pyradines [19]. We developed a method for direct assembly of aryl groups on silicon (n-doped, p-doped, undoped, single crystal, or polycrystalline) and gallium arsenide using aryl diazonium salts [20].

One can also construct molecules that have a preference for a given surface by judicious choice of the alligator clip. This surface-selective SAM formation could be a critical factor in determining the usefulness of SAMs for device placement, as heterogeneity in device patterns must be attained. In other words, an array of all AND logic gates would be useless. The heterogeneity is best if it is programmable in its design pattern. Using crystalline substrates that are composed of different atom types and yet are regular in their periodicity could provide a method for the predictable arrangement of surface molecules, but programmability to the array architecture would then be limited to one's ability to form tailored substrate crystals.

11.2.2. Molecular Bundle Switches

Reed developed a test-bed method, called the nanopore, to make reliable measurements on groups of molecules [21]. The nanopore consists of a small (30–50-nm diameter) surface of evaporated metal, most often gold or palladium, on which a SAM of the molecular wires or devices with ~ 1000 molecules is permitted to form. An upper metal (usually gold or titanium) contact is then evaporated onto the top of the SAM layer, forming a metal-SAM-metal sandwich through which I(V) measurements are recorded (Figure 11.2). Using such a small area for the SAM (~ 1000 molecules), we can potentially form defect-free SAMs because the entire area is smaller than the typical defect density of a SAM. This would eliminate electrical shorts that can occur when a larger, e.g., microscale, SAM is used.

We hypothesized that the conductivity of these OPE molecular scale wires and devices arises from electron transfer through the π-orbital backbone that extends over the entire molecule. When the phenyl rings of the OPEs are planar, the π-orbital overlap within the molecules is continuous. Thus transfer over the entire molecule is achieved: Electrons can freely flow between the two metal contacts, leading to maximum conductivity. But if the phenyl rings become perpendicular with respect to each other, the π-orbitals between the phenylene rings likewise become perpendicular. The discontinuity of the π-orbital network in the perpendicular arrangement minimizes electron transport through the molecular system, leading to greatly reduced conductivity [22].

We devised a method that would permit altering the degree of a molecule's π-orbital overlap through the use of a third electrode (gate). Thus molecules with orthogonal net dipoles could be controlled by use of a third electrode in the nanopore to modulate the conformation, and hence the current through the system. However, since nanopore devices with an electrode perpendicular to the SAM axis had not yet been fabricated, we simply began with the control experiments. Namely, to study the two-electrode nanopore made with molecules bearing dipolar groups.

Accordingly, the nitroaniline OPE was tested in the nanopore, in the absence of an orthogonal external electric field, to determine its electronic characteristics. Then a series of control experiments were first performed with alkanethiol-derived SAMs and systems containing no molecules. Both the Au-alkanethiolate-Au junctions and the Au-silicon nitride membrane-Au junctions showed current levels at the noise limit of the apparatus (<1 pA) for both bias polarities at both room and low temperatures. The Au-Au junctions gave ohmic I(V) characteristics with very low resistances. A device containing a SAM of the conjugated OPEs but not bearing the nitroaniline functionalities was fabricated

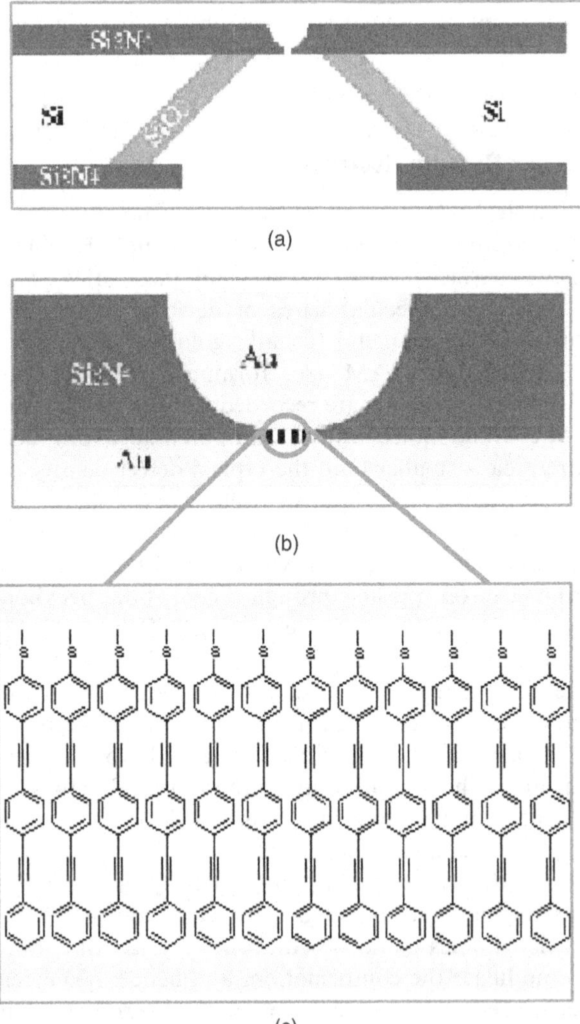

and measured in nearly identical conditions [23] and it exhibited essentially linear I(V) behavior within its noncrystalline temperature range.

Figure 11.3 shows the typical I(V) characteristics of an Au-(nitroaniline OPE)-Au device at 60 K [24, 25]. Positive bias corresponds to hole injection from the chemisorbed thiol-Au contact and electron injection from the evaporated contact. Unlike previous devices that also used molecules to form the active region, this device exhibits a robust and large NDR with a valley-to-peak ratio (PVR) of 1030:1 [24, 25]. We observed the NDR effect from the system up to 260 K but not beyond. Since then, room temperature NDR has been seen in the nanopores containing the SAM of the nitro OPEs [24].

Additionally, we demonstrated charge storage in a self-assembled nanoscale molecular device that operated as a molecular dynamic random access memory (mDRAM) with practical thresholds and output under ambient operation [25]. The memory device operates by the storage of a high or low conductivity state. Because added electrons dramatically affect the conductivity of the molecular system, the information state can be read by a conductivity check. Figure 11.4 shows the write, read, and erase sequence [25]. To further explore the mechanism of this mNDR and mDRAM phenomenon, we have synthesized several related compounds such as a nitroacetamide rather than a free nitroaniline moiety. After testing in the nanopore, we found that nitroacetamide exhibited the NDR effect, however, with a smaller PVR of 200:1 at 60 K.

Figure 11.5 shows the measured input and output of an mDRAM cell using the mononitro OPV in the nanopore at room temperature. To convert the stored conductivity to standard voltage conventions, the output of the device was

Figure 11.2. (a) The starting substrate for the device fabrication is a 250 μm-thick double side polished silicon wafer with 50 nm of low stress Si_3N_4. On the back surface, the nitride was removed in a 400 μm by 400 μm square. The exposed silicon was etched in an orientation-dependent anisotropic etchant through to the top surface to leave a suspended (40 μm by 40 μm) silicon nitride membrane. Reed then grew 100 nm of SiO_2 thermally on the Si sidewalls to improve electrical insulation. A single hole 30 to 50 nm in diameter was made through the membrane by electron beam lithography and reactive ion etching (RIE). Because of the constrained geometry, the RIE rates are substantially reduced so that the far side opening is much smaller than the actual pattern, thereby rendering the cross section bowl-shaped geometry. (b) and (c) A gold contact of 200 nm thickness was evaporated onto the topside of the membrane, filling the pore with Au. The sample was then immediately transferred into a solution to self-assemble the active electronic component, illustrated here with an unfunctionalized sulfur-tipped OPE. The sample was then rinsed, quickly loaded into a vacuum chamber, and mounted onto a liquid nitrogen cooling stage for the bottom Au electrode evaporation, where 200 nm Au was evaporated at 77 K at a rate of less than 1 Å/s. The devices were then diced into individual chips, bonded onto packaging sockets and loaded into a variable temperature cryostat and measured with a Semiconductor Parameter Analyzer.

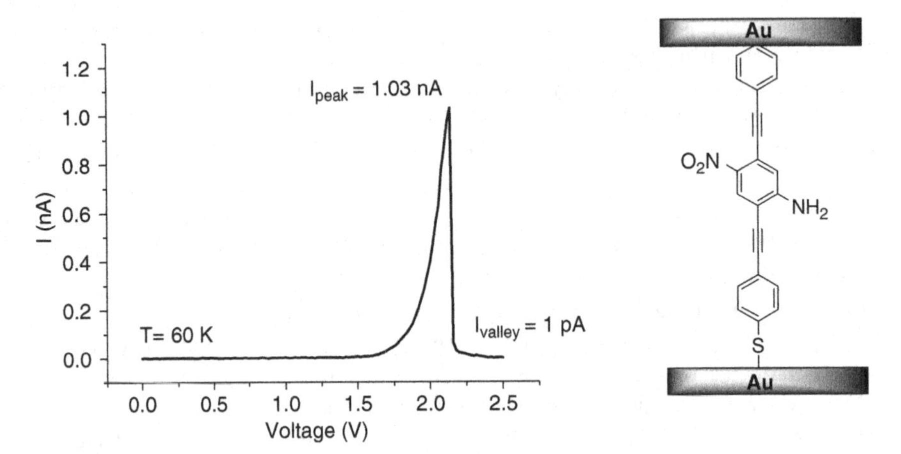

Figure 11.3. I(V) characteristics of an Au-(nitroaniline OPV)-Au device at 60 K in the nanopore.

dropped across a resistor, sent to a comparator, inverted, and gated with the "read" pulse. The upper trace shown in Figure 11.5 is an input waveform applied, and the lower is the mDRAM cell output. The first positive pulse configures the state of the cell by writing a bit, and the second and third positive pulses read the cell. The third pulse (and subsequent read pulses, not shown here for simplicity) demonstrates that the cell is robust and continues to hold the state (up to the limit of the bit retention time). The negative pulse erases the bit, resetting the cell. The second set of four pulses repeats this pattern, and many months of continuous operation have been observed with no degradation in performance [25].

This memory can be rationalized based upon conduction channels that change upon charge injection as studied by density functional theory (DFT) [26, 27]. These DFT studies further corroborate with the experimental results in that the unsubstituted OPE and the amine-substituted OPE would be inactive as devices (having linear I(V) curves) while nitroaniline and nitro OPE would both have switching states (exhibited by sharp nonlinear I(V) characteristics) due to the accepting of electrons during voltage application. Furthermore, the DFT calculations showed that nitroaniline would need to receive one electron in order to become conductive whereas the nitro OPE would be initially conductive ("on" in the mDRAM) and then become less conductive, "off", upon receipt of one electron [26, 27]. This is precisely the effect observed in the experiment [25].

Stoddart and Heath have synthesized molecular devices that would bridge the crossed nanowires and act as switches in memory and logic devices [28, 29]. The UCLA researchers have synthesized catenanes and rotaxanes [30] that can be switched between states using redox chemistry. For instance, Langmuir–Blodgett (LB) films were formed from the catenane, and the monolayers were deposited on polysilicon nanowires etched onto a silicon wafer through photolithography. A second set of orthogonal titanium nanowires were deposited through a shadow

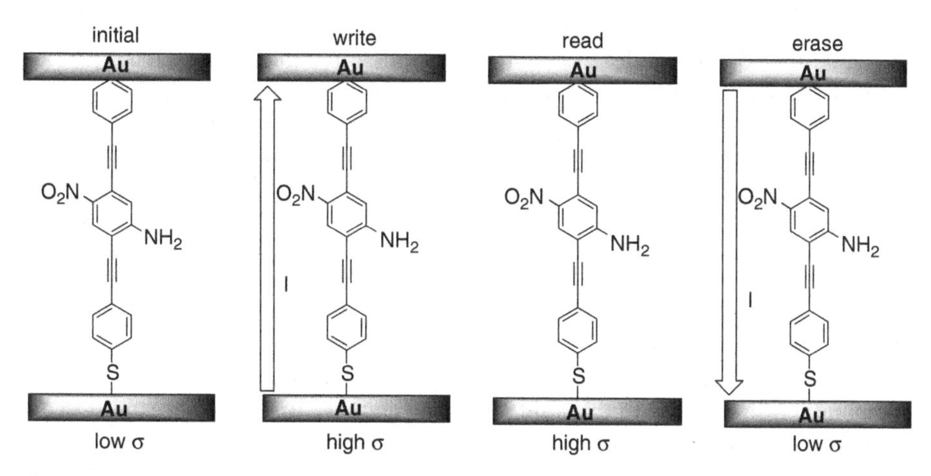

Figure 11.4. Write, read, and erase sequences for a nitroaniline OPV in the nanopore and its use as a 1-bit random access memory: An initially low conductivity state (low σ) is changed (written) into a high conductivity state (high σ) upon application of a voltage pulse at approximately 2 V. The direction of current that flows during this "write" pulse is diagrammed. The high σ state persists as a stored "bit," which is read in the low voltage region, approximately 0.3 V, as a nondestructive read. Again, this effect persisted up to 260 K. No particular mechanism, for example charge storage on the molecule vs. molecular tilting vs. molecular rotations, is implied by this scheme.

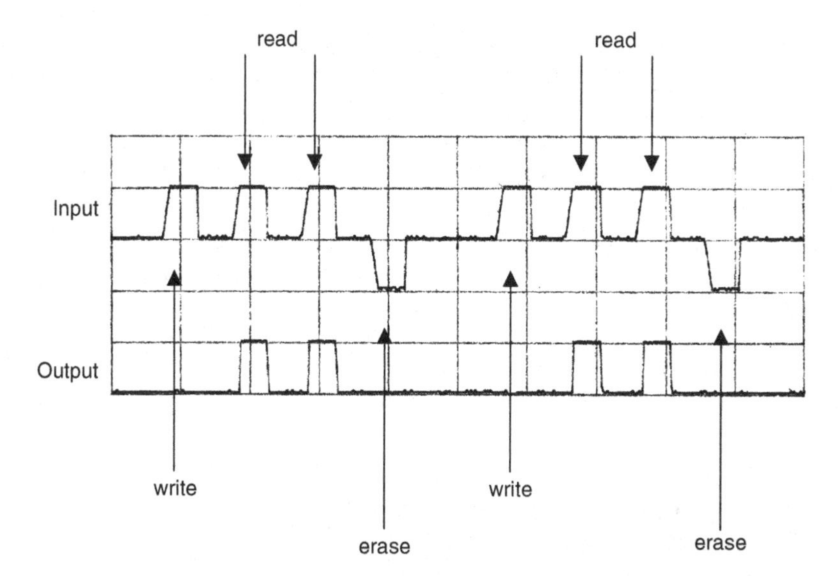

Figure 11.5. The mDRAM cell input and output that are constructed from the mononitro OPV in the nanopore.

mask, and the I(V) curve was determined. The data, when compared to controls, indicated that the molecules were acting as solid-state molecular switches [28, 29]. Small logic circuits [31] and, recently, large-scale memory units [32] have been demonstrated. Unlike the nanopore, Stoddart and Heath leveraged the nanoscale crosspoint formed between two orthogonal nanowires. In the most recent work [32], about 100 [2]rotaxane molecules were estimated to be sandwiched in the crosspoint of two nanowires.

A challenge to computing based on molecular switches is the impedance mismatch between the molecule and the metal contact, leading to a resistance barrier [16, 33–35]. Neither the selenium or tellurium alligator clip significantly reduced the barrier height [35]. But the use of an isonitrile as the contact between the organic molecular scale wire and a palladium probe would significantly reduce the conduction barrier, and would allow an increase in the conductivity of the molecular scale wires. Therefore molecular scale devices with isonitrile attachment moieties were synthesized [33, 36, 37].

11.3. CIRCUIT AND ARCHITECTURES IN MOLECULAR COMPUTING

Molecular computing requires not only molecular switches but also solutions to integrate them into microscale functional circuits. Two popular approaches have been taken. The first is based on quantum cellular automata (QCA) and related electrostatic information transfers [38–40]. It relies on electrostatic field repulsions to transport information throughout the circuitry. One major benefit of the QCA or electrostatics approach is that energy consumption is diminutive because only a few or fractions of an electron are used for each bit of information. Unfortunately, molecular QCA is based on the interconnection of individual molecules. While implementations of small QCA circuits have been reported, none are based on molecular quantum dots, due to our inability to position and interconnect individual molecules. The second approach is based on the crossbar of single-walled carbon nanotubes (SWNT) [41] or synthetic nanowires [42, 43] that sandwich molecular bundles as switching devices.

Many have investigated the integration of molecular circuits into large-scale computing systems. In particular, numerous architectural solutions have been proposed to combine the strength of conventional MOSFETs and molecular units. An orthogonal approach is to employ conventional lithography-based top-down approach to fabricate microscale patterns for large-scale integration while employing bottom-up nanoscale fabrication methods to fill microscale units with programmable molecular switches.

11.3.1. Quantum Cellular Automata (QCA) and Electrostatics Architectures

In the QCA approach toward molecular computing systems, four quantum dots in a square array are placed in a cell such that electrons are able to tunnel between the dots but are unable to leave the cell [44]. Coulomb repulsion will force the electrons to occupy dots on opposite corners. The two ground state polarizations are both energetically equivalent and can be labeled logic "0" or "1." Flipping the logic state of one cell, for instance by applying a negative potential to a lead near the quantum dot occupied by an electron, will result in the next door cell flipping ground states in order to reduce Coulomb repulsion. Amlani and co-workers have demonstrated experimental switching of six-dot QCA cells with Al islands on an oxidized Si surface [45, 46]. They later demonstrated a functioning majority gate with logic AND and OR operations [47].

While the use of quantum dots in the demonstration of QCA is a good first step, the ultimate goal is to use individual molecules to hold the electrons and pass electrostatic potentials down QCA wires. We have synthesized molecules that have the capability of transferring information from one molecule to another through electrostatic potentials (Fig. 11.6) [38]. The potentials only employ a millionth of an electron per bit of information, which is extreme attractive in terms of energy consumption.

Reshapes of the electron density due to the input signals produce electrostatic interactions. The electrostatic potential interactions between molecules could transport the information. They obviate electron currents or electron transfers as in present devices: A small change in the electrostatic potential of one molecule could suffice for intermolecular communication, leading to minute charge transfer, far less than one electron. External fields or excitations are able to change the boundary conditions of the molecule producing a change in the electrostatic potential generated by these sources. As an example, a charge or field on the left side of a molecule would reshape the electron density, providing a different potential at the output side. The change observed in the electrostatic potentials is in the range of values of nonbonded interactions, such as van der Waals interactions, which are easily detected by neighboring molecules. In fact, these are precisely the ranges of signal energies that would be attractive if we will ultimately utilize large-scale integration in very small areas for power density and heat dissipation considerations.

Although we synthesized molecules that included three-terminal molecular junctions, switches, and molecular logic gates to demonstrate the electrostatics methodology [38], none of the molecules were incorporated into an actual assembly. All results were based on simulation only because the QCA and electrostatics approach have major obstacles to overcome before even simple laboratory tests can be attempted. While relatively large quantum dot arrays can be fabricated using existing methods, a major problem is that placement of molecules in precisely aligned arrays at the nanoscale is very difficult to achieve. Another problem is that degradation of only one molecule in the array can cause failure of the entire circuit. Even small examples of two-dots have yet to be demonstrated using molecules because addressing of the molecular-sized inputs and the recording of a signal based on fractions of an electron make the hurdles enormous.

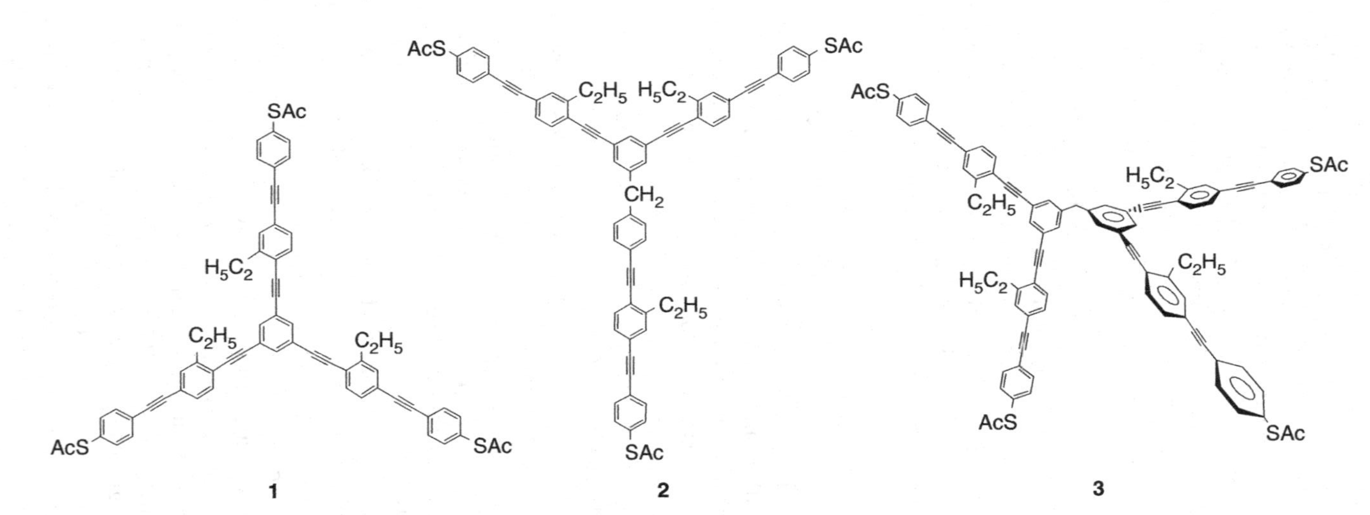

Figure 11.6. Three compounds synthesized for studying electrostatic interaction-based computing. (**1**) is a molecular three-terminal junction that could be used as a molecular interconnect. (**2**) is a molecular-sized switch with a source, drain and gate terminals similar to a bulk solid-state FET. (**3**) can be an active OR or a passive NOR gate if positive logic is used or an AND or NAND gate if negative logic is used [38].

11.3.2. Nanowire Crossbar Arrays

That QCA and electrostatics architectures require accurate positioning of individual molecules has become a major obstacle in their realization. On the contrary, the crossbar array architecture works for both individual molecules and molecular bundles. Therefore, prototype circuits of various scales have been demonstrated [32, 41, 42, 48, 49].

The basic crossbar array architecture employs two overlaid but orthogonal arrays of parallel nanowires. A molecular monolayer can be formed between the two arrays, particularly at the crosspoint of two orthogonal nanowires. Semiconductor nanowires, usually silicon nanowires and their variations, metallic nanowires, and SWNTs have been the most popular choices for the nanowires. The crossbar array architecture addresses the integration challenges by 1) leveraging a regular structure that seamlessly integrates devices and interconnects; 2) leveraging the nanometer diameter of nanowires to form nanoscale crosspoints, which can host nanoscale molecular bundles; and 3) allowing independent choices of interconnects and devices, especially the flexibility in the choice of devices. Moreover, the large number of identical nanowires and crosspoints provides the redundancy for critical fault tolerance. The flexibility of the basic crossbar array architecture has led to numerous variations of crossbar array architectures, depending on the choices of nanowires, devices in the crosspoints, and addressing methodologies. It is important to note that the crossbar array architecture has been traditionally employed by MOSFET programmable array logic (PAL) [50]. A FET can also be formed at the crosspoint by doping the silicon nanowire [51]; many circuits have been designed [52, 53] or fabricated [54] based on such FET crosspoints. However, we will focus on the solutions that are based on either the use of molecules, namely carbon nanotubes, as nanowires or the use of molecular bundles at the crosspoints.

11.3.2.1. Fabrication. Three approaches have been successfully employed to fabricate large-scale nanowire crossbar arrays. Lieber's group has pioneered the *self-assembly* approach, which grows nanowires and then assembles them into arrays. The diameter of nanowires can be controlled by seed catalysts, such as gold particles, and nanowires with diameters down to 3 nm have been demonstrated [63]. After nanowire fabrication, Langmuir–Blodgett (LB) flow techniques were developed to align them into a compact array [64, 65]. They were then repeated to form an orthogonal array to accomplish a crossbar array. The self-assembly approach separates the fabrication of nanowires and their assembly into arrays. Therefore, it can employ a variety of nanowires, e.g., Ge/Si nanowire heterostructures [66] and carbon nanotubes. On the other hand, the self-assembly approach is limited by the LB process to align the nanowires into an array. Essentially, the LB process cannot guarantee a uniform array; neither can it guarantee the pitch of the array (distance between adjacent nanowires) or the position of an individual nanowire.

The second approach is *nanoimprinting lithography* [67–69], which employs molds fabricated using costly but fine methods, e.g., electron beam lithography or the supperlattice nanowire pattern transfer (SNAP) method to be addressed below. Williams and Heath demonstrated an array of 150 Silicon nanowires with 15 nm wide at 34 nm pitch [68].

The third approach is the *superlattice nanowire pattern transfer (SNAP)* method [48, 49]. Similar to the nanoimprinting, the SNAP method does not separate nanowire growth and array formation clearly: Nanowires are formed at the same time as they are aligned. In this method, a GaAs/AlGaAs superlattice is first fabricated with molecular beam epitaxy (MBE). The AlGaAs of the superlattice is then etched to create parallel grooves between the GaAs layers. The superlattice is then tilted in order to deposit metal onto the edges of the grooves, forming metallic nanowires. The width of the metallic nanowires is defined by the thickness of the GaAs layers in the superlattice; the pitch, or separation, between the nanowires is defined by the thickness of the AlGaAs layers that are etched. The nanowire array formed along the grooves is then transferred to a silicon wafer by contacting the superlattice to an adhesive layer on top of the wafer. The metallic nanowire array is released on the wafer by etching the GaAs oxide layer between the nanowires and the GaAs layers. The metallic nanowires can be employed as etch masks to form semiconductor nanowires on the wafer, e.g., silicon-on-SiO_2 wafer. Crossbar arrays can be formed by overlaying orthogonal arrays. Using this method, Health's group has demonstrated a large-scale crossbar array, intended as a high-density memory module [32]. The metallic nanowire array can also be employed to fabricate molds for nanoimprinting [68], as discussed above.

Compared to the self-assembly approach, nanoimprinting and SNAP methods are limited in the nanowires that can be employed, primarily due to the fact that they form the nanowires and align them into arrays at the same time. Therefore, it would be extremely hard to use them to fabricate arrays using nanowires with heterostructures, such as the Ge/Si nanowires that were developed by Lieber [66]. However, compared to the self-assembly approach, nanoimprinting and the SNAP method enjoy a greater control on the pitch and the position of the nanowires, thereby producing much more uniform arrays.

11.3.2.2. Circuit and Architectural Solutions. Large-scale crossbar arrays intended as memory units have been fabricated; small logic circuit based on crossbar arrays have also been demonstrated. Inspired by these fabrications, computer engineering researchers have proposed numerous architectural solutions based on crossbar arrays. To achieve large scale integration, many have studied the possibility of a hybrid circuit with both traditional CMOS transistors and nanoscale crossbar arrays.

Lieber proposed a SWNT-based nonvolatile RAM device comprising a series of crossed nanotubes wherein one parallel layer of nanotubes is placed on a substrate and another layer of parallel nanotubes, orthogonal to the first set, is suspended above the lower nanotubes by placing them on a periodic array of supports [41]. The elasticity of the suspended nanotubes provides one energy minimum, wherein the contact resistance between the two layers is zero, and the switches (the contacts between the two sets of orthogonal nanowires) are "off". When the tubes are transiently charged to produce attractive electrostatic forces, the suspended tubes flex to meet the tubes directly below them, and a contact is made, representing the "on" state. The "on"/"off" state could be read by measuring the resistance at each junction, and could be switched by applying voltage pulses at the correct electrodes. This theory was tested by mechanically placing two sets of nanotube bundles in a crossed mode and measuring the I(V) characteristics when the switch was "off" or "on." Based on this nonvolatile RAM design, Zhang and Jha proposed a reconfigurable architecture that can achieve much better performance, reconfigurability, and density than existing field-programmable gate arrays (FPGA) [70].

While Lieber employed the SWNTs themselves as switches, Stoddart and Heath employed nanoimprinting and the SNAP method to fabricate large and fine nanowire crossbar arrays [32, 68]. These large and fine crossbar arrays (400 by 400 array and 33 nm pitch) have a monolayer of bistable [2]rotaxane molecules between the two orthogonal arrays [28, 29]. The monolayer is formed using the LB process. Stoddard and Heath intended the 400 by 400 crossbar array as a high-density memory unit. The promise of extremely high density memory units by nanowire crossbar arrays has also inspired sophisticated circuit and architecture designs based on nanowire crossbar memory units [71, 72]. However, none of these designs have been realized.

11.3.2.3. Challenges. The crossbar array architecture faces several hurdles before it can be useful. First, addressing an individual nanowire or crosspoint is difficult due to their miniature size. For example, although fine and large crossbar arrays have been demonstrated [32, 68], it is still difficult to address an individual crosspoint or make a contact with a single nanowire. For example, in [32], electrode contacts for crosspoint characterization can only be lithographically made to two to four adjacent nanowires. As a result, a small array of crosspoints had to be measured to infer the properties of a single crosspoint as a memory unit. The fundamental challenge is to select a single nanowire in a dense array through microscale metal wires. Most proposed solutions rely on lithographically overlaying an orthogonal array of microscale metal wire on top of the nanowire array to build a *decoder* or *demultiplexer*.

Williams and Kuekes patented a proposal that randomly deposits gold particles between the metal wire array and the nanowire array so that a random set of crosspoints becomes conductive when gold particles are deposited therein [73]. DeHon et al. employed modulation-doped nanowires to interface with microscale metal wires [74]. A modulation-doped nanowire has different dopings along its length, which can be controlled during growth. On the contrary, Savage et al. employed axial-doped nanowires, which have different dopings along the radius [75]. Both solutions seek to distinguish a nanowire from its neighbors in an array based on its unique doping heterogeneity sensed by the microscale array of metal wires. Like the gold particle-based solution, they are inherently probabilistic due to the inability to distinguish nanowires with different doping patterns during assembly or accurately position any given nanowire. Therefore, all three solutions can lead to low utilization of the nanowire array. So far fabrication has not yet been reported for any of these structures.

Lieber fabricated a small decoder based on chemically modifying lithographically selected crosspoints in the nanowire crossbar array [76]. The input nanowires interface with microscale metal wires through lithographically made contacts. Apparently, the decoder is limited by lithography. It does not really allow the selection of a nanowire from an array with a pitch beyond the resolution of lithography. Heath later demonstrated a decoder that selects a nanowire from an array of 150 silicon nanowires with a pitch of 34 nm [48]. The design is based on forming a binary tree pattern in dielectric between the microscale metal array and the nanowire array. The finest pattern is determined by the pitch of the nanowire array and hence e-beam lithography has to be employed. So far, no truly nanoscale addressing solution has yet enjoyed cost-effective, scalable fabrication.

Furthermore, signal strength degrades as it travels along the nanowires. Gain is typically introduced into circuits by the use of active devices, such as transistors. However, placing a transistor at each crosspoint is an untenable solution because doing so will eliminate the size advantage of the molecular-based system. Likewise, in the absence of a transistor at each cross point in the crossbar array, molecules with very large "on" : "off" ratios will needed. For instance, if a switch with a 10:1 "on" : "off" ratio was used, then 10 switches in the "off" state would appear as an "on" switch. Hence, isolation of the signal via a transistor is essential, but presently the only solution for the transistors' introduction would be for a large solid-state gate below each cross point, again defeating the purpose for the small molecules.

Additionally, if SWNTs are to be used as the crossbars, connection of molecular switches via covalent bonds introduces sp^3-hybridized carbon atom linkages at each junction, disturbing the electronic nature of the SWNT and possibly obviating the very reason to use the SWNTs in the first place. Noncovalent bonding of the device molecule to the SWNT will probably not provide the conductance necessary for the circuit to operate. Therefore, continued work is being done to devise and construct crossbar architectures that address these challenges.

11.4. MEET-IN-BETWEEN PARADIGM

The traditional silicon MOSFET fabrication is top-down, relying on lithography to accurately position each individual transistor, as specified by design. However, due to process variations, such a what-you-design-is-what-you-get assumption is no longer true for present deep submicron MOSFET fabrication, let alone truly nanoscale structures such as silicon nanowires. It is extremely hard, if at all possible, to position individual nanostructure, especially in a massive number. On the other hand, photolithography has been a successful and evolving technology for microscale patterning.

On the other hand, most nanoscale fabrication methods, including self-assembly, nanoimprinting, and the LB process, are bottom-up in nature, unable to accurately position a nanostructure effectively.[2] Although small logic and memory units have been demonstrated, it is expensive, if at all possible, to use the same methods to fabricate large-scale integrated systems. The bottom-up fabrication is only effective with making repetitive patterns, or homogeneous structures. It is extremely difficult to directly connect nanoscale units thus fabricated with each other, or even with microscale structures, as highlighted above, with certainty.

Combining the bottom-up and top-down approaches, or meet-in-between paradigm, is likely to provide solutions to such a dilemma. In this paradigm, instead of positioning each individual nanoscale element, traditional photolithography is employed to pattern microscale units and interconnects between them, which are potentially programmable. Each unit is then filled with nanoscale elements through bottom-up nanoscale fabrication methods. Finally, each unit is individually programmed through the microscale interconnection in order to produce functional circuits. Figure 11.7 illustrates this three-step process. The paradigm leverages the ever increasing computing power to relieve the limitations of fabrication. The meet-in-between paradigm is complementary to existing proposals to combine the strength of lithographically fabricated MOSFETs and nanoscale elements, including molecular switches [70, 77, 78]. In most of these proposals, MOSFETs are employed for logic and signal regeneration while nanoscale elements are employed for highly regular structures, such as memory units.

The meet-in-between paradigm is based on a trade off between design and fabrication capabilities. Instead of fabricating a prefixed design, it fabricates a prefixed but programmable microscale structure. It leverages the programmability of the nanoscale elements and existing computing power to work around the nanoscale uncertainty inside of the microscale units. Critical to the meet-in-between paradigm is a microscale unit made of nanoscale switching elements that are rapidly programmable and feature-rich. The two compelling candidates are nanowire crossbars and molecular nanocells [79]. As we have already addressed nanowire crossbar arrays, we describe the nanocell design and fabrication next.

11.4.1. Nanocell

We have developed a nanocell architecture that fits into the meet-in-between paradigm and takes advantage of the smallness in size of the molecules via lithographic tools [38, 79–81]. The nanocell architecture also offers enormous

[2] While nanoimprinting and the SNAP method can more or less accurately control the separation between nanowires, they are less capable of absolutely positioning a single nanowire.

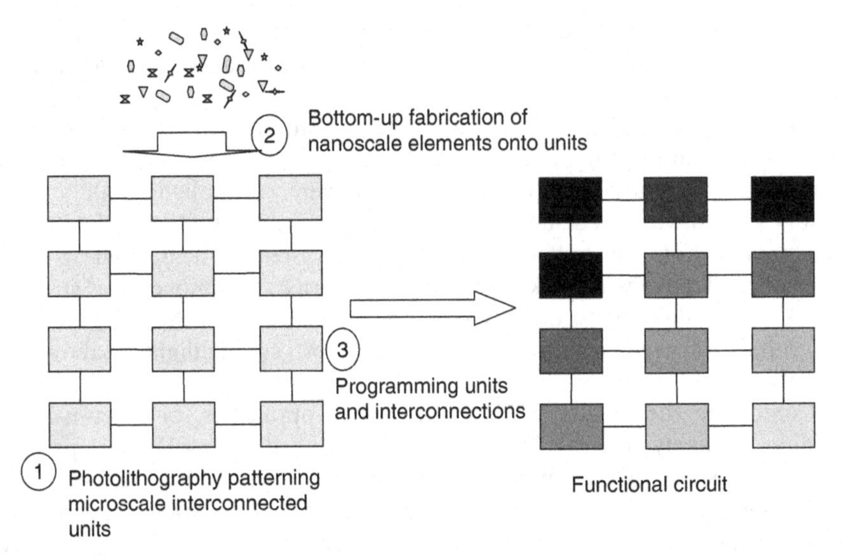

Figure 11.7. The meet-in-between paradigm based on programmable units of nanoscale basic elements, such as molecules, nanowires, and carbon nanotubes. For example, a unit can be a silicon nanowire crossbar array or a molecular nanocell [79].

defect tolerance and a fabrication simplicity that is unrivaled by other molecular electronic architectures.

A nanocell is a two-dimensional (three-dimensional models could also be considered) network of self-assembled metallic particles connected by molecules that show reprogrammable (can be turned "on" or "off") negative differential resistance (NDR) or even other switching and/or memory properties. The nanocell is surrounded by a small number of lithographically defined access leads at the edges of the nanocell. Figure 11.8 shows an assembled nanocell. Unlike typical chip fabrication, the nanocell is not constructed as a specific logic gate and the internal topology is, for the most part, disordered. Logic is created in the nanocell by training it post-fabrication, similar in some respects to a field-programmable gate array (FPGA). Even if this process is only a few percent efficient in the use molecular devices, very high logic densities will be possible. Moreover, the nanocell has the potential to be reprogrammed throughout a computational process via changes in the "on" and "off" states of the molecules, thereby creating a real-time dynamic reconfigurable hard-wired logic. The CPU of the computer would be comprised of arrays of nanocells wherein each nanocell would have the functionality of many transistors working in concert. A regular array of nanocells is assumed to manage complexity, and ultimately, a few nanocells, once programmed, should be capable of programming their neighboring nanocells through bootstrapping heuristics. Alternatively, arrays could be programmed one nanocell at a time via an underlying CMOS platform.

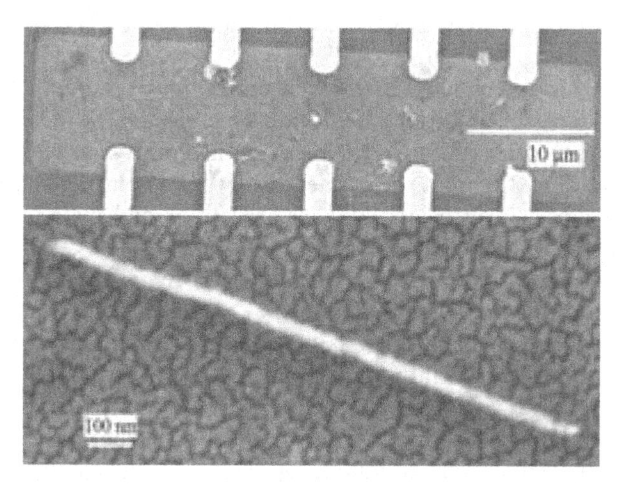

Figure 11.8. An assembled nanocell that exhibited memory with 10 pins (top); one of the metallic nanowires interspersed with electroactive molecules as the active elements (bottom) [79].

We have synthesized several types of room temperature-operable molecular switches and demonstrated them in nanopores and atop silicon-chip platforms. The functional molecular switches can be reversibly switched from an "off" state to an "on" state, and/or the reverse, based on stimuli such as voltage pulses. The number of nanoparticles (usually metallic or semiconducting) and the number of the interconnecting molecular switches can vary dramatically based on the chosen size of the nanocell and on the dimensions of the nanoparticles and molecules chosen.

Within the fabricated nanocell, the input and output leads could be repetitively interchanged based on the programming needs of the system, thereby demonstrating the pliability of the architecture. Naturally, issues of gain will eventually have to be addressed through either an underlying CMOS layer or clocked circuits programmed into the nanocell [82]. Even if one CMOS transistor was used for gain at the output from each nanocell, enormous space savings could be attained since a nanocell could possess the functionality of numerous transistors working in concert to produce a specified logic function. Furthermore, by capitalizing on the NDR properties of the molecular switches, internal gain elements based upon NDR/nanoparticle/NDR stacks (Goto pairs [83]) could be efficacious.

The object in programming or training a nanocell is to take a random, fixed nanocell and turn its switches "on" and "off" until it functions as a target logic device. The nanocell is then trained post-fabrication by changing the states, "on" or "off," of the molecular switches by imposing voltages at the surrounding input/output leads. Notice how we hope to address, in a broad sense, the internal molecular switches via the surrounding leads.

11.5. MOLECULAR GRAFTING FOR SILICON COMPUTING

We have described the intensive research endeavor in building switching and memory using molecules. Direct integration of molecular switches and memory units suffer from the uncertainty in chemical self-assembly. The works described in this chapter so far attempt to alleviate this problem through fabrication and design innovations. Nevertheless, all these solutions endeavor to replace semiconductor devices, at least partially, with molecular bundles. While lab demonstrations have show promise, these solutions remain ambitious and not immediately implementable in the industry, which has a considerable vested interest in semiconductor. In this last section, we examine a very different application of molecular computing: Control the properties of semiconductor for better computing with molecules, instead of the direct use of molecules or molecular bundles for switching and memory. The application is called molecular grafting. We showed that a monolayer of molecules can change the threshold voltage (V_T) of conventional silicon MOSFETs. Molecular grating thus provides a new mechanism to control process variations [84] and leakage current [85], which have been two of the most important challenges to modern CMOS-based circuits. We also showed that the mobility of intrinsic silicon nanowires can be dramatically changed through fluoride ion treatment. Molecule grafting can thus considerably improve the performance of silicon nanowires-based computing.

11.5.1. Controlled Modulation of Conductance in Silicon Devices

While many alternatives, including molecules, have been proposed to implement computing in the nanometer era, silicon remains the stalwart of the electronics industry. Generally, the behavior of silicon is controlled by changing the composition of the active region by impurity doping, while changing the surface (interface) states is also possible [86–89]. As scaling to the sub-20 nm-size region is pursued, routine impurity doping becomes problematic due to its resultant uncertainty of distribution [90, 91]. Doping uncertainty is a major contributor to process variations that have challenged deep-submicron silicon MOS circuit design [92]. Provided that back-end processing of future devices could be held to molecularly permissive temperatures (300–350°C) [93], it is attractive to seek controllable modulation of device performance through surface modifications, taking advantage of the dramatic increase in the surface-area-to-volume-ratios of nanoscale features.

Several techniques have been used to covalently attach molecules directly onto silicon surfaces [19, 20, 94, 95]. The Si-C bond formed using these methods is both thermodynamically and kinetically stable due to its high bond strength (3.5 eV) and low polarity [95, 96]. The majority of research in this area has focused on the grafting methods or the influence on the surface (or interface) properties of bulk semiconductors. So far little research has been conducted showing controlled modulation of semiconductor devices by grafting molecular layers onto oxide-free active device areas, and particularly via silicon—sp^2-hybridized-carbon bonds.

Since there is no intervening oxide between the π-rich molecules and the silicon, sequentially tuned molecular-structure changes can predictably regulate the device performances over a wide range.

We have been able to prepare an electronically controlled series of molecules, from strong π-electron donors to strong π-electron acceptors, and systematically covalently attach them as molecular monolayers onto the channel region of pseudo-MOSFETs (back gated). We have subsequently studied the device modulation.

We fabricated the pseudo-MOSFETs using a silicon-on insulator (SOI) wafer (Fig. 11.9). We used this simple back-gating design instead of a more complicated and potentially damaging top-gate fabrication in order to avoid destroying the grafted molecules and interfering with their influence. After etching the devices in an Ar-purged buffered oxide etch (BOE) for five minutes to remove the oxide layer and form the H-passivated silicon surface, we grafted the molecules **1–4** (Fig. 11.10) directly onto the active area in the channel region of the device by exposing the freshly etched samples to a 0.5 mM solution of the diazonium salt (**1–4**) in anhydrous acetonitrile (CH_3CN). The grafting time depends on the molecule that was used, and that was carefully calibrated. After grafting, we rinsed the samples thoroughly with CH_3CN and dried them with an N_2 flow.

We tested all the devices and measured the DC I(V) characteristics immediately after the BOE etching and before the molecular grafting. To get freshly cleaned surface for molecular grafting, we briefly etched the devices with BOE (30–60 s) and transferred them into the glove box for grafting. We then conducted a second DC $I(V)$ measurement after the grafting was completed. Devices with no molecules (H-passivated surface) were prepared and tested as the control samples. Both molecular grafting and testing were done at room temperature.

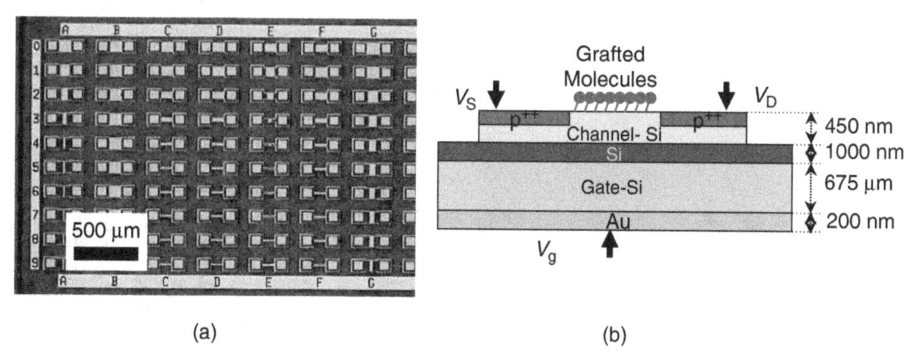

Figure 11.9. (a) An optical micrograph of pseudo MOSFETs on one chip. Boxed regions indicate the source and drain, between which sits the channel. The data shown in this contribution were collected with Row 0, for which both the length and width of the channel are 100 μm and the active area for molecular assembly is $110 \times 110\,\mu m^2$. (b) Schematic side-view representation (not to scale) of the device. The molecules were grafted between source and drain electrodes. V_S, V_D, and V_g refer to the bias applied on the source, drain, and gate, respectively.

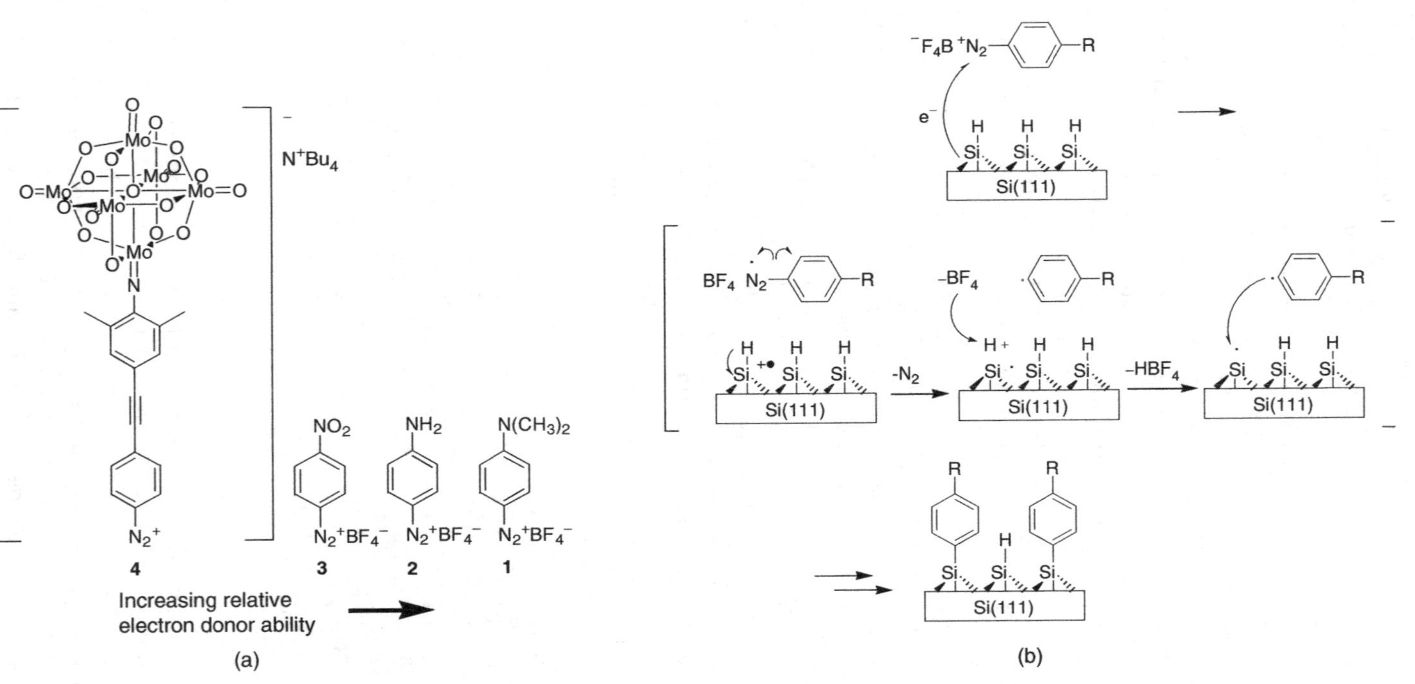

Figure 11.10. (a) Molecular structures used for grafting atop the pseudo-MOSFET channel. Structures of the starting molecules (**1–4**) used in the present contribution where (**1**) is the most electron rich system due to the dimethylamino substituent, (**2**) is slightly lower in its electron donation capability, followed by (**3**), and then finally (**4**) bears an extremely electropositive polymolybdate. (b). The grafting mechanism.

Figure 11.11 shows the typical transfer characteristics of the devices under test before and after the grafting of molecular monolayers. For compounds **1**–**3**, the drain current under the same gate bias decreased after the molecular grafting. The amplitude of this decrease is in the order of **1**>**2**>**3**. But I_D increased slightly after the grafting of **4**. The hystereses observed in the I_D–V_g curves are very similar in amplitude and shape before and after the molecular attachment (Fig. 11.11). Therefore these hysteretic effects are caused by the device itself, not by the molecular grafting.

The threshold voltage V_T of the pseudo-MOSFET can be extracted based on I_D, which represents the onset of significant drain current (Fig. 11.12a). The V_T values shown in Figure 11.12a were determined from the V_g axis intercept of the $I_{D,Sat}^{0.5}$-V_g characteristics linearly extrapolated [97–99]. We observed that the V_T became more negative (usually about $1 \sim 2$ V) after grafting **1**, **2** and **3** while less negative for **4**, relative to the control which was the Si-H surface before molecular grafting. Though there are slight differences between the values of ΔV_T when V_T was extracted from the forward and back scans (Fig. 11.12b) due to the inherent

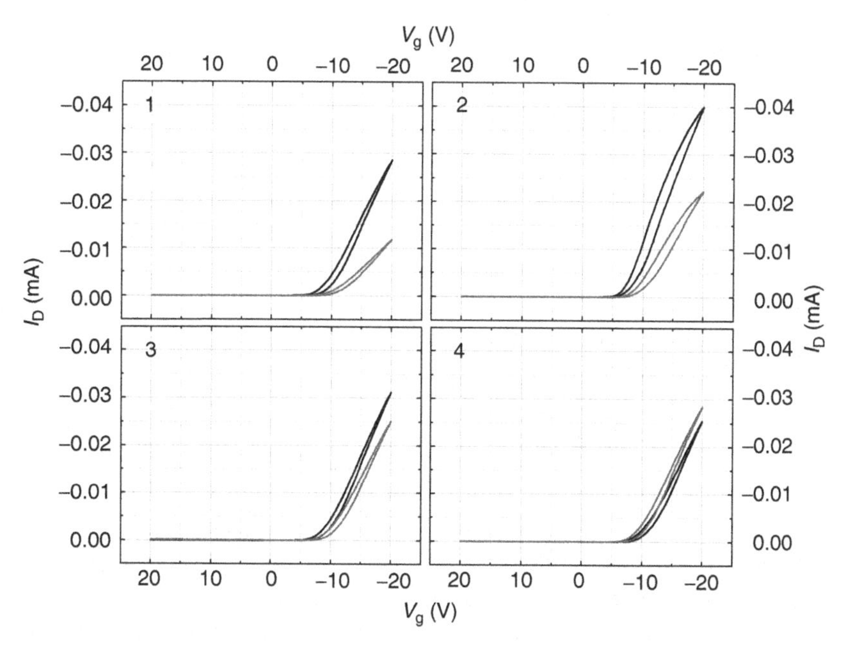

Figure 11.11. Transfer characteristics of the devices under test before and after the grafting of different molecular monolayers. The numbers (1, 2, 3, and 4) in the illustrations correspond to the compounds **1–4**, respectively. Data shown here were collected under −5 V of V_D and is the average value for 14 devices from Row 0 on one chip, as described in Figure 11.10a, and they are characteristic of the hystereses observed in all the devices. The gate bias was scanned first from +20 V to −20 V (forward) and then back from −20 V to +20 V.

hysteretic nature of the devices themselves (Fig. 11.11), the same trend in the change of V_T was observed. The control samples experienced the same treatment history as the devices under test, but had no molecules grafted on them, only the hydrogen passivation remained. For control samples, the V_T changed slightly (typically ≤ 0.3 V) from the first to the second etching. This supports our assertion that the V_T change in the devices under test is not caused by the etching but by the molecular grafting on the channel region which tracks directly with the electron donor ability of the molecules. The changes in V_T after the monolayer molecular grafting are consistent with those in I_D. Attaching compounds **1–3** led to a more negative V_T and less I_D while attaching **4** resulted in a less negative V_T and larger I_D. Therefore the channel conductances were reduced by grafting molecular monolayers of **1–3**, and they were increased by **4**, scaling directly with the relative electron donating ability of the molecules.

From the charge perspective, molecular grafting changes the channel conductance in a way similar to the impurity doping. The acceptor-like monolayer (more potent than the hydrogen atom of the H-passivated control) would decrease the V_T of the pseudo-MOSFET and the donors would increase the V_T. It implies that p-Si gained negative charges when modified by compounds **1–3**, while it gave up negative charges in the case of **4**.

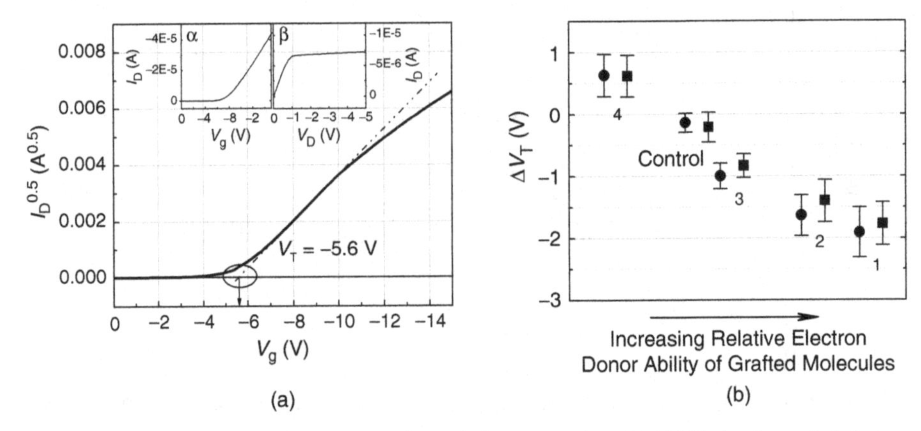

Figure 11.12. Electrical output results of the pseudo-MOSFET devices. (a) Extrapolation method for V_T used on the measured $I_{D,Sat}^{0.5}$-V_g characteristics. V_T was extracted at its maximum slope point. Inset α displays the typical transfer characteristics of the devices under test with an applied drain-source bias (V_D) of −5 V. Such a V_D was chosen for the V_T extraction according to Inset β, the typical output characteristics of the devices under test at $V_g > V_T$, to ensure the device operation is in the saturation region (Supporting Information). (b) Representative ΔV_T ($= V_{T\text{ (with molecules)}} - V_{T\text{ (without molecules)}}$) of the devices under test extracted from both the forward (solid circle) and back (solid square) scans after the grafting of different molecular monolayers (1–4), as well as on the control samples. Data shown here is the average value for 14 devices.

11.5.2. Molecular Grafting on Intrinsic Silicon Nanowires

Semiconductor nanowires [54, 64, 100, 101], including silicon nanowires, are promising as building blocks of nanoscale circuits, in particular those based on crossbar arrays addressed above [53, 78, 102, 103]. Because their carrier mobility is critical to the circuit performance, considerable efforts have been devoted to improving silicon nanowire performance [104, 105] with doping. However, Fernandez–Serra et al. [106] found that uniform doping may not be achievable in ultra small structures like silicon nanowires. As devices reach the sub-20-nm-sized regime, variations in doping profiles becomes a severe problem thereby generating inconsistencies between devices. Recent work done on ultra-thin silicon-on-insulator (SOI) revealed that its electronic conduction is determined not by bulk dopant but by the interaction of surface/interface electronic energy levels with the bulk band structure [87].

We have demonstrated that by controlling surface state densities with a very simple and rapid aqueous fluoride ion treatment, excellent electronic operation can be achieved on intrinsic silicon nanowires without bulk doping, thereby taking advantage of the increase in surface-to-volume ratios in these diminutive structures. Forming a fluoride-decorated (F-decorated) oxide surface significantly increases the conductivity and the mobility by more than three orders of magnitude over the oxide-free H-passivated surface. This provides a methodology that might sidestep the difficult-to-control impurity doping of nanodevices. Figure 11.13a shows the structure diagram of the fabricated silicon nanowire-FET test devices; Figure 11.13b shows a typical SEM image of the devices. Figure 11.13c illustrates the entire sequence: oxide-coated silicon nanowire to fluoride-decorated oxide coatings to H-passivated structures.

We also found that the silicon nanowire devices with F-decorated oxide surfaces exhibited large hystereses during the gate voltage sweeps, enabling the fabricated silicon nanowire FETs to serve as nonvolatile memory devices, although they are far from being useful in their present state in these electronic testbed structures. As shown in Figure 11.14a, the silicon nanowire has higher conductance, at the same gate voltage, during sweeping from 20 V to -20 V than the sweep in the opposite direction. This is caused by the change in surface charge density when applying a positive or negative gate bias. When applying a negative gate bias, the surface negative traps were discharged; when applying a positive gate voltage, the oxide traps were again charged. This makes the silicon nanowire-FET exhibit two reversible conductance states as shown in Figure 11.14b. We measured the memory performance of different devices and the results showed that the ratio of ON/OFF states ranged from 3–10. We used relatively long (5 s) pulse times in the write/erase operations in order to obtain the higher ON/OFF ranges from the silicon nanowire-FETs. Shorter write/erase pulses resulted in two different conductance states but with smaller ON/OFF ratios. For example, the ON/OFF ratio of a silicon nanowire-FET memory device was $\sim$4 with the 5 s write/erase pulsing but the ON/OFF ratio fell to $\sim$1.2 with 0.1 s write/erase pulsing. The shorter pulsing times apparently led to an unfinished charging/discharging process.

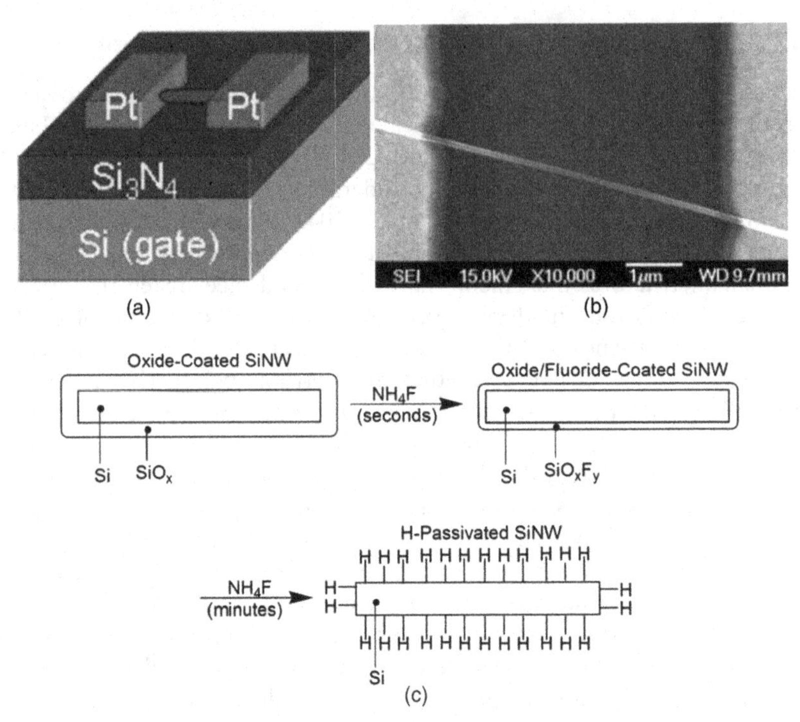

Figure 11.13. (a) Schematic of a silicon nanowire-FET in which Pt contacts are used as the source and drain electrodes, Si_3N_4 (300 nm) as the dielectric layer, and a highly doped (dopant: phosphorus at a level of $\sim 10^{20}\,cm^{-3}$) Si(100) substrate as the gate. (b) SEM image of a silicon nanowire-FET (1 μm bar). (c) A schematic of the sequence involving the treatment of an oxide-coated silicon nanowire with aqueous ammonium fluoride for seconds (typically 20–60 s) to generate the silicon oxide/fluoride surface (F-decoration) that shows the higher conductivity and mobility, and then after minutes (typically 10–12 min) of continued ammonium fluoride treatment, it forms the H-passivated silicon surface which results in low conductivity and mobility for the silicon nanowire.

Thinning of the gate nitride would reduce the voltage constraints in the system. Furthermore, the OFF state was never near zero, possibly because the surface charge could not be completely discharged.

11.6. CONCLUSIONS

This chapter presented an overview of the switching mechanisms in molecular bundles, as well as circuit and architecture proposals for their large-scale integration. Despite years of research endeavor, existing technologies still remain far from practical use and considerable research and development is necessary to

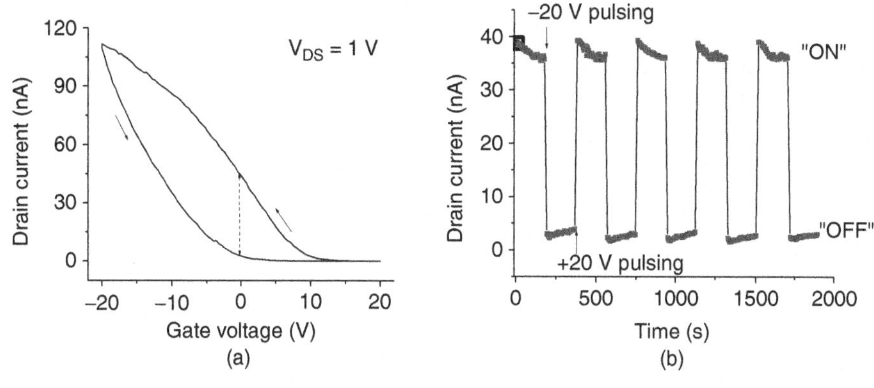

Figure 11.14. (a) I_D vs. V_G curve for a silicon nanowire with a F-decorated oxide surface recorded at V_{DS} of +1 V, showing a large hysteresis for different gate voltage sweeping directions. (b) The memory performance of the silicon nanowire-FET device showing distinct and stable ON/OFF state with write/erase operation. After each +20 V gate pulsing (write) or −20 V pulsing (erase) for 5 s, the device state was consecutively read 50 times with drain current at V_{DS} of +1 V.

realize large-scale integration for functional computing. In contrast, it was shown that molecules can be employed to change the behavior of silicon through surface grafting, instead of implementing switching themselves. This provides not only a new mechanism to enhance existing silicon-based process technologies but also a new modulation technology to implement silicon nanowire-based devices.

ACKNOWLEDGMENTS

We thank our collaborators and lab members for their contributions to the work described in this chapter; many if their names are cited in the references. The JMT program has been generously supported by DARPA and to a lesser extent by the ONR and AFOSR.

REFERENCES

1. J. M. Tour. *Molecular Electronics: Commercial Insights, Chemistry, Devices, Architecture and Programming*. Singapore: World Scientific, 2003.
2. G. Cuniberti, G. Fagas, and K. Richter. *Introducing Molecular Electronics: A Brief Overview* (Lecture Notes In Physics) New York: Springer, 2005.
3. J. R. Heath, et al. A defect-tolerant computer architecture: opportunities for nanotechnology. *Science*, 280(5370): p 1716, 1998.
4. M. Mishra and S. C. Goldstein. Scalable defect tolerance for molecular electronics. In: First Workshop on Non Silicon Computing, 2002.

5. K. Nikolic, A. Sadek, and M. Forshaw. Fault-tolerant techniques for nanocomputers. *Nanotechnology*, 13(3): pp 357–362, 2002.
6. J. Han and P. Jonker. A defect- and fault-tolerant architecture for nanocomputers. *Nanotechnology*, 14(2): pp 224–230, 2003.
7. M. B. Tahoori and S. Mitra. Defect and fault tolerance of reconfigurable molecular computing. In: 12th Annual IEEE Symposium on Field-Programmable Custom Computing Machines, 2004.
8. A. DeHon and H. Naeimi. Seven strategies for tolerating highly defective fabrication. *IEEE Design and Test of Computers*, 22(4): p 306, 2005.
9. E. Lörtscher, et al. Reversible and controllable switching of a single-molecule junction. *Small*, 2(8–9): pp 973–977, 2006.
10. D. Natelson, et al. Single-molecule transistors: electron transfer in the solid state. *Chemical Physics*, 324(1): p 267, 2006.
11. S. J. Tans, A. R. M. Verschueren, and C. Dekker. Room-temperature transistor based on a single carbon nanotube. *Nature*, 393(6680): p 49, 1998.
12. W. Liang, et al. Kondo resonance in a single-molecule transistor. *Nature*, 417(6890): p 725, 2002.
13. D. K. James and J. M. Tour. *Molecular Wires. Topics in Current Chemistry*, Vol. 257/2005. Berlin: Springer, p 33, 2005.
14. D. K. James and J. M. Tour. Organic synthesis and device testing for molecular electronics. *Aldrichimica Acta*, 39(2): pp 47–56, 2006.
15. B. A. Mantooth and P. S. Weiss. Fabrication, assembly, and characterization of molecular electronic components. *Proceedings of the IEEE*, 91(11): p 1785, 2003.
16. J. M. Tour. Molecular Electronics. Synthesis and testing of components. *Accounts of Chemical Research*, 33(11): p 791–804, 2000.
17. D. K. James and J. M. Tour. Analytical techniques for characterization of organic molecular assemblies in molecular electronics devices: A review. *Analytica Chimica Acta*, 568 pp 2–19, 2006.
18. T. He, et al. Controlled modulation of conductance in silicon devices by molecular monolayers. *Journal of American Chemistry Society*, 128(45): pp 14537–14541, 2006.
19. A. Ulman. Formation and structure of self-assembled monolayers. *Chemical Reviews*, 96(4): pp 1533–1554, 1996.
20. M. P. Stewart, et al. Direct covalent grafting of conjugated molecules onto Si, GaAs, and Pd Surfaces from Aryldiazonium Salts. *Journal of American Chemistry Society*, 126(1): pp 370–378, 2004.
21. M. A. Reed, et al. Conductance of a molecular junction. *Science*, 278(5336): pp 252–254, 1997.
22. C. Zhou, et al. Nanoscale metal/self-assembled monolayer/metal heterostructures. *Applied Physics Letters*, 71(5): p 611, 1997.
23. C. Zhou. *Atomic and Molecular Wires*. New Haven, CT: Yale University Press, 1999.
24. J. Chen, et al. Large on-off ratios and negative differential resistance in a molecular electronic device. *Science*, 286(5444): pp 1550–1552, 1999.

25. J. Chen, et al. Room-temperature negative differential resistance in nanoscale molecular junctions. *Applied Physics Letters*, 77(8): p 1224, 2000.
26. J. M. Seminario, A. G. Zacarias, and P. A. Derosa. Theoretical analysis of complementary molecular memory devices. *Journal of Physical Chemistry A*, 105(5): pp 791–795, 2001.
27. J. M. Seminario, A. G. Zacarias, and J. M. Tour. Theoretical study of a molecular resonant tunneling diode. *Journal of American Chemistry Society*, 122(13): pp 3015–3020, 2000.
28. C. P Collier, et al. A [2]Catenane-based solid state electronically reconfigurable switch. *Science*, 289(5482): pp 1172–1175, 2000.
29. A. R. Pease, et al. Switching devices based on interlocked molecules. *Accounts of Chemical Research*, 34(6): pp 433–444, 2001.
30. V. Balzani, M. Gomez-Lopez, and J. F. Stoddart. Molecular machines. *Accounts of Chemical Research*, 31(7): pp 405–414, 1998.
31. C. P. Collier, et al. Electronically configurable molecular-based logic gates. *Science*, 285(5426): pp 391–394, 1999.
32. J. E. Green, et al. A 160-kilobit molecular electronic memory patterned at 10^{11} bits per square centimeter. *Nature*, 445(7126): p 414, 2007.
33. J. M. Tour, et al. Synthesis and preliminary testing of molecular wires and devices. *Chemistry, A European Journal*, (7): pp 5118–5134, 2001.
34. M. A. Reed and J. M. Tour. Computing with molecules. *Scientific American*, (June): pp 86–93, 2000.
35. J. M. Seminario, A. G. Zacarias, and J. M. Tour. Molecular alligator clips for single molecule electronics. Studies of group 16 and isonitriles interfaced with Au contacts. *Journal of American Chemistry Society*, 121(2): pp 411–416, 1999.
36. J. Chen, et al. Molecular wires, switches, and memories. *Annals of the New York Academy of Sciences*, 960(1): pp 69–99, 2002.
37. J. Chen, et al. Electronic transport through metal-1, 4-phenylene diisocyanide-metal junctions. *Chemical Physics Letters*, 313: pp 741–748, 1999.
38. J. M. Tour, M. Kozaki, and J. M. Seminario. Molecular scale electronics: a synthetic/computational approach to digital computing. *Journal of American Chemistry Society*, 120(33): pp 8486–8493, 1998.
39. P. D. Tougaw and C. S. Lent. Logical devices implemented using quantum cellular automata. *Journal of Applied Physics*, 75(3): p 1818, 1994.
40. C. S. Lent and P. D. Tougaw. A device architecture for computing with quantum dots. *Proceedings of the IEEE*, 85(4): p 541, 1997.
41. T. Rueckes, et al. Carbon nanotube-based nonvolatile random access memory for molecular computing. *Science*, 289(5476): pp 94–97, 2000.
42. Y. Huang, et al. Directed assembly of one-dimensional nanostructures into functional networks. *Science*, 291(5504): pp 630–633, 2001.
43. S. W. Chung, J. Y. Yu, and J. R. Heath. Silicon nanowire devices. *Applied Physics Letters*, 76(15): p 2068, 2000.
44. G. L. Snider, et al. Quantum-dot cellular automata: Review and recent experiments (Invited). AIP: 1999.

45. G. H. Bernstein, et al. Observation of switching in a quantum-dot cellular automata cell. *Nanotechnology*, 10(2): p 166, 1999.
46. I. Amlani, et al. Demonstration of a six-dot quantum cellular automata system. *Applied Physics Letters*, 72(17): p 2179, 1998.
47. I. Amlani, et al. Digital logic gate using quantum-dot cellular automata. *Science*, 284(5412): pp 289–291, 1999.
48. R. Beckman, et al. Bridging dimensions: demultiplexing ultrahigh-density nanowire circuits. *Science*, 310(5747): pp 465–468, 2005.
49. N. A. Melosh, et al. Ultrahigh-density nanowire lattices and circuits. *Science*, 300(5616): pp 112–115, 2003.
50. J. M. Birkner and H. T. Chua, inventors. Monolithic Memories, Inc., assignees. Programmable array logic circuit. U.S. Patent. 1978.
51. Y. Cui, et al. High performance silicon nanowire field effect transistors. *Nano Letters*, 3(2): pp 149–152, 2003.
52. A. DeHon. Array-based architecture for FET-based, nanoscale electronics. *IEEE Transactions on Nanotechnology*, 2(1): pp 23–32, 2003.
53. G. Snider, P. Kuekes, and R. S. Williams. CMOS-like logic in defective, nanoscale crossbars. *Nanotechnology*, 15(8): pp 881–891, 2004.
54. Y. Cui and C. M. Lieber. Functional nanoscale electronic devices assembled using silicon nanowire building blocks. *Science*, 291(5505): pp 851–853, 2001.
55. K. D. Ausman, et al. Roping and wrapping carbon nanotubes. In: Electronic Properties of Molecular Nanostructures: XV International Winterschool/Euroconference, Kirchberg, Tirol (Austria), 2001: AIP.
56. J. L. Bahr, et al. Dissolution of small diameter single-wall carbon nanotubes in organic solvents? *Chemical Communications*, 193: p 194, 2001.
57. E. T. Mickelson, et al. Solvation of fluorinated single-wall carbon nanotubes in alcohol solvents. *Journal of Physical Chemistry B*, 103(21): pp 4318–4322, 1999.
58. J. Chen, et al. Dissolution of full-length single-walled carbon nanotubes. *Journal of Physical Chemistry B*, 105(13): pp 2525–2528, 2001.
59. J. L. Bahr, et al. Functionalization of carbon nanotubes by electrochemical reduction of aryl diazonium salts: a bucky paper electrode. *Journal of American Chemistry Society*, 123(27): pp 6536–6542, 2001.
60. J. L. Bahr and J. M. Tour. Highly functionalized carbon nanotubes using *in situ* generated diazonium compounds. *Carbon*, 313: pp 91–97, 2001.
61. J. L. Bahr and J. M. Tour. Covalent chemistry of single-wall carbon nanotubes. *Journal of Materials Chemistry*, 12: pp. 1952–1958.
62. Y. Wang, et al. Reversible water-solubilization of single-walled carbon nanotubes by polymer wrapping. *Chemical Physics Letters*, 342: pp 265–271, 2001.
63. Y. Cui, et al. Diameter-controlled synthesis of single-crystal silicon nanowires. *Applied Physics Letters*, 78(15): p 2214, 2001.
64. Y. Huang, et al. Logic gates and computation from assembled nanowire building blocks. *Science*, 294(5545): pp 1313–1317, 2001.
65. D. Whang, S. Jin and C. M. Lieber. Nanolithography using hierarchically assembled nanowire masks. *Nano Letters*, 3(7): pp 951–954, 2003.

66. J. Xiang, et al. Ge/Si nanowire heterostructures as high-performance field-effect transistors. *Nature*, 441(7092): p 489, 2006.
67. S. Y. Chou, P. R. Krauss, and P. J. Renstrom. Imprint lithography with 25-nanometer resolution. *Science*, 272(5258): pp 85–87, 1996.
68. G. Y. Jung, et al. Circuit fabrication at 17 nm half-pitch by nanoimprint lithography. *Nano Letters*, 6(3): pp 351–354, 2006.
69. M. Colburn, et al. Step and flash imprint lithography: a new approach to high-resolution patterning. *SPIE*: 1999.
70. W. Zhang, N. K. Jha, and L. Shang. A hybrid nanotube/CMOS dynamically reconfigurable architecture. *Nature*, 2006.
71. S. C. Goldstein and M. Budiu. Nanofabrics: Spatial computing using molecular electronics. In: Proceedings of the 28th Annual International Symposium on Computer Architecture, 2003.
72. A. DeHon, et al. Nonphotolithographic nanoscale memory density prospects. *IEEE Transactions on Nanotechnology*, 4(2): p 215, 2005.
73. P. J. Kuekes and R. S. Williams. Demultiplexer for a molecular wire crossbar network. Hewlett-Packard Company, 2001.
74. A. DeHon, P. Lincoln, and J. E. Savag. Stochastic assembly of sublithographic nanoscale interfaces. *IEEE Transactions on Nanotechnology*, 2(3): p 165, 2003.
75. J. E. Savage, et al. Radial addressing of nanowires. *ACM Journal on Emerging Technologies in Computing Systems (JETC)*, 2(2): pp 129–154, 2006.
76. Z. Zhong, et al. Nanowire crossbar arrays as address decoders for integrated nanosystems. *Science*, 302(5649): pp 1377–1379, 2003.
77. S. S. Gregory and R. S. Williams. Nano/CMOS architectures using a field-programmable nanowire interconnect. *Nanotechnology*, 18(3): p 035204, 2007.
78. A. DeHon. Nanowire-based programmable architectures. *ACM Journal on Emerging Technologies in Computing Systems (JETC)*, 1(2): pp 109–162, 2005.
79. J. M. Tour, et al. NanoCell Electronic Memories. *Journal of American Chemistry Society*, 125(43): pp 13279–13283, 2003.
80. J. M. Tour, et al. Nanocell logic gates for molecular computing. *IEEE Transactions on Nanotechnology*, 1(2): p 100, 2002.
81. C. P. Husband, et al. Logic and memory with nanocell circuits. *IEEE Transactions on Electron Devices*, 50(9): p 1865, 2003.
82. D. Nackashi and P. Franzon. Moletronics: a circuit design perspective. In: SPIE International Conference on Smart Electronics and MEMS, 2000, Melbourne Australia.
83. E. Goto. The parametron: a digital computing element which utilizes parametric oscillation. In: IRE, 1959.
84. OS Unsal, et al. Impact of parameter variations on circuits and microarchitecture. *IEEE Micro*, 26(6): p 30, 2006.
85. N. S. Kim, et al. Leakage current: Moore's law meets static power. *Computer*, 36(12): p 68, 2003.
86. G. Ashkenasy, et al. Molecular engineering of semiconductor surfaces and devices. *Accounts of Chemical Research*, 35(2): pp 121–128, 2002.

87. P. Zhang, et al. Electronic transport in nanometre-scale silicon-on-insulator membranes. *Nature*, 439(7077): p 703, 2006.
88. J. Yang, et al. Controlling the threshold voltage of a metal–oxide–semiconductor field effect transistor by molecular protonation of the Si:SiO[sub 2] interface. *Journal of Vacuum Science and Technology B: Microelectronics and Nanometer Structures*, 20: pp 1706–1709, 2002.
89. J. J. Boland. Semiconductor physics: transport news. *Nature*, 439(7077): p 671, 2006.
90. T. Shinada, et al. Enhancing semiconductor device performance using ordered dopant arrays. *Nature*, 437(7062): p 1128, 2005.
91. S. Roy and A. Asenov. Applied physics: Where do the dopants go? *Science*, 309(5733): pp 388–390, 2005.
92. T. Sugii, et al. Doping process issues for sub-0.1 μm Generation MOSFETs. In: Materials Research Society Symposium Proceedings, Warrendale, Pennsylvania, 2001; Materials Research Society; 1999.
93. Z. Liu, et al. Molecular memories that survive silicon device processing and real-world operation. *Science*, 302(5650): pp 1543–1545, 2003.
94. R. J. Hamers, et al. Cycloaddition chemistry of organic molecules with semiconductor surfaces. *Accounts of Chemical Research*, 33(9): pp 617–624, 2000.
95. J. M. Buriak. Organometallic chemistry on silicon and germanium surfaces. *Chemical Reviews*, 102(5): pp 1271–1308, 2002.
96. M. A. Brook. *Silicon in Organic, Organometallic, and Polymer Chemistry*. New York: Wiley, 2000.
97. K. Terada and S. Okamoto. Zero-point correction of the carrier density in the measurement of MOS inversion-layer mobility. *Solid-State Electronics*, 47(9): p 1457, 2003.
98. D. K Schroder. *Semiconductor Material and Device Characterization*. New York: Wiley, 1990.
99. A. Ortiz-Conde, et al. A review of recent MOSFET threshold voltage extraction methods. *Microelectronics Reliability*, 42: p 583, 2002.
100. J. Y. Yu, S. W. Chung, and J. R. Heath. Silicon nanowires: preparation, device fabrication, and transport properties. *Journal of Physical Chemistry B*, 104(50): pp 11864–11870, 2000.
101. J. Goldberger, et al. Silicon vertically integrated nanowire field effect transistors. *Nano Letters*, 6(5): pp 973–977, 2006.
102. M. M. Ziegler and M. R. Stan. CMOS/nano co-design for crossbar-based molecular electronic systems. *IEEE Transactions on Nanotechnology*, 2(4): pp 217–230, 2003.
103. K. K. Likharev and D. B. Strukov. CMOL: Devices, circuits, and architectures. *Introducing Molecular Electronics*, pp 447–477, 2005.
104. Y. Cui, et al. Doping and electrical transport in silicon nanowires. *Journal of Physical Chemistry B*, 104(22): pp 5213–5216, 2000.
105. K. Byon, et al. Synthesis and postgrowth doping of silicon nanowires. *Applied Physics Letters*, 87(19): p 193104, 2005.
106. M. V. Fernandez-Serra, C. Adessi, and X. Blase. Surface segregation and backscattering in doped silicon nanowires. *Physical Review Letters*, 96(16): p 166805, 2006.

12

SELF-ASSEMBLY OF SUPRAMOLECULAR NANOSTRUCTURES: ORDERED ARRAYS OF METAL IONS AND CARBON NANOTUBES

Mario Ruben

The use of molecular units within information processing algorithms and devices involves the controlled handling of molecules within the nanoworld. Working at the lower nanometer limit, the combination of scanning probe techniques with bottom-up self-assembly concepts has been proven to be a pivotal tool for such handling. Because "nano-handling" deals intrinsically with the investigation of physical phenomena close to their intrinsic correlation lengths, new ideas in terms of computing concepts may spring up from such scientific work. The perspectives emerging from nonbinaric logics by using ion dots instead of the larger quantum dots are discussed: the use of molecules within the cellular automata scheme; propositions for molecular quantum computing concepts; and the impact of surface-confined self-assembly schemes. All concepts overlap structurally in specific 2D arrangements, called metal ion assemblies (MIAs) and networks (MINs), which represent the material platform for alternative computation approaches discussed herein. Finally, the self-assembly of carbon nanotubes (CNTs) and their use as interface between nanoscopic molecular devices and macroscopic environment will be reviewed.

Bio-Inspired and Nanoscale Integrated Computing. Edited by Mary Mehrnoosh Eshaghian-Wilner

12.1. INTRODUCTION

12.1.1. Molecules: Nanoscale and Computing

Living organisms can be considered as information processing entities designed on the base of molecules (among other views). Taking a closer look at the information processing capacities of living beings, it becomes clear that their performances are (still) superior to that of silicon-based devices. This advantage is basically founded on design principles involving, in particular, massive parallelism, high integration depth, and network-like organization (e.g., brain, nerve cells, receptors, recognition schemes); all implicating molecules. However, the difficulties in mastering the built-in complexity set the main obstacle for in the emergence of purely molecularly and biologically based computing devices. First, rational steps moving along this tempting road map have raised considerable interest, mainly using DNA molecules as structural and functional base (see also Chapter 13) [1]. At this point, it has to be mentioned that the herein reviewed field of molecular computing is still in its early infancy and so marked by several teething problems, e.g., the need for a complete redesign of the communication interface with the environment and the still widely unexplored integration of molecules into electrical circuits. In addition, the use of basic building blocks of living systems in technological devices raises moral and ethical questions, which have to be addressed in parallel with the progressing scientific development of the field.

The use of molecular concepts is not only restricted to biological material. Beyond this scope, the introduction of simpler, organic molecules into already established information processing paradigms of silicon technology seems to be an attractive trade off. Driven by the recent emergence and massive application of new technologies working at the lower nano-regime, a length scale which matches perfectly with the intrinsic size of organic molecules, is a concept that seems to be a possible shortcut for the device development of the nearer future.

Throughout this chapter, the term "biologically inspired" will be used in the sense that it stands for the involvement of molecules in both conventional and advanced information processing schemes (e.g., storage, logic operations, bistability). The use of molecular units within information processing algorithms and devices represents herein the introduction of molecules into the nanoworld; these units can be addressed selectively by applying suitable techniques and methods exhibiting nanometer resolution (e.g., scanning probe techniques). For this to occur, several premises have to be fulfilled on the molecule side: The (i) molecular components have to possess a physical switching property (e.g., redox, magnetic, and/or spin state transitions), and (ii) their geometry should enable a smooth arrangement in two-dimensions (2D) in order to warrant (iii) precise addressing of the single molecule components. The degree of fulfilment of these premises will set the stage for the successful integration and implementation of molecular switching units into nanostructured or even nanoscaled devices. Thus, device architectures possessing addressable nanopatterned switching units promise to serve as platforms for breakthrough solutions in view of the continuously increasing demand for increased storage capacity performances [2].

The introduction of physical switching properties into molecules is a genuine task of chemical design, while the realization of molecular 2D arrangement relies on default of top-down construction principles within the lower nanometer regime and on bottom-up approaches such as supramolecular self-assembly techniques. Single molecule addressing and switching will take advantage of cutting-edge scanning probe and other techniques provided by physicists. In this way, nanoscale molecular information storage represents an archetypical example for the need of cross-border research.

Cutting edge research in the lower nanometer regime may be suspicious of scientific discoveries at the fundamental level of knowledge because the research deals with the investigation of physical phenomena close to their intrinsic correlation lengths. Within this respect, we will discuss in the following four candidates of "more-than-Moore" strategies, standing for technologies that exceed the conventional state-of-the-art technological scheme [3]. The following perspectives are discussed: the gloaming of nonbinaric logics by using ion dots instead of the larger quantum dots [4, 5]; the use of molecules within the cellular automata scheme [6], propositions for molecular quantum computing concepts [7]; and the impact of surface-confined self-assembly schemes. All four concepts materialize structurally in specific 2D arrangements, called metal ion assemblies (MIAs) and networks (MINs), which represent the platform for the alternative computation approaches. Finally, the self-assembly of carbon nanotubes (CNTs) and their use as interface between nanoscopic molecular devices and macroscopic environment will be reviewed.

12.1.2. The Ion Dot Concept

By analogy to semiconductor quantum dots [8], a metal ion coordinated by organic ligands can be considered as a natural quantum dot, i.e., the artificial atom is replaced by a real one (Fig. 12.1). Typically, the feature size of a single quantum dot is at the very lower limit, around 10 nm. However, most are much larger. So called *ion dots*, defined as single metal ion centers surrounded by their organic ligands, are at least more than one order of magnitude smaller in size [4]. Whereas the electronic level structure of a semiconductor quantum dot is merely controlled by its geometry, the electronic levels of a coordinated metal ion are tailored through the type and the geometrical arrangement of its organic ligands.

The local distribution of electron density around the respective metal ion can be used effectively as the structural and functional base for the design of nanoscaled molecular devices. Consequently, the genuine redox, electronic, or spin states of metal ions and their explicit variability (beyond the two-level scheme) will set the base for novel kinds of multistate (meaning beyond binary logics) digital information storage and processing concepts [10].

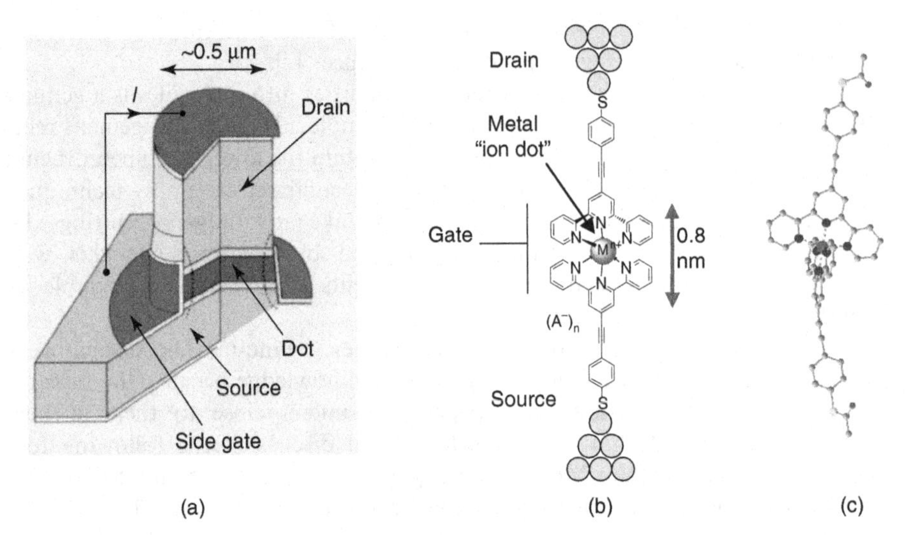

Figure 12.1. (a) Schematic representation of a typical semiconductor quantum dot showing a pillar-like structure with typical dimension of ca. 0.5 μm [8]. (b) By analogy, the same representation exhibiting the concept of a metal ion dot. The ion dot region has an extension of ca. 0.8 nm as shown in (c) the X-ray structure of the respective Ruthenium (II) compound [9].

12.1.3. Cellular Automata Based on Molecules

Further interest in molecularly based devices has risen from an alternative encoding concept called cellular automata (see Chapter 4). Herein, molecules do not act as 0/1 switches but instead as substructured charge containers with changing local electron density distribution [10]. First envisioned for 2D assemblies of quantum dots, binary information is encoded in the internal charge distribution of appropriated dot ensembles or suited molecules. Such type of molecules should possess a small number of differently charged redox centers in a geometrically well defined intramolecular configuration. In case of half-charging, each molecule (here called "cell") exhibits degenerated ground states that can be interconverted by changing external field parameter triggers. Finally, the degeneracy of the ground states of a single cell/molecule is lifted by the electrostatic interaction with the neighbouring cells (arranged in 1D, 2D, or even 3D). The removal of the degeneracy results in "1" and "0" states, which are used for computation in the usual binary way, although the logic operation depends directly on a structural parameter, i.e., the supramolecular 2D arrangement (Fig. 12.2). The critical parameter of the encoding and processing of information here is the degree of intramolecular electronic interaction between the redox centers within the cells: They should be able to communicate, but be neither completely delocalized nor too localized. Additionally, and perhaps an even bigger challenge is the fine tuning of the intermolecular, electrostatic Coloumb interactions between the cells/molecules. Here, surface effects involving

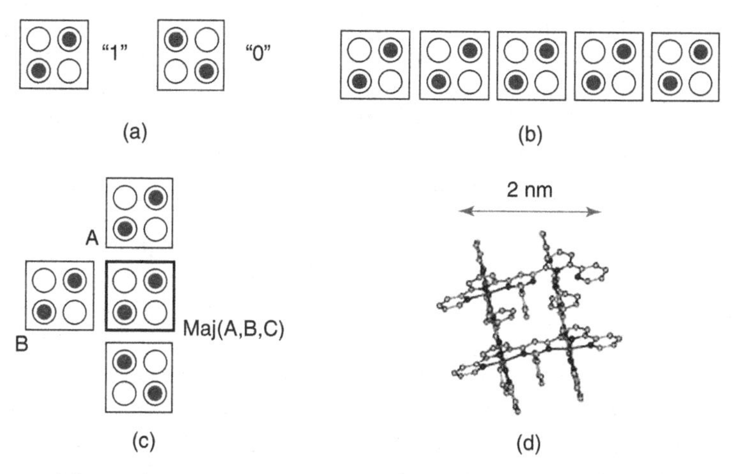

Figure 12.2. Schematic representation of the general working principle of cellular automata: (a) Coulomb repulsion keeps the electron density (dark) at antipodal sites resulting in the degenerated "1" and "0" state. (b) A wire of cellular automata can be formed by a one-dimensional arrangement of cells. The intercellular Coulomb interaction lifts the degeneracy of the 0/1 and forces all encoding units into the same state. (c) Working principle of a majority logic gate consisting of three inputs (A, B, C) which converge on an output (Maj(A;B;C;)) [6]. (d) X-ray structure of a molecule comprised of four metal ions in a array-like configuration, which might be suited for the formation of molecule-based cellular automata by controlled formation of supramolecular 2D arrangements [10].

the underlying substrate may be very helpful to make the molecular cells communicate in the right degree. Remarkably, once the molecules are correctly charged and the intermolecular communication established, such a computing scheme does not involve any current flow.

12.1.4. Molecular Qubits

More recently, the use of molecules as quantum bits in quantum computing algorithms was proposed [7]. Quantum computers could potentially speed up certain kinds of mathematical operations by using elementary units based on quantum bits, so called qubits, instead classical binary bits. Owing to its quantum nature, a single qubit can exist in states spanning any combination of two basic wave functions $|0>$ and $|1>$. Consequently, an operation on qubit causes simultaneous operations on each of the combination's components. Thus, a single operation on a multi-qubit system can affect a huge amount of information; this is called quantum parallelism. In view of molecules, the use of either electronic states or nuclear spin states in quantum computing systems was proposed [11], whereby the primary challenge for scientists in the field lies in the maintenance of the coherence, i.e., the ability to protect the molecular information from deterioration. Among the multitude of evoked quantum computing schemes and algorithms, the implementation of Grover''s algorithm in high-spin molecules exhibiting a slow relaxation of magnetism seems to be a realistic option [12].

12.1.5. Surface-Confined Self-Assembly

There are main two ways to apply supramolecular self-assembly and self-organization techniques in order to steer molecules on surfaces (Fig. 12.3**A,B**): **A**(i), the molecules can be pre-assembled from their components (organic compounds, metal ions, anions, etc.) under bulk conditions. **A**(ii) Subsequently, the so formed supramolecules will be deposited on the surfaces in a second, separated step. (**B**) In a one-step-procedure, the same molecular components can be deposited altogether onto the respective surface by a surface-assisted self-assembly protocol. There, reversible, weak intermolecular forces (hydrogen bonding, metal coordination, van der Waals forces, etc.) will lead, if the molecular components are sufficiently instructed, to highly ordered surface-confined self-assembled network structures (right in Fig. 12.3). Both concepts will be discussed separately in the following two sections, whereby only highly ordered, matrix-like assemblies will be presented due to their general interests in computing concepts. It is beyond the scope of this chapter to review the more general field of surface-deposition of molecules [13]. In both approaches, straightforward scanning

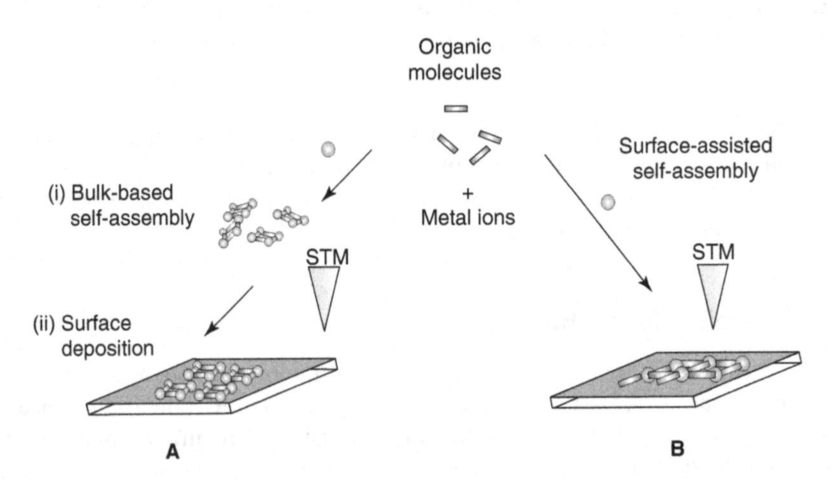

Figure 12.3. Schematic representation of the two general concepts steering the surface-confined organization of molecules and metal ions on surfaces: (**A**) The two-tiered approach involving (i) bulk self-assembly and (ii) surface deposition. (**B**) The one-step procedure of surface-assisted self-assembly by depositing organic ligands (rods) and metal ions directly on surfaces. In both approaches, scanning tunneling microscopy (STM) provides direct access to the nanodimension of the supramolecular self-organization processes.

tunneling microscopy techniques will be used to investigate the surface-confined molecular architectures. Complementary to these two concepts, some examples exhibiting the surface organization of carbon nanotubes (CNTs), which can be considered as extended 1D molecule with high aspect ratios, will be discussed.

12.2. METAL ION ARRAYS (MIAS)

12.2.1. Synthesis and Properties of Metal Ion Arrays (MIAs)

Grid-like metal ion arrays (MIAs) represent a class of coordination compounds, in which a set of metal ions is held in a matrix-like arrangement by a second set of tailor-made organic ligands. The well defined, two-dimensional (2D) arrangement of an exact number of metal ions strongly resembles the binary coded matrices and cross bar architectures used in information storage and processing technology. The metal ions at the crossing points of the finite network architecture can be supplied with well defined redox, magnetic, and spin-state transitions (all properties susceptible to use as switching parameters, either globally or locally) in a controlled manner. Furthermore, due to their very distinguished rectangular geometry, such metal ion arrays might self-assemble on surfaces into extended 2D ensembles.

The design of such metal ion arrays rests on the direction of the coordination instructions, which are based on the cross-over coordination algorithms of both involved metal ions and the ligand's coordination sites (e.g., nature, geometry, positions of the donor atoms). Therefore, the design of rectangular metal ion arrays requires perpendicular arrangements of the ligand planes at each metal center leading (depending on the topocity n and m of the ligands) to array structures of the $n \times m$ type (for $n = m$ to squares). According to this general procedure, metal ion arrays can be prepared in principle by careful prearrangement of the subunits using any set of metal ions and organic ligands, which opens access to a high structural and functional flexibility.

The synthesis of the metal ion arrays follows a mixed synthetic/self-assembly protocol. The organic ligands (rods in Fig. 12.3) are synthesized by conventional synthetic procedures, mainly heterocyclic chemistry. It is important that these well established techniques allow for the deliberate choice of combinations of donor atoms (e.g., N, O, P, etc.) positioned in different overall geometric environments, a necessary perquisite for the perpendicular metal ion coordination. The organic ligands are coordinated by the respective metal ions in solution-based self-assembly processes yielding the metal ion array molecules (MIAs) as bulk after removal from the solution (see step (i) in Fig. 12.3, left). Using such a general procedure, a multitude of square-like $[n \times n]$ arrays with n up to 5, but also rectangular $[n \times m]$ and more differentiated $[p \times [n \times m]]$ metal ion architectures are currently accessible (Fig. 12.4).

Aside from their synthetic accessibility and broad variability, metal ion arrays (MIAs) show very interesting physical properties in light of their potential to

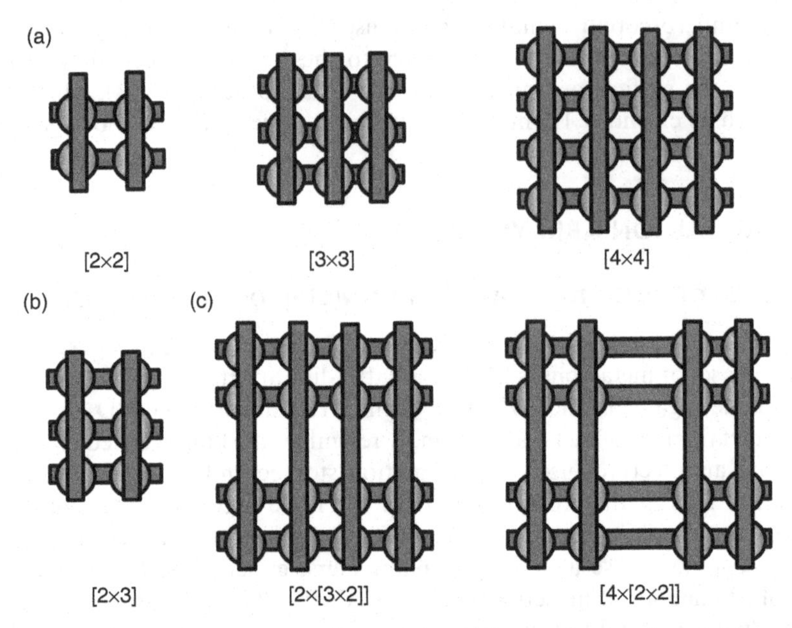

Figure 12.4. Schematic representation of the different types of metal ion arrays (MIAs): (a) [2 × 2], [3 × 3], and [4 × 4] squares, (b) [2 × 3] rectangles and (c) some examples of more complex architectures of the [2 x [4 × 2]] and [4 x [2 × 2]] type (reprinted from [10]).

enable bi- or even multistability at the nanometer scale. Among the promising molecular parameters, the redox and the magnetic behavior of some [2 × 2] and [3 × 3] metal ion arrays have been studied in detail.

In their bulk, some [2 × 2] Co_4^{II} and [3 × 3] Mn_9^{II} arrays have been proven to be very efficient electron reservoirs exhibiting multiple and oxidation steps in diluted solutions [14]. The [2 × 2] Co_4^{II} metal ion array exhibits very extraordinary reduction behavior in multiple (up to twelve), well resolved single electron steps with wave separations between 20 to 40 mV (Fig. 12.5a). Rather low redox potentials (below 3 V) guarantee robust cycling for some Co_4^{II} compounds; this is a necessary prerequisite for low fatigue rates in future device stability. Note the stability of the recuced species can be directly correlated to the nature of the organic ligands and the metal ions, since replacing the Co^{2+} ions by Fe^{2+}, Zn^{2+}, Mn^{2+} decreases the cyclability considerably [15]. In view of the realization of molecular cellular automata (see Chapter 4), two basic results dealing with metal ion arrays of the [2 × 2]Co family might be of interest: (i) The metal ion arrays can be ordered along step edges of graphite into 1D infinite chains (Fig. 12.5b) [16]. (ii) The step-wise synthesis of a [2 × 2]$Co_2^{II}Co_2^{III}$ metal ion array yields molecular species, which show exactly the diagonally localized electron density distribution required for the internal cell structure of such automata devices (Fig. 12.5c) [17].

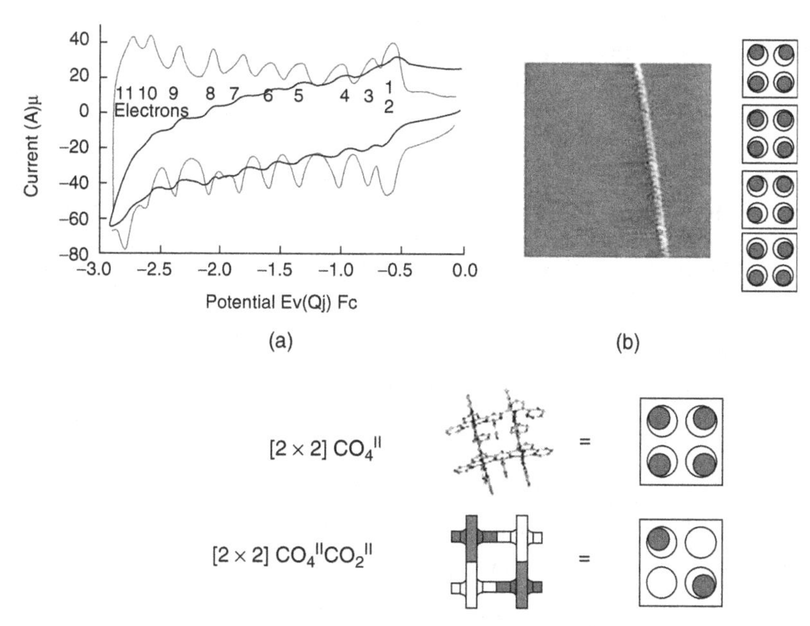

Figure 12.5. (a) Solution cyclovoltammogramm showing the well-resolved single electron reductions of a [2 × 2] Co_4^{II} metal ion array in potential window between 0 and 3 V. Below, the X-ray structure and its translation into the cellular automata symbol are shown [15a]. (b) STM image showing the aligning of the [2 × 2] Co_4^{II} metal ion arrays along the step edges of a graphite substrate after drop casting [16]. (c) Constitution and symbol of the mixed valence [2 × 2] $Co_2^{II}Co_2^{III}$ metal ion array [17].

However, questions concerning the degree of intramolecular delocalization and the strength of the intermolecular Coloumb interaction remain to be investigated before the utility of the metal ion arrays within the frame of the cellular automata concept can be proven. In addition, on the apparatus' side, single molecule techniques discriminating between the two charge states of a cell rest to be developed as feasible read-out tools.

Among the physical molecule properties that may be considered for magnetic, molecular data storage systems, the spin transition (ST) phenomenon featuring the transition between the low spin (LS) and the high spin (HS) states of Fe^{II} ions is an attractive process [18]. Molecular ST systems possess a unique concomitance of possible "write" (temperature, pressure, light) and "read" (magnetic, optical) parameters [19]. Investigations along these lines revealed spin transition behavior in several [2 × 2] Fe_4^{II} metal ion arrays. The internal spin states of the incorporated Fe^{II} ions can be switched successively between their diamagnetic Fe^{II}(LS) and the paramagnetic Fe^{II}(HS) spin states by applying external field triggers (temperature, pressure, light) on macroscopic samples (Fig. 12.6) [10, 20].

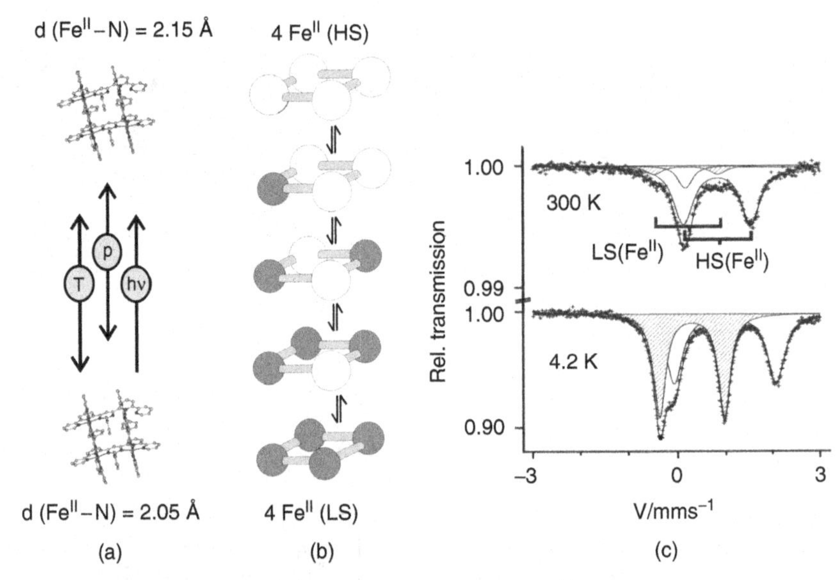

Figure 12.6. Spin transition scheme of the [2 × 2] Fe_4^{II} metal ion arrays exhibiting switching between the diamagnetic low spin (LS) and the paramagnetic high spin state. (a) and (b) The spin state change can be triggered by temperature, pressure and light; the averaged Fe-N bond lengths in the 3 HS/1 LS and in the 1 HS/3 LS states are given. (c) The Moessbauer spectra at two different temperatures showing the spin transition between the two Fe(II) species [20].

In addition, the magnetic anisotropy in a series of [2 × 2] Co_4^{II} metal ion arrays was investigated by single-crystal magnetization measurements at low temperatures. The magnetization data exhibit metamagnetic-like behavior and are explained by the weak-exchange limit of a minimal-spin Hamiltonian including Heisenberg exchange, easy-axis ligand fields, and Zeeman terms [21].

12.2.2. Surface Organization of Metal Ion Arrays (MIAs)

Supramolecular chemistry, with its characteristic control of the self-assembly process and its intrinsic defect tolerance, is a very efficient synthetic tool to achieve ordered arrangements of metal ions with subnanometer precision [22]. In order to study supramolecular entities under restricted spatial dimensions, in particular on flat surfaces, scanning tunneling microscopy (STM) is the investigation method of choice due to its excellent real-space imaging and manipulation capabilities. Recent advances in scanning probe techniques of molecules have enabled imaging and also manipulation with partially sub-molecular resolution [13c]. Thus, after bulk self-assembly synthesis, monolayers of [2 × 2] Co_4^{II} metal ion arrays can be generated simply by drop casting the molecules on an atomically flat graphite surface (HOPG) (following way **A** in Fig. 12.3). The highly ordered supramolecular surface structures consisting of densely packed flat-lying [2 × 2] Co_4^{II} metal ion arrays (MIAs) of rectangular shape are formed spontaneously from dilute acetone solution as almost defect-free domains of up to 0.5 μm^2 (Fig. 12.7). The domain growth proceeds outwards from single nucleation points, a process which might be considered as two-dimensional crystallization.

Certain substitution patterns at the organic ligands provoke on-the-edge orientation of the metal ion arrays with respect to the surface. But, most of the observed metal ion [2 × 2] arrays result in flat tiles forming a "grid-of-grids" superarray, in which the presence of the [2 × 2] metal ion grids is reflected by the 2.5 × 2.4 nm periodicity in agreement with the molecule sizes determined by X-ray crystallography [23].

It was shown that films of the pure [2 × 2] Co_4^{II} metal ion arrays are poorly conductive, but doping them with an excess of cadmium (II) ions could increase the conductivity up to $10^{-2}\,S\,cm^{-1}$ [24]. In similar ways, [3 × 3] Mn_9^{II} metal ion arrays were deposited on both HPOG and ^{II}Au(111) surfaces [25]. By application of a voltage pulse through the STM-tip on the monolayer of metal ion arrays, a single [2 × 2] Co_4^{II} metal ion array could be lifted, leaving a square-like hole of the dimension of the molecule (see Fig. 12.7b). The migration rate of the hole was measured to be 200 times slower than in a monolayer of cycloalkanes, reflecting the degree of adsorption of the molecules to the graphite surface [23].

Further insight into the intramolecular electronic situation of isolated single [2 × 2] Co_4^{II} metal ion arrays at room temperature and ambient conditions could

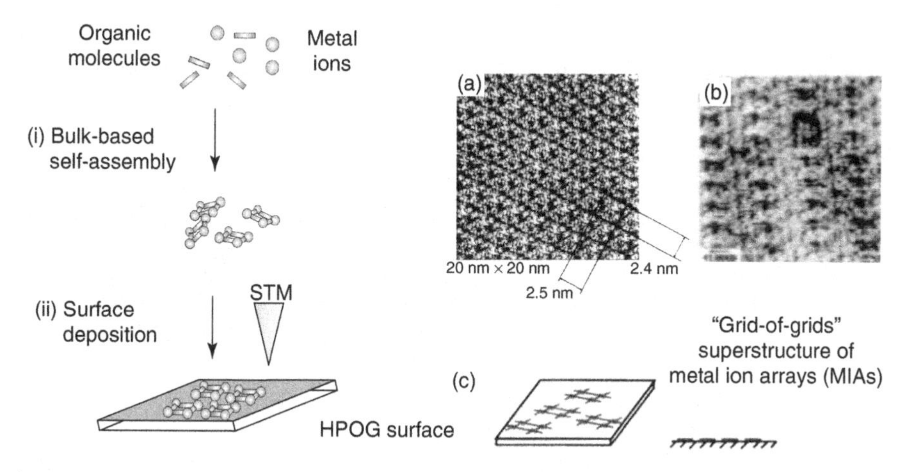

Figure 12.7. Schematic representation of the deposition of metal ion arrays (MIAs) on surfaces: (a) STM image of the monolayer of [2 × 2] Co_4^{II} metal ion arrays on graphite. The periodicity of the grid-of-grids network is shown with 2.5 × 2.4 nm. (b) A hole in the supramolecular monolayer is produced by potential induced lifting of a single [2 × 2] Co_4^{II} molecule with the STM tip. (c) Schematic representation of the disposition of the MIAs at the surface (top and side view) [23].

be gained by a scanning tunneling spectroscopy technique called current induced tunneling spectroscopy (CITS) [26]. The experiments allowed the localization of the positions of the incorporated Co^{II} ions by a selective mapping of the highest occupied molecular orbitals (HOMOs). Principle density functional theory calculations confirmed that in this type of molecules the HOMOs possess a large d-character, such that they are strongly localized around the positions of the metal ions. Consequently, the projection (of the CITS maps at certain negative tunneling biases) reveals electronically the cornerstone positions of the four Co^{II} metal ions (Fig. 12.8) [10]. The same technique was successfully applied to the higher homologous $[3 \times 3]$ Mn_9^{II} and $[4 \times 4]$ Mn_{16}^{II} MIAs aligning respectively 9 and 16 manganese metal ions. The obtained CITS maps mirror the structural situation within the metal ion arrays; although very regularly arranged, the metal ions display in these higher homologues a more lozenge-like structure (Fig. 12.8). This structural deviation from the optimal square-like arrangement can be attributed to the "pinching-in" of the organic ligands during metal ion coordination, reflecting the importance and the consequences of sufficiently instructed metal–ligand interactions for the outcome the self-assembly processes [25].

In conclusion, the formation of highly ordered 2D monolayers of metal ion arrays on surfaces represents a two-tiered self-assembly process: (i) The $[n \times n]$ metal ion arrays are formed in a bulk self-assembly step in solution from their molecular components (organic ligands and metal ions). Subsequently, (ii) the $[n \times n]$ metal ion arrays are self-assembled themselves into densely packed domains of monolayers on the graphite surface. The first self-assembly process relies on the read-out of the coordination instructions stored in the ligands and the metal ions, while the second is steered by van der Waals forces between the metal ion arrays on one side and between arrays and graphite surface on the other side. Due to the flat, square-like geometry of the metal ion arrays, this second

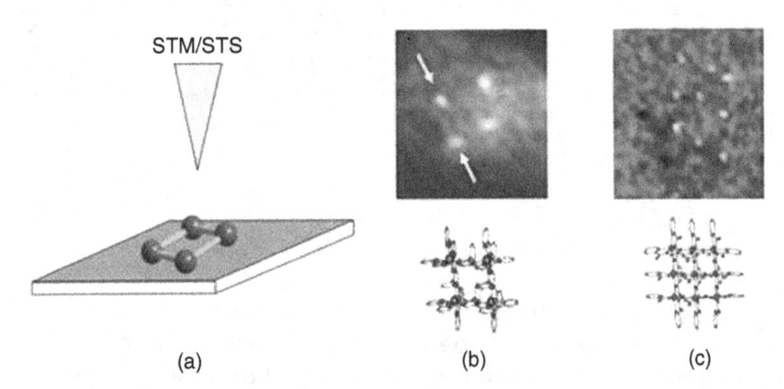

Figure 12.8. (a) Schematic principle showing the metal ion array on a graphite surface. (b) and (c) Show the results of the locally resolved current-induced tunneling spectroscopy (CITS) measurements of a $[2 \times 2]$ Co_4^{II} and $[3 \times 3]$ Mn_9^{II} indicating the position and arrangement of the respective metal ions [26].

self-assembly process under 2D confinement automatically results in a highly ordered "grid-of-grids" superstructure. Consequently, monolayers of $[n \times n]$ MIAs exhibit a two-fold supramolecular matrix structure, (i) internally by the ligand-directed coordinative positioning of the metal ions and (ii) externally by the van der Waals directed formation of the "grid-of-grids" superstructure. The addressing of single metal ion arrays at the single array level was achieved by removing one molecule from the monolayer with help of the scanning tunneling microscopy (STM) tip. Single metal ion addressing inside of isolated metal ion arrays could be achieved electronically through the use of scanning tunneling spectroscopy technique (CITS) [26].

12.3. METAL ION NETWORKS (MINs)

12.3.1. Surface-Confined Assembly of Metal Ion Networks (MINs)

Method **B** in Fig. 12.3 introduces a shortened conceptual alternative to the above described multistep approach in achieving extended highly ordered MIAs on surfaces. Instead to the three-tiered self-assembly approach of method **A**, metal ions and organic ligands are now ordered in only one self-assembly step under immediate 2D confinement of the surfaces [22]. Experimentally, the organic ligand molecules are deposited by organic molecular beam epitaxy on a metallic surface under ultra high vacuum conditions. A more or less complete organic monolayer is formed, on which the respective metal ions are subsequently co-deposited by electron beam evaporation. A short annealing period (typically several minutes) supplies the necessary mobility of the molecular components, guaranteeing the required kinetics for accomplishing the surface-assisted self-assembly process [23]. Depending on the applied molecular ligand/metal couples, regular network structures of different geometries, with domain sizes up to several hundred nanometers, are formed. Within these extended, polymeric 2D network structures, the metal ions are positioned at the crossing points in very regular distances of a few nanometers. Scanning tunneling microscopy (STM) has been proven to be the ideal tool to obtain structural information of the formed metal ion networks (MINs) structures (Fig. 12.9).

One example of such a metal ion network is the self-assembly of linear dicarboxylic acid ligands with Fe(0) metal ions on a Cu(100) surface [28]. At a Fe/ligand ratio of 0.5/1, rectangular molecular assemblies could be obtained consisting of dimeric iron nods that are interconnected by an organic backbone of orthogonally arranged ligand linkers (Fig. 12.10). Within the resolution limit of the STM, the length of the Fe–O bond of the metal ion to the coordinating carboxylates was determined with d(Fe–O) = 1.9–2.3 Å (an expected range for a Fe(II) species). The Fe–Fe distance within the dimeric nods was shown to be between 4.5 and 5.0 Å, a value that is considerably larger than in comparable bulk structures [28]. The coordination geometry around each iron ion can be interpreted as square-planar, which is quite unusual for Iron(II) ions. However,

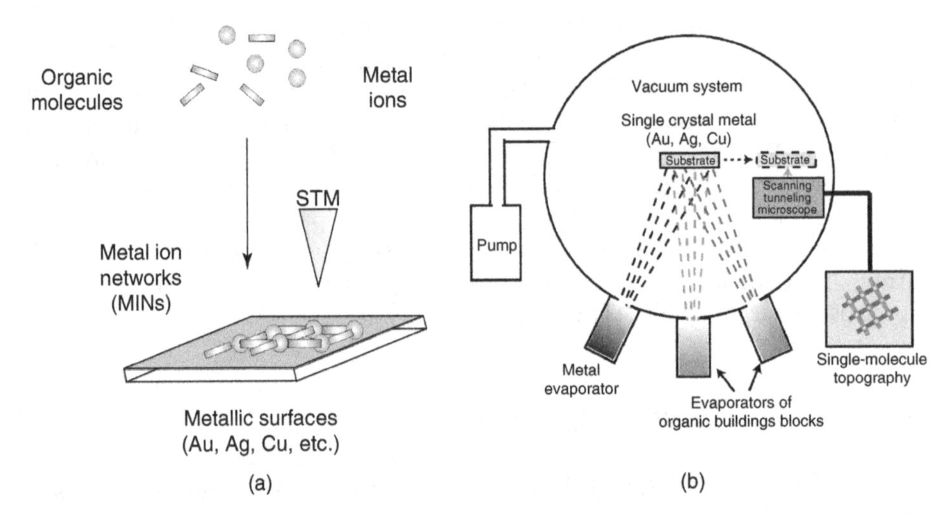

Figure 12.9. (a) Schematic representation of the experimental setup of the UHV-based codeposition of the molecular components (metal atoms and organic ligands) on metallic substrates. (b) The components self-assemble on metallic surfaces under UHV 2D confinement to extended metal ion networks (MINs), which can be investigated *in situ* by scanning probe techniques [27].

the electronic coupling of the coordinated metal centers to the metallic substrate or possible counterbalancing of the charges of the assembled molecular structures by mirror charges within the upper layers of the metallic substrate have to be taken into consideration to explain the electronic properties of surface-deposited metal atoms or ions [29]. Looking at the distinct metal/ligand stoicheometry around the dimeric Fe_2-nodes, it seems reasonable to believe that the principle of electro-neutrality persists under near-surface conditions.

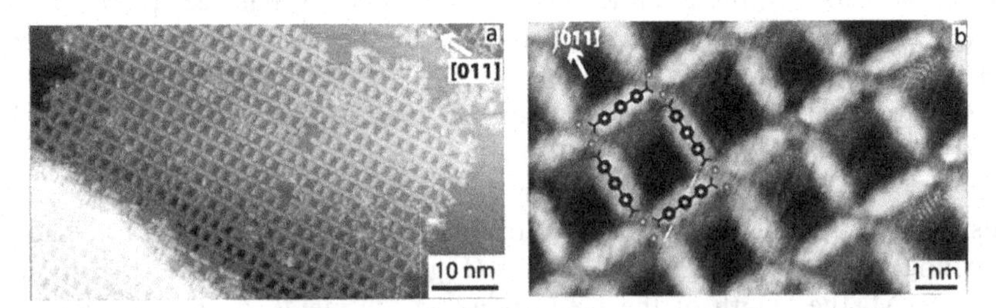

Figure 12.10. (a) STM image showing the 2D topography of the extended Fe-metal ion dicarboxylic acid network on a Cu(100) surface; (b) high resolution image of the same network indicating the positions of the organic ligand dicarboxylic acid backbone and the interconnecting dimeric Fe-nod structures [28].

Such view would lead to the definition of two Iron (II) metal ions surrounded by four deprotonated, negatively charged carboxylic groups of the interconnecting ligands. The formation of electro-neutral 4+/4-units will depend critically on the smoothness of the deprotonation reaction of the carboxylic acid groups. Alternative to the spontaneous deprotonation of carboxylic acid groups on copper surfaces, a redox reaction involving the reduction of four carboxylic protons under simultaneous oxidation of the two Iron(0) centers might be considered. The formed gaseous hydrogen could easily migrate into the UHV environment so favoring the accomplishment of the redox reaction. Furthermore, it has been mentioned that the principle of maximal occupancy of the coordination sites might not be strictly valid under 2D-UHV confinement, since the ligands tend to layer down onto the surface and additional solvent molecules, present in conventional reaction conditions, are not present to fill open coordination sites at the metal ion.

On the ligand side, it was shown that the lengths of linear rod-like ligands can be two, three, or four phenyl units without losing their self-assembly ability. Thus, the distance between the positions of two Fe-dimers within the metal ion network can be deliberately chosen between 1.2 and 2.0 nm. Near-edge, X-ray adsorption fine structure studies have shown that the aromatic backbones of the ligands are adsorbed with their phenyl rings almost parallel to the Cu(1 0 0) surface plane [28]. Besides changing the lengths, the introduction of photoactive double bond structures into the ligand backbone is possible. Interestingly, the introduction 2D prochirality on the ligand side leads to less-ordered metal ion network structures,

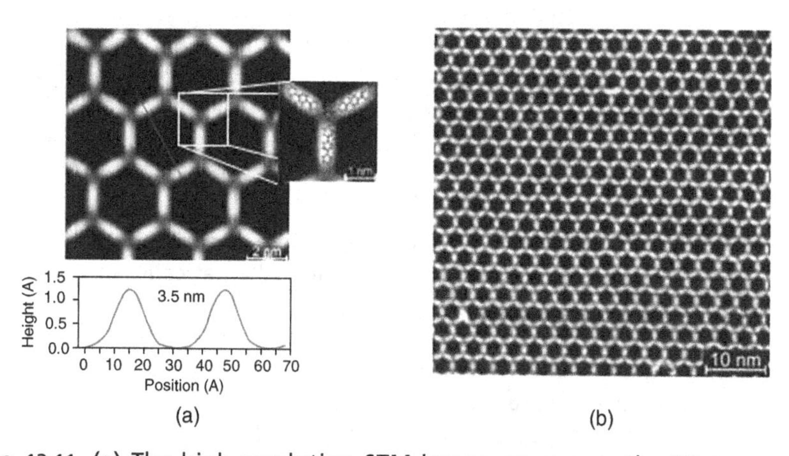

Figure 12.11. (a) The high resolution STM image represents the 2D topography and the internal structure of the extended hexagonal Co-metal ion dicarbonitrile terphenyl network indicating three-fold carbonitrile coordination around monomeric Co-centers at the crossing points. Internal cavities of 3.5 A diameter are generated on the Ag(111) surface. (b) STM image showing the extended regular metal ion network topology [31].

where the energy gain during the metal ion coordination directed self-assembly is reduced; this leads to considerably smaller domain sizes on comparable substrates [27]. Moreover, it was shown that heteroleptic metal ion networks are also accessible by replacing two dicarboxylic acid ligands with two linear pyridines [30].

The same surface-assisted self-assembly strategy can be applied to different metal ion–ligand couples. Replacing the dicarboxylic acids with linear biphenols leads to hexagonal metal ion networks now exhibiting single ion arrangement at the crossing points. By the same token, the combination of Cobalt ions with linear bis-dicarbonitrile ligands generates hexagonal networks with domains up to several hundred nanometer domain sizes (Fig. 12.11) [31]. The symmetry of the evolving metal ion networks (MINs) has been proven to be independent from the symmetry of the substrate, so excluding mere templating effects. It was shown that intrinsic metal ion network symmetry is predominant with respect to the substrate symmetry. The match or mismatch between substrate and metal ion network symmetry is expressed by the evolving domain size and defect probability [31].

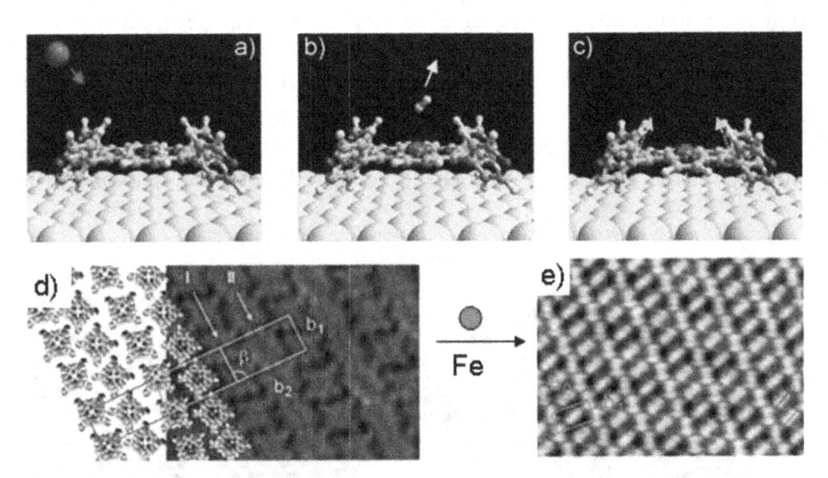

Figure 12.12. (a–c) The three main processes during the metallation of surface-deposited porphyrins on a Ag(111) surface are schematically illustrated: (a) A pure protonated porphyrin layer is exposed to an iron monomer from an atomic beam. (b) The pyrrolic protons are reduced to molecular hydrogen which desorbs while the codeposited Fe(0) is oxidized and incorporated into the dianionic porphyrinato core. (c) This process is associated with a nonplanar deformation of the porphyrin macrocycle. (e) STM image and model of the densely packed monolayer of empty *meso*-porphyrins on Ag(111) exhibiting the two different orientations. The models including the unit cell (b_1 = 13.9 Å, b_2 = 27.4 Å) highlight the structure and facilitate the identification of the molecular moieties. (e) STM image of the metalated monolayer showing the unaffected unit cell [32].

12.3.2. Filling Macrocycle Arrays

Not only linear ligands but also macrocylic ligands as porphyrines and phthalocyanines can be used in the surface-assisted self-assembly approach. Thus, *meso*-tetrapyridyl porphyrines form on Ag(111) surfaces' densely packed monolayers containing two orientations of the macrocycles (due to packing effects). Co-depositing of iron atoms onto this monolayer leads to selective incorporation of the metal atoms into the porphyrin macrocycles whereby the template structure is strictly preserved (Fig. 12.12) [32]. The immobilization of the molecular reactants allows the identification of single metalation events in a novel reaction scheme.

This "filling" approach opens up appealing opportunities, especially because it seems to be easily applied to a large variety of porphyrin or related macrocycle species organized on surfaces (which can be metalated by iron and other metal centers). Specifically, the "filling approach" allows the formation of low-dimensional metallo-porphyrin architectures by using preorganized immobilized macrocycle template arrangements that are subsequently functionalized by metallation. Moreover, novel porphyrin compounds can be created because procedures can be conceived where the addition of a metal center enters as a final step. Current work addresses the elucidation of the coordination characteristics of the involved metal centers, which will also set the base for the investigation of the physical properties (e.g., the magnetism) of single metal centers in the formed extended metal ion networks.

12.4. SELF-ASSEMBLY OF CARBON NANOTUBES (CNTs)

Remarkable electronic properties make carbon nanotubes (CNTs) promising building blocks for molecular or nanoscale devices. In comparison with molecules, CNTs represent an ideal link between the nanoscopic molecular world and their macroscopic implementation. CNTs can fulfill this role because of their typical dimensions involving both nanometer diameter and micrometer lengths. To use them in this interlinking role, CNTs need to be assembled with nanometer precision into hierarchical arrays over large scales of areas; at the same time, they have to be connected to partially macroscopic device components (e.g., electrodes) [33]. At the nanoscale, different techniques are pursued to achieve highly ordered structures. Thus, most of the available methods rely on fabrication of CNTs on prepatterned substrates or catalysts [34]. In a different approach, self-assembly techniques take use of the capillary forces leading to three-dimensional micro-patterns of aligned CNT films [35]. Thereby, a water-spreading method on prepatterned substrates is used to direct the growth of highly ordered CNT films. Although this process is still restricted to the micrometer length scale, improvement of the feature resolutions seems possible (Fig. 12.13).

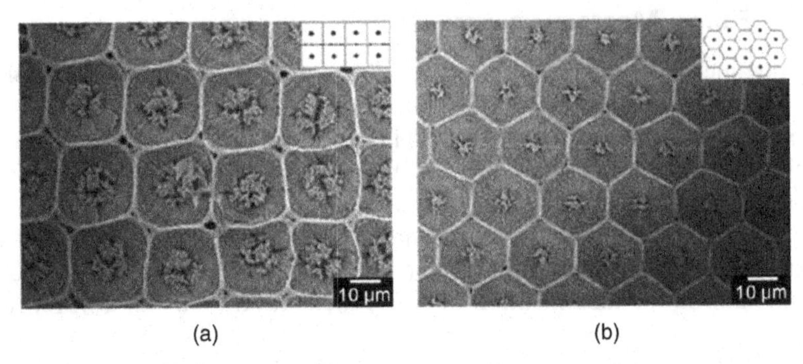

Figure 12.13. Highly ordered carbon nanotube film patterns—(a) cubic and (b) hexagonal—are prepared by applying capillary forces on growing carbon nanotubes on prepatterned surfaces [35].

Within the nanometer regime, different arrangements of carbon nanotubes have achieved on nanometer-scale electrodes of metals. Such controlled arrangement foreshadows the implementation of nanotubes into device architectures that interface the nano world with the macrosopic world. Some early examples make use of individual carbon nanotubes as simple quantum wires by interconnecting a pair of electrodes as depicted in Figure 12.14a [36, 37]. Based on this motif, more elaborate device structures like transistors or SQUID loops were reported (Fig. 12.14b,c) [38, 39]. The different kinds of CNT-based devices recently produced represent a tool kit to interface functional molecules directly within devices, enabling the exploitation of functionalities of molecules within the

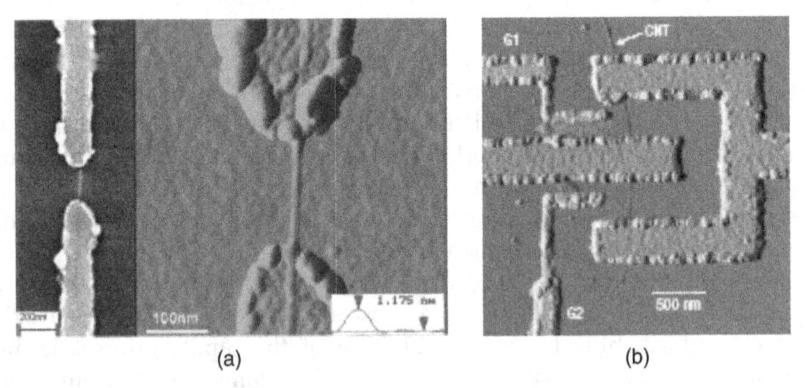

Figure 12.14. (a) Scanning electron microscope (SEM) (left) and STM images (right) of an individual single wall carbon nanotube deposited onto a pair of electrodes. Inset: an AFM profile across a carbon naotube showing its thickness of 1.2 nm [37] (b) AFM image of a typical device geometry of the carbon nanotube CNT-SQUID [39].

nanoregime. To move towards this goal, it will be necessary to control the binding interaction of the respective molecules with the carbon nanotube surface selectively via suited linkers (e.g., pyrene groups).

12.5. CONCLUSIONS

Highly ordered supramolecular metal ion arrays (MIAs) and metal ion networks (MINs) can be constructed on surfaces by using two different general approaches: (A) a three-tiered hierarchical synthesis/self-assembly/deposition protocol or (B) by one-step 2D confined coordination of the metal ions/atoms through organic ligands directly under assistance of the surface. Both strategies lead to nanometer sized, highly symmetric regular arrangements of metal ions on surfaces, whereby the components (organic ligands and metal ions) can be changed in their size, identity, and electronic properties. This modular approach enables a deliberate choice of the nature as well as the relative positioning of the metal ions, in addition to the fine-tuning of their electronic environment. Remarkably, individual metal ions can be effectively imaged and addressed by scanning tunneling microscopy (STM) and spectroscopy (STS) techniques. In the future, direct addressing of electronic properties of the metal ion states (redox, spin states, magnetic anisotropy) might lead to the exploitation of bi- or even multistability at the single ion level, e.g., within the 1 nm regime. Such possibilities posit the ion dot concept in analogy to the well established quantum dot approach. In addition to the possibility of controlled nanopatterning, new horizons in (molecular) data storage are opened. We are given access to completely new avenues in view of alternative information processing technologies (e.g., cellular automata, quantum computing) [11]. If one considers each metal ion within the extended network as addressable (still) bistable data point with an averaged metal ion distance of ca. 2 nm, a functional MIN nanostructure would easily lead to data storage capacities in the several hundreds of Tb/in^2. This is clearly above the actual state-of-the-art of storage devices (below 1 Tb/in^2), but it is also more than what might be achievable by 2D monodomain cluster deposition (without organic molecules) on Au (788) surfaces (26 Tb/in^2) [40]. The additional use of the intrinsic multistability of metal ions through the ion dot concept would go beyond these numbers; this might be a strong motivation to implement molecular metal ion components into the nanoscale devices. However, the implementation and controlled exploitation of metal ion arrays (MIAs) and networks (MINs) within device architectures must find a way to bridge the gap between the molecular nano- and the environmental macrodimensions. Highly integrated carbon nanotube-electrode structures can act as the appropriate connector tool, linking the nanoworld of supramolecular metal ion assemblies with the macroscopic device environments. In sum, such molecule-based device geometries comprising (i) metal ion assemblies, (ii) carbon nanotubes, and (iii) nanostructured metallic electrodes would combine the advantages of both bottom-up self-assembly concepts and cutting-edge top-down fabrication techniques for the development of future information processing applications.

REFERENCES

1. N. C. Seemann. *Nature*, 421: pp 427–431, 2003.
2. R. F. Service. *Science*, 314: pp 1868–1870, 2006.
3. ENIAC-Strategic Research Agenda, European Technology Platform Nanoelectronics. Nov 2005. http://www.cordis.lu/ist/eniac.
4. J. M. Lehn. *Supramolecular Chemistry: Concepts and Perspectives*. Weinheim: VCH, 1995, Ch. 9; p 200.
5. J. Repp, G. Meyer, F. E. Olsson, and M. Person. *Science*, 305: pp 493–497, 2004.
6. C. S. Lent, B. Isaksen, and M. Lieberman. *Journal of American Chemistry Society*, 125: pp 1056–1063, 2003. M. Lieberman, S. Chellamma, B. Varughese, Y. Wang, C. Lent, G. H. Bernstein, G. Snider, and F. C. Peiris. *Annals of the New York Academy of Sciences*, 960: pp 225–239, 2002. A. O. Orlov, I. Amlani, G. H. Bernstein, C. S. Lent, and G. L. Snider. *Science*, 277: pp 928–930, 1997.
7. F. Meier, J. Levy, and D. Loss. *Physical Review Letters*, 90: p. 047901, 2003.
8. L. Kouwenhoven and C. Marcus. *Physics World*: pp 36–39, June 1998.
9. M. Ruben, A. Landa, E. Lörtscher, H. Riel, M. Mayor, H. Görls, H. B. Weber, A. Arnold, and F. Evers. *Small*, to be published. 2008 (DOI: 10.1002/smll.200800390).
10. M. Ruben, J. Rojo, F. J. Romero-Salguero, L. H. Uppadine, and J. M. Lehn. *Angewandte Chemie International Edition*. 43: pp 3644–3662, 2004.
11. S. Bertaina, S. Gambarelli, A. Tkashuk, I. N. Kurkin, B. Malkin, A. Stepanov, and B. Barbara. *Nature Nanotechnology*: 2: pp 39–42, 2007.
12. M. N. Leuenberger and D. Loss. *Nature*, 410: pp 789–794, 2001.
13. See some recent examples: (a) S. D. Feyter, and F. C. D. Schryver. *Chemical Society Reviews*, 32: pp 139–150, 2003; (b) P. Samorí. *Chemical Society Reviews*, 34: pp 551–561, 2005; (c) J. V. Barth, G. Costantini, and K. Kern. *Nature*, 437: pp 671–684, 2005.
14. (a) M. Ruben, E. Breuning, J. P. Gisselbrecht, and J. M. Lehn. *Angewandte Chemie International Edition*, 39: pp 4139–4142, 2000; (b) L. Zhao, C. J. Matthews, L. K. Thompson, and S. L. Heath. *Chemical Communications*: pp 265–266, 2000.
15. (a) M. Ruben, E. Breuning, M. Barboiu, J. P. Gisselbrecht, and J. M. Lehn. *Chemistry, A European Journal*, 9: pp 291–299, 2003; (b) D. M. Bassani, J. M. Lehn, S. Serroni, F. Puntoriero, and S. Campagna. *Chemistry, A European Journal*, 9: pp 5936–5946, 2003.
16. M. S. Alam, S. Strömsdorfer, V. Dremov, P. Müller, J. Kortus, M. Ruben, and J. M. Lehn. *Angewandte Chemie International Edition*, 117: pp 8109–8113, 2005.
17. L. H. Uppadine, J. P. Giesselbrecht, N. Kyritsakas, K. Nättinen, K. Rissanen, and J. M. Lehn. *Chemistry, A European Journal*, 11: pp 2549–2565, 2005.
18. O. Kahn and J. C. Martinez. *Science*, 279: pp 44–48, 1998.
19. P. Gütlich, A. Hauser, and H. Spiering. *Angewandte Chemie International Edition*, 33: pp 2024–2054, 1994.

20. (a) M. Ruben, U. Ziener, J. M. Lehn, V. Ksenofontov, P. Gütlich, and G. B. M. Vaughan. *Chemistry, A European Journal*, 11: pp 94–100, 2005; (b) M. Ruben, E. Breuning, J. M. Lehn, V. Ksenofontov, F. Renz, P. Gütlich, and G. B. M. Vaughan. *Chemistry, A European Journal*, 9: pp 4422–4429, 2003; (c) E. Breuning, M. Ruben, J. M. Lehn, F. Renz, Y. Garcia, V. Ksenofontov, P. Gütlich, E. Wegelius, and K. Rissanen. *Angewandte Chemie International Edition*, 39: pp 2504–2507, 2000.
21. O. Waldmann, M. Ruben, U. Ziener, P. Müller, and J. M. Lehn. *Inorganic Chemistry*, 45: pp 6535–6540, 2006.
22. M. Ruben. *Angewandte Chemie International Edition*, 44: pp 1594–1596, 2005.
23. (a) A. Semenov, J. P. Spatz, M. Möller, J. M. Lehn, B. Sell, D. Schubert, C. H. Weidl, and U. S. Schubert. *Angewandte Chemie International Edition*, 38: pp 2547–2550, 1999; (b) U. Ziener, J. M. Lehn, A. Mourran, and M. Möller. *Chemistry, A European Journal*, 8: pp 951–957, 2002.
24. J. Hassmann, C. Y. Hahn, O. Waldmann, E. Volz, H. J. Schleemilch, N. Hallschmid, P. Müller, G. S. Hanan, D. Volkmer, U. S. Schubert, J. M. Lehn, H. Mauser, A. Hirsch, and T. Clark. *Materials Research Society Symposium Proceedings*, 488: pp 447–452, 1998.
25. (a) L. Weeks, L. K. Thompson, J. G. Shapter, K. J. Pope, and Z. Xu. *Journal of Microscopy*, 212: pp 102–106, 2003; (b) V. A. Milway, S. M. Tareque Abedin, V. Niel, T. L. Kelly, L. N. Dawe, S. K. Dey, D. W. Thompson, D. O. Miller, M. S. Alam, P. Müller, and L. K. Thompson. *Dalton Transactions:* pp 2835–2851, 2006.
26. M. Ruben, J. M. Lehn, and P. Müller. *Chemical Society Reviews*, 35: pp 1056–1067, 2006.
27. N. Lin, S. Stepanow, F. Vidal, K. Kern, M. S. Alam, S. Stromsdörfer, V. Dremov, P. Müller, A. Landa, and M. Ruben. *Dalton Transactions*: pp 2794–2800, 2006.
28. N. Lin, S. Stepanow, F. Vidal, K. Kern, M. S. Alam, S. Stromsdörfer, V. Dremov, P. Müller, A. Landa, and M. Ruben. *Dalton Transactions*: pp 2794–2800, 2006.
29. A. P. Seitsonen, M. A. Lingenfelder, H. Spillmann, A. Dmitriev, S. Stepanow, N. Lin, K. Kern, and J. V. Barth. *Journal of American Chemistry Society*, 128: pp 5634–5635, 2006.
30. L. Lin, S. Stepanow, F. Vidal, J. V. Barth, and K. Kern. *Chemical Communications*, 13: pp 1681–1683, 2005.
31. S. Stepanow, N. Lin, D. Payer, U. Schlickum, F. Klappenberger, G. Zappellaro, M. Ruben, H. Brune, J. V. Barth, and K. Kern. *Angewandte Chemie International Edition*, 46: pp 710–713, 2007.
32. W. Auwärter, A. Weber-Bargioni, S. Brink, A. Riemann, A. Schiffrin, M. Ruben, and J. V. Barth. *ChemPhysChem*, 8: pp 250–254, 2007.
33. Y. Yan, M. B. Chan-Park, and Q. Zhang. *Small*, 3: pp 24–42, 2007.
34. (a) S. Sun, C. B. Murray, D. Weller, L. Folks, and A. Moser. *Science*, 287: pp 1989–1992, 2000; (b) Y. Yang, S. Huang, H. He, A. W. H. Mau, and L. Dai. *Journal of American Chemistry Society*, 121: pp 10832–10833, 1999.
35. H. Lui, S. Li, J. Zhai, H. Li, Q. Zheng, L. Jiang, and D. Zhu. *Angewandte Chemie International Edition*, 43: pp 1146–1149, 2004.
36. S. J. Tans, M. H. Devoret, H. J. Dai, A. Thess, R. E. Smalley, L. J. Geerlings, and C. Dekker. *Nature*, 386: pp 474–477, 1997.

37. R. Krupke, F. Hennrich, H. B. Weber, M. M. Kappes, and H. von Loehneysen. *Nano Letters*, 3: pp 1019–1022, 2003.

38. P. Jarillo-Herrero, J. A. van Dam, and L. P. Kouwenhouven. *Nature*, 439: pp 953–956, 2006.

39. J. P. Cleuziou, W. Wernsdorfer, V. Bouchiat, T. Ondarcuhu, and M. Monthiuox. *Nature Nanotechnology*, 1: pp 53–59, 2006.

40. N. Weiss, T. Cren, M. Epple, S. Rusponi, G. Baudot, S. Rohart, A. Tejeda, V. Repain, S. Rousset, P. Ohresser, F. Scheurer, P. Bencok, and H. Brune. *Physical Review Letters*, 95: pp 157–204, 2005.

13

DNA NANOTECHNOLOGY AND ITS BIOLOGICAL APPLICATIONS

John H. Reif and Thomas H. LaBean

This chapter presents an overview of the emerging research area of DNA nanostructures and biomolecular devices. We discuss work involving the use of synthetic DNA to self-assemble DNA nanostructure devices. Recently, there have been a series of quite astonishing experimental results that have taken the technology from a state of intriguing possibilities into demonstrated capabilities of quickly increasing scale. We particularly emphasize molecular devices that are programmable and autonomous. By programmable, we mean the tasks executed can be modified without entirely redesigning the nanostructure. By autonomous, we mean that the steps are executed with no exterior mediation after starting. We discuss such programmable molecular-scale devices that achieve various capabilities, including computation, 2D patterning, amplified sensing, and molecular transport.

13.1. INTRODUCTION

13.1.1. DNA Nanotechnology and its Use to Assemble Molecular-Scale Devices

The particular molecular-scale devices that are the topic of this article are known as DNA nanostructures. We will explain how DNA nanostructures have some unique advantages among nanostructures: They are relatively easy to design, fairly predictable in their geometric structures, and have been experimentally implemented in a growing number of labs around the world. They are constructed primarily of synthetic DNA. A key principle in the study of DNA nanostructures is the use of self-assembly processes to actuate the molecular assembly. Since self-assembly operates naturally at the molecular scale, it does not suffer from the limitation in

Bio-Inspired and Nanoscale Integrated Computing. Edited by Mary Mehrnoosh Eshaghian-Wilner

scale reduction that so restricts lithography or other more conventional top-down manufacturing techniques. Other surveys of DNA nanotechnology and devices have been given by LaBean [1], Mao [2], Reif [3], and Seeman [4].

In attempting to understand these modern developments, it is worth recalling that mechanical methods for computation date back to the very onset of computer science (for example to the cog-based mechanical computing machine of Babbage). Lovelace stated in 1843 that Babbage's "analytical engine weaves algebraic patterns just as the Jacquard loom weaves flowers and leaves." In some of the recently demonstrated methods for biomolecular computation described here, computational patterns were essentially woven into molecular fabric (DNA lattices) via carefully controlled and designed self-assembly processes.

In general, nanoscience research is highly interdisciplinary. In particular, DNA self-assembly uses techniques from multiple disciplines such as biochemistry, physics, chemistry, and material science, as well as computer science and mathematics. We will observe that many of these self-assembly processes are computational-based and programmable, and it seems likely that a variety of interdisciplinary techniques will be essential to the further development of this emerging field of biomolecular computation.

13.1.2. The Topics Discussed in this Chapter

While a high degree of interdisciplinarity makes the topic quite intellectually exciting, it also makes it challenging for a typical reader. For this reason, this article was written with the expectation that the reader has little background knowledge of chemistry or biochemistry. We define a few relevant technical terms in Section 13.3.1. In Section 13.3.2 we list some known enzymes used for manipulation of DNA nanostructures. In Section 13.3.3 we list some reasons why DNA is uniquely suited for assembly of molecular-scale devices.

In many cases, the self-assembly processes are programmable in ways analogous to more conventional computational processes. We present an overview of theoretical principles and techniques (such as tiling assemblies and molecular transducers) developed for a number of DNA self-assembly processes that have their roots in computer science theory (e.g., abstract tiling models and finite state transducers). Computer-based design and simulation are also essential to the development of many complex DNA self-assembled nanostructures and systems. Error-correction techniques for correct assembly and repair of DNA self-assemblies are also discussed.

The area of DNA self-assembled nanostructures and robotics is by no means simply a theoretical topic—many dramatic experimental results have already been demonstrated, and a number of these will be discussed. The complexity of these demonstrations has been increasing at an impressive rate (even in comparison to the rate of improvement of silicon-based technologies). This chapter discusses the accelerating scale of complexity of DNA nanostructures (such as the number of addressable pixels of 2D patterned DNA nanostructures) and provides some predictions for the future.

Molecular-scale devices using DNA nanostructures have been engineered to have various capabilities, ranging from (i) the execution of molecular-scale computation, (ii) use as scaffolds or templates for the further assembly of other materials (such as scaffolds for various hybrid molecular electronic architectures or perhaps high-efficiency solar-cells), (iii) robotic movement and molecular transport, and (iv) exquisitely sensitive molecular detection and amplification of single molecular events, and (v) transduction of molecular sensing to provide drug delivery.

13.2. INTRODUCTORY DEFINITIONS

13.2.1. A Brief Introduction to DNA

Single stranded DNA (denoted ssDNA) is a linear polymer consisting of a sequence of DNA bases oriented along a backbone with chemical directionality. By convention, the base sequence is listed starting from the 5-prime end of the polymer and ending at the 3-prime end (these names refer to particular carbon atoms in the deoxyribose sugar units of the sugar-phosphate backbone, the details of which are not critical to the present discussion). The consecutive nucleotide bases (monomer units) of an ssDNA molecule are joined through the backbone via covalent bonds. There are four types of DNA bases: adenine, thymine, guanine, and cytosine, typically denoted by the symbols A, T, G, and C, respectively. These bases form the alphabet of DNA; the specific base sequence comprises DNA's information content. The bases are grouped into *complementary pairs* (G, C) and (A, T).

The most basic DNA operation is *hybridization*, where two ssDNA oriented in opposite directions can bind to form a double stranded *DNA helix* (dsDNA) by pairing between complementary bases. DNA hybridization occurs in a buffer solution with appropriate temperature, pH, and salinity. A dsDNA helix is illustrated in Figure 13.1.

Since the binding energy of the pair (G, C) is approximately half-against the binding energy of the pair (A, T), the association strength of hybridization depends on the sequence of complementary bases, and can be approximated by

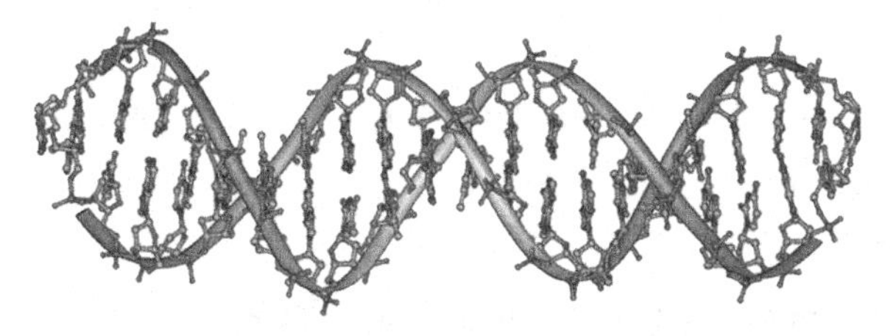

Figure 13.1. Structure of a *DNA double helix* (Created by Michael Ströck and released under the GNU Free Documentation License (GFDL).)

known software packages. The *melting temperature* of a DNA helix is the temperature at which half of all the molecules are fully hybridized as double helix, while the other half are single stranded. The kinetics of the DNA hybridization process is quite well understood; it occurs in a (random) zipper-like manner, similar to a biased one-dimensional random walk.

Whereas ssDNA is a relatively flexible molecule, dsDNA is quite stiff (over lengths of less than 150 or so bases) and has the well characterized, double helix structure. There are about 10.5 bases per full rotation on this helical axis. The exact geometry of the double helix depends slightly on the base sequence in a way readily computed by existing software. A *DNA nanostructure* is a multimolecular (supramolecular) complex consisting of a number of ssDNA that have partially hybridized, as designed, along their subsegments.

13.2.2. Manipulation of DNA

In addition to the hybridization reaction, there are a wide variety of known enzymes and other proteins used for manipulation of DNA nanostructures and have predictable effects. (Interestingly, these proteins were discovered in natural bacterial cells and tailored for laboratory use.) These Include:

- *Restriction enzymes* can cut (double-strand break) or nick (single-strand break) a DNA backbone at specific locations determined by short base sequences.
- *Ligase enzymes* can heal or repair DNA nicks by forming covalent bonds in the sugar-phosphate backbone.
- *Polymerase* can extend an ssDNA by covalently coupling further complementary bases, as dictated by a template ssDNA, thus forming a longer sequence of dsDNA.

Besides their extensive use in other biotechnology procedures, the above reactions, together with hybridization, are often used to execute and control DNA computations and DNA molecular robotic operations. The restriction enzyme reactions are programmable in the sense that they are site specific, only executed as determined by the appropriate DNA base sequence. The latter two reactions, using ligase and polymerase, require the expenditure of energy via consumption of ATP molecules and can thus be controlled by ATP concentration.

13.2.3. Why Use DNA to Assemble Molecular-Scale Devices?

There are many advantages of DNA as a material for building things at the molecular scale.

(a) The advantages from a design perspective are:
- The basic geometric and thermodynamic properties of dsDNA are well understood and can be modeled by available software systems. The

structure of a large number of more complex DNA nanostructures can be predicted by a number of prototype software systems from details like the sequence composition, temperature, and buffer conditions (which are the key relevant parameters).

- Design of DNA nanostructures can be assisted by software. To design a DNA nanostructure or device, one needs to design a library of ssDNA strands with specific segments that hybridize to (and only to) specific complementary segments on other ssDNA. There are a number of software systems for this combinatorial sequence design task and for design of DNA nanostructures with desired structures.

(b) The advantages from an experiments perspective are:

- The chemical synthesis of ssDNA is now routine and inexpensive; a test tube of ssDNA consisting of any specified short sequence of bases (< 150) can be obtained from commercial sources for modest cost (at this time, about half a U.S. dollar per base); it will contain a very large number (typically at least 10^{12}) of identical ssDNA molecules. The synthesized ssDNA can have errors (premature termination of the synthesis is the most frequent error), but can be easily purified by well known techniques (e.g., electrophoresis, mentioned below).
- The assembly of DNA nanostructures is a very simple experimental process; in many cases, one simply combines the various component ssDNA into a single test tube with an appropriate buffer solution at an initial temperature above the expected melting temperature of the most stable base-pairing structure, and then slowly cools the test tube below the melting temperature.
- The assembled DNA nanostructures can be characterized by a variety of techniques. One such technique is electrophoresis. It provides information about the relative molecular mass of DNA molecules, as well as some information regarding their assembled structures, depending on what type of electrophoresis (denaturing or native, respectively) is used. Other techniques like atomic force microscopy (AFM) and transmission electron microscopy (TEM) provide images of the actual assembled DNA nanostructures on 2D surfaces.

13.3. ADELMAN'S INITIAL DEMONSTRATION OF A DNA-BASED COMPUTATION

13.3.1. Adleman's Experiment

The field of DNA computing began in 1994 with a laboratory experiment described in [5, 6]. The goal of the experiment was to find, within a given directed graph, a Hamiltonian path, which is a path that visits each node exactly once. To solve this problem, a set of ssDNA was designed based on the set of edges of the graph. When combined in a test tube and cooled, they self-assembled into dsDNA.

Each of these DNA nanostructures was a linear DNA helix that corresponded to a path in the graph. If the graph had a Hamiltonian path, then one of these DNA nanostructures encoded the Hamiltonian path. Through conventional biochemical extraction methods, Adelman was able to isolate only DNA nanostructures encoding Hamiltonian paths and, by determining their sequence, the explicit Hamiltonian path. It should be mentioned that this landmark experiment was designed and demonstrated experimentally by Adleman alone, a computer scientist with limited training in biochemistry.

13.3.2. The Nonscalability of Adelman's Experiment

While this experiment founded the field of DNA computing, it was not scalable in practice, since the number of different DNA strands needed increased exponentially with the number of nodes of the graph. Although there can be an enormous number of DNA strands in a test tube (10^{15} or more, depending on solution concentration), the size of the largest graph that could be solved by his method was limited to at most a few dozen nodes. This is not surprising, since finding the Hamiltonian path is an NP complete problem, whose solution is likely to be intractable using conventional computers. Even though DNA computers operate at the molecular scale, they are still equivalent to conventional computers (e.g., deterministic Turing machines) in computational power. This experiment taught a healthy lesson to the DNA computing community (which is now well recognized): Carefully examine scalability issues and judge any proposed experimental methodology by its scalability.

13.3.3. Autonomous Biomolecular Computation

Shortly following Adleman's experiment, there was a burst of further experiments in DNA computing, many of which were quite ingenious. However, almost none of these DNA computing methods were autonomous, and instead required many tedious laboratory steps to execute. In retrospect, one of the most notable aspects of Adleman's experiment was that the self-assembly phase of the experiment was completely autonomous—it required no exterior mediation (the bulk of the labor was in the nonautonomous molecular sorting steps). The strategy can be termed generate-and-sort, since all possible answers are created and incorrect solutions are subsequently discarded. Maximizing molecular autonomy makes an experimental laboratory demonstration much more feasible as the scale increases. The remaining sections mostly discuss autonomous devices for biomolecular computation based on self-assembly.

13.4. SELF-ASSEMBLED DNA TILES AND LATTICES

13.4.1. Computation By Self-Assembly

The most fundamental way computer science ideas have impacted DNA nanostructure design is via the pioneering work by theoretical computer scientists on a formal model of 2D tiling due to Wang (in 1961), which culminated in a proof by Berger in 1966 that universal computation could be done via tiling assemblies. Winfree [7] was the first to apply the concepts of computational tiling assemblies to DNA molecular constructs. His core idea was to use tiles composed of DNA to perform computations during the process of self-assembly, where only valid solutions to the computation are allowed to assemble. To understand this idea, we will need an overview of DNA nanostructures, as presented in the next section.

13.4.2. DNA Nanostructures

Recall that a DNA nanostructure is a multimolecular complex consisting of a number of ssDNA that have partially hybridized along their subsegments. The field of DNA nanostructures was pioneered by Seeman [4].

Particularly useful types of motifs often found in DNA nanostructures include stem-loops and sticky ends, as illustrated below.

Figure 13.2a illustrates a *stem-loop* where ssDNA loops back to hybridize on itself; that is, one segment of the ssDNA (near the 5′ end) hybridizes with another segment further along (nearer the 3′ end) on the same ssDNA strand. The shown stem consists of the dsDNA region with sequence CACGGTGC on the bottom

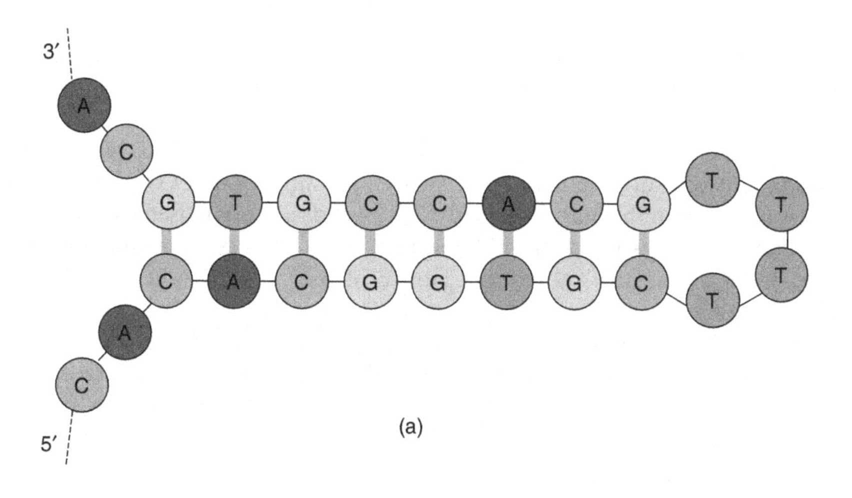

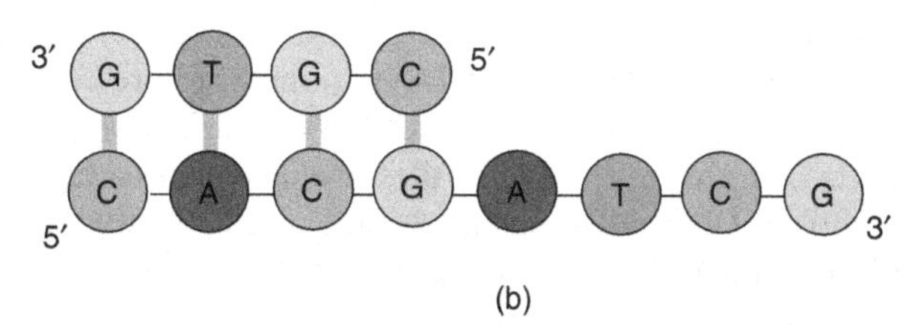

Figure 13.2. A stem-loop and a sticky end.

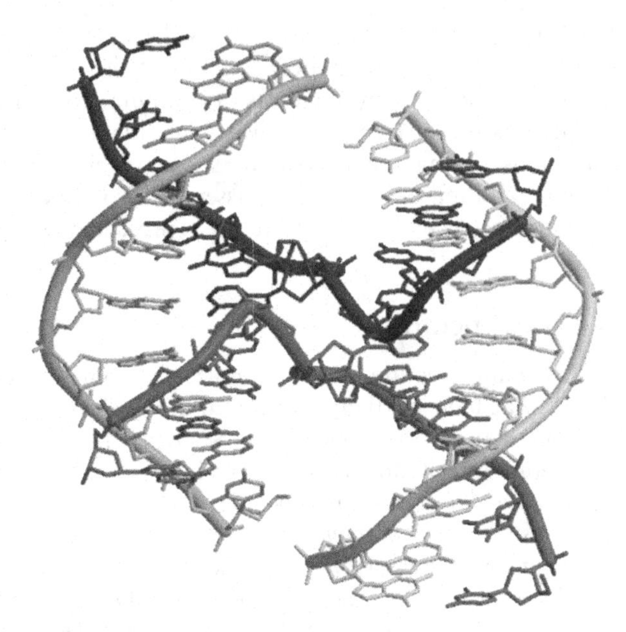

Figure 13.3. A Holliday junction. (Image created by Miguel Ortiz-Lombardía, CNIO, Madrid, Spain and used with permission.)

strand. The loop in this case consists of the ssDNA region with sequence TTTT. Stem-loops are often used as markers for visualizing programmed patterning on DNA nanostructures.

Figure 13.2b illustrates a *sticky end* where unhybridized sDNA protrudes from the end of a double helix. The sticky end shown (ATCG) protrudes from dsDNA (CACG on the bottom strand). Sticky ends are often used to combine two DNA nanostructures together via hybridization of their complementary ssDNA. Figure 13.2b shows the antiparallel nature of dsDNA with the 5′ end of each strand pointing toward the 3′ end of its partner strand.

Figure 13.3 illustrates a *Holliday junction*, where two adjacent DNA helices form a junction with one strand of each DNA helix crossing over to the other DNA helix. Holliday junctions are often used to tie together various parts of a DNA nanostructure.

13.4.3. DNA Tiles and Lattices

A *DNA tile* is a DNA nanostructure that has a number of sticky ends on its sides, which are termed *pads*. A DNA lattice is a DNA nanostructure composed of a group of DNA tiles that are assembled together via hybridization of their pads. Generally, the strands composing the DNA tiles are designed to have a melting temperature above those of the pads, ensuring that when the component DNA molecules are combined together in solution, first the DNA tiles assemble, and only then, as the solution is further cooled, do the tiles bind together via

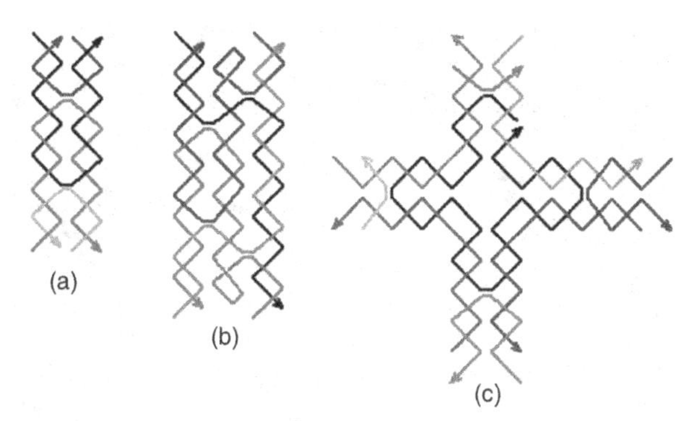

Figure 13.4. DNA tiles.

hybridization of their pads. A number of prototype computer software systems have been developed for the design of the DNA sequences composing DNA tiles, and for optimizing their stability.

To program a tiling assembly, the pads of tiles are designed so that tiles assemble together as intended. Proper designs ensure that only the adjacent pads of neighboring tiles are complementary so only those pads hybridize together. Figure 13.4 illustrates some principal DNA tiles.

Winfree and Seeman [7] developed a family of DNA tiles known collectively as DX tiles (see left tile illustrated in Fig. 13.4a) that consisted of two parallel DNA helices linked by two immobile Holliday junctions. They demonstrated that these tiles formed large 2D lattices, as viewed by AFM. Subsequently, other DNA tiles were developed to provide for more complex strand topology and interconnections, including a family of DNA tiles known as *TX tiles* (see tile illustrated in Fig. 13.4b) [8] composed of three DNA helices linked by four crossover junctions. Both the DX tiles and the TX tiles are rectangular in shape, where two opposing edges of the tile have pads consisting of ssDNA sticky ends. In addition, TX tiles have topological properties that allow for strands to propagate in useful ways though tile lattices. (This property is often used for aid in patterning DNA lattices as described below). Other DNA tiles known as *cross tiles* (see Fig. 13.4c) [8] are shaped roughly square and have pads on all four sides, allowing for binding of the tile directly with neighbors in all four directions in the lattice plane. Figure 13.5 gives an AFM image of a 2D lattice using cross tiles.

13.5. AUTONOMOUS FINITE-STATE COMPUTATION USING LINEAR DNA NANOSTRUCTURES

13.5.1. The First Experimental Demonstration of Autonomous Computations Using Self-Assembly of DNA Nanostructures

The first experimental demonstrations of computation using DNA tile assembly was [9], which demonstrated two-layer, linear assemblies of TX tiles that executed

Figure 13.5. AFM image of 2D lattice of cross tiles.

a bit-wise cumulative XOR computation. (Given n bits as input, each *i*th output is the XOR of the first i input bits, which is the computation occurring when one determines the output bits of a full-carry binary adder circuit.) The experiment [9] is described further in Figure 13.6.

Figure 13.6 shows a unit TX tile (a) and the sets of input and output tiles (b) with geometric shapes conveying sticky-end complementary matching. The input layer and corner condition tiles were designed to assemble first [see example computational assemblies in Fig. 13.6 (c) and (d)]. The output layer would then assemble specifically starting from the bottom left using the inputs from the input layer. The tiles were designed such that an output reporter strand ran through all the n tiles of the assembly by bridges across the adjoining pads in input, corner, and output tiles. This reporter strand was pasted together from the short ssDNA sequences within the tiles using ligation enzyme mentioned previously. When the solution was warmed, this output strand was isolated and identified. The output data was read by experimentally determining the sequence of cut sites (as described below). In principle, the purified output strands could be used for subsequent computations.

This experiment [9] provided answers to a basic question:

- *How can one provide data input to a molecular computation using DNA tiles?*

In this experiment the input bits (1's and 0's) were encoded on two different tile types with specific sticky-ends and specific endonuclease cleavage sites

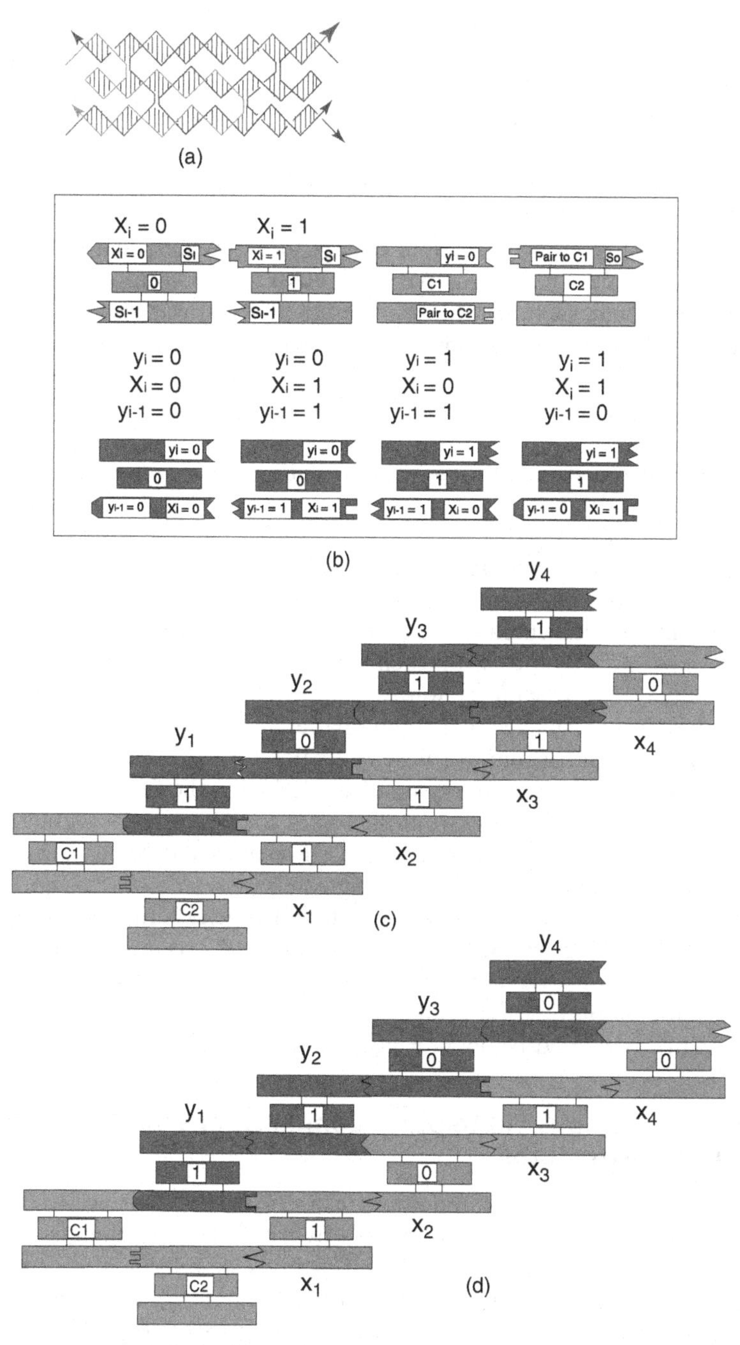

Figure 13.6. Sequential Boolean computation via a linear DNA tiling assembly. (Figure adapted with permission from [9].)

(subsequences at which protein enzymes can cut the DNA backbone). The input sequence was defined by specific sticky-ends that assembled a specific input layer (Figure 13.6).

Here is the next question of concern:

- *How can one execute a step of computation using DNA tiles?*

To execute steps of computation, the TX tiles were designed to have pads at one end that encoded the cumulative XOR value. Also, since the reporter strand segments ran though each tile, the appropriate input bit was also provided within its structure. These two values implied that the opposing pad on the other side of the tile be the XOR of these two bits.

This is the final question of concern:

- *How can one determine and/or display the output values of a DNA tiling computation?*

The output in this case was read by determining which of two possible cut sites (endonuclease cleavage sites) were present at each position in the tile assembly. This was executed by first isolating the ligated reporter strand, then digesting separate aliquots with each endonuclease separately and then two together; finally, these samples were examined by gel electrophoresis and the output values were displayed as banding patterns on the gel.

Another method for output (presented below) is the use of AFM observable patterning. Such patterning can be made by designing the tiles computing a bit 1 to have a stem loop protruding from the top of the tile or by providing a site for binding of a marker protein. Sequences of such molecular patterning are clearly viewable under appropriate AFM imaging conditions.

Although they are quite simple computations, the experiments of [9] and [10] did demonstrate pioneering methods for autonomous execution of a sequence of finite-state operations via algorithmic self-assembly, as well as for providing inputs and for outputting the results. Further DNA tile assembly computations [11, 12] will be presented below in Figure 13.11.

13.5.2. Autonomous Finite-State Computations via Disassembly of DNA Nanostructures

An alternative method for autonomous execution of a sequence of finite-state transitions was subsequently developed by [13]. Their technique essentially operated in the reverse of the assembly methods described above and instead was based on disassembly. They began with a linear DNA nanostructure whose sequence encoded the inputs; then they executed series of steps that digested the DNA nanostructure from one end. On each step, a sticky-end at one end of the nanostructure encoded the current state, and the finite transition was determined by hybridization of the current sticky end with a small "rule" nanostructure encoding the finite-state transition rule. Then a restriction enzyme, which recognized the sequence encoding the current input as well as the current state, cut the appended end of the linear DNA nanostructure to expose a new a sticky-end encoding the next state. The hardware–software complex is composed of dsDNA with an ssDNA overhang (shown at top left ready to bind with the input molecule) and a protein restriction enzyme (shown as gray pinchers). See Figure 13.7 for further details. This ingenious design is an excellent demonstration that there is often more than one way to do any task at the molecular scale.

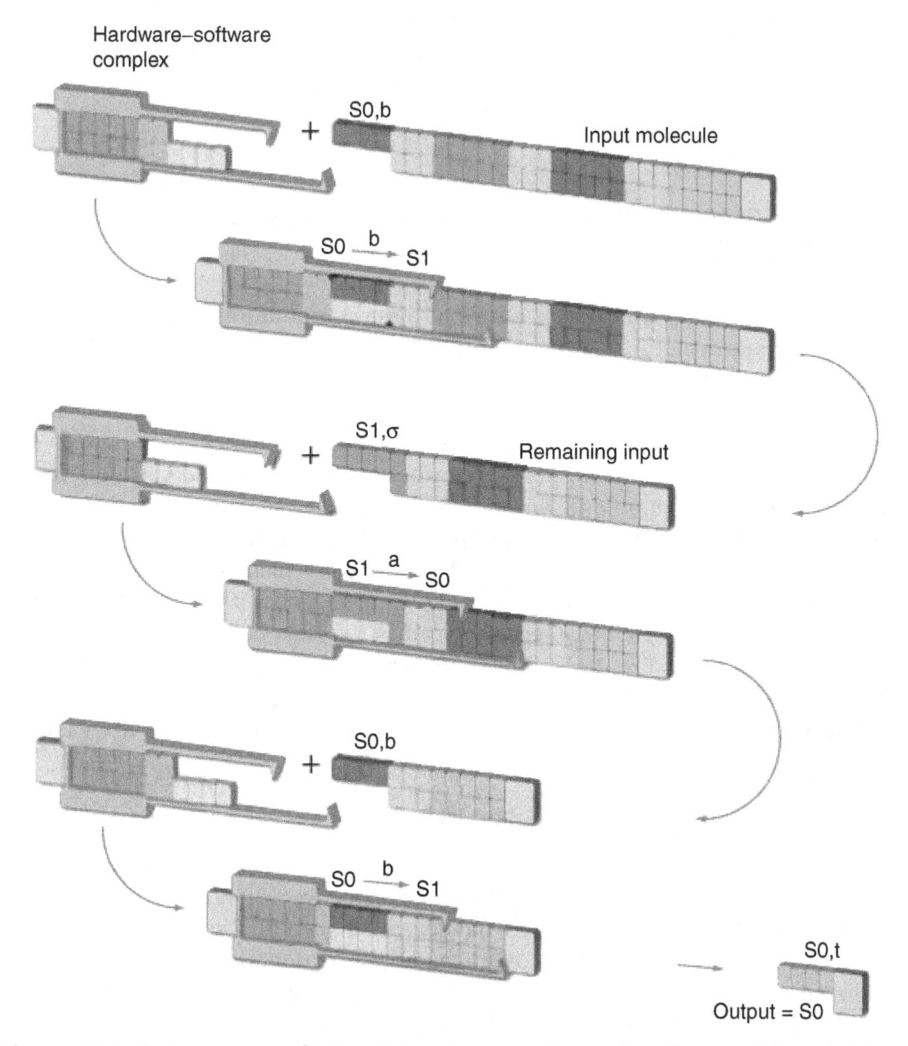

Figure 13.7. Autonomous finite-state computations via disassembly of DNA nanostructures. (Figure adapted with permission from [13]).

13.5.3. Applications of Autonomous Finite-State Computations at the Molecular Scale

Even very simple operations, such as the above Boolean or finite-state transitions, operating at the molecular-scale could have important potential applications, for example, for drug mediation [13]. The idea is for the DNA nanostructures to take as input a set of RNA sequences, whose level of expression (or lack of expression) within a cell indicates a particular disease state. Then the execution of simple Boolean operations executable by finite-state transitions can determine that a disease exists, and execute a response (e.g., the release of RNA sequences which provide a remediation of the disease by altering the expression of proteins expressed by the cell). While such a scheme was demonstrated by [13] in the test tube, it remains to be demonstrated in the much more challenging environment of a cell. Another class of applications is for control of molecular robotic devices, described in Section 13.7.

13.6. ASSEMBLING PATTERNED AND ADDRESSABLE 2D DNA LATTICES

One of the most appealing applications of tiling computations is their use to form patterned nanostructures to which other, perhaps functional, materials can be selectively bound.

An *addressable* 2D DNA lattice is one that has a number of sites with distinct ssDNA. This provides a superstructure for selectively attaching other molecules at addressable locations. The input layer for the computational assembly described in Figure 13.6 is an example of an addressable system, since unique ssDNA pads defined the tile locations. Other examples will be presented in the following text discussion. As Many types of molecules exist for which we can attach DNA. Known attachment chemistry allows them to be tagged with a given sequence of ssDNA. Each of these DNA-tagged molecules can then be assembled by hybridization of their DNA tags to a complementary sequence of ssDNA located within an addressable 2D DNA lattice. In this way, we can program the assembly of each DNA-tagged molecule onto a particular site of the addressable 2D DNA lattice.

13.6.1. Attaching Materials to DNA

Many materials can be made to directly or indirectly bind to specific segments of DNA using a variety of known attachment chemistries. Materials that can directly bind to specific segments of DNA include other (complementary) DNA, RNA, proteins, peptides, and various other materials. Materials that can be made to indirectly bind to DNA include a variety of metals (e.g., gold) that bind to sulfur-labeled compounds, carbon nanotubes (via various attachment chemistries), etc. These attachment strategies provide molecular-scale "Velcro" for attaching heterogeneous materials to DNA nanostructures. For example, they can potentially be

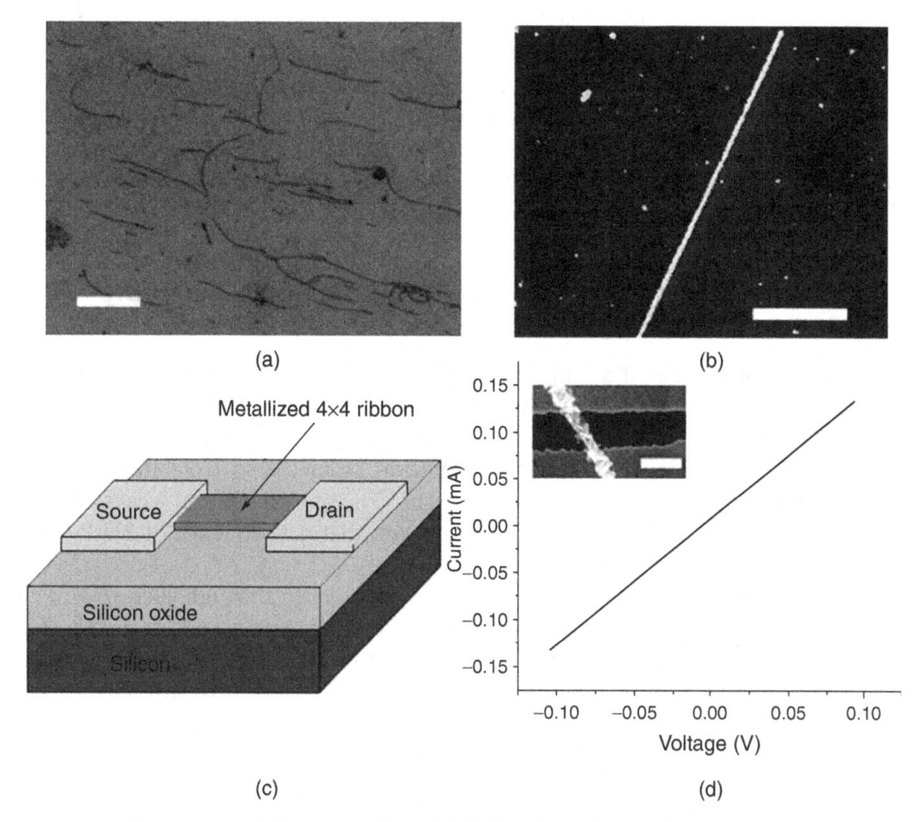

Figure 13.8. DNA-templated fabrication of metal nanowires.

used for attaching molecular electronic devices to 2D or 3D DNA nanostructures. See Figure 13.8 [14] for an example of conductive wires composed of self-assembled DNA tubes covered with gold or silver.

Figure 13.8a is a SEM image of bare self-assembled DNA nanotube on silicon oxide surface (scale bar equals 500 nm). Figure 13.8b is a SEM image of a gold-coated DNA nanotube on silicon oxide surface (scale bar equals 500 nm). Figure 13.8c is a schematic representation of the measured device (inset of Figure 13.8d) showing the tempated nanowire and source and drain electrodes fabricated by electron beam lithography. Figure 13.8d gives the current/voltage curve of a gold nanowire. Smaller, smoother silver nanowires are presented in [14, 15].

13.6.2. Methods for Programmable Assembly of Patterned 2D DNA Lattices

The first experimental demonstration of 2D DNA lattices by Winfree and Seeman provided very simple patterning by repeated stripes determined by a stem loop projecting from every DNA tile on an odd column. This limited sort of patterning needed to be extended to large classes of patterns.

In particular, the key capability needed is a programmable method for forming distinct patterns on 2D DNA lattices, without having to completely redesign the lattice to achieve any given pattern. There are at least three methods [16] for assembling patterned 2D DNA lattices that now have been experimentally demonstrated, as described in the next sections.

13.6.3. Use of Scaffold Strands for Programmable Assembly of Patterned 2D DNA Lattices

The first published use of a scaffold strand, a long ssDNA around which shorter ssDNA assemble to form structures larger than individual tiles, is given in [1]. Scaffold strands were used to demonstrate programmable patterning of 2D DNA lattices in [17] by propagating 1D information from the scaffold into a second dimension to create AFM observable patterns. The scaffold strand weaves though the resulting DNA lattice to form a desired and distinct sequence of 2D barcode patterns (Figure 13.9).

In this demonstration, identical scaffold strands ran through each row of the 2D lattices, using short stem loops extending above the lattice to form pixels. This determined a bar code sequence of stripes over the 2D lattice that was viewed by AFM. In principle, this method may be extended to allow for each row's

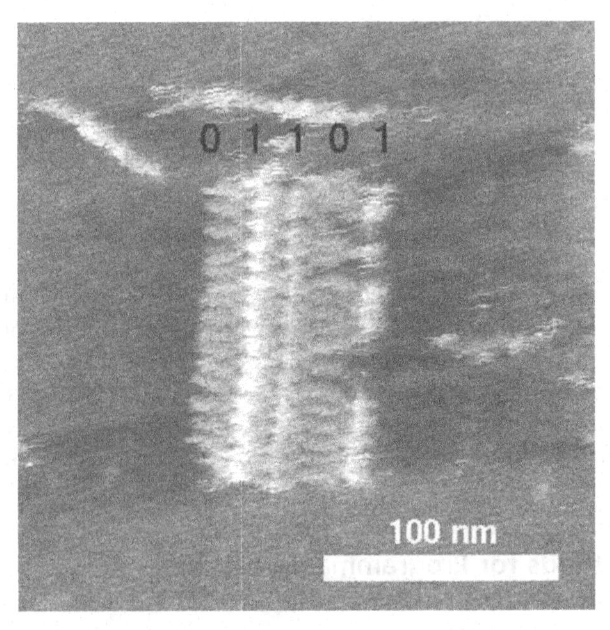

Figure 13.9. Use of scaffold strands for programmable assembly of barcode patterned 2D DNA lattices.

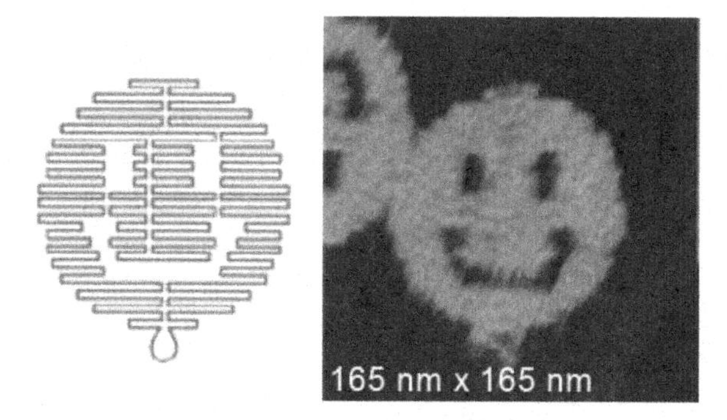

Figure 13.10. Use of DNA origami for programmable assembly of patterned 2D DNA lattices.

patterning to be determined by a distinct scaffold strand, defining an arbitrary 2D pixel image. A spectacular experimental demonstration of patterning via scaffold strand is also known as DNA origami [18]. This approach makes use of a long strand of scaffold ssDNA (such as from the sequence of a viral phage) that has only weak secondary structure and no long repeated or complementary subsequences. A large number of relatively short "staple" ssDNA sequences to this, with subsequences complementary to certain subsequences of the scaffold ssDNA. These staple sequences are chosen so that they bind to the scaffold ssDNA by hybridization, and induce the scaffold ssDNA to fold together into a DNA nanostructure. A schematic trace of the scaffold strand is shown in Figure 13.10 (left panel) and an AFM image of the resulting assembled origami is shown in Figure 13.10 (right panel). This landmark work of Rothemund [18] very substantially increases the scale of 2D patterned assemblies to hundreds of molecular pixels (e.g., stem loops viewable via AFM) within square area less than 100 nm on a side. In principal this "molecular origami" method with staple strands can be used to form arbitrary complex 2D patterned nanostructures as defined.

13.6.4. Use of Computational Assembly for Programmable Assembly of Patterned 2D DNA Lattices

Another very promising method is to use the DNA tile's pads to program a 2D computational assembly. Recall that in the 1970s, computer scientists showed that any computable 2D pattern can be so assembled. [11] and [12] have experimentally demonstrated two distinct and quite interesting 2D computational assemblies, and furthermore provided AFM images of the resulting nanostructures as illustrated in Figure 13.11.

In Figure 13.11, each tile determines and outputs to neighborhood pads the XOR of two of the tile pads. Example AFM images of the assembled structures are shown in the three panels (scale bars = 100 nm).

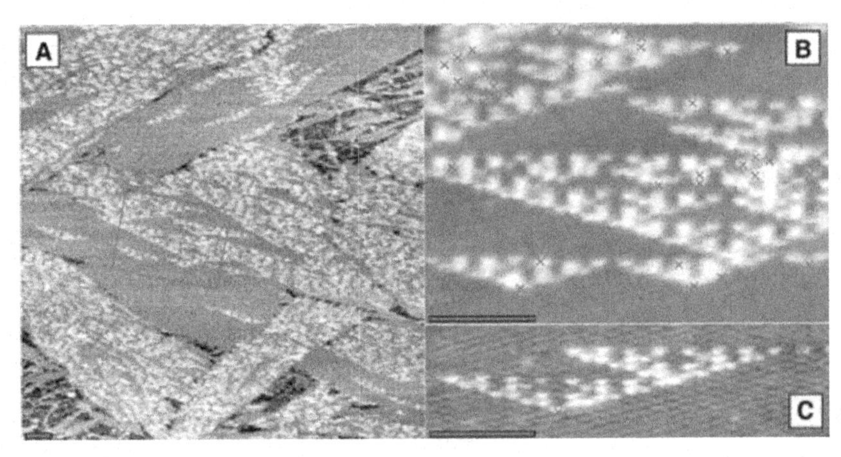

Figure 13.11. A modulo-2 version of Pascal's Triangle, known as the Sierpinski Triangle [11]. (Figure adapted with permission from [11].)

Figure 13.12 gives Winfree's design for a self-assembled binary counter [12], starting with 0 at the first row, and on each further row being the increment by 1 of the row below. The pads of the tiles of each row of this computational lattice were designed in a similar way to that of the linear XOR lattice assemblies described in the prior section. The resulting 2D counting lattice is found in MUX designs for address memory, and so this patterning may have major applications for patterning molecular electronic circuits.

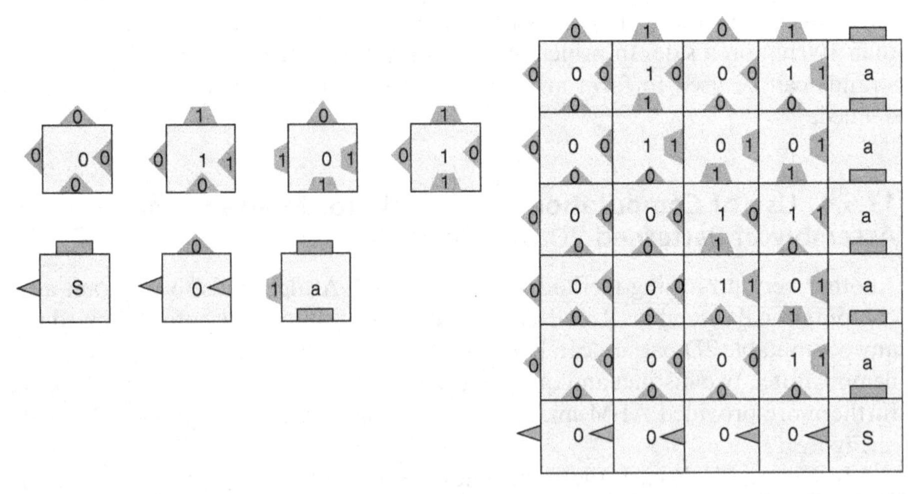

Figure 13.12. A modulo-2 version of Pascal's Triangle (known as the Sierpinski Triangle) [11].

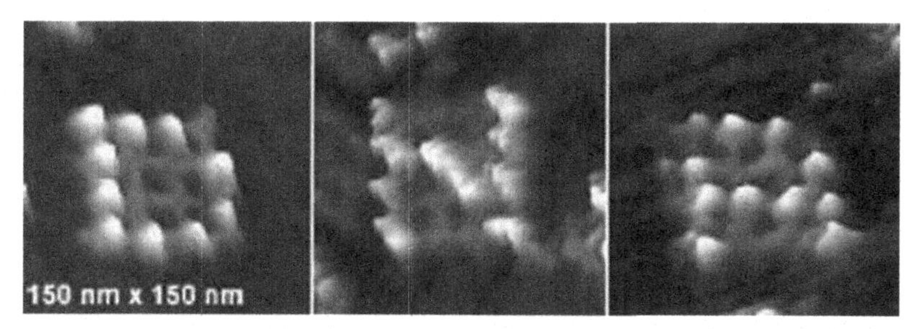

Figure 13.13. 2D patterns by hierarchical assembly. (Figure adapted with permission from [19].)

13.6.5. Use of Hierarchical Assembly for Programming of Patterned 2D DNA Lattices

A further approach is to assemble DNA lattices in a hierarchical fashion [19]. Figure 13.13 gives three examples of preprogrammed patterns displayed on addressable DNA tile lattices. Tiles are assembled prior to mixing with other preformed tiles. Unique ssDNA pads direct tiles to designed locations. White pixels are "turned on" by binding a protein (streptavidin) at programmed sites as determined in the tile assembly step by the presence or absence of a small molecule (biotin) appended to a DNA strand within the tile. Addressable, hierarchical assembly has been demonstrated for only modest size lattices to date, but has considerable potential particularly in conjunction with the above methods for patterned assembly.

13.7. ERROR CORRECTION AND SELF-REPAIR AT THE MOLECULAR SCALE

In many of the self-assembled devices described here, there can be significant levels of error. These errors occur both in the synthesis of the component DNA and in the basic molecular processes that are use to assemble and modify the DNA nanostructures, such as hybridization and the application of enzymes. There are various purification and optimization procedures developed in biochemistry for minimization of many of these types of errors. However, there remains a need for development of methods for decreasing the errors of assembly and for self-repair of DNA tiling lattices comprised of a large number of tiles. A number of techniques have been proposed for decreasing the errors of a DNA tiling assembly by providing increased redundancy.

Winfree [20] developed a "proofreading" method of replaced each tile with a subarray of tiles that provide sufficient redundancy to quadratically reduce errors, but increase the size of the assembly. This scheme is given in Figure 13.14 (top), with the original tiles in Figure 13.15 (top) and the modified tiles in Figure 13.14 (bottom).

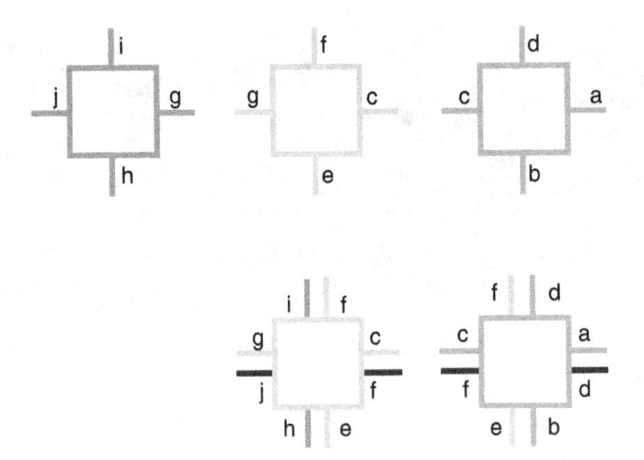

Figure 13.14. Winfree's error-resilient tiles.

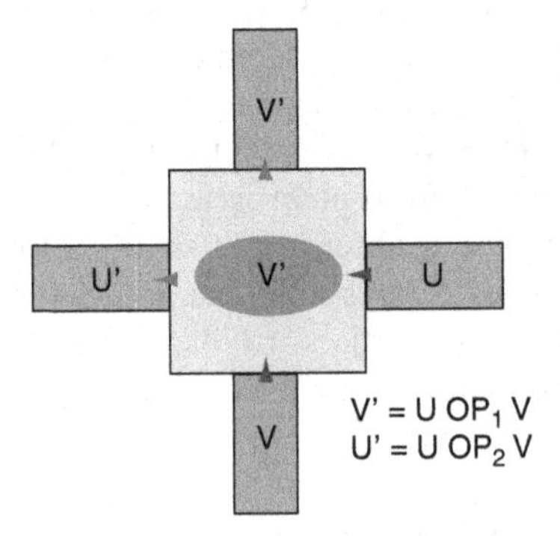

Figure 13.15. Reif's compact error-correction scheme: an original tile.

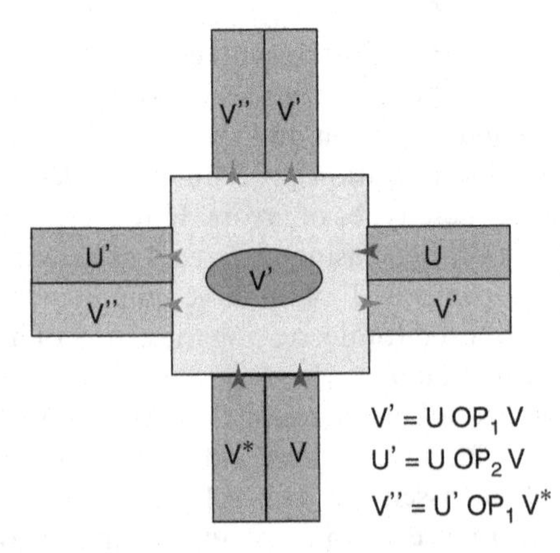

Figure 13.16. Reif's compact error-correction scheme: the modified tile.

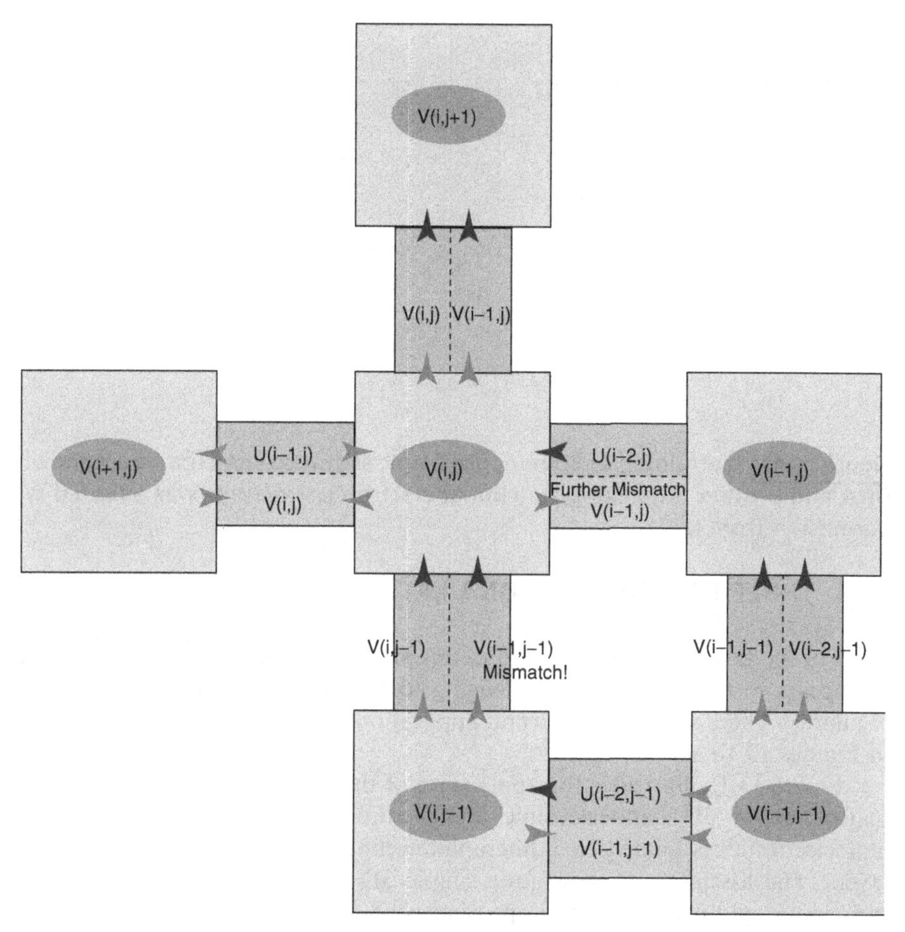

Figure 13.17. Reif's compact error-correction scheme: the error propagation process.

Reif et al. [21] provides a more compact method for decreasing assembly errors. This compact method takes each original tile (as illustrated in Fig. 13.15) and modifies the pads of each tile, to form a modified error-resultant tile (as illustrated in Fig. 13.16).

The result is that essentially each tile executes both the original computation required at that location as well as the computation of a particular neighbor. An illustration of the error propagation process is illustrated in Figure 13.17.

This provides a quadratic reduction of errors without increasing the assembly size. The experimental testing of these and related error-reduction methods is ongoing. It seems possible that other error-correction techniques developed in computer science may also be utilized.

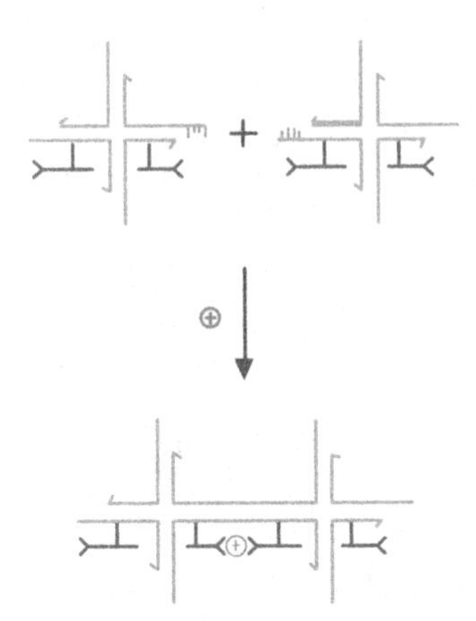

Figure 13.18. Scaffolding of 3D nanoelectronic architectures. (The figure showing DNA and protein organizing functional electronic structures was adapted with permission from [22].)

13.8. THREE-DIMENSIONAL DNA LATTICES

Most of the DNA lattices described in this article have been limited to 2D sheets. It appears to be much more challenging to assemble 3D DNA lattices of high regularity. There are some important applications if this can be done, as described in Figures 13.18 and 19.

Figure 13.18 illustrates the application of three-dimensional DNA lattices to scaffolding of 3D nanoelectronic architectures. The density of conventional nanoelectronics is limited by lithographic techniques to only a small number of layers. The assembly of even quite simple 3D nanoelectronic devices, such as memory, would provide much improvement in density.

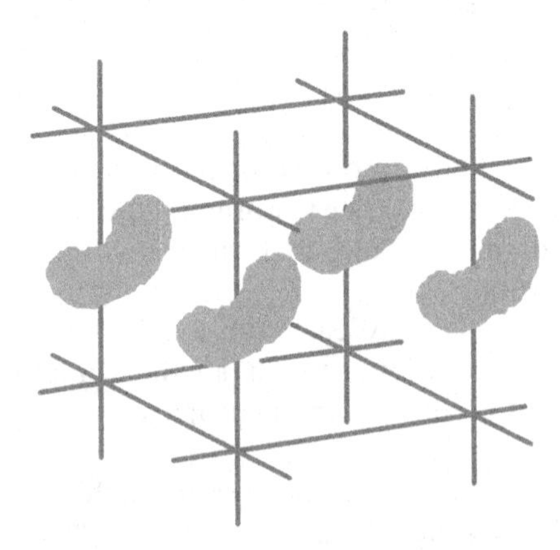

Figure 13.19. Application of 3D DNA lattices to scaffolding of proteins into regular 3D arrays.

Figure 13.19 illustrates the application of three-dimensional DNA lattices to scaffolding of proteins into regular 3D arrays. It has been estimated that at least one half of all natural proteins cannot be readily crystallized and have unknown structure; determining these structures would have a major impact in the biological sciences. Suppose a 3D DNA lattice can be assembled with sufficient regularity and with regular interstices (say, within each DNA tile comprising the lattice). Then a given protein might be captured within each of the lattice's interstices, allowing it to be in a fixed orientation at each of its regularly spaced locations in 3D. This would allow the protein to be arranged in 3D in a regular way to allow for X-ray crystallography studies of its structure. This visionary idea is due to Seeman. So far there has been only limited success in assembling 3D DNA lattices, and they do not yet have the degree of regularity (down to 2 or 3 Å) required for the envisioned X-ray crystallography studies. However, given the successes up to now for 2D DNA lattices, this seems eventually achievable.

13.9. AUTONOMOUS MOLECULAR TRANSPORT DEVICES SELF-ASSEMBLED FROM DNA

There are a number of other tasks that can be done at the molecular scale that would be considerably aided by this technology. For example, many molecular-scale tasks may require the transport of molecules. The cell uses protein motors fueled by ATP to do this. While a number of motors composed of DNA nanostructures have been demonstrated, they do not operate autonomously, and instead require some sort of externally mediated changes (such as temperature-cycling) on each work-cycle of the motor.

Peng et al. [23] experimentally demonstrated the first autonomously operating device composed of DNA providing transport as described in Figures 13.20 and 21.

First a linear DNA nanostructure (the "road") with a series of attached ssDNA strands (the "steps") was self-assembled, as illustrated in Figure 13.20. A fixed-length segment of DNA helix (the "walker") with short sticky ends (it's

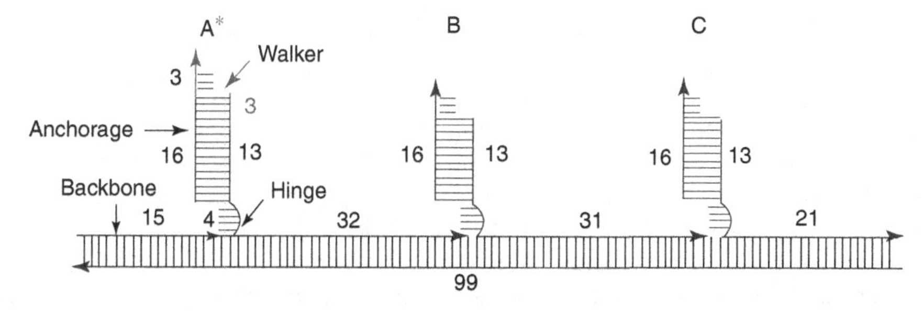

Figure 13.20. Design of autonomous molecular transport devices self-assembled from DNA.

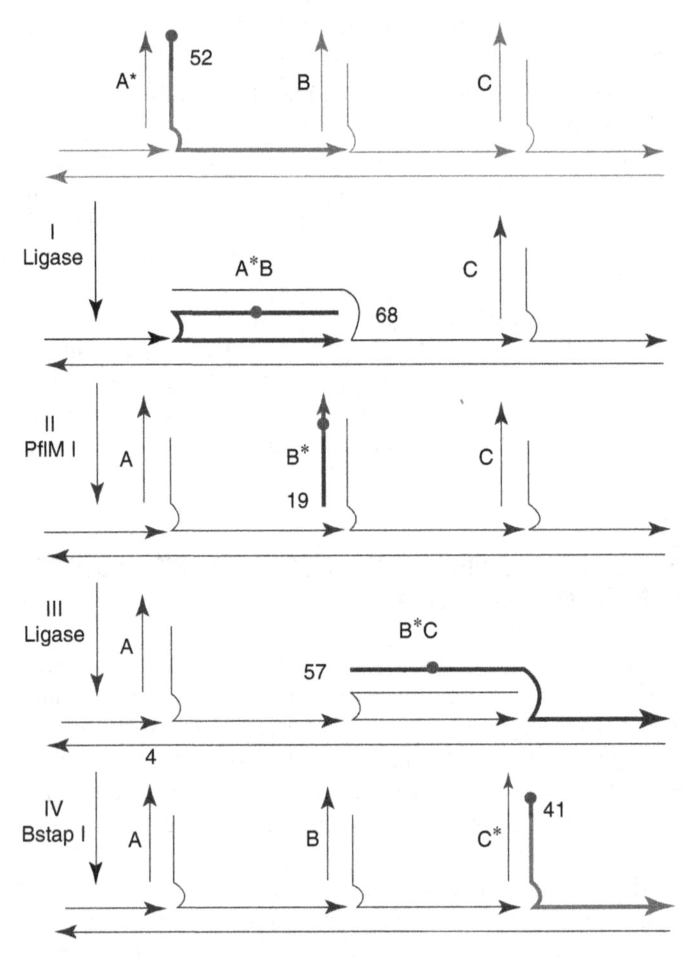

Figure 13.21. Operation of autonomous molecular transport devices self-assembled from DNA.

"feet") assembled on one end of the road, with the feet of the walker hybridized to the first two steps of the road.

Then, as illustrated in Figure 13.21, the walker proceeded to make sequential movement along the road, where at the start of each step, the feet of the walker are hybridized to two consecutive steps of the road. Then a restriction enzyme cuts the DNA helix where the backward foot is attached, exposing a new sticky-end forming and a new replacement foot that can hybridize to the next step that is free, which can be the step just after the step where the other foot is currently attached. A somewhat complex combinatorial design for the sequences composing the steps and the walker ensures that there is unidirectional forward motion along the road.

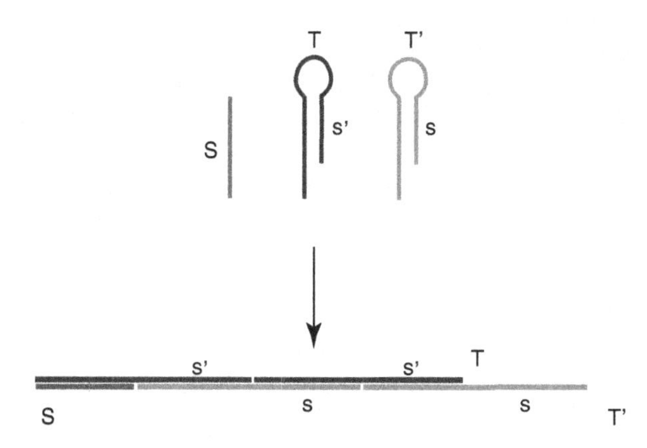

Figure 13.22. Autonomous molecular cascade for signal amplification.

13.10. AUTONOMOUS MOLECULAR CASCADE DEVICES FOR MOLECULAR SENSING

Another type of task that can be done at the molecular scale (and would be considerably aided by this technology) is to sense a particular molecule and amplify a response signal to achieve detection with extremely few starting target molecules. There are a number of protocols, such as PCR, used to detect and amplify a given sequence of DNA, but most of these require a repeated temperature-cycling and so are not autonomous [24]. Demonstrated an autonomous system using DNA nanostructures that initiated a hybridization cascade reaction in response to detection of a given ssDNA sequence S. It is described in Figure 13.22.

The experiment made use of multiple copies of two distinct DNA nanostructures T and T′ that are initially added to a test tube.

When ssDNA sequence S is added to the test tube, S initially has a hybridization reaction with a part of T, thus exposing a second ssDNA S′ that had been previously hidden within the nanostructure of T.

Next, S′ has a hybridization reaction with a part of T′, thus exposing a second copy of S that had been previously hidden within the nanostructure of T′. That other copy of S then repeats the process of other similar (but so far unaltered) copies of T and T′, allowing a cascade effect to occur completely autonomously. Such autonomous molecular cascade devices have applications to a variety of medical applications, where a larger response (e.g., a cascade response) is required in response to one of multiple molecular detection events.

13.11. CONCLUSIONS

We have provided an overview of a number of methods for assembling computational patterns within the molecular fabric of DNA lattices. We have surveyed the varied interdisciplinary techniques for carefully designing and controlling these self-assembly processes. Many of these self-assembly processes are computational-based and programmable; it seems likely that interdisciplinary techniques will be essential to other emerging subfields of nanoscience and biomolecular computation. We have also discussed a number of key challenges still confronting this emerging field on DNA nanostructures, including the need for error correction and the challenge and applications of constructing three-dimensional DNA lattices.

ACKNOWLEDGMENTS

This research in this chapter was supported by NSF grants CCF-0523555, CCF-0432038, and CCF-0432047. Thanks to N. Gopalkrishnan, U. Majumder, and S. Sahu for their very useful comments on this chapter.

REFERENCES

1. T. H. LaBean, E. Winfree, and J. H. Reif. Experimental progress in computation by self-assembly of DNA tilings. Proceedings of the 5th DIMACS Workshop on DNA Based Computers. in: E. Winfree and D. K. Gifford, (eds), DIMACS Series in Discrete Mathematics and Theoretical Computer Science, *Volume 54*. Cambridge, MA: MIT Press, pp 123–140.
2. Z. Deng, Y. Chen, Y. Tian, and C. Mao. A fresh look at DNA nanotechnology. In: J. Chen, N. Jonoska and G. Rozenberg, (eds), *Science and Computation*. Berlin: Springer, 2006, pp 23–34.
3. J. H. Reif and T. H. LaBean. Autonomous programmable biomolecular devices using self-assembled DNA nanostructures. *Communications of the ACM (CACM), Special Section: New Computing Paradigms*. T. Munakata, editor. To appear 2007.
4. N. C. Seeman. Nanotechnology and the double helix. *Scientific American*, 290(6): pp 64–75, June 2004.
5. L. Adleman. Molecular computation of solutions to computational problem. *Science*, 266: pp 1021–1024, 1994.
6. L. Adleman. Computing with DNA. Scientific American, 279(2): pp 34–41, Aug 1998.
7. E. Winfree, X. Yang, and N. C. Seeman. Universal computation via self-assembly of DNA: some theory and experiments. In: *DNA Based Computers II*: pp 191–213, 1998.
8. T. H. LaBean, H. Yan, J. Kopatsch, F. Liu, E. Winfree, J. H. Reif, and N. C. Seeman. The construction, analysis, ligation and self-assembly of DNA triple crossover complexes. *Journal of American Chemistry Society*, 122: pp 1848–1860, 2000.
9. C. Mao, T. H. LaBean, J. H. Reif, and N. C. Seeman. Logical Computation using algorithmic self-assembly of DNA triple-crossover molecules. *Nature*, 407: pp 493–495, Sep 28, 2000.
10. H. Yan, L. Feng, T. H. LaBean, and J. Reif. DNA nanotubes, parallel molecular computations of pairwise exclusive-or (XOR) using DNA "string tile" self-assembly. *Journal of American Chemistry Society*, 125(47): pp 14246–14247, 2003.

11. P. W. K. Rothemund, N. Papadakis, and E. Winfree. Algorithmic self-assembly of DNA Sierpinski triangles. *PLoS Biology*, 2(12): Dec 2004.
12. R. D. Barish, P. W. K. Rothemund, and E. Winfree. Two computational primitives for algorithmic self-assembly: copying and counting. *Nano Letters*, 5(12): pp 2586–2592, 2005.
13. E. Shapiro and Y. Benenson. Bringing DNA computers to life. *Scientific American*: pp 45–51, May 2006.
14. H. Yan, S. Ha Park, G. Finkelstein, J. H. Reif, and T. H. LaBean. DNA-templated self-assembly of protein arrays and highly conductive nanowires. *Science*, 301: pp 1882–1884, Sep 26, 2003.
15. S. H. Park, M.W. Prior, T. H. LaBean, and G. Finkelstein. Optimized fabrication and electrical analysis of silver nanowires templated on DNA molecules. *Applied Physics Letters*, 89: p. 033901, 2006.
16. T. H. LaBean, K. V. Gothelf, and J. H. Reif. Self-Assembling DNA Nanostructures for Patterned Molecular Assembly In: C. A. Mirkin and C. M. Niemeyer, (eds), *Nanobiotechnology*. Hoboken, NJ: Wiley, 2006.
17. H. Yan, T. H. LaBean, L. Feng, and J. H. Reif. Directed nucleation assembly of barcode patterned DNA lattices. *Proceedings of the National Academy of Sciences*, 100(14): pp 8103–8108, July 8, 2003.
18. P. W. K. Rothemund. Folding DNA to create nanoscale shapes and patterns. *Nature*, 440: pp 297–302, March 16, 2006.
19. S. H. Park, C. Pistol, S. J. Ahn, J. H. Reif, A. R. Lebeck, C. Dwyer, and T. H. LaBean. Finite-size, fully addressable DNA tile lattices formed by hierarchical assembly procedures. *Angewandte Chemie International Edition*, 45(5): pp 735–739, Jan 23, 2006.
20. J. H. Reif, S. Sahu, and P. Yin. Compact Error-Resilient Computational DNA Tiling Assemblies. In: N. Jonoska, (ed), *Nanotechnology: Science and Computation*. Springer Verlag Series in Natural Computing. Berlin: Springer, pp 79–104, 2006 .
21. E. Winfree and R. Bekbolatov. Proofreading tile sets: error correction for algorithmic self-assembly. *DNA Computers 9, LNCS*, 2943: pp 126–144, 2004.
22. B. H. Robinson and N. C. Seeman. *Protein Engineering*, 1: pp 295–300, 1987.
23. P. Yin, Hao Yan, X. G. Daniel, A. J. Turberfield, and J. H. Reif. A unidirectional DNA walker moving autonomously along a linear track. *Angewandte Chemie International Edition*, 43(37): pp 4906–4911, Sep 20, 2004.
24. R. M. Dirks and N. A. Pierce. Triggered amplification by hybridization chain reaction. *Proceedings of the National Academy of Sciences*, 101(43): pp 15275–15278, 2004.

14

DNA SEQUENCE MATCHING AT NANOSCALE LEVEL

Mary Mehrnoosh Eshaghian-Wilner, Ling Lau, Shiva Navab, and David D. Shen

In this chapter, we show how to form partial-order multiple-sequence alignment graphs on two types of reconfigurable mesh architectures. The first reconfigurable mesh is a standard microscale one that uses electrical interconnects, while the second one can be implemented at a nanoscale level and employs spin waves for interconnectivity. We consider graph formations for two cases. In one case, the number of distinct variables in the data sequences is constant. In the other case, it can be as much as $O(N)$. We show that given $O(N)$ aligned sequences of length L, we can combine the sequences to form a graph in $O(1)$ time, using either architecture if there is a constant number of distinct variables in the sequence. Otherwise, it will take $O(1)$ time if we use the spin-wave model and $O(N)$ time if we use the standard VLSI version.

14.1. INTRODUCTION

Research in molecular biology has been moving at an astonishing pace in recent years. The rapid accumulation of biological data has necessitated more robust databases and data processing algorithms. Consequently, new areas of computational biology are being created rapidly, combining the biological and informational sciences. To illustrate the complexities of some of the problems in computational biology, let us consider the genetic material in all living organisms, DNA. DNA is a polymer of nucleotides, where each nucleotide contains one of four bases: A, G, T, or C. An average gene in a human genome has 30,000

The authors of this chapter are listed alphabetically.

Bio-Inspired and Nanoscale Integrated Computing. Edited by Mary Mehrnoosh Eshaghian-Wilner

basepairs. With 30,000 genes estimated in each human genome, there are roughly three billion basepairs. A challenging task in computational biology deals with the finding of the "best" alignment amongst a given set of sequences. This is denoted as multiple sequence alignment (MSA). The domain of MSA is not limited to DNA sequences. It could be for proteins [1] and could become even more complex in sequence splicing problems [2].

The alignment of two sequences of length L can be solved via dynamic programming in $O(L^2)$ time. Extending it to N sequences, it will take $O(L^N)$ time, and in some cases the MSA problem is NP-complete [3]. Recent work includes CLUSTLW, T-COFFEE, etc. In one of the pioneering papers by Chris Lee, he demonstrates the advantages of applying a graph theoretical approach to the MSA problem (partial ordered alignment); and many graph-based techniques have been proposed since then as explained in [3–5], and [6]. Lee proposed the partial-ordered multiple-sequence alignment (PO-MSA) and demonstrated its application in pair-wise alignment, as well as its application in progressive pair-wise alignment [4, 5]. The results are comparable or even faster than some of the best known algorithms.

In this chapter, we study the complexities of the process of forming a PO-MSAG. Assuming there are $O(N)$ sequences of length $O(L)$ that are already aligned, it will take at least $O(NL)$ time to simplify the sequences sequentially into a PO-MSAG. Here, we present a set of fast and parallel algorithms to generate a PO-MSAG for any given set of aligned sequences at the hardware architecture level. The hardware can then be fabricated as a low cost chip that can be used as a coprocessor with a more powerful processor.

We will consider two reconfigurable mesh architectures on which to implement our solution: the nanoscale spin wave and the microscale electrical VLSI. We separate the problem into two cases: data sequences with a fixed amount of data variations and those with an arbitrary amount up to $O(N)$. In the fixed variations case, the spin wave and the VLSI architectures can construct the PO-MSAG in $O(1)$ time, as shown in [7, 8]. As for arbitrary $O(N)$ variations, the spin-wave architecture will construct the PO-MSAG in $O(1)$ time whereas the VLSI one takes $O(N)$ time. An example of an input sequence set is shown in Figure 14.1.

A - C - G - T - T - A - C - T
A - G - T - T - G - G - G - C - T
(unsorted)

⇓

A - C - G - T - T - A - * - * - C - T
A - * - G - T - T - G - G - G - C - T
(sorted)

Figure 14.1. Example of a partial-order multiple-sequence alignment.

The rest of this chapter will be organized as follows: Section 14.2 gives some background information on reconfigurable meshes; Section 14.3 shows how the two architectures as well as the multiscale architecture can solve the MSA problem with a fixed amount of data variables; Section 14.4 shows how to solve the problem with up to $O(N)$ data variables; and the final sections will discuss our concluding remarks and list our references.

14.2. PRELIMINARIES

For some basic information about reconfigurable meshes and our reasons for choosing this architecture to implement graph formations of partial-order multiple-sequence alignments algorithm, please refer to Chapter 7.

14.3. PO-MSAG FORMATION FOR CONSTANT VARIATION

The sequence output in this section is limited to at most a finite (constant) number of variables. To make the problem more concrete, we focus on DNA sequences, which have four variables (nucleotides) and possibly a special character, "*", representing a blank space, as our example (Figure 14.2).

14.3.1. Mapping on the Spin-Wave Architecture

INITIAL MAPPING. In our first step, we map the given data sequences (length: L; number of sequences: N) onto $N \times L$ processors (we shall address these processors as nodes) each with its own memory capable of storing any six of the five possible data variables (in this case, the nucleotides A, T, G, C, and the blank *) by placing each data variable of all sequences into the first memory slot of each respective processor. This takes constant time $O(1)$. So, for example, the first four nodes in rows and columns one and two will be mapped into memory as follows (Figure 14.3):

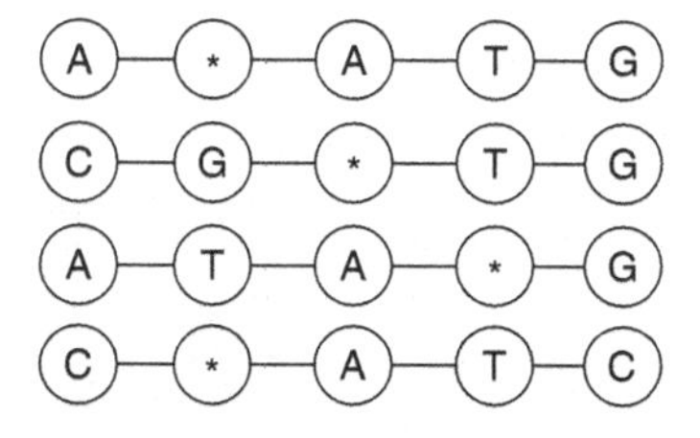

Figure 14.2. Example of a PO-MSA output of DNA.

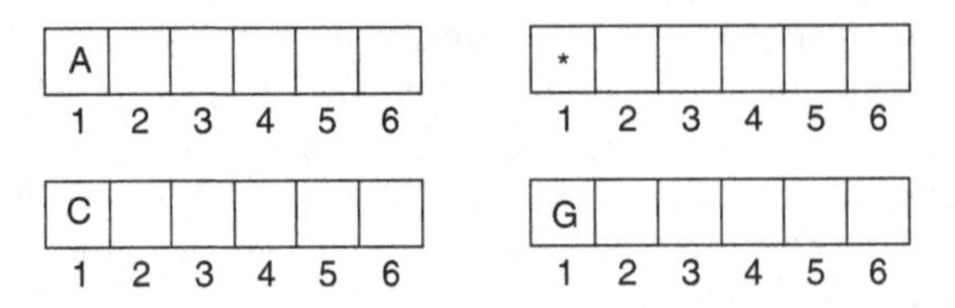

Figure 14.3. Example of the memory space of the nodes.

Neighbor Recording. Next, we let each node store its right neighbor in its second memory slot. This step still takes constant time since retrieving the information is $O(1)$ and also since all nodes are performing this task simultaneously on parallel processors. So once again, the first four nodes in rows and columns one and two will be mapped into memory as shown here (Figure 14.4):

Node Elimination. Now comes the grouping of the similar nodes in each column.

A. We start by turning off (disconnect the channel) all the switches above each node to force the signal sent by each node downwards.
B. Then we let all nodes simultaneously send out a signal downward in its own frequency and turn on all switches (reconnecting the channel), thereby lettings all signals pass through the bus. Then the nodes that receive a signal in their own frequencies will be repeated nodes. We disabled these nodes.

One node (the first or uppermost) of each type in that column that did not receive a signal remains active and represents the rest of its type. To better understand this procedure, we can take a look at the third column of our example. Each of the A nodes will first send out a signal downward in Frequency A. Then, there will be one—the first or uppermost—A node left that has not received any signals in Frequency A. The rest of the A's will be obsolete, since the rest will now be represented by that first A node. So for example, just for the A nucleotide in the third column, this process will be as follows (Figure 14.5):

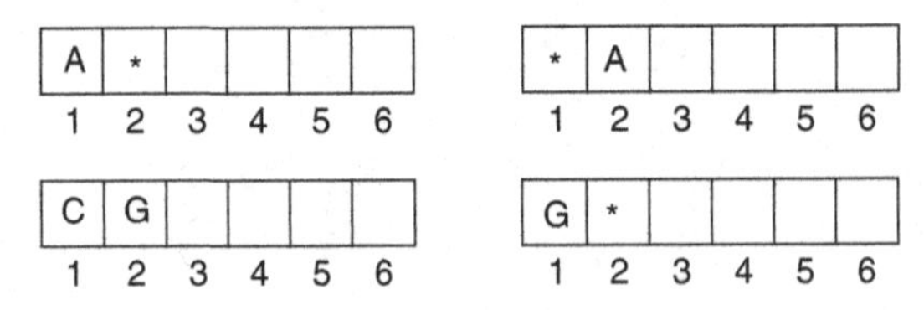

Figure 14.4. Memory spaces of the nodes after the set-up stage.

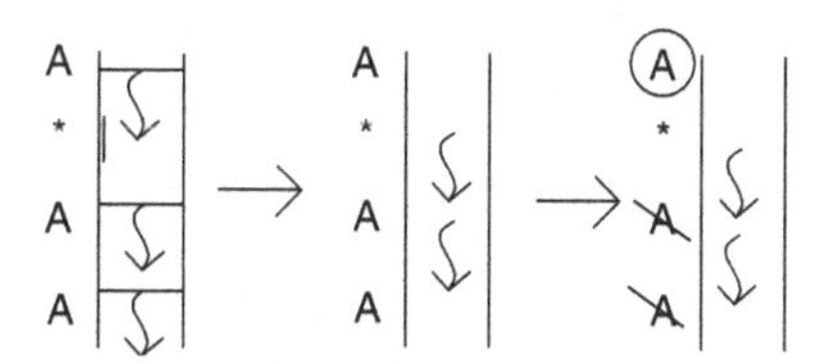

Figure 14.5. Disabling nodes using spin wave.

We perform this procedure with all nodes of all types simultaneously, each type on its own frequency. This process also has a run time of $O(1)$ because in the spin-wave mesh, similar nodes can communicate to each other in constant time.

MEMORY UPDATE. Now, we must update the memory of the remaining nodes.

A. We first turn off all switches below each node to force all sent signals upward. This process takes constant time.
B. Then we ask all disabled nodes whether or not their right neighbor (at memory slot 2) is an A. If so, that disabled node will send a signal upward to its respective representative node and turn on all switches. This part still runs at constant time since all disabled nodes can perform this step simultaneously.
C. Finally, the representative node receives that signal and checks its right neighbor memory (memory slots 2–6) for an A. If an A is not already there, the representative node will place an A in the next available slot. This process take constant time since at most, the representative node will need to look all five of its right neighbor memory slots.
D. We repeat steps *A*, *B*, and *C* four more times, each time requesting the disabled nodes to check for a different right neighbor: T, G, C, and *. Since steps *A*, *B*, and *C* are constant, 4*$O(1)$ is still $O(1)$. All disabled nodes of each type can perform these steps simultaneously because each type can communicate in its own frequency. Thus the overall performance of the Memory Update stage is still $O(1)$ (Figure 14.6).

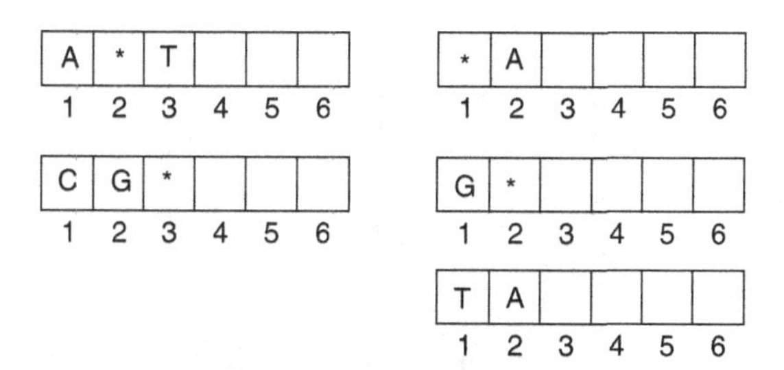

Figure 14.6. Memory space after the memory update stage.

14.3.2. Mapping on the Electrical VLSI Architecture

Initial Mapping. To implement our algorithm using electrical VLSI mesh, the initial mapping to memory of each data variable is identical to the spin-wave technique and will thus run in constant time, $O(1)$.

Neighbor Recording. Next we store each node's right neighbor in its second memory slot. This is still the same as in the spin-wave architecture, taking $O(1)$ time.

Node Elimination. Here is the main difference between spin-wave and VLSI meshes. In the electrical VLSI, only one signal can be sent through the channel (wire) at a time. However we can still perform the Node Elimination stage the same way as we did previously, as long as we do it for one type of node. We can then sequentially repeat this procedure for the other four node types. Since it took constant time for one node type, it will still take constant time for four more node types.

We use the third column again as our example. We turn off all switches to force sent signals downward. We then have all the A nodes send a signal downward and then we turn on all switches to let the signals pass. The A nodes that do receive a signal are disabled, while the one A node that does not receive any signals (the first and uppermost) is chosen as the representative node (Figure 14.7).

We repeat this process sequentially with each of the other variables (the T, G, and C nucleotides, and the blank *). This process still runs in constant time.

Memory Update. The last stage of updating all representative nodes' right neighbor memory with its respective disabled nodes' right neighbors is the same as the one shown in the spin-wave mesh. The only difference is that since each node type does not have its own frequency, everything must be done sequentially. However since there are only five node types, the entire stage is still completed in constant time. Thus, the overall runtime of the graph formation using electric VLSI will also run in constant time, $O(1)$.

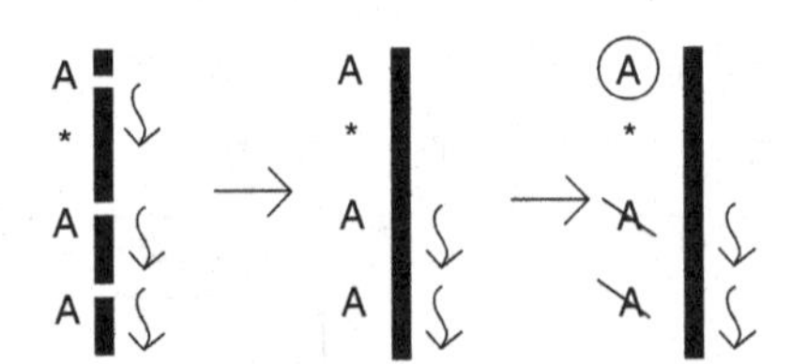

Figure 14.7. Disabling repeated nodes using VLSI.

14.4. PO-MSAG FORMATION FOR O(N) VARIATION

The nature of the problem is the same (N sequences of length L on a reconfigurable mesh). However, the data considered in this section can have as many as N different variations.

14.4.1. O(N) Variables on the Spin-Wave Version

Before we begin, we must make some assumptions. First, the size of our spin-wave reconfigurable mesh must be increased to $2N \times 2L$. This is still acceptable because it is only a constant increase. Next, we must assume that all nodes can access their own row index very quickly in constant time. Finally, we assume that when the nodes are mapped initially, they are done so in such a way that there is an empty column to the right of each data column. These empty columns will be known as graph columns.

Initial Mapping. This step is exactly the same as shown previously.

Neighbor Recording. This step is also exactly the same as shown previously.

Node Elimination. This step is exactly the same as shown previously as well.

Graph Formation. We will form the graph and store it in the graph columns. The overall idea here is that we place each representative node in the graph column to its right and all its right connections beneath it.

A. Counting disabled nodes. When we store the representative nodes in the graph columns, we need to leave the correct number of empty graph column slots beneath each representative node. These empty slots will be used to store the right neighbors of the representative nodes as well as of that of their respective disabled nodes. Therefore, we must first count the number of disabled nodes of each type within a column. To do this, we utilize a property of spin waves: superposition. The main idea behind superposition is that when nodes transmit signals in the same frequency, the signals will superpose in amplitude when they meet. So we can have all disabled nodes of each type transmit a signal with amplitude 1 in their own frequencies and have their respective representative node receive in its own frequency. We tell the representative node to keep receiving for a constant, certain amount of time (the time taken for a signal to propagate through the entire channel). When the representative is done receiving, it will have received the total amplitude of all the superposed signals sent by its disabled nodes. This amplitude received by the representative node is the number of its disabled nodes. We use our third column again as an example. The two disabled A nodes send up signals with amplitude 1 in their own frequency. The superposed signal of amplitude 2 is received by the representative A (Figure 14.8).

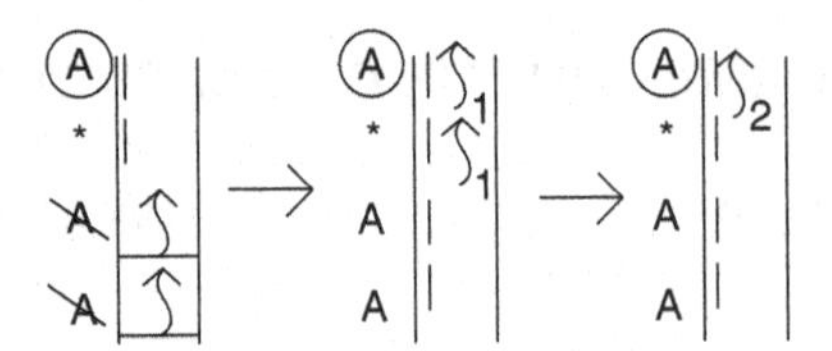

Figure 14.8. Counting disabled nodes using spin wave.

B. Indexing representative nodes. Now, we must find the appropriate location to place each representative node in its right graph column. In other words, we must find each representative node's row index in the graph column. We begin by telling all the representative nodes (disregarding its type) within a column to communicate with each other, through a common frequency (i.e., Frequency Active). Now we find the first or uppermost representative node of any type in a column in constant time by using a method similar to the Node Elimination procedure. Then we turn on all switches in the channel and have each representative node send a signal with an amplitude equal to two (in order to leave two extra slots in the graph column for itself and its own right neighbor) plus the number of disabled nodes of that type (obtained from the previous step), and immediately turn off the switch above it to direct its generated wave downwards. Note that for synchronization purposes, each representative node generates its spin wave signal after a short time to compensate the spin-wave propagation delay in the ferromagnetic channel. The time period should be equal or greater than the distance between adjacent nodes over speed of spin waves, which is in the order of $10^{-9}/10^{4} = 10^{-13}$. Considering the fact that the frequency is in the order of GHz or even THz, this whole process takes the same amount of time (constant time) as it would for one signal to travel the entire bus. Lastly, the amplitude of the superposed wave received by each representative node will be its row index in the graph column. We illustrate this using the third column of our example. The first representative node, A, sends a signal of amplitude 4 which is 2 (one for itself and one for its own right neighbor) + 2 (the number of disabled A nodes in that column). The next representative node, *, sends $2+0=2$ amplitude, which will be superposed with the first signal, making the total signal have an amplitude of 6. Since the A representative node did not receive any signals, its row index in the graph column will be 0 whereas the * representative node received a signal with an amplitude of 4 and thus that will be its row index in the graph column (Figure 14.9).

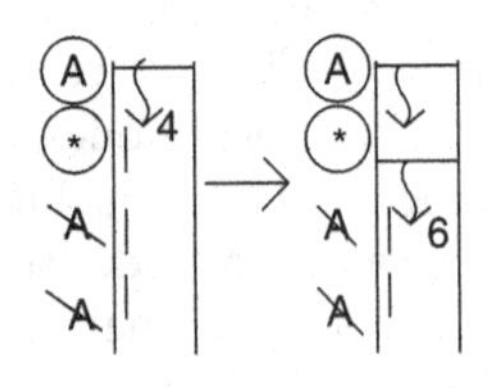

Figure 14.9. Calculating row indices in graph column using spin wave.

C. Indexing disabled nodes' right neighbors. Now, we find the row index in the graph column for each disabled node. We begin by turning on all switches and having each representative node sent a signal with an amplitude equal to its row index in the graph column plus two in its own frequency. Since all switches are on, the transmitted signals will travel in two opposite directions up and down the bus. Then each disabled node that receives a signal in its own frequency will take the amplitude of the received signal as its row index in the graph column, sends a signal with amplitude of 1, and then finally disables its receiver (by changing its receiving frequency to a dummy frequency), so that the node will not receive any signals from below. Similar to the previous procedure, this procedure is also done in constant time. We show how this works using the third column of our example. The representative nodes A and $*$ send signals of $0+2=2$ and $4+2=6$, respectively. Then the first disabled A node that receives the 2 sends a 1 making the total signal 3. Then last disabled A node that receives the 3 sends a 1 making the total signal 4. If there were another disabled A node, it would receive that 4. If there were a disabled * node, it would receive 6. The signals received by the disabled nodes are their respective row indices in the graph column. The first and last disabled A nodes will have row indices in the graph column of 2 and 3, respectively. Since each disabled node has its row index in the graph column now, it can retrieve its own right neighbor and copy it to the appropriate slot in the graph column based its row index in constant time.

D. Eliminating repeated right connections. Finally, we eliminate the repeated right neighbors under each representative node in each graph column. We first turn off the switch above each representative node and then perform the Node Elimination procedure under each representative node. This is done in constant time. Our graph is now formed and store in the graph columns (Figure 14.10).

Table 14.1 lists the time complexities of each stage of the process. The total time complexity of the process will be $O(1)$.

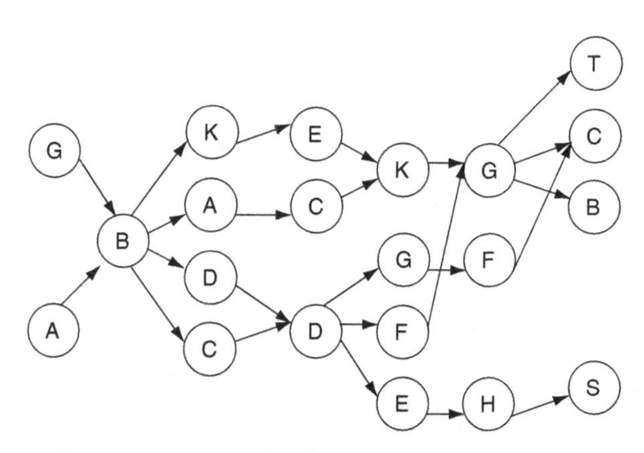

Figure 14.10. Result of the example sequence.

TABLE 14.1. Time Analysis: Spin Wave

1: Neighbor Recording	$O(1)$
2: Node Elimination	$O(1)$
3: Graph Formation	$O(1)$

14.4.2. O(N) Variables on the Electrical Version

The VSLI reconfigurable mesh architecture shares many similarities with the spin-wave reconfigurable mesh; however, multiple nodes cannot send messages at the same time. This poses major challenges when the application is communication-intensive. The graph formation algorithms we presented in the previous section, was optimized for a spin-wave reconfigurable mesh where we could benefit from different properties of waves such as simultanious multiple data transmission and superposition. The graph formation algorithm for the VLSI reconfigurable mesh is exactly the same as the one we presented for the constant number of variables. The steps are performed for each type of node sequentially.

Initial Mapping. As explained in the constant number of variables, this step is done in constant time.

Neighbor Recording. In this step, as explained previously, each node's right neighbor is stored in its second memory slot in constant time.

Node Elimination. The elimination process will again be handled via disabling the repeated nodes in each column, as discussed previously. This step is done sequentially for each type of node, so it takes $O(N)$ time.

Graph Formation. This step is also done exactly the same as memory update step in constant number of variations case. Since this step is done sequentially for each of $O(N)$ type of nodes, it is performed in $O(N)$.

Table 14.2 lists the time complexities of each stage of the process. The total time complexity of the process will be $O(1) + O(N) + O(N) = O(N)$.

To summarize the time complexities for both arbitrary variation and constant variation of the PO-MSAG formation problem solved with two reconfigurable mesh architectures is Table 14.3.

TABLE 14.2. Time Analysis: VLSI

1: Neighbor Recording	$O(1)$
2: Node Elimination	$O(N)$
3: Graph Formation	$O(N)$

TABLE 14.3. Summarized Time Analysis

	Arbitrary variation	Constant variation
VLSI	$O(N)$	$O(1)$
Spin Wave	$O(1)$	$O(1)$

14.4.3. PO-MSAG on a Hierarchical Multiscale Reconfigurable Mesh

The algorithm we presented works for a limited data size of $O(NL)$, for N sequences of length L. In some situations, however, the database size might be larger than this limit. For instance, the sequences might be longer than L or there might exist more than N sequences depending on the nature of the data and the size of the dataset. Let's assume we have a reconfigurable mesh of a fixed $N \times L$ size, and the dataset size exceeds the $N \times L$ limit of the reconfigurable mesh. We can approach this problem by using multiple $N \times L$ reconfigurable meshes and connecting each module in a multiscale fashion. We propose using the Hierarchical Multi-scale Reconfigurable Mesh described in Chapter 7 or [9] to solve this problem.

Assume there are $N \times N$ nanoscale reconfigurable meshes in this hierarchical multi-scale architecture. All these N^2 reconfigurable meshes can communicate simultaneously at the micro-level in unit time as long as there is only one read or write from or to each module. Data can now flow freely between reconfigurable mesh modules, so multiple reconfigurable meshes can be employed to solve PO-MSAG algorithm for a dataset larger than the $N \times L$ limit posed by each individual reconfigurable mesh.

In the algorithm explained in Section 14.3, during the phase in which the nodes record their right and left neighbors, the nodes at the edge of one reconfigurable mesh will have to record their left or right neighbors that are on another reconfigurable mesh through the microlevel interconnects. Since there can only be one receiving or transmitting at any given time, for the N or L nodes trying to transmit and receive data, the system time complexity degrades to the maximum of $O(N)$ and $O(L)$.

The inevitable time complexity degradation from $O(1)$ or $O(\log N)$ to $\max[O(N), O(L)]$ might seem to be disadvantageous. However, there can actually be an advantage in certain cases. Let us assume that we increase the size of our dataset to $A \times N$ sequences each having length $B \times L$, where A and B are some constants. We also assume that the reconfigurable meshes we normally use are only of size $N \times L$. So first of all, the advantage of the multiscale hierarchy is that there is no need to build a new reconfigurable mesh of size $AN \times BL$. In the case of the $AN \times BL$ dataset (or as a matter of fact, for any given size dataset), our multiscale architecture would simply just use multiple reconfigurable meshes of fixed size $N \times L$. However, without using the multiscale architecture, we would first have to obtain a new reconfigurable mesh or somehow increase the size of the standard one

TABLE 14.4. Summarized Time Analysis of Standard and Multiscale Architectures

	Arbitrary variation	Constant variation
STANDARD VLSI	$O(AN)$	$O(1)$
Spin Wave	$O(1)$	$O(1)$
MULTISCALE VLSI	$\max[O(N),O(L)]$	$\max[O(N),O(L)]$
Spin Wave	$\max[O(N),O(L)]$	$\max[O(N),O(L)]$

to $AN \times BL$, a costly operation. Also, there can be an advantage in the case where we use the VLSI reconfigurable for arbitrary variation. With an $AN \times BL$ dataset, we would expect the run-time to be $O(AN)$ for this case. However, in the multiscale architecture, this same case (VLSI with up to $O(AN)$ variation) will still run at $\max[O(N),O(L)]$. So as long as we make $A>1$ (when $N>L$) or $A>L/N$ (when $L>N$), we will have gained a run-time advantage (Table 14.4).

14.5. CONCLUSIONS

This chapter, demonstrated several techniques for forming graphs representing partial-order multiple-sequence alignment of a given set of N-aligned sequences, using two types of reconfigurable mesh architectures (the spin-wave version [10] and the VLSI version [11, 12]). It showed that for a constant number of variables, the run-times of both architectures are the same, $O(1)$. However for an arbitrary number of variables, the spin-wave architecture will have an $O(1)$ time complexity as opposed to an $O(N)$ time complexity using the other version. The algorithms described in this chapter belong to one of the first sets of algorithms currently under study in the area of bioinformatics. These results can be extended to large-scale graph databases. Furthermore, other applications will also benefit from such graphical representation of data, such as those in the areas of biological pathways and sequence splicing. Such areas will continue to demand even more efficient computing tools. The scale of our solution can be expanded by using clusters of mainframes to aid in the sequence data processing. Mapping tools such as Cluster-M, introduced in [13], can be used to handle the mapping and scheduling of graph data bases. All of these will serve as preliminary steps towards coming up with paradigms that could satisfy the computational needs of bioinformatics tasks.

REFERENCES

1. M. M. Eshaghian-Wilner. Integrated architectural solutions for protein sequence-structure alignment. In: Proceedings of the Sixth World Multi-Conference on Systemics, Cybernetics, and Informatics, SCI2002, July 2002.

2. X. Zhang and T. Kahveci. A New Approach for Alignment of Multiple Proteins. Pacific Symposium on Biocomputing, Maui. 11: pp 339–350, 2006.
3. C. Grasso and C. Lee. Combining partial order alignment and progressive multiple sequence alignment increases alignment speed and scalability to very large alignment problems. *Bioinformatics*, 20(10): pp 1546–1545, 2004.
4. C. Lee, C. Grasso, and M. F. Sharlow. Multiple Sequence Alignment Using Partial Order Graphs. *Bioinformatics,* Oxford, England, 18(3): 452–464, 2002.
5. Y. Ye and A. Godzik. Multiple flexible structure alignment using partial order graphs. *Bioinformatic*, 21(10): pp 2362–2369, 2005.
6. M. M. Eshaghian-Wilner, L. Lau, S. Navab, and D. Shen. Graph formations of partial-order multiple-sequence alignments using nano and micro-scale reconfigurable meshes. In: Proceedings of the 2006 International Conference on Computer Design CDES: pp 242–250, May 2006.
7. M. M. Eshaghian-Wilner, L. Lau, S. Navab, and D. Shen. Parallel graph formations of partial-order multiple-sequence alignments using nano-, micro-, and multi-scale reconfigurable meshes. Submitted to the *IEEE Transaction on NanoBioScience,* 2008.
8. M. M. Eshaghian-Wilner, A. Khitun, S. Navab, and K. Wang. Hierarchical Multi-Scale Architectures with Spin Waves. Proceedings of the 2006 International Conference on Computer Design CDES 2006, Las Vegas, Nevada: pp 220–226, June, 2006.
9. M. M. Eshaghian-Wilner, A. Khitun, S. Navab, and K. Wang. A nano-scale reconfigurable mesh with spin waves. In: Proceedings of the ACM International Conference on Computing Frontiers: pp 65–70, 2006.
10. M. M. Eshaghian-Wilner and R. Miller. The systolic reconfigurable mesh. *Journal of Parallel Processing Letters*, 14(3–4): pp 337–350, 2004.
11. Miller, Russ, V. K. Prasanna-Kumar, D. L. Reisis, and Q. F. Stout. Parallel computations on reconfigurable meshes. *IEEE Transactions on Computers*, 42(6): pp 678–692, 1993.
12. M. M. Eshaghian-Wilner. Mapping arbitrary heterogeneous task graphs onto arbitrary heterogeneous system graphs. *International Journal on Foundation of Computer Science*, 12(5): pp 599–628, 2001.

15

COMPUTATIONAL TASKS IN MEDICAL NANOROBOTICS

Robert A. Freitas, Jr.

Nanomedicine is the application of nanotechnology to medicine: the preservation and improvement of human health, using molecular tools and molecular knowledge of the human body. Medical nanorobotics is the most powerful form of future nanomedicine technology. Nanorobots may be constructed of diamondoid nanometer-scale parts and mechanical subsystems including onboard sensors, motors, manipulators, power plants, and molecular computers. The presence of onboard nanocomputers would allow *in vivo* medical nanorobots to perform numerous complex behaviors which must be conditionally executed on at least a semiautonomous basis, guided by receipt of local sensor data and constrained by preprogrammed settings, activity scripts, and event clocking, and further limited by a variety of simultaneously executing real-time control protocols. Such nanorobots cannot yet be manufactured in 2007 but preliminary scaling studies for several classes of medical nanorobots have been published in the literature. These designs are reviewed with an emphasis on the basic computational tasks required in each case, and a summation of the various major computational control functions common to all complex medical nanorobots is extracted from these design examples. Finally, we introduce the concept of nanorobot control protocols which are required to ensure that each nanorobot fully completes its intended mission accurately, safely, and in a timely manner according to plan. Six major classes of nanorobot control protocols have been identified and include operational, biocompatibility, theater, safety, security, and group protocols. Six important subclasses of theater protocols include locational, functional, situational, phenotypic, temporal, and identity control protocols.

Bio-Inspired and Nanoscale Integrated Computing. Edited by Mary Mehrnoosh Eshaghian-Wilner

15.1. INTRODUCTION

Nanotechnology is the engineering of molecularly precise structures and, ultimately, molecular machines. Nanomedicine [1, 2] is the application of nanotechnology to medicine: the preservation and improvement of human health, using molecular tools and molecular knowledge of the human body. Nanomedicine encompasses at least three types of molecularly precise structures: nonbiological nanomaterials, biotechnology materials and organisms, and nonbiological devices including inorganic nanorobotics. In the near term, the molecular tools of nanomedicine will include biologically active nanomaterials and nanoparticles having well-defined nanoscale features. In the mid-term (5–10 years), knowledge gained from genomics and proteomics will make possible new treatments tailored to specific individuals, new drugs targeting pathogens whose genomes have been decoded, and stem cell treatments. Genetic therapies, tissue engineering, and many other offshoots of biotechnology will become more common in therapeutic medical practice. We may also see biological robots derived from bacteria or other motile cells that have had their genomes re-engineered and reprogrammed, along with artificial organic devices that incorporate biological motors or self-assembled DNA-based structures for a variety of useful medical purposes.

In the farther term (2020s and beyond), the first fruits of molecular nanorobotics [3]—the most efficacious of the three classes of nanomedicine technology, though clinically the most distant and still mostly theoretical—should begin to appear in the medical field. These powerful therapeutic instrumentalities will become available once we learn how to design [4–10] and construct [3, 11–13] complete artificial nanorobots composed of diamondoid nanometer-scale parts [12a] and onboard subsystems including sensors [1a], motors [12b], manipulators [12b, 12c], power plants [1c], and molecular computers [1d, 12d]. The presence of onboard computers is essential because *in vivo* medical nanorobots will be called upon to perform numerous complex behaviors which must be conditionally executed on at least a semiautonomous basis, guided by receipt of local sensor data and constrained by preprogrammed settings, activity scripts, and event clocking, and further limited by a variety of simultaneously executing real-time control protocols.

The development pathway for diamondoid medical nanorobots will be long and arduous. First, theoretical scaling studies [4–10] are used to assess basic concept feasibility. These initial studies would then be followed by more detailed computational simulations of specific nanorobot components and assemblies, and ultimately full systems simulations, all thoroughly integrated with additional simulations of massively parallel manufacturing processes from start to finish consistent with a design-for-assembly engineering philosophy. Once nanofactories implementing molecular manufacturing capabilities become available, experimental efforts may progress from component fabrication and testing to component assembly and finally to prototypes and mass manufacture of medical nanorobots, ultimately leading to clinical trials. By 2007 there was some limited experimental work with microscale-component microrobots [14–18] but progress on nanoscale-component nanorobots remains largely at the concept feasibility stage. Since 1998, the author has published seven theoretical nanorobot scaling studies [4–10], several of which are briefly summarized below. Such studies are not intended to produce an actual engineering design for a future nanomedical product. Rather, the purpose is merely to examine a set of appropriate design constraints, scaling issues, and reference designs to assess whether or not the basic idea might be feasible, and to determine key limitations of such designs, including the many issues related to biocompatibility of medical nanorobots as extensively discussed elsewhere [2].

Complex medical nanorobots probably cannot be manufactured using the conventional techniques of self-assembly. As noted in the final report [19] of the recently completed congressionally-mandated review of the U.S. National Nanotechnology Initiative by the National Research Council (NRC) of the National Academies and the National Materials Advisory Board (NMAB): "For the manufacture of more sophisticated materials and devices, including complex objects produced in large quantities, it is unlikely that simple self-assembly processes will yield the desired results. The reason is that the probability of an error occurring at some point in the process will increase with the complexity of the system and the number of parts that must interoperate."

The opposite of self-assembly processes is positionally controlled processes, in which the positions and trajectories of all components of intermediate and final product objects are controlled at every moment during fabrication and assembly. Positional processes should allow more complex products to be built with high quality and should enable rapid prototyping during product development. Positional assembly is the norm in conventional macroscale manufacturing (e.g., cars, appliances, houses) but has not yet been seriously investigated experimentally for nanoscale manufacturing. Of course, we already know that positional fabrication will work in the nanoscale realm. This is demonstrated in the biological world by ribosomes, which positionally assemble proteins in living cells by following a sequence of digitally encoded instructions (even though ribosomes themselves are self-assembled). Lacking this positional fabrication of proteins controlled by DNA-based software, large, complex, digitally specified organisms would probably not be possible and biology as we know it would cease to exist.

The most important inorganic materials for positional assembly may be the rigid covalent or "diamondoid" solids, since these could potentially be used to build the most reliable and complex nanoscale machinery. Preliminary theoretical studies have suggested great promise for these materials in molecular manufacturing. The NMAB/NRC Review Committee recommended [19] that experimental work aimed at establishing the technical feasibility (or lack thereof) of positional molecular manufacturing should be pursued and supported: "Experimentation leading to demonstrations supplying ground truth for abstract models is appropriate to better characterize the potential for use of bottom-up or molecular manufacturing systems that utilize processes more complex than self-assembly." Making complex nanorobotic systems requires manufacturing techniques that can build a molecular structure by positional assembly [3]. This will involve picking and placing molecular parts one by one, moving them along controlled trajectories much like the robot arms that manufacture cars on automobile assembly lines. The procedure is then repeated over and over with all the different parts until the final product, such as a medical nanorobot, is fully assembled.

The positional assembly of diamondoid structures, some almost atom by atom, using molecular feedstock has been examined theoretically [12, 20] via computational models of diamond mechanosynthesis (DMS). DMS is the controlled addition of individual carbon atoms, carbon dimers (C_2), or single methyl (CH_3) groups to the growth surface of a diamond crystal lattice in a vacuum manufacturing environment. Covalent chemical bonds are formed one by one as the result of positionally constrained mechanical forces applied at the tip of a scanning probe microscope apparatus. Programmed sequences of carbon dimer placement on growing diamond surfaces *in vacuo* appear feasible in theory [20–23]. Diamond mechanosynthesis is being sought [11, 21] but has not yet been achieved experimentally; in 1999 Ho and Lee [24] demonstrated the first site-repeatable site-specific covalent bonding operation of a two diatomic carbon-containing molecules (CO), one after the other, to the same atom of iron on a crystal surface. In 2003, Oyabu et al. [25] vertically manipulated single silicon atoms from the Si(111)–(7 × 7) surface, using a low-temperature near-contact atomic force microscope to demonstrate removal of a selected silicon atom from its equilibrium position without perturbing the (7 × 7) unit cell and also the deposition of a single Si atom on a created vacancy, both via purely mechanical processes. Efforts are currently underway to achieve DMS with carbon atoms experimentally [11].

To be practical, molecular manufacturing must also be able to assemble very large numbers of medical nanorobots very quickly. Approaches under consideration include using replicative manufacturing systems [13] or massively parallel fabrication, employing large arrays of scanning probe tips all building similar diamondoid product structures in unison [13]. For example, simple mechanical ciliary arrays consisting of 10,000 independent microactuators on a 1 cm^2 chip have been made at the Cornell National Nanofabrication Laboratory for microscale parts transport applications and similarly at IBM for mechanical data storage applications [26]. Active probe arrays of 10,000 independently actuated microscope tips have been developed by Mirkin's group at Northwestern University for dip-pen nanolithography [27] using DNA-based ink. Almost any desired 2D shape can be drawn using 10 tips in concert. Another microcantilever array manufactured by Protiveris Corp. has millions of interdigitated cantilevers on a single chip [28]. Martel's group has investigated using fleets of independently mobile wireless instrumented microrobot manipulators called NanoWalkers to collectively form a nanofactory system that might be used for positional manufacturing operations [29]. Zyvex Corp. (www.zyvex.com) of Richardson, Texas received a $25 million, five-year, National Institute of Standards and Technology (NIST) contract to develop prototype microscale assemblers using microelectromechanical systems.

Ultimately, medical nanorobots will be manufactured in nanofactories efficiently designed for this purpose. One possible rough outline for a combined experimental and theoretical program to explore the feasibility of nanoscale positional manufacturing techniques, starting with the positionally controlled mechanosynthesis of diamondoid structures using simple molecular feedstock and progressing to the ultimate goal of a desktop nanofactory appliance able to manufacture macroscale quantities of molecularly precise product objects according to digitally defined blueprints, is available at the Nanofactory Collaboration website: http://www.MolecularAssembler.com/Nanofactory/Challenges.htm.

The purpose of this chapter is to examine the wide range of computational tasks that might need to be performed by various representative classes of medical nanorobots. Only a sampling of such tasks will be presented because time and space do not permit an exhaustive survey. The discussion starts with descriptions of several classes of medical nanorobots for which preliminary scaling studies have already been published in the literature. Basic computational tasks are described in each case. Following a more general discussion of the various major functions common to many or all complex medical nanorobots, we introduce the concept of nanorobot control protocols which are required to ensure that each nanorobot fully completes its intended mission accurately, safely, and in a timely manner according to plan.

15.2. EXEMPLAR MEDICAL NANOROBOT DESIGNS

Preliminary scaling studies of several classes of medical nanorobots have been published in the literature and include respirocytes (artificial mechanical red cells) [4], microbivores (artificial white cells) [5], clottocytes (artificial platelets) [6], and chromallocytes (chromosome exchanging nanorobots) [7]. These studies are summarized here, with emphasis on the computational tasks and requirements for such devices.

15.2.1. Respirocytes

15.2.1.1. Nanorobot Description. The first theoretical design study of a medical nanorobot ever published in a peer-reviewed medical journal (in 1998) described an artificial mechanical red blood cell or “respirocyte” [4] made of 18 billion precisely arranged atoms (Fig. 15.1)—a bloodborne, spherical 1-micron diamondoid 1000-atmosphere pressure vessel [1e] with active pumping [1f] powered by endogenous serum glucose [1g], able to deliver 236 times more oxygen to the tissues per unit volume than natural red cells and to manage carbonic acidity, controlled by gas concentration sensors [1h] and an onboard nanocomputer [1d, 12d].

In the exemplar design, onboard pressure tanks can hold up to three billion oxygen (O_2) and carbon dioxide (CO_2) molecules. Molecular pumps are arranged on the surface to load and unload gases from the pressurized tanks. Tens of thousands of these individual pumps, called molecular sorting rotors [1f], cover a large fraction of the hull surface of the respirocyte. Molecules of oxygen (O_2) or

Figure 15.1. An artificial red cell: the respirocyte [4]. Designer Robert A. Freitas, Jr.; illustration by E-spaces. (© 2002 E-spaces and Robert A. Freitas, Jr.)

carbon dioxide (CO_2) may drift into their respective binding sites on the exterior rotor surface and be carried into the respirocyte interior as the rotor turns in its casing. Once inside, a small pin is inserted into the binding site, forcibly ejecting the bound molecule into the interior tank volume. There are 12 identical pumping stations laid out around the equator of the respirocyte, with oxygen rotors on the left, carbon dioxide rotors on the right, and water rotors in the middle. Temperature [1i] and concentration [1j] sensors tell the devices when to release or pick up gases. Each station has special pressure sensors [1k] to receive ultrasonic acoustic messages [1m], so doctors can tell the devices to turn on or off, or change the operating parameters of the devices, while the nanorobots are inside a patient.

The basic operation of respirocytes is straightforward. These nanorobots, still entirely theoretical, would mimic the action of the natural hemoglobin-filled red blood cells. In the tissues, oxygen is pumped out of the device by the sorting rotors on one side. Carbon dioxide is pumped into the device by the sorting rotors on the other side, one molecule at a time. Half a minute later, when the respirocyte reaches the patient's lungs in the normal course of the circulation of the blood, these same rotors reverse their direction of rotation, recharging the device with fresh oxygen and dumping the stored CO_2, which diffuses into the lungs and can then be exhaled by the patient. The onboard nanocomputer and numerous chemical and pressure sensors enable complex device behaviors remotely reprogrammable by the physician via externally applied ultrasound acoustic signals. There is also a large internal void surrounding the nanocomputer which can be a vacuum, or can be filled or emptied with water. This allows the device to control its buoyancy very precisely (Section 2.1.2) and provides a crude but simple method for removing respirocytes from the blood using a centrifuge, a procedure called nanapheresis [1n].

Various sensors are needed to acquire external data essential in regulating gas loading and unloading operations, tank volume management, and other special protocols. It is also convenient to include internal pressure sensors [1k] to monitor O_2 and CO_2 gas tank loading, ullage (container fullness) sensors [1o] for ballast and glucose fuel tanks, and internal/external temperature sensors [1i] to help monitor and regulate total system energy output. The attending physician can broadcast signals to molecular mechanical systems deployed inside the human body most conveniently using modulated compressive pressure pulses received by mechanical transducers embedded in the surface of the respirocyte. Converting a pattern of pressure fluctuations into mechanical motions that can serve as input to a mechanical computer requires transducers that function as pressure-driven actuators [1p, 12e]. Data transmitted at ~ 10 MHz ($\sim 10^7$ bits/sec) using peak-to-trough 10-atm pressure pulses (the same as medical pulse-echo diagnostic ultrasound systems [30]) should attenuate only $\sim 10\%$ per 1 cm of travel [1q], so whole-body broadcasts should be feasible even in emergency field situations. Internal communications within the respirocyte itself can be achieved by impressing modulated low-pressure acoustical spikes [1r] on the hydraulic working fluid of the power distribution system [1s], or via simple mechanical rods and couplings [1t].

Onboard power is provided by a mechanochemical engine [1u] or fuel cell [1v] that exoergically combines ambient glucose and oxygen to generate mechanical or electrical energy to drive molecular sorting rotors and other subsystems, as demonstrated in principle in a variety of biological motor systems [1w]. Sorting rotors absorb glucose directly from the blood and store it in an internal fuel tank. Oxygen is tapped from onboard storage. The power system is scaled such that each glucose engine can fill the primary O_2 tank from a fully empty condition in 10 seconds, requiring a peak continuous output of 0.3 pW. The glucose fuel tank is scaled such that one tankful of fuel drives the glucose engine at maximum output for 10 seconds, consuming 5% of the O_2 gas stored onboard and releasing a volume of waste water approximately equal to the volume of the glucose consumed. Power is transmitted mechanically or hydraulically using an appropriate working fluid, and can be distributed as required using rods and gear trains [1x] or using pipes and mechanically operated valves, controlled by the nanocomputer.

A 5 cc therapeutic dose of 50% respirocyte saline suspension containing 5 trillion nanorobots would exactly replace the gas carrying capacity of the patient's entire 5.4 liters of blood. If up to 1 liter of respirocyte suspension can safely be added to the human bloodstream [2a], this could keep a patient's tissues safely oxygenated for up to 4 hours, even if a heart attack caused the heart to stop beating or if there was a complete absence of respiration or no external availability of oxygen. Primary medical applications of respirocytes would include emergency revival of victims of carbon monoxide poisoning at the scene of a fire; transfusable blood substitution; partial treatment for anemia and various lung and perinatal/neonatal disorders; enhancement of cardiovascular/neurovascular procedures, tumor therapies, and diagnostics; prevention of asphyxia; artificial breathing (e.g., underwater, high altitude, etc.); sports applications (e.g., ability to sprint at Olympic speed for up to 15 minutes without breathing); and a variety of other applications in veterinary medicine, military science, and space exploration.

15.2.1.2. Computational Tasks. An onboard nanocomputer is necessary to provide precise control of all basic respirocyte operations. These operations include respiratory gas loading and unloading, sensor field management and data handling, molecular rotor field and storage/ballast tank control, glucose engine throttling, management of glucose fuel supply, powerplant effluent regulation, control of power distribution, and interpretation of sensor data and commands received from the outside. The nanocomputer is also responsible for basic self-diagnosis monitoring and activation of failsafe shutdown protocols, and ongoing real-time revision or correction of protocols for *in vivo* devices.

With efficient programming, a 10^4 bit/sec computer can probably meet all computational requirements, given the simplicity of analogous chemical process control systems in factory settings [31, 32]. That's roughly the computing capacity of a transistor-based 1960-vintage IBM 1620 computer, or about 1/50th the capacity of a 1976-vintage Apple II microprocessor-based PC. Both the IBM 1620 and the Apple II used $\sim 10^5$ bits of internal memory, but even the early PCs typically had access to 10^6 bits (0.1 megabyte) of external floppy drive memory. Assuming ~ 500 bits/sec/nm^3 and 10^{18} bits/sec/watt for nanomechanical computers [12d], and ~ 5 bits/nm^3 for nanomechanical mass storage systems [1y, 12f], each 10^4 bit/sec CPU is allocated a volume of $\sim 10^4$ nm^3 and consumes ~ 0.01 pW (3% of the power output of one glucose engine), while 500 kilobits of memory requires $\sim 10^5$ nm^3. The use of reversible logic could significantly reduce power consumption [33–35].

Respirocyte behavior is initially governed by a set of default protocols which can be modified at any time by the attending physician. Basic protocols will exist for operating molecular sorting rotors at various speeds and directions in response to sensor data. For example, ballast water pumping will normally be driven by internal ullage and temperature sensors. O_2 rotors may load tanks at $P_{O2} > 95$ mmHg (alveolar) and unload at $P_{O2} < 40$ mmHg (tissues). CO_2 rotors may fill at $P_{CO2} > 46$ mmHg and unfill at $P_{CO2} < 40$ mmHg, and may incorporate other sensor data including P_{O2}, temperature, etc., to fine-tune pressure thresholds and enhance reliability. Gas loading parameters may be precisely specified in an individualized onboard lookup table provided by the physician for his patient, as for instance to adjust for declining arteriovenous oxygen gradient at high altitudes [36].

Respirocytes, like natural hemoglobin, may also participate in the elimination of CO and in NO-mediated vascular control [37] if appropriate sorting rotors and onboard tankage are provided.

One important protocol described in the original paper [4] is the initial warmup procedure for empty respirocytes that must be infused into a patient. Respirocytes comprising a single augmentation dosage are stored as a dry powder, tanks empty, in sealed plastic drip bags with two hose couplings. With no batteries to run down, consumables to age, vapors to outgas or organic matter to decompose, the product should have a long shelf life. To use the product, the bag is filled with ice-cold 0.13 M glucose (23 gm/liter) solution, plus any salts, minerals, vitamins, proteins or other substances the physician deems appropriate. The powder floats on the liquid surface. An external source of O_2 gas (and CO_2, if required) is provided through the second coupling. Respirocyte sensors detect the presence of glucose and begin pumping fuel into the glucose tanks. As these tanks fill, each device loads its oxygen tank to rated capacity. The powder still floats on the surface. Finally, the respirocytes load ballast tanks and sink to the bottom of the bag. (Powder remaining on the surface evidences malfunction.) During this ~30 sec charging process, the respirocytes absorb 53.7 liters of O_2 @ STP, store 0.78 liters of CO_2 waste, release enough energy to warm bag contents to 42°C, and leave behind a 0.005 M glucose solution, closely matching normal blood concentration and temperature. Upon receiving the correct activation command, broadcast acoustically using an ultrasound transmitter device pressed against the bag, the respirocytes blow sufficient ballast water to achieve neutral buoyancy, creating a perfect suspension (after agitation) ready for IV drip. The suspension is ~300 times plasma viscosity (~castor oil or canola oil at 37°C), still permitting ready plug flow [38].

Once a therapeutic purpose is completed, it may be desirable to extract artificial devices from circulation, requiring activation of a respirocyte filtration protocol. During this protocol, called nanapheresis [1n], blood to be cleared may be passed from the patient to a specialized centrifugation apparatus where ultrasound acoustic transmitters command respirocytes to establish neutral buoyancy. No other solid blood component can maintain exact neutral buoyancy, hence those other components precipitate outward during gentle centrifugation and are drawn off and added back to filtered plasma on the other side of the apparatus. Meanwhile, after a period of centrifugation, the plasma, containing mostly suspended respirocytes but few other solids, is drawn off through a 1-micron filter, removing the respirocytes. Filtered plasma is recombined with centrifuged solid components and returned undamaged to the patient's body. The rate of separation is further enhanced either by commanding respirocytes to empty all tanks, lowering net density to 66% of blood plasma density, or by commanding respirocytes to blow a 5-micron O_2 gas bubble to which the device may adhere via surface tension, allowing it to rise at 45 mm/hour in a normal gravitational environment.

Respirocytes can be programmed with even more sophisticated behaviors. Detection of $P_{CO2} < 0.5$ mmHg and $P_{O2} > 150$ mmHg, indicating direct exposure to atmosphere and a high probability that the device has been bled out of the body, should trigger a prompt gas venting and failsafe device shutdown procedure. This is particularly important in the case of oral bleeding, since respirocytes could explode if crushed between two hard planar surfaces with no avenue of escape as might occur during dental grinding [1z]—tooth enamel is the hardest natural substance in the human body, and a patient with an oral lesion could spread respirocyte-impregnated blood over the teeth. Of course, single-device explosions of a micron-size pressure tank caused by dental grinding would probably be undetectable by the patient.

Self-test algorithms monitoring tank filling rates, unaccounted pressure drops (indicating a leak), clutch responses, etc. may detect significant device malfunction, causing the respirocyte to place itself in standby mode ready to respond to an acoustic command to execute the filtration protocol for nanapheresis. Basic self-diagnostic routines should be able to detect simple failure modes such as jammed rotor banks, plugged flues, gas leaks, and so forth, and to use backup systems to place the device into a fail-safe dormant mode pending removal by filtration. Dormant mode can also be triggered by thermal shutoff protocols which activate upon self-detection of any of at least four failure scenarios: (1) A normally functioning respirocyte is deposited in a relatively dry (e.g., osteal or cartilaginous) location, losing contact with the aqueous heat sink; (2) one or more glucose engines become jammed at full open throttle; (3) inbound sorting rotors overload pressure vessel; or (4) device fragmentation or combustion.

Using onboard $(21\ \text{nm})^3$ pressure transducers, respirocytes can detect, record, or respond to changes in heart rate or blood pressure, since arterial pressure is normally 0.1–0.2 atm ($\sim 100\ \text{zJ}/(21\ \text{nm})^3$ sensor) and systolic/diastolic differential is 0.05–0.07 atm ($40\text{–}60\ \text{zJ}/(21\ \text{nm})^3$ sensor), both well above the mean thermal noise limit of $k_B T \sim 4\ \text{zJ}$ at human body temperature. Outmessaging [1aa] protocols could allow the population of respirocytes to communicate systemwide status (e.g., "low oxygen," "low glucose," "under immune attack," "cyanide detected" [39]) directly with the patient by inducing recognizable physiological cues (fever, shivering, gasping) or by transmitting the information to implanted GUIs accessible to the patient [1bq], or with the physician by generating subtle respiratory patterns requiring diagnostic equipment to detect, either automatically or in response to an acoustically transmitted global inquiry initiated by patient or physician. Attempted phagocytosis of bloodborne respirocytes by the RES [2b], particularly Kupffer cells in the liver [2c], bloodborne white cells, macrophages, or other natural phagocytic cells could, upon detection by the respirocyte, activate appropriate phagocyte avoidance and escape protocols (Section 15.4.2) using techniques detailed elsewhere [2d].

Other useful control protocols may enable *in vivo* respirocytes to be commanded to cease or resume operating locally/globally, to enable/disable one or another class of sorting rotors, to alter sensor sensitivities, to blow tanks or to run all engines continuously, and so forth. Note that the long-term bloodstream presence of a therapeutic dose of working respirocytes would preclude tissue hypoxia, possibly reducing EPO secretion as low as 1% of normal levels [40], thus suppressing erythropoiesis. It may be possible to avoid the resulting decimation of the natural erythrocyte population by adjusting respirocyte P_{O2}/P_{CO2} response thresholds so that these devices activate only when red cells are stressed. Command messages of these and similar types should require lengthy authorization codes (Section 15.4.5) to prevent accidental or malicious triggering of respirocyte behaviors having potentially harmful consequences.

15.2.2. Microbivores

15.2.2.1. Nanorobot Description. A second theoretical design study of a medical nanorobot describes an artificial mechanical white cell or "microbivore"—an oval-shaped device (Fig. 15.2) measuring a few microns in size and made of diamond and sapphire—that would seek out and digest unwanted bloodborne pathogens [5]. One main task of natural white cells is to absorb and consume microbial invaders in the bloodstream, called phagocytosis. Microbivore nanorobots would also perform phagocytosis, but would operate much faster, more reliably, and under human control.

The microbivore is an oblate spheroidal nanomedical device measuring 3.4 microns in diameter along its major axis and 2.0 microns in diameter along

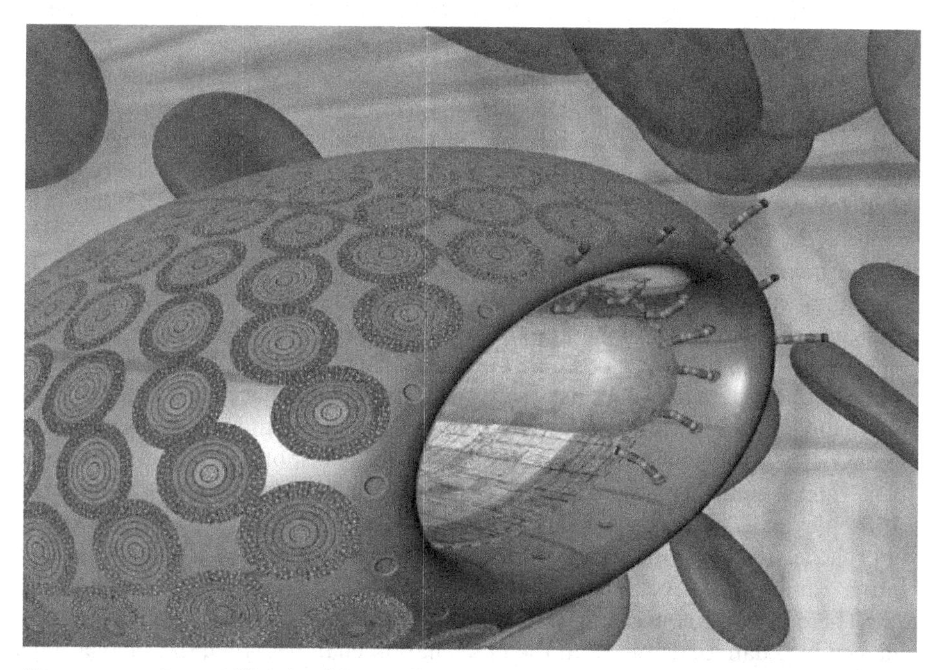

Figure 15.2. An artificial white cell: the microbivore [5]. Designer Robert A. Freitas, Jr.; illustration and additional design Forrest Bishop. (© 2001 Zyvex Corp. and Robert A., Freitas Jr.)

its minor axis, consisting of 610 billion precisely arranged structural atoms in a gross geometric volume of 12.1 micron^3 and a dry mass of 12.2 picograms. This size helps to ensure that the nanorobot can safely pass through even the narrowest of human capillaries and other tight spots in the spleen (e.g., the interendothelial splenofenestral slits [2e]) and elsewhere in the human body [2f]. The microbivore has a mouth with an irising door, called the ingestion port, where microbes are fed in to be digested. The microbivore also has a rear end, or exhaust port, where the completely digested remains of the pathogen are harmlessly expelled from the device. The rear door opens between the main body of the microbivore and a tail-cone structure. The device may consume up to 200 pW of continuous power while completely digesting trapped microbes at a maximum throughput of 2 micron^3 of organic material per 30-second cycle, which is large enough to internalize a single microbe from virtually any major bacteremic species in a single gulp. This "digest and discharge" protocol [1ab] is conceptually similar to the internalization and digestion process practiced by natural phagocytes, except that the artificial process should be much faster and cleaner. For example, it is well-known that macrophages release biologically active compounds during bacteriophagy [41], whereas well designed microbivores need only release biologically inactive effluent.

The first task for the bloodborne microbivore is to reliably acquire a pathogen to be digested. If the correct bacterium bumps into the nanorobot surface, reversible species-specific binding sites on the microbivore hull can recognize and weakly bind to the bacterium. A set of 9 distinct antigenic markers should be specific enough [1ac], since all 9 must register a positive binding event to confirm that a targeted microbe has been caught. There are 20,000 copies of these 9-marker receptor sets, distributed in 275 disk-shaped regions across the microbivore surface. Inside each receptor ring are more rotors to absorb ambient glucose and oxygen from the bloodstream to provide nanorobot power. At the center of each 150-nm diameter receptor disk is a grapple silo. Once a bacterium has been captured by the reversible receptors, telescoping robotic grapples [1ad] rise up out of the microbivore surface and attach to the trapped bacterium, establishing secure anchorage to the microbe's plasma membrane. The microbivore grapple arms are about 100 nm long and have various rotating and telescoping joints that allow them to change their position, angle, and length. After rising out of its silo, a grapple arm can execute complex twisting motions, and adjacent grapple arms can physically reach each other, allowing them to hand off bound objects as small as a virus particle. Grapple handoff motions can transport a large rod-shaped bacterium from its original capture site forward to the ingestion port at the front of the device. The captive organism is rotated into the proper orientation as it approaches the open microbivore mouth, where the pathogen cell is internalized into a 2 micron^3 morcellation chamber.

There are two concentric cylinders inside the microbivore. The bacterium is minced into nanoscale pieces in the morcellation chamber [1ae], the smaller inner cylinder, then the remains are pistoned into a separate 2 micron^3 digestion chamber, a larger outer cylinder. In a preprogrammed sequence, ~40 different engineered digestive enzymes are successively injected and extracted six times during a single digestion cycle, progressively reducing the morcellate to monoresidue amino acids, mononucleotides, glycerol, free fatty acids and simple sugars, using an appropriate array of molecular sorting rotors. These basic molecules are then harmlessly discharged back into the bloodstream through the exhaust port at the rear of the device, completing the 30-second digestion cycle. When treatment is finished, the doctor may transmit an ultrasound signal to tell the circulating microbivores that their work is done. The nanorobots then exit the body through the kidneys and are excreted with the urine in due course.

The microbivore needs a variety of external and internal sensors to complete its tasks. External sensors include chemical sensors for glucose, oxygen, carbon dioxide, and so forth, up to 10 different molecular species with 100 sensors per molecular species. Pressure sensors for acoustic communication are mounted within the nanorobot hull to permit the microbivore to receive external instructions from the attending physician during the course of *in vivo* activities. Ten (redundant) internal temperature sensors capable of detecting 0.3°C temperature change [1i] are positioned near each of the 10 independent powerplants.

Microbivores use bloodstream glucose for power like respirocytes, but also use blood-dissolved oxygen which is available only at much lower concentrations than in the lung capillary bed, so many more molecular sorting rotors for O_2 must be present on the microbivore hull. The microbivore is scaled for a maximum power output of 200 pW, about 1000 times higher than for the respirocyte. Diamondoid mechanical cables may transmit internal mechanical energy at power densities of $\sim 6 \times 10^{12}\,W/m^3$ [1x]. To connect every powerplant with each of its 9 neighbors via power cables, permitting rapid load sharing among any pair of powerplants inside the device, requires 45 power cables; assuming 1000 internal power cables to accommodate additional power distribution tasks and for redundancy, total power cable volume is 0.05 micron^3. By varying the cable rotation rate, the same power cables can simultaneously be used to convey necessary internal operational information [1t] including sensor data traffic and control signals from the computers.

A human neutrophil, the most common type of leukocyte or white cell, can capture and engulf a microbe in a minute or less, but complete digestion and excretion of the organism's remains can take an hour or longer. Thus our natural white cells—even when aided by antibiotics—can sometimes take weeks or months to completely clear bacteria from the bloodstream. By comparison, a single terabot (10^{12}-nanorobot) dose of microbivores should be able to fully eliminate bloodborne pathogens in just minutes or hours, even in the case of locally dense infections. This is accomplished without increasing the risk of sepsis or septic shock because all bacterial components (including all cell-wall LPS) are internalized and fully digested into harmless nonantigenic molecules prior to discharge from the device. No matter that a bacterium has acquired multiple drug resistance to antibiotics or to any other traditional treatment—the microbivore will eat it anyway. Microbivores would be up to $\sim$1000 times faster-acting than antibiotic-aided natural leukocytes. The nanorobots would digest $\sim$100 times more microbial material than an equal volume of natural white cells could digest in any given time period, and would have far greater maximum lifetime capacity for phagocytosis than natural white blood cells. Besides intravenous bacterial, viral, fungal, and parasitic scavenging, microbivores or related devices could also be used to help clear respiratory or cerebrospinal bacterial infections, or infections in other nonsanguinous fluid spaces such as pleural [42], synovial [43], or urinary fluids; eliminate bacterial toxemias and biofilms [44]; eradicate viral, fungal, and parasitic infections; disinfect surfaces, foodstuffs, or organic samples; and help clean up biohazards and toxic chemicals. Related nanorobots with enhanced tissue mobility could be programmed to quickly recognize and digest even the tiniest aggregates of early tumor cells with unmatched speed and surgical precision, eliminating cancer. Similar nanorobots of this class could be programmed to remove circulatory obstructions in just minutes, quickly rescuing even the most compromised stroke victim from near-certain ischemic brain damage, and could have other uses in various veterinary and military applications.

15.2.2.2. Computational Tasks. An onboard nanocomputer is required to provide precise control of all basic microbivore operations. These operations include most of the basic computational tasks already described for the respirocyte (Section 2.1.2) such as onboard tank volume control, sensor coordination, data and power management, and nanapheresis protocols. Other tasks unique to the microbivore include management of multiple reversible microbial binding sites to ensure accurate identification of the targeted microbial species, control of hundreds of independent grapple elevator and segment rotation mechanisms, coordination of the grapple motion field to transport trapped microbes in a controlled manner across the nanorobot surface or for nanorobot locomotion, and sequence control for morcellation and digestion activities (ingestion port opening/closing, mincing, transfer pistoning, enzyme cycling, effluent ejection pistoning, and exhaust port opening/closing).

The microbivore computer is scaled as a 0.01 micron^3 mechanical nanocomputer [1y, 12d], in principle capable of >100 megaflops but normally operated at ~1 megaflop to hold power consumption to ~60 pW. Assuming ~5 bits/nm^3 for nanomechanical data storage systems [12d] and a read/write cost of ~10 zJ/bit at a read/write speed of ~10^9 bits/sec [1af, 12d], then a comparison with other software systems of comparable complexity (Table 15.1) suggests that the ~5 megabits of mass nanomechanical memory needed to hold the microbivore control system displaces a volume of 0.001 micron^3 and draws ~10 pW while in continuous operation. The baseline microbivore design includes 10 duplicate computer/memory systems for redundancy (with only one of the ten computer/memory systems in active operation at a time), displacing a total of 0.11 micron^3 and consuming $\leq$70 pW.

As with respirocytes, microbivore behavior is initially governed by a set of default protocols many of which can be modified at any time by the attending physician. Basic protocols have already been described for the respirocyte (Section 2.1.2). Since virtually every medical nanorobot placed inside the human

TABLE 15.1. Lines of Compactly Written Low-Error Software Code Required to Control Complex Semiautonomous Machines

Control software for device	Estimated lines of code	Bits of code (~100 bits/line)	Ref.
Voyager Spacecraft Software	3,000	300,000	[45]
Viking Lander Software	—	432,000	[46]
Respirocyte Control System (est.)	—	~500,000	[4]
Galileo Spacecraft Software	8,000	800,000	[45]
Cassini Spacecraft Software	32,000	3,200,000	[45]
Microbivore Control System (est.)	—	~5,000,000	[5]
Ariane Flight Control Software	90,000	9,000,000	[47]
Airbus 340 Flight Warning System	100,000	10,000,000	[48]
Mars Pathfinder Spacecraft	160,000	16,000,000	[45]
Space Shuttle Software	500,000	50,000,000	[49]
Boeing 777 and Airbus 340	3,000,000	300,000,000	[50]

body will encounter natural phagocytic cells many times during its mission [2d], microbivores, like respirocytes, may incorporate any of several possible phagocyte avoidance and escape techniques [2d], possibly including, for example, surface-tethered phagocyte chemorepellent molecules [2g, 51] or phagocyte engulfment inhibitors [2h, 52], as well as more proactive approaches to phagocytosis avoidance [2d].

Many special protocols will be needed by microbivores but a complete enumeration is beyond the scope of this chapter. Biocompatibility-related protocols (Section 15.4.2) provide examples ranging from the simple to the more complex. At the simple end of the range is the microbial tail protocol [5]. Free releases of bacterial flagella into the bloodstream could produce inflammation or elicit various immune system responses and thus should be avoided. Complete internalization of tail may be ensured by specialized operational routines (e.g., forced end-over-end rotation of an internalized microbe while inside the morcellation chamber, thus completely spooling the tail into the microbivore before fully sealing the ingestion port door) or by specialized mechanical tools or jigs (e.g., a counterrotating interdigitated-knobbed capstan-roller pair).

An example of a more complex protocol arises from the fact that all bloodborne nanorobots larger than ~1 micron in all three physical dimensions are subject to possible geometrical trapping in the fenestral slits of the splenic sinusoids in the red pulp of the spleen [2e]. A small percentage of blood is forced to circulate through a physical filter in the spleen requiring passage through slits measuring 1–2 microns in width and ~6 microns in length [2e]. Microbivores which become pinned to a slit face-on, or which become stuck edge-on during an attempted passage, can detect that they have become trapped by measuring various blood component concentration and pressure differentials across their surfaces. The nanorobot then activates its automatic splenofenestral escape protocol [5], which involves the extension and patterned ciliation of surface grapples until sensor readings reveal that passage through the slit has been achieved, after which the grapples are retracted.

15.2.3. Clottocytes

15.2.3.1. Nanorobot Description. Another theoretical design study describes an artificial mechanical platelet [6] or "clottocyte" that would allow complete hemostasis in as little as ~1 second, even in moderately large wounds. This response time is on the order of 100–1000 times faster than the natural hemostatic system. The baseline clottocyte is conceived as a serum oxyglucose-powered spherical nanorobot ~2 microns in diameter (~4 micron^3 volume) containing a fiber mesh that is compactly folded onboard. Upon command from its control computer, the device promptly unfurls its mesh packet in the immediate vicinity of an injured blood vessel—following, say, a cut through the skin. Soluble thin films coating certain parts of the mesh dissolve upon contact with plasma water, revealing sticky sections (e.g., complementary to blood group antigens unique to red cell surfaces [1ag]) in desired patterns. Blood cells are immediately trapped in the overlapping artificial nettings released by multiple neighboring activated clottocytes, and bleeding halts at once.

The required blood concentration n_{bot} of clottocyte nanorobots required to stop capillary flow at velocity $v_{cap} \sim 1$ mm/sec [1ah] in a response time $t_{stop} = 1$ sec, assuming $n_{overlap} = 2$ fully overlapped nets each of area $A_{net} = 0.1\ mm^2$, is $n_{bot} \sim n_{overlap}/(A_{net}\ t_{stop}\ v_{cap}) = 20\ mm^{-3}$, or just ~110 million clottocytes in the entire 5.4-liter human body blood volume possessing ~11 m^2 of total deployable mesh surface. This total dose is ~0.4 mm^3 of clottocytes, which produces a negligible serum nanocrit [1ai] or "Nct" of ~0.00001% (nanorobot/blood volume ratio). During the 1 second hemostasis time, an incision wound measuring 1 cm long and 3 mm deep would lose only ~6 mm^3 of blood, less than one-tenth of a single droplet. There are 2–3 red cells per deployed 1 micron^2 mesh square, more than enough to ensure that the meshwork will be completely filled, allowing complete blockage of a breach.

Total natural bleeding time, as experimentally measured from initial time of injury to cessation of blood flow, may range from 2–5 minutes [53] up to 9–10 minutes [54, 55] if even small doses of anticoagulant aspirin are present [56], with 2–8 minutes being typical in clinical practice. With a ~1 sec response time, artificial mechanical platelets appear to permit the halting of bleeding 100–1000 times faster than natural hemostasis. While 1–300 platelets might be broken and still be insufficient to initiate a self-perpetuating clotting cascade, even a single clottocyte, upon reliably detecting a blood vessel break, can rapidly communicate this fact to its neighboring devices, immediately triggering a progressive controlled mesh-release cascade. Clottocytes may perform a clotting function that is equivalent in its essentials to that performed by biological platelets—but at only ~0.01% of the bloodstream concentration of those cells. Hence clottocytes appear to be ~10,000 times more effective as clotting agents than an equal volume of natural platelets.

15.2.3.2. Computational Tasks. Besides the nanocomputers and control systems similar to those previously described for other nanorobots, clottocytes crucially require special control protocols to ensure that these nanorobots cannot release their mesh packets in the wrong places inside the body, or at an inappropriate time. These protocols will demand that carefully specified constellations of sensor readings must be observed before device activation is permitted. Reliable communications protocols will be required to control co-ordinated mesh releases from multiple neighboring devices and to regulate the maximum multidevice-activation radius within the local clottocyte population.

Detection of the "bled out of body" condition will be an especially important component of these protocols. Atmospheric concentrations of gases such as carbon dioxide and oxygen are different than the concentrations of those same gases in blood serum. As clottocyte-rich blood enters a breach in a blood vessel, nanorobot onboard sensors can rapidly detect the change in partial pressures, often indicating that the nanodevice is being bled out of the body. At a nanorobot whole-blood concentration of $20\,mm^{-3}$, mean device separation is 370 microns. If the first device to be bled from the body lies 75 microns from the air–serum interface, oxygen molecules from the air can diffuse through serum at human body temperature (310 K) from the interface to the nanodevice surface in $\sim$1 second [1aj]. Detection of this change can be rapidly broadcast to neighboring clottocytes using $\sim$1500 m/sec waterborne acoustic pulses that are received by devices up to several millimeters away in times on the order of microseconds, allowing rapid propagation of a carefully controlled device-enablement cascade. Similarly, air temperature is normally lower than body temperature. The thermal equilibration time [1ak] across a distance L in serum at 310 K is $t_{EQ} \sim (6.7 \times 10^6)\, L^2$, hence a device that lies 75 microns from the air–plasma interface can detect a change in temperature in $t_{EQ} \sim 40$ millisec. Other relevant sensor readings may include blood pressure profiles, bioacoustic monitoring, bioelectrical field measurements, optical and ultraviolet radiation detection, and sudden shifts in pH or other ionic concentrations. At some cost in rapidity of response, clottocytes also could eavesdrop [1am] on natural biological platelet control signals, using sensors with receptors for the natural prostaglandins produced by endothelial cells that normally induce or inhibit platelet activation, and then take appropriate action upon receipt of those natural biochemical signals.

Biocompatibility [2] requirements engender additional needs for specialized clottocyte control protocols. For instance, the rapid mechanical action of clottocytes could interfere with the much slower natural platelet adhesion and aggregation processes, or disturb the normal equilibrium between the clotting and fibrinolytic systems [2i]. Thus it may be necessary for artificial platelets to release quantities of various chemical substances that will encourage the remainder of the coagulation cascade to proceed normally or at an accelerated pace, including timed localized vasodilation and vasoconstriction, control of endothelial cell modulation of natural platelet action, and finally clot retraction and fibrinolysis much later during tertiary hemostasis.

Yet another biocompatibility-related complication is that the bare tissue walls of a wound will continue to exude fluid [2j], and may begin to desiccate, if only the capillary termini are sealed but the rest of the tissue is left exposed to open air. Since clottocytes may remain attached to their discharged nets, and can communicate with each other via acoustic channels [1m], it should be possible to precisely control the development of a larger artificial mesh-based clot via coordinated mesh extensions or retractions within the clot. Alternatively, clottocytes could allow blood fluids to flood small incised or avulsed wound volumes, allowing exposed tissue walls to be bathed in fluids but casting a watertight sealant net across the wound opening flush with the epidermal plane of the wound cavity. The operational protocols that trigger and direct these behaviors will need to be rather complex to ensure safe and fully biocompatible operation.

Additional protocols are required to allow clottocytes to prevent accidental but potentially fatal catastrophic natural clotting cascades such as disseminated intravascular coagulation or DIC [57]. One solution is to equip clottocytes with sensors to detect decreased serum levels of fibrinogen, plasminogen, alpha$_2$-antiplasmin, antithrombin III, factor VII, and protein C, and elevated levels of thrombin and various fibrin/fibrinogen-derived degradation products [2i]. If DIC conditions arise, nanorobots might respond by absorbing and metabolizing the excess thrombin (which trigger clotting), or by releasing thrombin inhibitors such as antithrombin III, hirudin, argatroban, or lepirudin [58] or anticoagulants that reduce thrombin generation such as danaparoid [58] to interrupt the cascade [58, 59]. For example, a $\sim$0.02% Nct concentration of nanorobots, suitably activated according to physician-approved parameters, could replace the entire depleted natural bloodstream content of antithrombin III from onboard stores in seconds.

Clottocytes will require still more sophisticated operational protocols if they are intended to assist platelets participating in the sealing of internal blood vessel lesions, in order to avoid inadvertently blocking the lumen of the entire vessel, e.g., in the case of minor internal bleeding. Similarly, prevention of bleeding at vascular anastomoses, hemarthroses, internal bruising, "blood blisters," and larger tissue hematomas, as well as forced local coagulation in tumors or in intracerebral aneurysms, may also require more advanced protocols, possibly including integration with preexisting *in vivo* navigation systems [1an]. For some of these applications, modified clottocytes possessing the mobility of microbivores may be required in tandem with a graduated recruitment response depending upon how many (intercommunicating) devices appear to be involved in the event.

15.2.4. Chromallocytes

15.2.4.1. Nanorobot Description. The chromallocyte [7] is a hypothetical mobile cell-repair nanorobot whose primary purpose is to perform chromosome replacement therapy (CRT). In CRT, the entire chromatin content of the nucleus in a living cell is extracted and promptly replaced with a new set of prefabricated chromosomes which have been artificially manufactured as defect-free copies of

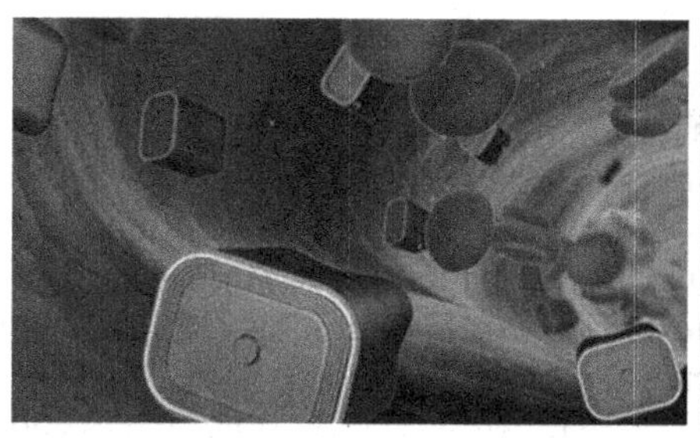

Figure 15.3. Artist's conceptions of the basic chromallocyte [7] design: devices walking along luminal wall of blood vessel (left); telescoping funnel assembly and proboscis manipulator in extended position (right). [© 2006 Stimulacra LLC (www.stimulacra.net) and Robert A. Freitas, Jr. (www.rfreitas.com)]

the originals. The chromallocyte (Fig. 15.3) will be capable of limited vascular surface travel into the capillary bed of the targeted tissue or organ, followed by diapedesis (exiting a blood vessel into the tissues) [1ao], histonatation (locomotion through tissues) [1ap], cytopenetration (entry into the cell interior) [1aq], and complete chromatin replacement in the nucleus of the target cell. The CRT mission ends with a return to the bloodstream and subsequent extraction of the device from the body at the original infusion site. Replacement chromosomes are manufactured in a desktop *ex vivo* chromosome sequencing and manufacturing facility, then loaded into the nanorobots for delivery to specific targeted cells during CRT. A single lozenge-shaped 69 micron^3 chromallocyte measures 4.18 microns and 3.28 microns along cross-sectional diameters and 5.05 microns in length, typically consuming 50–200 pW in normal operation and a maximum of 1000 pW in bursts during outmessaging, the most energy-intensive task. Treatment of an entire large human organ such as a liver, involving simultaneous CRT on all 250 billion hepatic tissue cells, might require the localized infusion of a $\sim$1 terabot (10^{12} devices) $\sim$69 cm^3 chromallocyte dose in a 1-liter 7% saline suspension during a $\sim$7 hour course of therapy.

The chromallocyte includes an extensible primary manipulator 4 microns long and 0.55 microns in diameter called the Proboscis that is used to spool up chromatin strands via slow rotation when inserted into the cell nucleus. After spooling, a segmented funnel assembly is extended around the spooled bolus of DNA, fully enclosing and sequestering the old genetic material. The new chromatin is then discharged into the nucleus through the center of the Proboscis by pistoning from internal storage vaults, while the old chromatin that is sequestered inside the sealed watertight funnel assembly is forced into the storage vaults as space is vacated by the new chromatin that is simultaneously being pumped out. The chromallocyte includes a mobility system similar to the microbivore grapple system, along with a solvation wave drive [1ar] that is used to ensure smooth passage through cell plasma and nuclear membranes.

15.2.4.2. Computational Tasks. Besides the usual sensor, mobility, data and power management tasks common to other nanorobots, chromallocyte computers must also control Proboscis extension and variable rotation rates, adjustment of adhesioregulatory surfaces [2k] on the Proboscis, pumping and fluid gating through the Proboscis, funnel assembly extension and retraction motions, and valving of materials into and out of the storage vaults. Locomotion also requires protocols for diapedesis through capillary walls, ECM (extracellular matrix) brachiation [1as] or grapple-mediated ciliary swimming [1at] through acellular tissues spaces (including emergency use of the Proboscis), transit through plasma and nuclear membranes [1aq], and nanapheresis [1n] or other nanorobot extraction procedures. (Chromallocytes must also be capable of emergency auto-excretion through the renal tubules as long as external acoustic power is still being supplied to the patient, or they must have a capture protocol in which they can allow themselves to be harmlessly phagocytosed and transported to lymph nodes [2m] for subsequent removal by other nanorobots in a separate procedure.) The position of the apical terminus of the Proboscis must be controlled during new chromatin discharge to allow placement of new chromosomes near their optimum territorial regions inside the nucleus, requiring integration of Proboscis position information with sensor-derived information about the nanorobot's position relative to the cell nucleus. Precise control, timing, and coordination with sensor data is required to manage the release and subsequent recovery, via molecular sorting rotors, of dozens of small-molecule engineered biological reagents that are deployed in waves by the nanorobot to temporarily suppress various natural processes such as mechanotransduction, apoptosis, and inflammation. The computer must also guide nanorobot navigation [1an] to and from the target cell by coordinating locomotion with real-time positional information possibly received from a navigational microtransponder network [1au] that has been preinstalled in the patient's body. (The navigational system itself, including its installation and management, are a separate nanorobotic instrumentality that is operated independently of the chromallocytes.)

Chromallocyte onboard computation and control is provided by a computer system similar to that employed in the microbivore [5]. This includes a tenfold redundant 0.01 micron^3 CPU throttled back to a $\sim$1 megaflop processing rate to conserve energy, giving a total computer volume of 0.1 micron^3. However, the chromallocyte incorporates a tenfold redundant mass memory system that is ten times larger (50 megabits, 0.01 micron^3) than for the microbivore (5 megabits, 0.001 micron^3), giving a total data storage volume of 0.1 micron^3. This increased memory allocation is justified by (1) the increased complexity of a CRT mission as compared to an antimicrobial mission, and (2) the need for greater reliability, safety, and certainty of result in the case of CRT, where a mission failure could have more serious medical consequences. There is sufficient unallocated volume in the current chromallocyte design to permit significantly increased onboard data storage if required.

The overall CRT mission requires completing a five-phase procedure. One of these phases, called "chromosome replacement," involves at least semiautonomous nanorobot activities and includes a specific 26-step sequential process for performing whole-genome chromosome exchange nanosurgery on a living cell using the chromallocytes [7]. Each step must be verified as complete before the next step can be initiated. Failure to reliably and provably complete any given step should trigger a safe mission abort protocol in which the nanorobot must appropriately reverse any steps already completed, thus restoring the cell to its original condition, then withdraw safely from the cell and subsequently from the patient's body, while reporting all details of this failure to the attending physician. Autonomous control of a 26-step chromosome exchange operation will include numerous sensor-driven checkpoints. Motions and speeds of sensored mobile components must also be restrained to within safe operating envelopes. Other possible operating protocols of this highly sophisticated medical nanorobot are outlined in Section 15.4.

Specialized protocols required for chromallocytes could be particularly complex. Significantly modified procedures will be required for numerous unusual cases including (1) proliferating, pathological, multinucleate, and karyolobate cells, (2) cells in locations where access is difficult such as brain, bone, or mobile cells, (3) cells expressing genetic mosaicism, and (4) alternative missions such as mitochondrial DNA replacement [7].

15.3. COMMON FUNCTIONS REQUIRING ONBOARD COMPUTATION

Functions which may be common to many different classes of medical nanorobots and must be controlled by the onboard nanocomputer include thosc listed below.

1. *Pumping*. Single-molecule recognition, sorting, and pumping via molecular sorting rotors [1f] to allow molecule-by-molecule exchanges with the environment. The typical $\sim$1 micron nanorobot might have $\sim 10^4$–10^5 sorting rotors requiring individual control, though many can probably be operated in banks or clusters of 10–100 rotors. Molecular pumps would be a primary system in nanorobots such as pharmacytes [10] which are intended solely to dispense drugs or other reagents.
2. *Sensing*. Chemical sensors [1j], pressure [1k] and temperature [1i] sensors, ullage sensors (include onboard pressure tank management) [1o], electrical, magnetic and optical sensors [1av], position/orientation sensors [1aw], gravity sensors [1ax], and molecular recognition sites [1ay, 1az]; and coordination and interpretation of macrosensing data providing onboard information from external acoustic [1ba], proprioceptive [1aw], electric/magnetic [1bb], or optical [1bc] sources, or direct neural traffic eavesdropping [1bd]. The typical $\sim$1 micron nanorobot might employ $\sim 10^4$–10^5 sensors of various kinds requiring individual control.
3. *Configuration*. Control of device shape [1be]; gas-driven extensible bumpers to maintain physical contact among adjacent devices during

stationkeeping [1bf]; control of internal ballasting for nanapheresis [1n]; and control of chemical ligands for hull displays, flags, or semaphores [1bg], or for controlled adhesioregulation of external surfaces [2k].

4. *Energy.* Control of onboard power generation or power receiver systems including thermal, mechanical, acoustic, chemical, electrical, photonic, or nuclear sources; management of onboard energy storage; controlling the transduction, conditioning, and conversion of rf or tethered energy sources; and control of internal power distribution and load balancing throughout a nanorobotic device [1c].
5. *Communication.* Control of communications hardware including receivers and transmitters, whether chemical, acoustic, electromagnetic, or other modality; interpretation of received signals as new commands from the physician; replacement of existing operating parameters with new ones (while the devices are *in vivo*), such as the changing of respirocyte operating parameters at high altitudes or deep underwater at high pressures; coordination of internal fiber or mobile networks; control of inmessaging (signals from external sources that are directed to *in vivo* nanorobots) and outmessaging (signals from *in vivo* nanorobots that are directed to external recipients), including coordination of nanorobot populations to accurately transfer information directly to or from the patient (e.g., directly via somesthetic, kinesthetic, auditory, gustatory, olfactory, ocular or dermal [1bq] displays, or indirectly via artificial symptoms); and routing of data signals internally throughout the nanorobot [1bh].
6. *Navigation.* Establishing absolute or relative physical position across many regimes including bloodstream, tissues, organs, and cells; positional navigation by dead reckoning, cartotaxis, macro/microtransponder networks (including stationkeeping protocols among neighboring intercommunicating devices and self-correcting calibration protocols allowing ~3 micron positioning accuracies in ~1000-device stacks maintaining ~100-micron interdevice separations across 10-cm tissue columns); functional navigation using thermographic, barographic, chemographic, or microbiotic modalities; and validation of the identity of encountered tissue types [1an].
7. *Manipulation.* Deployment and actuation of manipulators including ciliary, pneumatic, or telescoping systems; stowage, retrieval, selection, installation, use, and detachment of tooltips and other end-effectors; management of tool and manipulator garages; management of coordinated manipulator arrays; and control of onboard disposal or disassembly systems including morcellation, grinding, sonication, thermal or chemical decomposition systems [1b].
8. *Locomotion.* Control of specific *in vivo* locomotion systems [1bi] including ciliary or grapple systems, surface deformation, inclined planes/screws, volume displacement, and viscous anchoring systems; control of locomotion across cell-coated tissue surfaces (e.g., vascular lumen) via

legged ambulation, tank-tread rolling, amoeboid motion or inchworm locomotion; control of histonatation (tissue swimming) and cytopenetration as required; cytocarriage (nanorobots controlling movement of natural cells such as leukocytes from intracellular berths); and *ex vivo* locomotion [1bj] including dental walking, epidermal locomotion, and airborne nanoflight.

9. *Computation.* Control of all onboard computer systems [1d] including CPUs, memory, and internal data transmission lines, and procedures for switching between redundant computers or computer components; and the management, synchronization and calibration of onboard clocks and calendars [1bk]. Nanocomputers that employ standard dissipative architectures (as opposed to energy-saving reversible computing architectures [33–35]) will generate a high power density of waste heat, hence nanocomputers in larger nanorobots will commonly be throttled down from maximum processing speeds during normal operations both to save energy and to avoid excessive localized device heating. This is a principal design limitation on medical nanorobot computers. Consequently, a good design philosophy is to offload as many computational tasks as possible to *in vivo* data processing devices located external to the nanorobot (e.g., fixed-location tissue-embedded *in vivo* nanorobotic implanted nodes), or preferably to large *ex vivo* computers embedded in the physician's office, the surgical suite, or the hospital infrastructure, with data transferred in and out of the patient's body through fixed or mobile communications networks [1bm].
10. *Redundancy Management.* Acceptable system reliability for populations of trillions of cooperating medical nanorobots will require extensive subsystem and component redundancy. Typically, tenfold redundancy among mission-critical components appears sufficient to ensure acceptable mission reliability [12g] in therapeutic applications. For example, a good nanorobot design [4–8] may specify 10 times more sorting rotors, sensors, or appendages for locomotion than are strictly necessary, with the large surplus held in reserve as spares and backups. This implies that another important computational function will be redundant systems management [60, 61] and modeling [62]. Onboard computers must continuously monitor the performance of all redundant components and subsystems to determine whether or not any have failed, and if so, to decide which backup system to swap in to replace the function performed by the failed system. Studies are needed to define optimal control protocols for redundant systems management. As a simple example, in the case of the respirocytes which employ tenfold-redundant onboard glucose-oxygen power stations, reliability simulations may be useful to determine whether all ten power stations ideally should be run: (a) at peak power on a rotating schedule, (b) at partial power on a continuous basis, or (c) one at a time until failure, thereafter switching to

the next backup. During any component switchovers, the respirocyte computer must control the real-time distribution of power that is transmitted hydraulically to local station subsystems and also along a dozen independent interstation trunk lines that allow stations to pass hydraulic power among themselves as required, permitting load shifting and balancing. Redundancy management also applies to onboard computers, which are themselves multiply redundant. Further analysis is needed to determine the best techniques for safely switching among them without sacrificing system reliability.

11. *Flawless Compact Software.* Another fundamental limit to onboard computing is the minimum practical volume that may be occupied by registers or other physical media capable of storing or executing onboard data or instructions. For example, a 1 micron^3 storage volume could contain $\sim 10^{10}$ bits ($\sim$1 GB assuming 8-bit words) on a hydrofluorocarbon memory tape with a maximum $\sim$10 sec access time, or $\sim 10^7$ bits ($\sim$1 MB) in a tightly packed block of mechanical registers comprised of nanoscale diamond rods with a maximum $\sim$1 millisec access time [1af]. Given the presumed requirements for tenfold redundancy and the need for non-memory computer components, maximum allowable onboard data storage may often be an order of magnitude smaller than the above figures, or less. Given such limited availability of onboard nanorobot computer memory, it seems inevitable that efficient ultracompact software and space-efficient algorithms will again come into vogue. Such software was commonplace in the early days of PC development when total RAM memories were often limited to 48 KB or less—a meager allocation that was nonetheless sufficient to hold an entire word processing program. High software reliability must also be a major design criterion [63] because medical nanorobots must be extremely reliable. People have already been killed or almost killed [64] by software bugs in conventional medical devices. Policies for medical and product liability insurance will almost certainly incorporate strict requirements for provable software reliability and system noncrashability as measured by formalized industry metrics because of the potential significant risks to human health if a nanorobot control malfunction due to software error occurs during therapeutic use—a stringent requirement that should, at long last, make the formalized production of "error-free" code economically viable. The related supervisory software must also be designed with a simple user interface to minimize the possibility of error even when operated by weary, distracted, and fallible human medical personnel [65].

15.4. NANOROBOT CONTROL PROTOCOLS

A nanorobot control protocol is a specialized sequence of nanorobot actions that may be executed in carefully specified circumstances in order to ensure that the defined mission is successfully completed in an accurate, safe, and timely manner according to plan. Protocol-driven actions may be obligate (mandatory) or facultative (permissive) as required. Such actions may be initiated in response to sensor input (ranging from simple inputs such as ambient temperature or chemical concentration, to more complex inputs such as detection of engulfment by a phagocyte), clock or calendar datemarks, completion of intermediate mission benchmarks, or external commands.

Six general classes of nanorobot control protocols have been identified and are described below. All nanorobots will require some of these protocols, and some may require most or all of them.

15.4.1. Operational Protocols

Operational protocols control the simplest and most basic procedures that nanorobots must perform. Such protocols may control the operation of individual or groups of sensors or sorting rotors, or may regulate internal power transfer and energy storage using logic-branched feedback loops. In the case of the respirocyte, a proper analysis of operational protocols would involve drawing up an operational flowchart for its normal behaviors, including alternate modes, operational options, parameter lookup tables, communications interfaces, load levels, switching among backups, internal state monitoring, sensor integration with sorting rotor function, engine throttling, and device-specific operational modes—e.g., resting, exercising, asphyxiating, underwater, high altitude, and so forth. Also included here are standardized user interfaces and user protocols to make it as convenient as possible for doctors to access and reprogram the nanodevices while ensuring patient safety.

The nanorobot must also be able to follow its activity script and recognize when its mission has ended, and then be conveniently shut down. In the case of the microbivore, the end of the mission is not clearly defined because another target microbe might always be found, hence the termination of this nanorobot's open-ended mission most likely will be determined by a clock or will be externally commanded by the attending physician. In other cases, like the chromallocyte, the mission sequence is clearly defined and has a specific end state because the device can track its progress through the preprogrammed activity script checklist as a function of completed tasks and external events. Activity scripts may be the most complex form of operational protocol.

15.4.2. Biocompatibility Protocols

Most nanorobots may require active processes to maintain biocompatibility [2]. These processes will require computer control and may range from simple protocols such as controlling alterable surface chemical modifications (e.g., adhesioregulation [2k]) and the microbial tail protocol (Section 2.2.2) to relatively complex processes such as phagocyte avoidance and escape protocols [2d] or the splenofenestral escape protocol [5] (for bloodborne nanorobots above a certain size), activation of which will initiate a stereotyped set of locomotive movements that result in breaking free from splenic captivity and returning to the bloodstream (Section 2.2.2).

Numerous other biocompatibility situations requiring specialized protocols are identified elsewhere [2]. One interesting example involves the commandeering of natural motile cells by medical nanorobots, a procedure known as cytocarriage [1bn], as an alternative mode of *in vivo* transport. During cytocarriage, one or more medical nanorobots may enter a motile cell, ride or steer the cell to a desired destination inside the human body, then vacate the cell upon arrival. When macrophages and other leukocytic cells become infected, they express B7 on their membrane surface which can be recognized by a T-cell CD28 receptor protein, triggering an immunologic response. In the "infected cytovehicle protocol," nanorobot pilots would inspect the cell surface of a prospective cytovehicles for B7 and similar flags prior to cytopenetration to avoid choosing an infected cell for cytocarriage that could then spread the infection. If the cell subsequently becomes infected and begins expressing B7 or other warning substances during the journey, the protocol would direct the nanorobot pilot to abort the mission and steer the cytovehicle to a nearby disposal site, or implement immediate therapeutic measures; failing this, the pilot should abandon the vehicle at once.

15.4.3. Theater Protocols

A theater protocol is a control process by which the nanorobot verifies that it is performing its tasks in the intended location or in the desired circumstances, as prescribed by the attending physician. If the nanorobot determines that its present location is inappropriate or the necessary conditions to enable action no longer exist, the protocol triggers an appropriate corrective or defensive response. When a nanorobot detects an out-of-theater condition (e.g., bled out of body), the most appropriate response may often be to safe the device and then shut down. Subsequent reentry into theater after shutdown (e.g., bled-out devices find their way back into the bloodstream) normally should not allow reactivation of the device because during its absence from theater, the nanorobot may have been subjected to unknown forces, chemicals, radiation, or even reprogramming that could render it harmful to the patient upon reentry, if reactivated. Even if a full pre-restart self-diagnostic routine could be performed after an out-of-theater excursion, unwanted modifications may be too subtle to detect via internal sensors and the risk of malicious tampering is too great. One possible exception is where the violated theater of operation is a functional condition such as "patient is asleep," "patient is drunk," or "patient is sexually aroused." In these cases, it is less likely that switching on or off in response to theater protocols would pose significant risk either to the devices or to the patient because these conditions are normally cyclical volitional states that define inherently ephemeral or periodic theaters of action.

The longer a therapeutic nanorobotic medical procedure takes to complete, the more times a theater protocol should be executed during the mission. Protocols might be checked more or less frequently depending upon their mission criticality, their impact on safety, the likelihood of a rapid change in status, their computational overhead (the more often the protocol is run, the fewer computer resources are available for primary tasks), and data acquisition time (e.g., detecting a very rare molecule in the bloodstream might take many seconds or minutes; the reprotocol cycle time should be significantly longer than the maximum practical sensing period). Some protocols might be checked continuously—biochemical processes may occur in milliseconds, other biological responses may require seconds or minutes (e.g., automobile fuel gauges are rechecked electronically about once every second.)

It is worth noting that theater protocols are found in natural human cell biology. For example, most tissue cells require attachment to the ECM and subsequent spreading on the ECM, for proper growth, function, and even survival, a feature known as anchorage dependence [66]. Upon losing this anchorage and departing from the desired theater of operation, cells often die by undergoing apoptosis or programmed cell death. (Only the cells circulating in the blood are designed to survive without attachment and spreading—some tumor cells acquire this ability and leave their original tissue site to form metastases, a pathological state.) In this theater protocol, the cell is designed to default to suicide, but defers this action as long as a survival signal is received, indicating continued residence in the permitted theater of operations. The survival signal is generated as long as the cell maintains integrin attachment to a basement surface or ECM. This protocol trigger is called an apoptotic switch [67]. T-cells employ a similar theater protocol via the "immunological synapse" [68].

There are at least six basic classes of theater protocols.

1. *Locational.* Nanorobot activities may be locationally restricted to certain specific regions of the body (e.g., heart, liver, epidermis, skull, ear canal, right arm, big toe, or bloodstream). They may be restricted negatively—allowed to operate only as long as they are not present in a particular location—or may have a combination such that they exhibit two distinct modes of behavior (e.g., location A behavior vs. location B behavior, or location A versus location not-A behaviors). Thus a nanorobot that has been bled out of the body or has been surreptitiously syringed from one body to another would automatically deactivate. Since nanorobots can contain reliable onboard geographic location and gravity sensors [1ax], nanorobots could vary their activities as the patient moved across the surface of the Earth, and astronaut-implanted nanorobots could vary their behaviors according to whether the human is (a) on the ground, (b) experiencing high-acceleration during boost phase, or (c) experiencing on-orbit weightlessness.
2. *Functional.* Nanorobots may include theater protocols to restrict their actions according to functional criteria, such as where a set of specified conditions exists (e.g., only in a cancer tumor secreting certain biochemical factors), or only where blood composition lies within certain preset limits, or only in bloodstreams in which large amounts of alcohol, nicotine,

hallucinogenic drugs, or other bioactive substances are present (or absent). Nanorobots might be restricted to operations only when blood pressure moves above or below certain preset thresholds, or when respiration or heart rate is too fast or too slow, or within a defined range. Measured progesterone level could serve as one indicator of pregnancy, and a high blood concentration of the hormone relaxin could indicate that the adult patient has recently undergone childbirth. Other devices might be programmed to operate only during periods of sexual arousal (as determined operationally by a set of critical body-wide physiological parameters), or only when the patient is hungry, or is having an epileptic seizure, or is in REM-phase sleep, or only when the patient is talking, sneezing, defecating, or urinating (all of which have unique and distinct physiological adjuncts that can be detected, measured, and acted upon by a sufficiently discriminating acoustic macrosensing system [1ba]). Another functional protocol that might be useful in some cases is a determination as to whether the patient was functionally "alive" or "dead"—as provisionally determined by a molecular analog to the Karnofsky Performance Score [69] using a constellation of sensor readings from an *in vivo* nanorobot population including blood velocity, tissue oxygenation or carbonation levels, glucose depletion, monitoring of mechanical body noises and body temperature, presence of chemical poisons or forensic biochemical death markers, absence of electrical or other neural activity, dysfunctional ionic balances, and so forth.

3. *Situational.* Nanorobots could be programmed with narrowly specialized situational protocols driven by complex multiple sensor data designed to detect rare events. For example, one situational theater might be defined as the simultaneous receipt of sensor data from nanorobots stationed throughout the body indicating (1) high concentrations of adrenalin in the bloodstream, (2) a sudden transition from a gravity-detected to a low-gravity (e.g., free-fall) environment as recorded by acceleration and gravity sensors, (3) whistling noises detected by nanorobots stationed at the cochlear nerves, (4) continuous rotational movements detected by the vestibular apparatus, (5) drop in ambient air pressure in the lungs, and (6) kinesthetic sensors detecting flailing motions of arms and legs. The simultaneous occurrence of all these sensory events might suggest that the patient was falling through the air, and might provide the unique theater of operation of nanorobots designed to automatically respond to this event. Other more common situations might also be recognized using situational protocols. For instance, determination that the patient is wearing clothing, and the approximate extent of coverage of the epidermis by that clothing, may be detectable by a sophisticated whole-body nanorobot population by sampling dermal, neural, pressure, and temperature receptor output, by measuring variations in transdermal light transmission, or by mapping variations in sweat gland cell metabolic rate [1bo].

4. *Phenotypic*. Nanorobots might be allowed to operate only in patients that have (or lack) specific detectable phenotypes. Membership in such classifications might be assessed biochemically by identifying specific sets of bloodborne substances. For example, determining the ratio of estrogen to testosterone might be one of several coordinated tests for gender. Biological age might be roughly estimated by combining measurements of glucosamine levels (which increase with age), testosterone levels (which often decrease with age), racemization of proteins, and other biochemical markers of aging. Obesity might be detectable by sampling the ratios of specific classes of lipid carriers or monitoring leptin and other biochemicals in the blood. It might be possible to detect certain behavioral traits such as thrill-seeking, shyness, psychopathology, or schizophrenia, all of which are known or suspected to be associated with significant biochemical or hormonal variations that are in principal detectable by bloodborne nanorobots. Hair or skin color might also be measurable, if the pigmentation generates detectable bloodborne adjuncts. Recent research indicating possible physiological and biochemical correlates of sexual preference [70–72] may put this into the class of phenotypes that could be selectable as permissible or restrictive theaters of nanorobotic operation, though reliability may be a significant challenge in the implementation of certain phenotypic protocols.
5. *Temporal*. The theater of nanorobot activity may be further limited by temporal restrictions to operations performed at certain specific times or with measured durations. Examples of temporal protocols might include restrictions to operation only at night (e.g., when the patient is likely to be asleep) or at a specific time of day (e.g., between 4 AM and 5 AM in the morning). Activity may be event-sequenced—permitted only at specific points in the daily circadian rhythm, or in synchronization with other natural rhythms of the body such as brain waves, gastric oscillations, cardiac impulses, respiratory cycling, daily cortisol variations, menstrual cycles, or coughing stacatta. Nanorobots could be restricted to operating for only one minute out of every hour, whether simply to save power or alternatively to perform a lengthy procedure in smaller increments for medical reasons. Temporal protocols might apply only for a set period of time followed by termination of activity, or may be keyed to the calendar, activating, for example, only at midnight on New Years Eve or annually on one's marital anniversary. (These latter instances may presume a longer-lived nanorobotic residence *in vivo*.)
6. *Identity*. Nanorobots may also be subject to identity protocols in which activities are permitted only if the device can confirm that it is present within the body of a specific person and no one else. This sort of control may be enabled via biochemical biometrics in which onboard sensors are used to detect specific cytochemical or plasma membrane self-markers [1ag] such as HLA specificities in MHC Class I molecules on cell

membrane, blood group antigens on red cells surfaces, and other individual biochemical signatures other than genetic differences (which are not directly accessible to most nanorobot species). An alternate method for uniquely identifying individual people is to insert an artificial gene into all cells of each person from birth whose sole purpose is to cause cells to synthesize a harmless simple chemical marker, perhaps a digitally encoded short carbohydrate chain [73]—even a small hexasaccharide can encode $\sim 10^{12}$ unique combinations [74]—that is released into the bloodstream and is readily detectable and decoded by theater-restricted medical nanorobots. The deployment of such self-renewable *in vivo* fiducial chemical markers could serve as a convenient substitute for less-intrusive phenotypic measurements but might be regarded as controversial much like the 1999 USPTO-granted patent for barcoding humans [75].

15.4.4. Safety Protocols

Safety engineering involves making sure things do not fail in the presence of random faults. As Dorner [76] notes that

> learning theory tells us [that] breaking safety rules is usually reinforced, which is to say, it pays off. Its immediate consequence is only that the violator is rid of the encumbrance the rules impose and can act more freely. Safety rules are usually devised in such a way that a violator will not be instantly blown sky high, injured, or harmed in any other way but will instead find that his life is made easier. The positive consequences of violating safety rules reinforce our tendency to violate them, so the likelihood of a disaster increases. And when one does in fact occur, the violator of safety rules may not have another chance to modify his behavior in the future.

Since the risk of harm is great if medical nanorobots are misused, future designers should try to make it as hard as possible to disable the safety features. To the greatest degree practical, these features should be permanently embedded in hardware to minimize the probability of circumvention.

Nanorobot control systems should employ failsafe designs which may incorporate parallelism (dividing tasks among a large number of simple systems), specialization (individual systems optimized for particular tasks), and redundancy (comparing the output of multiple systems to improve reliability of the results). These built-in safeguards are then enhanced by the use of safety protocols which are active device behaviors designed to further enhance and reinforce safety. Failsafe designs must ensure that even a total system failure will not lead to death or serious injury of the patient. Safety protocols should be able to recognize various catastrophic internal failure states including compromised physical structure or software/data corruption, necessitating localized, intermediate, or even whole system shutdown, with entry into a safe-harbor mode that may permit the nanorobot to engage in actions that will safely remove it from the body since it can no longer fulfill its mission. One simple failsafe mechanism widely employed in high-reliability programmable devices is the watchdog timer—a counter that can shut down the computer if it ever reaches zero, but which is continually reset by a correctly operating program so that it never reaches zero as long as the program continues functioning (analogous to the biological "apoptotic switch" described earlier). In the case of medical nanorobots that require active supervision, the reset command could be periodically rebroadcast to the nanodevices *in vivo* by the attending physician via ultrasound messaging.

Safety protocols may range from relatively simple procedures, such as self-diagnostic routines, to very complex procedures such as dental protocols which instruct mobile nanorobots bled into the mouth to avoid the hard grinding surfaces by retreating to lower positions on the teeth [1z], or flight protocols for aerial nanorobots including self-enforcement of no-fly zones near nose and mouth along with other active anti-inhalation, inhale-safe, and post-inhalation extraction protocols [1bp]. Other safety protocols may incorporate a wide variety of user-set locks and limits, command blocking, limits on access to functions by non-physicians, and perhaps some equivalent to the humorously termed "shame blocker" (a recent telephone gimmick in which the user dials 333 and a number, and then his phone won't let him call that number, e.g., his ex-wife when he's drunk).

15.4.5. Security Protocols

Even the best security cannot prevent all harm but can help avoid significant harm while allowing the system to continue operating normally. While most features are useful for what they do, security features within products are useful because of what they don't allow to be done. Security engineering helps to ensure that the nanodevice will not fail "in the presence of an intelligent and malicious adversary who forces faults at the worst time and in the worst way." Security protocols are required to ensure that incoming commands originate from trusted and authorized sources, and may employ checksums, signed and certified programs, and formally proofed systems without trapdoors. These protocols should normally refuse to accept commands that could cause the device to exhibit behaviors that would harm the patient. Communication protocols might include TCP/IP protocol stacks combined with typical security technologies such as firewalls, packet filtering, intrusion detection, and secure procedures for flow control and authentication (e.g., passwords, biometrics, and public-key cryptography) to prevent "body-hacking." Depending on circumstances, nanorobots may need blockers for viruses, worms, and spam, and protocols linked to sensors that monitor structural integrity to ensure tamper resistance.

Other security protocols may be required to authorize reading and writing data into personal medical record caches implanted in the patient's body—multiply redundant caches $\sim 1\ mm^3$ in size could hold >1000 TB of fast-access mechanical memory, capacious enough to store a lifetime of detailed medical data, thus simplifying and perfecting medical recordkeeping while making lifesaving data more quickly accessible to emergency medical personnel. Security protocols must safeguard not just these internal data stores, but also access to the patient's implanted internal communications network and any other *in vivo* nanorobotry that might be present. Other security protocols enforcing digital rights management (DRM), audit trails, process accounting, expiration dates and related schemes may be employed on commercial systems to enforce patent rights. However, any regulations requiring implementation of embedded government-sponsored controls or content filtering should be regarded with deep suspicion.

15.4.6. Group Protocols

Group protocols may be required to control the collective behaviors of large populations of simultaneously interacting *in vivo* medical nanorobots. It is hoped that relatively simple individual nanorobot behaviors [77] can be programmed that will give rise to more complex desired group behaviors. Multirobot control algorithms are a major research field today with precedents in agoric algorithms [78–81], stigmergy [82], swarm computing [83, 84], and agent-based systems [80].

15.5. CONCLUSIONS

Medical nanorobots may be constructed of diamondoid nanometer-scale parts and subsystems including onboard sensors, motors, manipulators, power plants, and molecular computers. The presence of onboard nanocomputers will allow *in vivo* medical nanorobots to perform numerous complex behaviors which must be conditionally executed on at least a semiautonomous basis, guided by receipt of local sensor data, constrained by preprogrammed settings, activity scripts, and event clocking, and further limited by a variety of simultaneously executing real-time control protocols.

Such nanorobots cannot yet be manufactured, but preliminary scaling studies for several classes of medical nanorobots including respirocytes, microbivores, clottocytes and chromallocytes have been published in the literature. These designs allow an analysis of basic computational tasks and a summation of major computational control functions common to all complex medical nanorobots. These functions include the control and management of pumping, sensing, configuration, energy, communication, navigation, manipulation, locomotion, computation, and the use of redundancy management and flawless compact software.

Nanorobot control protocols are required to ensure that each nanorobot completes its intended mission accurately, completely, safely, and in a timely manner according to plan. Six major classes of nanorobot control protocols have been identified and include operational, biocompatibility, theater, safety, security, and group protocols. Six important subclasses of theater protocols include locational, functional, situational, phenotypic, temporal, and identity control protocols.

ACKNOWLEDGMENTS

Thanks are due to Forrest Bishop for the respirocyte and microbivore images, Zyvex Corp. for the use of the microbivore image, and John Luu and his colleagues at Stimulacra LLC for the chromallocyte images. Research grants from Alcor Foundation, Life Extension Foundation, Kurzweil Foundation, and the Institute for Molecular Manufacturing supported this work.

REFERENCES

1. R. A. Freitas, Jr. *Nanomedicine, Volume I: Basic Capabilities.* Georgetown, TX: Landes Bioscience, 1999; http://www.nanomedicine.com/NMI.htm; (a) Chapter 4, (b) 9.3, (c) Chapter 6, (d) 10.2, (e) 10.3, (f) 3.4.2, (g) 6.3.4, (h) 4.2.1, (i) 4.6, (j) 4.2, (k) 4.5, (m) 7.2.2, (n) 10.3.6, (o) 4.5.4, (p) 6.3.2, (q) 6.4.1, (r) 7.2.5.3, (s) 6.4.3, (t) 7.2.5.4, (u) 6.3.4.4, (v) 6.3.4.5, (w) 6.3.4.2, (x) 6.4.3.4, (y) 10.2.1, (z) 9.5.1, (aa) 7.4.6, (ab) 10.4.2.4.2, (ac) 8.5.2.2, (ad) 9.3.1.4, (ae) 9.3.5.1, (af) 7.2.6, (ag) 8.5.2.1, (ah) 8.2.1.1, (ai) 9.4.1.4, (aj) 3.2, (ak) 10.5.4, (am) 7.4.5.2, (an) Chapter 8, (ao) 9.4.4.1, (ap) 9.4.4, (aq) 9.4.5, (ar) 9.4.5.3, (as) 9.4.4.2, (at) 9.4.2.5.1, (au) 8.3.3, (av) 4.7, (aw) 4.9.2, (ax) 4.9.2.4, (ay) 4.2.8, (az) 8.5.2, (ba) 4.9.1, (bb) 4.9.3, (bc) 4.9.4, (bd) 4.9.5, (be) Chapter 5, (bf) 5.4, (bg) 5.3.6, (bh) Chapter 7, (bi) 9.4, (bj) 9.5, (bk) 10.1, (bm) 7.3, (bn) 9.4.7, (bo) 8.4.1.3, (bp) 9.5.3.6, (bq) 7.4.6.7.
2. R. A. Freitas, Jr. *Nanomedicine, Volume IIA: Biocompatibility.* Georgetown, TX: Landes Bioscience, 2003; http://www.nanomedicine.com/NMIIA.htm; (a) 15.6.2, (b) 15.4.3.1, (c) 15.4.3.2.3, (d) 15.4.3.6, (e) 15.4.2.3, (f) 15.4.2, (g) 15.4.3.6.1, (h) 15.4.3.6.4, (i) 15.2.5, (j) 15.5.2.2, (k) 15.2.2.4, (m) 15.4.3.4.
3. R. A. Freitas, Jr. Current status of nanomedicine and medical nanorobotics (Invited Survey). *Journal of Computational and Theoretical Nanoscience*, 2: pp 1–25, Mar 2005; http://www.nanomedicine.com/Papers/NMRevMar05.pdf.
4. R. A. Freitas, Jr. Exploratory design in medical nanotechnology: A mechanical artificial red cell. *Artificial Cells, Blood Substitutes, and Immobilization Biotechnology*, 26: pp 411–430, 1998; http://www.foresight.org/Nanomedicine/Respirocytes.html.
5. R. A. Freitas, Jr. Microbivores: Artificial mechanical phagocytes using digest and discharge protocol. *Journal of Evolution and Technology*, 14: pp 1–52, Apr 2005; http://www.jetpress.org/volume14/freitas.html.
6. R. A. Freitas, Jr. Clottocytes: Artificial Mechanical Platelets. Foresight Update No. 41: pp 9–11, June 30, 2000; http://www.imm.org/Reports/Rep018.html.
7. R. A. Freitas, Jr. The ideal gene delivery vector: chromallocytes, cell repair nanorobots for chromosome replacement therapy. *Journal of Evolution and Technology*, 16: pp. 1–97, June 2007; http://jetpress.org/v16/freitas.pdf.
8. R. A. Freitas, Jr. Christopher J. Phoenix. Vasculoid: a personal nanomedical appliance to replace human blood. *Journal of Evolution and Technology*, 11: pp 1–139, Apr 2002; http://www.jetpress.org/volume11/vasculoid.pdf.

9. R. A. Freitas, Jr. Nanodentistry. *The Journal of the American Dental Association*, 131: pp 1559–1566, Nov 2000; http://www.rfreitas.com/Nano/Nanodentistry.htm.
10. R. A. Freitas Jr. Pharmacytes: An ideal vehicle for targeted drug delivery. *Journal of Nanoscience and Nanotechnology*, 6: pp 2769–2775, Sep–Oct 2006; http://www.nanomedicine.com/Papers/JNNPharm06.pdf.
11. Nanofactory Collaboration website. Accessed Jan 2007. http://www.MolecularAssembler.com/Nanofactory.
12. K. E. Drexler. *Nanosystems: Molecular Machinery, Manufacturing, and Computation*. New York: Wiley, 1992; (a) Chapter 10, (b) Chapter 11, (c) 13.4, (d) Chapter 12, (e) 16.3.2, (f) 12.4, (g) 6.7.2.
13. R. A. Freitas Jr. and R. C. Merkle. *Kinematic Self-Replicating Machines*. Georgetown, TX: Landes Bioscience, 2004; http://www.molecularassembler.com/KSRM.htm.
14. K. Ishiyama, M. Sendoh, and K. I. Arai. Magnetic micromachines for medical applications. *Journal of Magnetism and Magnetic Materials*, 242–245: pp 1163–1165, 2002.
15. D. D. Chrusch, B. W. Podaima, and R. Gordon. Cytobots: intracellular robotic micromanipulators. In: W. Kinsner, A. Sebak, editors. Conference Proceedings, 2002 IEEE Canadian Conference on Electrical and Computer Engineering. Winnipeg, Canada: 2002.
16. J. B. Mathieu, S. Martel, L. Yahia, G. Soulez, and G. Beaudoin. MRI systems as a mean of propulsion for a microdevice in blood vessels. *Biomedical Materials and Engineering*, 15: pp 367, 2005.
17. K. B. Yesin, P. Exner, K. Vollmers, and B. J. Nelson. Biomedical micro-robotic system. 8th International Conference on Medical Image Computing and Computer Assisted Intervention (MICCAI 2005), Palm Springs, California, Oct 26–29, 2005; www.miccai2005.org; p 819.
18. Micro-robots take off as ARC announces funding. Monash University, Oct 11, 2006; http://www.monash.edu.au/news/newsline/story/1038.
19. Committee to Review the National Nanotechnology Initiative, National Materials Advisory Board (NMAB), National Research Council (NRC). *A Matter of Size: Triennial Review of the National Nanotechnology Initiative*. Washington DC: The National Academies Press, 2006; http://www.nap.edu/catalog/11752.html#toc.
20. R. C. Merkle and R. A. Freitas Jr. Theoretical analysis of a carbon-carbon dimer placement tool for diamond mechanosynthesis. *Journal of Nanoscience and Nanotechnology*, 3: pp 319–324, Aug 2003; http://www.rfreitas.com/Nano/JNNDimerTool.pdf.
21. R. A. Freitas Jr., inventor. A simple tool for positional diamond mechanosynthesis, and its method of manufacture. U.S. Provisional Patent Application No. 60/543,802, filed Feb 11, 2004; U.S. Patent Pending, Feb 11, 2005; http://www.MolecularAssembler.com/Papers/DMSToolbuildProvPat.htm.
22. J. Peng, R. A. Freitas Jr., R. C. Merkle, J. R. Von Ehr, J. N. Randall, and G. D. Skidmore. Theoretical analysis of diamond mechanosynthesis. Part III. Positional C2 deposition on diamond C(110) surface using Si/Ge/Sn-based dimer placement tools. *Journal of Computational and Theoretical Nanoscience*, 3: pp 28–41, Feb 2006; http://www.MolecularAssembler.com/Papers/JCTNPengFeb06.pdf.
23. B. Temelso, C. D. Sherrill, R. C. Merkle, and R. A. Freitas Jr. High-level *ab initio* studies of hydrogen abstraction from prototype hydrocarbon systems. *Journal of Physical*

Chemistry A, 110: pp 11160–11173, Sep 28, 2006; http://www.MolecularAssembler.com/Papers/TemelsoHAbst.pdf.

24. H. J. Lee and W. Ho. Single bond formation and characterization with a scanning tunneling microscope. *Science*, 286: pp 1719–1722, 1999; http://www.physics.uci.edu/%7Ewilsonho/stm-iets.html.
25. N. Oyabu, O. Custance, I. Yi, Y. Sugawara, and S. Morita. Mechanical vertical manipulation of selected single atoms by soft nanoindentation using near contact atomic force microscopy. *Physical Review Letters*, 90: p 176102, May 2, 2003; http://link.aps.org/abstract/PRL/v90/e176102.
26. P. Vettiger, G. Cross, M. Despont, U. Drechsler, U. Duerig, B. Gotsmann, W. Haeberle, M. Lantz, H. Rothuizen, R. Stutz, and G. Binnig. The millipede: nanotechnology entering data storage. *IEEE Transactions on Nanotechnology*, 1: pp 39–55, Mar 2002.
27. D. Bullen, S. Chung, X. Wang, J. Zou, C. Liu, and C. Mirkin. Development of parallel dip pen nanolithography probe arrays for high throughput nanolithography. Symposium LL: Rapid Prototyping Technologies, Materials Research Society Fall Meeting; Dec 2–6, 2002; Boston, Proceedings of the MRS, Vol. 758, 2002; http://mass.micro.uiuc.edu/publications/papers/84.pdf.
28. Microcantilever Arrays. Protiveris Corp., 2003; http://www.protiveris.com/cantilever_tech/microcantileverarrays.html.
29. S. Martel and I. Hunter. Nanofactories based on a fleet of scientific instruments configured as miniature autonomous robots. Proceedings of the 3rd International Workshop on Microfactories; Sep 16–18, 2002; Minneapolis, pp 97–100, 2002.
30. G. R. Ter Haar. Biological effects of ultrasound in clinical applications. In: K. S. Suslick, editor. *Ultrasound: Its Chemical, Physical and Biological Effects*. New York: VCH Publishers, 1988.
31. E. Ayers. An automatic chemical plant. *Scientific American*, 187: pp 82–88; 1952.
32. H. D. Luke. *Automation for Productivity*. New York: Wiley, 1972.
33. C. Bennett. The thermodynamics of computation – A review. *International Journal of Theoretical Physics*, 21: pp 905–940, 1981.
34. R. C. Merkle. Reversible electronic logic using switches. *Nanotechnology*, 4: pp 21–40, 1993.
35. J. S. Hall. Nanocomputers and reversible logic. *Nanotechnology*, 5: pp 157–167, 1994.
36. W. H. Weihe, editor. *Physiological Effects of High Altitude*. New York: Macmillan, 1964
37. L. Jia, C. Bonaventura, J. Bonaventura, and J. S. Stamler. S-nitrosohaemoglobin: a dynamic activity of blood involved in vascular control. *Nature*, 380: pp 221–226, 1996.
38. H. L. Goldsmith and V. T. Turitto. Rheological aspects of thrombosis and haemostasis: basic principles and applications. *Journal of Thrombosis and Haemostasis*, 55: pp 415–435, 1986.
39. M. H. Smit and A. E. G. Cass. Cyanide detection using a substrate-regenerating, peroxidase-based biosensor. *Analytical Chemistry*, 62: pp 2429–2436, 2006.
40. A. J. Erslev. *In vitro* production of erythropoietin by kidneys perfused with a serum-free solution. *Blood*, 44: pp 77–85, 1974.
41. E. F. Fincher, L. Johannsen, L. Kapas, S. Takahashi and J. M. Krueger. Microglia digest staphylococcus aureus into low molecular weight biologically active compounds. *American Journal of Physiology*, 271: pp R149–R156, July 1996.

42. C. Strange and S. A. Sahn. The definitions and epidemiology of pleural space infection. *Seminars in Respiratory and Critical Care Medicine*, 14: pp 3–8, Mar 1999.
43. L. Carreno Perez. Septic arthritis. *Baillieres Best Practice and Research Clinical Rheumatology*, 13: pp 37–58, Mar 1999.
44. J. W. Costerton, P. S. Stewart, and E. P. Greenberg. Bacterial biofilms: a common cause of persistent infections. *Science*, 284: pp 1318–1322, May 21, 1999.
45. D. Goldin. Remarks at the 15th Annual NASA Continual Improvement and Reinvention Conference. Apr 27, 2000; http://rk.gsfc.nasa.gov/richcontent/Speeches/goldin_remarks_5ps.pdf.
46. N. Leveson. Information and Computer Science Department, University of California, Irvine. Viking Lander. *The Risks Digest* 3: Oct 1, 1986; http://128,240.150.127/Risks/3.72.html#subj1.
47. P. Marks. Dependence day. *New Scientist*: July 31, 1999; http://www.newscientist.com/article/mg16321971.000-dependence-day.html.
48. Ada Information Clearinghouse. Ada in Airbus 340 Flight Warning System. ITT Research Institute, 1994; http://www.adahome.com/Ammo/Success/aerofws.html.
49. H. Lin. The development of software for ballistic-missile defense. *Scientifc American*, 253: p 52, Dec 1985.
50. V. Cortellessa, B. Cukic, D. Del Gobbo, A. Mili, M. Napolitano, M. Shereshevsky, and H. Sandhu. Certifying adaptive flight control software. Proceedings of ISACC 2000: The Software Risk Management Conference. Reston, VA, Sep 2000; http://www.isacc.com/presentations/3c-bc.pdf.
51. D. E. Van Epps and B. R. Andersen. Streptolysin O inhibition of neutrophil chemotaxis and mobility: nonimmune phenomenon with species specificity. *Infection and Immunology*, 9: pp 27–33, Jan 1974.
52. A. Odegaard and J. Lamvik. The effect of phenylbutazone and chloramphenicol on phagocytosis of radiolabeled candida albicans by human monocytes cultured *in vitro*. *Acta Pathologica et Microbiologica Scandinavica. Section C, Immunology*, 84: pp 37–44, Feb 1976.
53. R. Kumar, J. E. Ansell, R. T. Canoso, and D. Deykin. Clinical trial of a new bleeding-time device. *American Journal of Clinical Pathology*, 70: pp 642–645, Oct 1978.
54. L. R. Hertzendorf, L. Stehling, A. S. Kurec, and F. R. Davey. Comparison of bleeding times performed on the arm and the leg. *American Journal of Clinical Pathology*, 87: pp 393–396, Mar 1987.
55. S. E. Lind. Chapter 33. The Hemostatic System In: R. I. Handin, T. P. Stossel, and S. E. Lux, editors. *Blood: Principles and Practice of Hematology*. Philadelphia: Lippincott, 1995 pp 949–972.
56. L. Ardekian, R. Gaspar, M. Peled, B. Brener, and D. Laufer. Does low-dose aspirin therapy complicate oral surgical procedures? *Journal of the American Dental Association*, 131: pp 331–335, Mar 2000.
57. R. L. Bick, B. Arun, and E. P. Frenkel. Disseminated intravascular coagulation. Clinical and pathophysiological mechanisms and manifestations. *Haemostasis*, 29: pp 111–134, 1999.
58. T. E. Warkentin. Heparin-induced thrombocytopenia: a ten-year retrospective. *Annual Review of Medicine*, 50: pp 129–147, 1999.

59. D. B. Brieger, K. H. Mak, K. Kottke-Marchant, and E. J. Topol. Heparin-induced thrombocytopenia. *Journal of the American College of Cardiology*, 31: pp 1449–1459, June 1998.

60. J. R. Sklaroff. Redundancy management technique for space shuttle computers. *IBM Journal of Research and Development*, 20: pp 20–28, Jan 1976; http://www.research.ibm.com/journal/rd/201/ibmrd2001E.pdf.

61. M. Malek. Survivable algorithms and redundancy management in NASA's distributed computing systems. Final Technical Report, NAS 1.26:189827, NASA/NTIS, 1992.

62. G. Huszerl and I. Majzik. Modeling and analysis of redundancy management in distributed object-oriented systems by using UML statecharts. 27th Euromicro Conference, 2001: A Net Odyssey (Euromicro'01): p 200, 2001.

63. C. Fishman. They write the right stuff. *Fast Company*, no. 6: p 95, Dec 1996; http://www.fastcompany.com/magazine/06/writestuff.html.

64. M. A. Coppess, J. M. Miller, D. P. Zipes, and W. J. Groh. Software error resulting in malfunction of an implantable cardioverter defibrillator. *Journal Cardiovascular Electrophysiology*, 10: pp 871–873, June 1999.

65. T. P. Ozahowski, M. L. Greenberg, P. Mock, P. T. Holzberger, B. Gerling, C. Zalinger, and C. Perry. Implantable cardioverter defibrillator clinic casualties: inadvertent reprogramming during routine implantable cardioverter defibrillator follow-up. *Pacing and Clinical Electrophysiology*, pp 1524–1525, Oct 19, 1996.

66. E. Ruoslahti. Stretching is good for a cell. *Science*, 276: pp 1345–1346, May 30, 1997.

67. C. S. Chen, M. Mrksich, S. Huang, G. M. Whitesides, and D. E. Ingber. Geometric control of cell life and death. *Science*, 276: pp 1425–1428, May 30, 1997.

68. A. Grakoui, S. K. Bromley, C. Sumen, M. M. Davis, A. S. Shaw, P. M. Allen, and M. L. Dustin. The immunological synapse: a molecular machine controlling T cell activation. *Science*, 285: pp 221–227, July 9, 1999.

69. D. A. Karnofsky and J. H. Burchenal. The clinical evaluation of chemotherapeutic agents in cancer In: C. M. MacLeod, editor. *Evaluation of Chemotherapeutic Agents*. New York: Columbia University Press, 1949, p 196.

70. D. F. Swaab and M. A. Hofman. Sexual differentiation of the human hypothalamus in relation to gender and sexual orientation. *Trends in Neurosciences*, 18: pp 264–270, June 1995.

71. W. H. James. Biological and psychosocial determinants of male and female human sexual orientation. *Journal of Biosocial Science*, 37: pp 555–567, Sep 2005.

72. M. J. Sergeant, T. E. Dickins, M. N. Davies, and M. D. Griffiths. Women's hedonic ratings of body odor of heterosexual and homosexual men. *Archives of Sexual behavior*: Dec 22 2006.

73. N. Sharon and H. Lis. Carbohydrates in Cell Recognition. *Scientific American*, 268: pp 82–89, Jan 1993.

74. R. A. Laine. A calculation of all possible oligosaccharide isomers, both branched and linear yields 1.05 × 1012 structures for a reducing hexasaccharide: the isomer barrier to development of single-method saccharide sequencing or synthesis systems. *Glycobiology*, 4: pp 1–9, 1994.

75. T. W. Heeter, inventor. Method for verifying human identity during electronic sale transactions. United States Patent 5,878,155, Mar 2 1999; http://patents.uspto.

gov/cgi-bin/ifetch4?ENG + PATBIB-ALL + 0 + 967198 + 0 + 7 + 25907 + OF + 1 + 1 + 1 + PN%2f5878155.

76. D. Dorner. *The Logic of Failure, Metropolitan Books.* (English translation). New York: Henry Holt, 1996.

77. A. Cavalcanti and R. A. Freitas Jr. Autonomous multi-robot sensor-based cooperation for nanomedicine. *International Journal of Nonlinear Sciences and Numerical Simulation,* 3: pp 743–746, Aug 2002; http://lipari.usc.edu/~pal/cs5xx/Cavalcanti02.pdf.

78. M. S. Miller and K. E. Drexler. Markets and computation: agoric open systems. In: B. A. Huberman, editor. *The Ecology of Computation.* Amsterdam: Elsevier, 1988; http://www.agorics.com/Library/agoricpapers/aos/aos.0.html.

79. O. Guenther, T. Hogg, and B. A. Huberman. Market organizations for controlling smart matter In: R. Conte, R. Hegselmann, and P. Terna, editors. *Simulating Social Phenomena, Lecture Notes in Economics and Mathematical Systems.* Berlin: Springer, 1997, pp 241–257.

80. J. S. Hall, L. Steinberg, and B. D. Davison. Combining agoric and genetic methods in stochastic design. *Nanotechnology*, 9: pp 274–284, Sep 1998; http://www.cs.rutgers.edu/~davison/pubs/chsmith/chsmith.html.

81. T. Toth-Fejel. Agents, assemblers, and ANTS: scheduling assembly with market and biological software mechanisms. *Nanotechnology*, 11: pp 133–137, June 2000; http://www.erim.org/cec/papers/NanoAgents.pdf.

82. E. Bonabeau, S. Guerin, D. Snyers, P. Kuntz, and G. Theraulaz. Three-dimensional architectures grown by simple 'stigmergic' agents. *BioSystems*, 56: pp 13–32, 2000.

83. E. Bonabeau, M. Dorigo, and G. Theraulaz. *Swarm Intelligence: From Natural to Artificial Systems.* Santa Fe Institute Sciences of Complexity Series. New York: Oxford University Press, 1999.

84. G. Theraulaz and E. Bonabeau. Modeling the collective building of complex architectures in social insects with lattice swarms. *Journal of Theoretical Biology*, 177: pp 381–400, 1995.

16

HETEROGENEOUS NANOSTRUCTURES FOR BIOMEDICAL DIAGNOSTICS

Bio-Inspired Multiscale Structures with NEMS and MEMS Components

Hongyu Yu, Mahsa Rouhanizadeh, Lisong Ai, and Tzung K. Hsiai

NEMS (nanoelectromechanical systems) and MEMS (microelectromechanical systems) have enthused both biological and physical scientists as well as engineers and physicians to engage in cross-disciplinary research. Interfacing nano- and microscale entities has ushered in a paradigm shift for life science applications. This chapter introduces several heterogeneous NEMS and MEMS devices dovetailed for biomedical diagnostics. The chapter will start with the biologically inspired nano- and microfluidic components that have impacted the design and invention of micro total analysis systems (μ-TAS) applications. Next, nano- and microbiosensors will be introduced to demonstrate the feasibility of translating bench lab research to *in vivo* testing with a potential to close the lab-to-patient gap. The chapter will conclude with the emerging bio fuel cells that have unfolded the integration of energy source and functional components for the fully implantable nano-, micro-biosystems. The authors hope that this chapter will take readers through a bioinspired journey in the micro- and nanodomain.

16.1. INTRODUCTION

Bioinspired nanoengineering and sciences involve a dynamic range of length and time scales. While classical Newtonian mechanics intimately governs the design

Bio-Inspired and Nanoscale Integrated Computing. Edited by Mary Mehrnoosh Eshaghian-Wilner

rules, surface tension and quantum mechanics are emerging as the operating principle for the micro- and nanoscale devices [1]. Nanoscale structures have materialized phenomena otherwise undiscovered, such as Coulomb blockade (or single-electron tunneling) and quantized or ballistic conductance to metal insulator transition. Quantum confinement of electrons by the potential wells at the nanoscale structures has provided a basis for the electrical, optical, magnetic and thermoelectric properties of a solid-state functional material. The advent of microelectromechanical systems (MEMS) and nanoelectromechanical systems (NEMS) has enabled researchers to interface physical phenomena with biochemical and molecular interactions, thereby challenging the existing engineering and biomedical paradigms.

The last decade has attested a paradigm shift in research activities from that of the individualized endeavor to cross-disciplinary efforts. In 330 B.C., Aristotle recognized cross-fertilization as a means to "search for truth "and he wrote:

> The search for truth is one way hard another easy, for no one can master it fully nor miss it fully, each adds a little knowledge to our nature, and from all things assembled there arises a certain grandeur.

In this chapter, readers will embark on a journey to cross boundaries in search of the bioinspired heterogeneous structures.

16.2. MICROSCALE COMPONENTS

16.2.1. Nature-Inspired Surface Tension Force

Nature reveals surface tension force at its best. Inspiring phenomena are the water striders walking on the surface of water, a raindrop forming a spherical shape, and blood flow in the capillary bed. The common property of these three examples lies in the enormous surface to volume ratio. Surface tension σ is defined as the ratio of the magnitude of force F tangentially applied to the surface of a liquid to the length L along which the force exerts as follows:

$$\sigma = F/L \tag{16.1}$$

Using the jointed stilt-like legs, water striders flit to and fro across the unflustered surfaces of water up to 1.5 m/s [2] (Fig. 16.1). Stationary water striders rest their weight on all six legs. The minimal requirement for static stability on the surface is defined as [3]

$$M_c = \frac{Mg}{\sigma P} = 1, \tag{16.2}$$

where M denotes mass, g force of gravity, σ surface tension, and P the curvature parameter. When $M_c = 1$, the body force Mg is equal to surface tension σP.

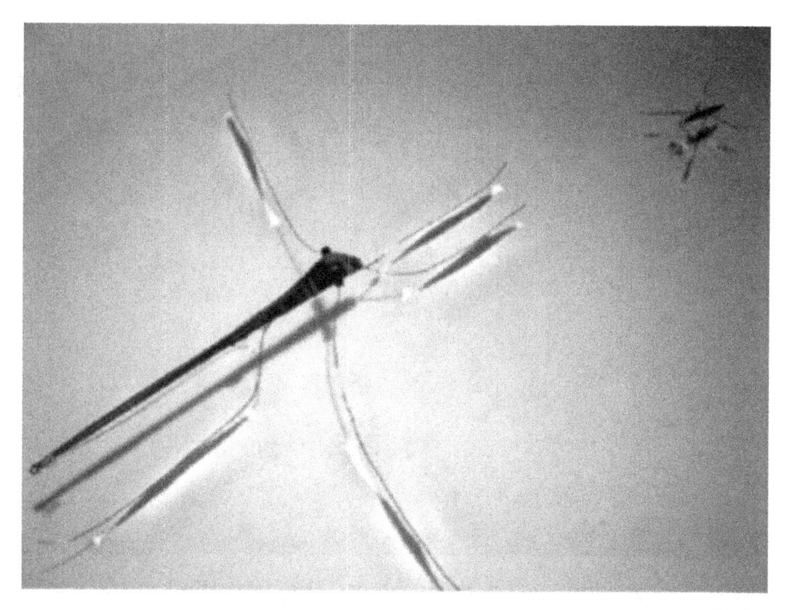

Figure 16.1. Water robot strider in relation to a water strider. (http://www.math.mit.edu/~dhu/Striderweb/striderweb.htm)

Water striders are able to remain stationary on water with a M_c ratio less than 1. In contrast, geckos are capable of dashing across the surface of water at a M_C ratio greater than 1 (Fig. 16.2). The attractive forces that hold geckos to the surface are van der Waals interactions between the finely divided setac and the surface [4].

While the precise mechanical principles by which large animals such as geckos and geese propel themselves against the surface of water remain unknown, investigators have been inspired by nature. The endless possibilities to manipulate the relation between body force and surface tension have generated tremendous enthusiasm to develop novel microscale devices.

16.2.2. Surface Tension-Inspired Microfluidics

The top-to-down approach to design small devices has encountered an increasing surface to volume ratio. While surface tension is a potential obstacle to mobilize mechanical components, researchers have employed various driving mechanisms, namely, thermal actuation [5], piezoelectrical property [6], electrolysis and electrostatic forces [7], acoustic actuation [8], and alike. Recently, Eun Sok Kim et al. have developed a focusing acoustic ejector [9] to overcome surface tension. The droplet emerges from the liquid surface as a result of convergence of piezoelectrically generated acoustic waves (Fig. 16.3).

Rather than avoiding surface tension, Jin (CJ) Kim's group at University of California, Los Angeles, developed a digital droplet production utilizing the

Figure 16.2. A crocodile gecko darts across the unruffled surface of a pond despite an $M_c > 1$.

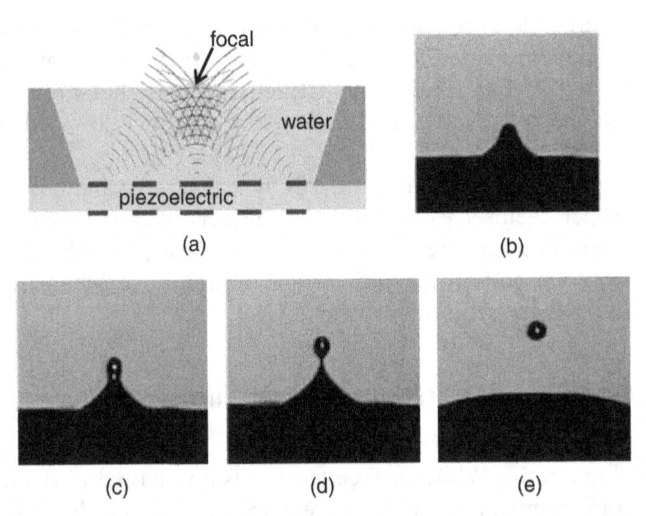

Figure 16.3. The droplet formation by breaking the surface tension. (a) Scheme of piezoelectric transducer-generated focal point providing sufficient energy to generate a discrete liquid droplet despite the surface tension. (b) A real-time photograph of a focal point emerging from the surface. (c) The droplet is taking into shape. (d) The droplet is about to be separated from the focal point. (e) The droplet is ejected into the mid-air, reaching to 230 μm in height. The diameter of this droplet is 70 μm.

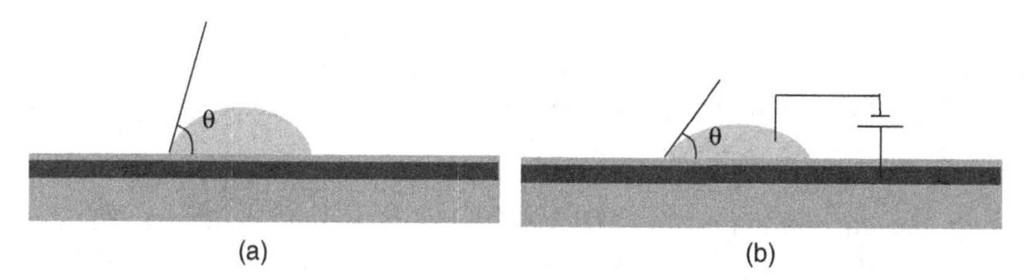

Figure 16.4. Contact angle (θ) between the liquid droplet and surface. (a) Contact angle in the absence of electrical filed. (b) Contact angle in the presence of applied electrical field.

property of electrowetting. Inspired by Lippmann's work [10], Kim's group incorporates electrostatic charge to modify the capillary force at the liquid and solid interface. The electrostatic force is directly related to the liquid surface tension. When a liquid droplet is placed on a dielectric layer (Fig. 16.4), various contact angles develop due to different hydrophobicities of the surface. The angle represents a quantitative measurement of the wetting ability of a liquid to the surface. When an electric field is generated between the liquid droplet and an electrode underneath the dielectric layer, the contact angle is altered, giving rise to a phenomenon known as electrowetting.

Based on Lippmann's equation, the tension γ_{SL} in the interface between the dielectric layer and liquid is defined as

$$\gamma_{SL}(V) = \gamma_{SL}|_{V=0} - \frac{CV^2}{2}, \tag{16.3}$$

where V represents the electrical potential difference between liquid and the electrode layer, and C denotes the specific capacitance of the dielectric layer.

By controlling the interface tension, E. S. Kim's group has developed digital microfluidic device containing a glass cover (top) and patterned electrode underneath (Fig. 16.5). An asymmetric electrostatic field is generated to create differential wettabilty between the front and rear sides of the droplet, thereby developing pressure difference between the front and the rear to drive the droplet to migrate forward. This driving mechanism is characterized as an electrowetting-on-dielectric (EWOD) actuation.

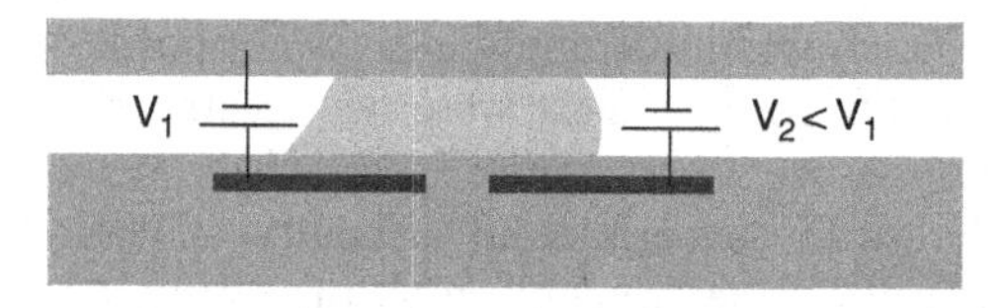

Figure 16.5. Asymmetric pressure develops due to the asymmetric electrical field. The droplet migrates to the left when V_2 is less than V_1.

The principle of EWOD has made possible the droplet formation, droplet separation, and droplet coalesce. The integration of these manipulations has materialized the possibility of digital microfluidics with relevance to biomedical applications such as DNA microstamping.

On the nanoscale, the electrowetting force is sufficient to drive the fluid into or through the carbon nanotube. The electrostatic potential across single-walled carbon nanotubes [11] overcomes the capillary resistance. This breakthrough translates electrowetting into the nanoscale technology such as nanolithography and zepto (10^{-21}) liters of liquid handling.

16.3. NANOSCALE COMPONENTS

As the surface tension becomes dominant over the force of gravity in the micro- and nanoscale domain, design issues such as static friction between two separated plates in vacuum requires the understanding of quantum effects. Chan et al. experimentally demonstrated the significance of such a phenomenon at the submicron scales [12]. Quantum electrodynamics predicts that even in a vacuum at a temperature of absolute zero, particles exist in the context of Heisenberg uncertainty principle.

When two parallel, uncharged conducting plates are positioned close together, the plates will start to move towards each other by a force. The magnitude of this force is a function of the area A of the plates, and inverse to the fourth power of the distance of plate separation:

$$\frac{F_c}{A} = -\frac{\pi^2 \hbar c}{240} \frac{1}{z^4}, \tag{16.4}$$

where c is the speed of light, $\hbar$ is called the reduced Plank constant which is equal to Planck constant h divided by 2π, and z is the separation between the plates. This phenomenon is called the Casimir effect, first predicted in 1948 [13]. When the separation between the surfaces decreases, the Casimir pressure increases rapidly, reaching about 1 atmosphere at $z \sim 10\,\text{nm}$.The quantum-mechanical effect of a vacuum fluctuation of the electromagnetic field or the Casimir effect is not limited to parallel plates, but it occurs in any two conducting materials in close juxtaposition.

Chan et al. designed an elegantly simple experiment [12]. Using MEMS fabrication technique, the authors etched a 3.5 μm thick, 500 μm^2-doped poly-silicon square plate, anchored in the middle of two opposite sides by small rods so that the plate was free to rotate. There were two electrodes underneath the plate, which operate as sensors. A separation of 2 μm existed between the plate and the electrodes. Considering the experimental difficulty of testing the Casimir effect with two plates parallel in extremely close proximity, the group at Georgia Institute of Technology measured the Casimir effect between their micromachined plate and a gold-coated Styrofoam ball with a radius of 100 μm [14]. This ball was

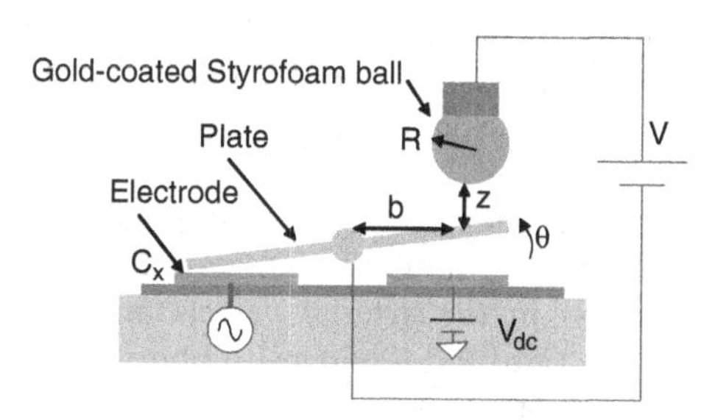

Figure 16.6. A cross-section schematic of the device with the electrical connections. The capacitance C_x between the top plate and the left electrode is a function of dc voltage V_{dc} applied to the electrode on the right. Angle of rotation θ of the top plate and the attractive electrostatic force is a function of separation (z). The voltage V applied to the sphere is 289 mV. The scheme is not drawn to scale.

glued to a wire whose displacement was precisely controlled to separate less than 2 μ from the plate. The Casimir effect produced a torque on the plate towards the Styrofoam ball. As the distance between the plate and the electrodes beneath it (increasing the distance to one of them, and decreasing it to the other) changes, the capacitances between the electrodes and the plate would be altered. By measuring the capacitance variations, the angular displacement of the plate was measured (Fig. 16.6).

Chan et al. showed that the dependence of the rotation angle on the separation between the surfaces is in agreement with calculations of the Casimir force. This experiment demonstrated that quantum-electrodynamical effects play a key role in MEMS and NEMS components when the separation between components is in the nanometer range.

16.4. BIOMEDICAL-INSPIRED MICRO- AND NANOCOMPONENTS FOR CARDIOVASCULAR MEDICINE

16.4.1. Integrating Micro- and Nanosensors: Biomedical Implication

Recent advances in molecular biology and dramatic progress in micro- and nanotechnology have occurred concomitantly. Development of specialized biomedical interfaces involves the exploration of biocompatible materials, ultra-low-noise microelectronics to sense biological signals without disturbing or interfering with the living tissue, as well as possible application of nanoscale components such as carbon nano tubes (CNTs). It is anticipated that the developed technologies in this field would be instrumental for a wide range of implanted and portable analytic devices, such as DNA and protein sensors, diagnostic micro Lab-on-a-Chip devices, and microelectronic interfaces for therapeutic applications. The MEMS and NEMS sensors offer an entry point to address biomedical issues. Their special and temporal resolution provides a basis to tackle the micro- and nanoscale domains otherwise difficult with the conventional technology.

In clinical medicine, coronary artery disease or atherosclerosis remains the leading cause of death in the industrialized nations. Atherosclerosis involves complex plaque formation in the arterial vessels and is considered to be an inflammatory disease [15]. Oxidized low-density lipoprotein cholesterol (oxLDL) has been considered important in the development of these inflammatory processes with the seminal observations that LDL must be oxidatively modified for it to be taken up by the inflammatory cells that are present in the blood and tissues [16] (Fig. 16.7). In this context, the development of nanowire sensors provides a new avenue to understand the fundamentals of oxidative modification of LDL in the arterial circulation with unprecedented sensitivity and selectivity.

Biomechanical forces, especially fluid and solid shear stresses, have significant effects on the arterial wall [17–19]. Fluid shear stress, the tangential drag force of blood passing along the surface of the endothelium [20], has metabolic as well as mechanical effects on endothelial cell (EC) function [21–23]. At the arterial bifurcations or branching points, endothelial cells become hyperpermeable in the presence of hyperlipoproteinemia that may favor intimal uptake and retention of LDL, resulting in local oxidative degradation of trapped LDL [24–26]. A complex flow profile develops at bifurcations, namely, flow separation and migrating stagnation points that create low and oscillating shear stress (Fig. 16.8).

At the lateral walls of arterial bifurcations, disturbed flow, including oscillatory flow (bidirectional with no net forward flow), is considered to be an inducer of oxidative stress that favors the pathogenesis of atherosclerotic plaque, whereas in the medial wall of bifurcation, pulsatile flow develops and down-regulates adhesion molecules, inflammatory cytokines, and oxidative stress [27, 28]

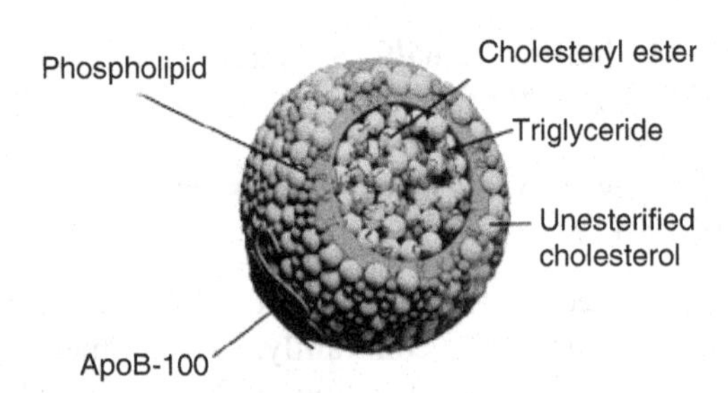

Figure 16.7. A three-dimensional representation of low density lipoprotein cholesterol (LDL). LDL particle contains phospholipids with their hydrophobic tails exposed to the outer surface. Embedded by these phospholipids are cholesteryl ester, and triglyceride and unesterified cholesterol. Surrounded the phospholipids are ApoB-100 protein. A typical LDL particle has a typical diameter from 25 to 27 nm.

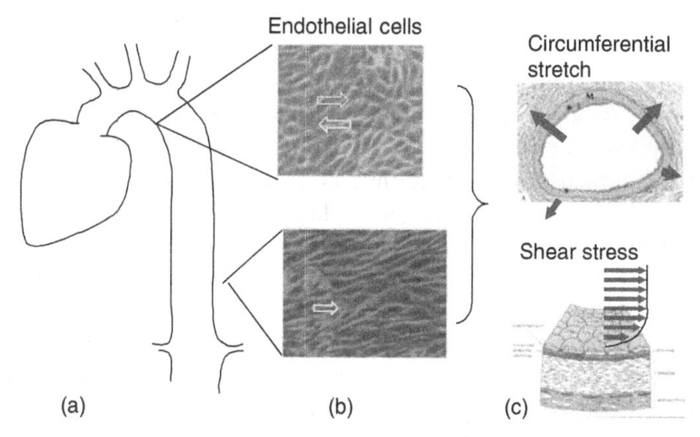

Figure 16.8. (a) Schematic drawing that shows the regions of curvatures and straight segments of the aorta. (b) Endothelial cells (EC) at the aortic arch experience disturbed flow, assuming random morphology, whereas in the straight portion, EC line in the direction of blood flow. (c) The arterial wall is exposed to circumferential stretch, wall shear stress, and hydrostatic force that acts perpendicularly on the inner wall.

(Fig. 16.9). Thus, direct measurement of shear stress is clinically relevant to the athero-prone regions.

The understanding of lipid biochemistry leads to the development of indium oxide (In_2O_3) nanowire sensors to investigate whether LDL particles preferentially undergo oxidative modification [29] at the arterial bifurcations where flow

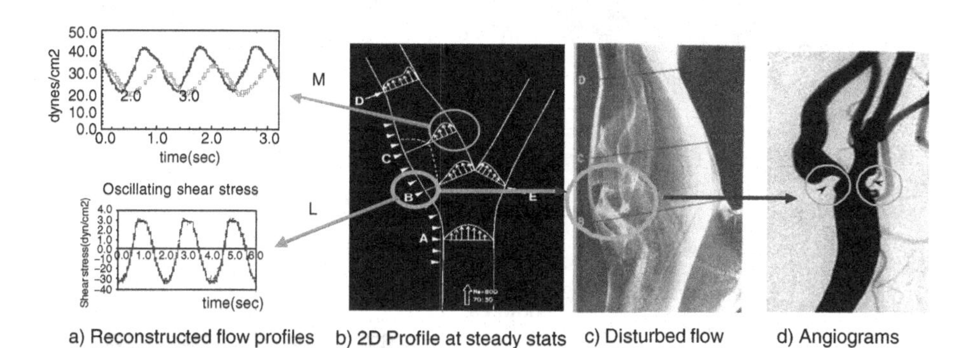

Figure 16.9. (a) Shear stress profiles at the lateral (L) and medial (M) walls of arterial bifurcations. (b) Pulsatile shear stress develops at the medial wall (M), whereas oscillating flow develops at the lateral (L) wall where migrating stagnation point occurs. (c) The lateral wall is the region where flow separation and disturbed flow develop. (d) Arterial angiogram shows the predilection sites for plaque formation.

separation occurs. In parallel, shear stress sensors [30] were developed to localize arterial sites where the uptake and local oxidative degradation of trapped LDL particles occur. The prototypes of nano- and microscale sensors will be discussed. Both sensors provide a basis to translate model of *in vitro* investigation to *in vivo* assessment in animal models.

Nanoscale biosensors based on individual semiconductive nanowires take the advantage of the enormous surface-to-volume ratio of such nanostructures and hold great promise to offer unprecedented sensitivity and response time, as well as the capability to directly convert biological signals to electrical signals. The feasibility of detecting LDL particles by individual In_2O_3 (Indium Oxide) nanowire-based FETs *in vitro* has been demonstrated by Zhou and Hsiai [31, 32]. The possibility of translating nanowire sensors to detect the oxidatively modified LDL trapped in the atherosclerotic lesions will be a paradigm to detect, predict, and treat individuals who are at risk for acute coronary syndromes.

16.4.2. Nanowire and Nanotube Networked Transistors

Individual nanowire and nanotube transistors offer superior electronic characteristics; however, device-to-device variation has been consistently observed and may cause difficulty in the biosensing experiments. A second family of sensors—nanowire/nanotube networked sensors—has been developed to address the above issues. These devices have been made by controlling the density of as-synthesized nanowires/nanotubes by tuning the catalyst density (Fig. 16.10). In_2O_3 nanowires-based FETs

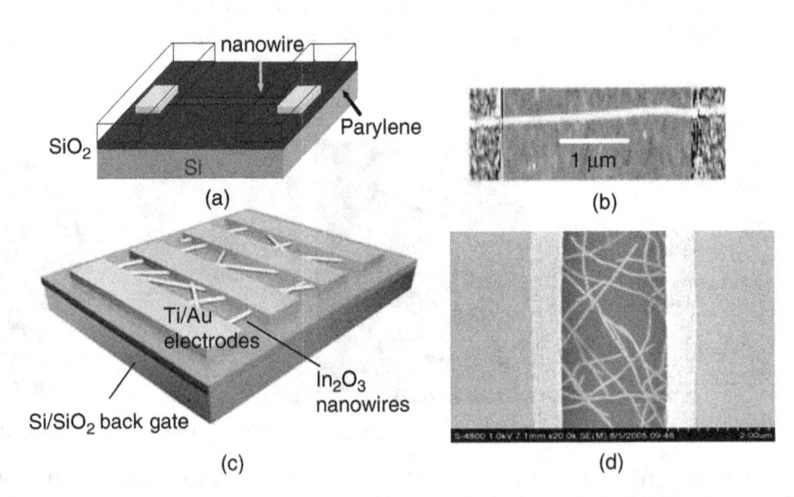

Figure 16.10. Nanowire-based field effect transistor (FET). (a) In_2O_3 nanowire transistor with metal electrode passivated by parylene. (b) Scanning electron microscope (SEM) of nanowire-based FET. (c) Concept of In_2O_3 nanowire networked transistor. (d) SEM image of the active area of one device showing multiple In_2O_3 nanowires bridging a pair of electrodes.

have been demonstrated as a sensitive chemical sensor for NO_2 and NH_3 [33]. In_2O_3 functions as the n-type semiconductor as a result of oxygen vacancy doping. In this context, the redox property of LDL cholesterol permits the nanosensors to accumulate or deplete electrons.

Indium oxide nanowires were synthesized through the chemical vapor deposition by laser ablating InAs as a target in a high pressurized quartz furnace. This process was performed on a silicon-silicon dioxide substrate that was patterned with 10 nm monodispersed gold nanoparticles. Individual In_2O_3 nanowires at 10 nm width were then sonicated into isopropyl alcohol suspension, followed by their deposition onto a degenerately doped silicon wafer coated with 500 nm SiO_2. Between the source and drain electrodes at 500 μm apart, the individual transistor contained multiple nanowires/nanotubes as the conductive channel. Therefore, the overall transistor and sensing characteristics were averaged over an ensemble of nanowires or nanotubes, and the device-to-device variation is significantly suppressed [32].

In_2O_3 represents a wide band gap transparent conductor with a direct band gap of about 3.6 eV and an indirect band gap of about 2.6 eV [34]. With these properties, In_2O_3 nanowires were selected to test whether LDL particles in the oxidized and reduced states modify the conductivity of the filed effect transistor based on In_2O_3 nanowires devices.

Selective detection of oxLDL was established by functionalization of the nanowire/nanotube surfaces with antibodies specific for LDL species. First, the nanowire/nanotube transistor surface was decorated with a layer of poly (ethylene imine) (PEI) and poly (ethylene glycol) (PEG). The surface chemistry was achieved via (a) the attachment of the antibodies to the PEI/PEG surface, and (b) interaction of the carboxyl group at the Y-end of the antibodies and the primary amines available in PEI (Fig. 16.11).

Figure 16.12 demonstrates the feasibility of developing In_2O_3 nanowires to detect redox state of protein molecules. Two factors may account for the increased electron concentration in the nanowires. The first is that the amino groups carried by the ApoB-100 protein in LDL particles may function as reductive species and hence, donate electrons to the nanowires. The second concomitant factor is due to positive charges carried by the amino groups, which can function as a positive gate bias to our nanowires, thus leading to the enhanced carrier concentration [35, 36]. Rouhanizadeh et al. reported a In_2O_3 nanowire-based FET as an emergent sensor to detect oxLDL [36]. Using both the I_D-V_{DS} (current versus drain-source voltage) and I_D-V_{GS} (current versus gate-source voltage), the investigators found that the sample which contained more oxidized LDL particles and consequently more free electrons, increased the conductivity of the nanowire-based FET. This increase in conductivity was distinct from the presence of ferrocytochrome-C.

16.4.3. Microscale Shear Stress Sensors

The emerging MEMS systems provide a spatial resolution comparable to the size of the endothelial cells (EC) and temporal resolution in the kHz range to

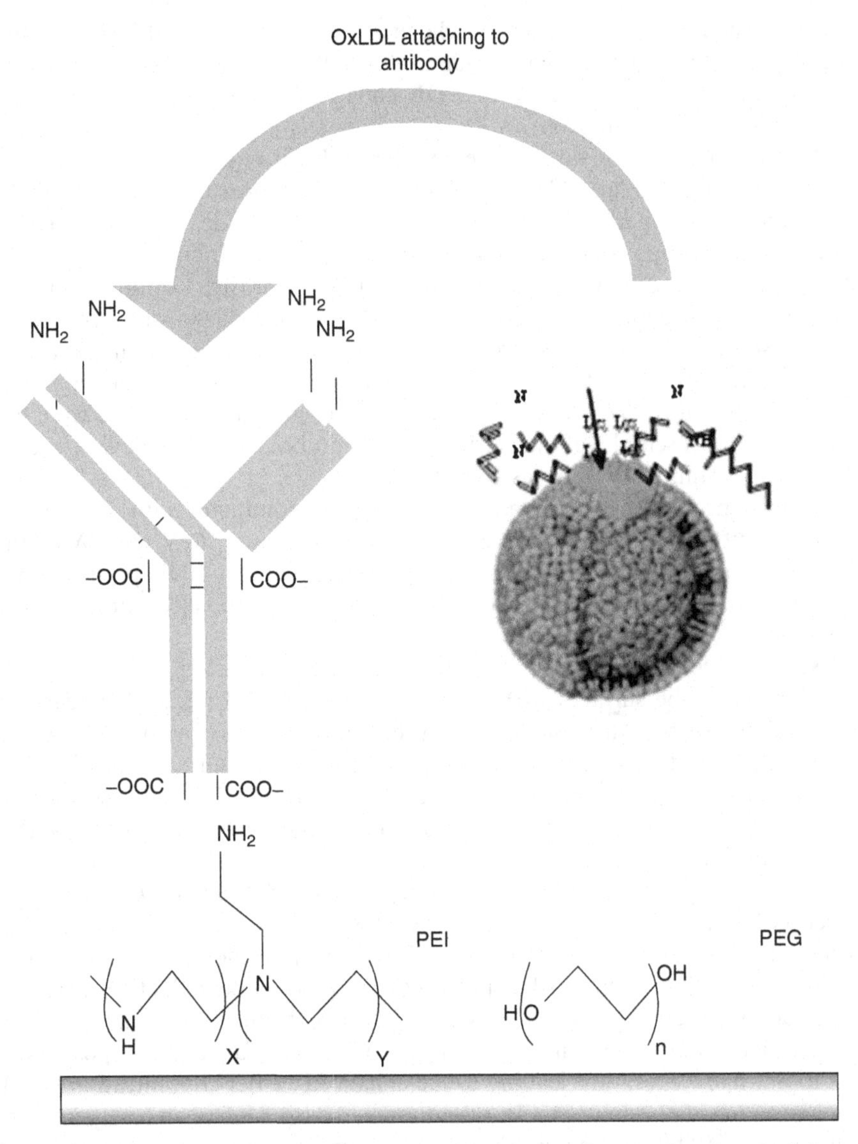

Figure 16.11. The nanosensor was coated with a layer of PEI/PEG polymer, followed by the attachment of antibodies to the primary amine of PEI. This sensor was then exposed to various biospecies, including oxLDL and oxidized cytochrome C. Selectivity is achieved via the selective binding of oxLDL to the antibodies.

investigate the mechanisms by which the characteristics of shear stress regulates the biological activities of endothelial cells at the complicated arterial geometry.

MEMS shear stress sensors are operated based on the heat transfer principle [37] by which the voltage output of the MEMS sensors is responsive to the

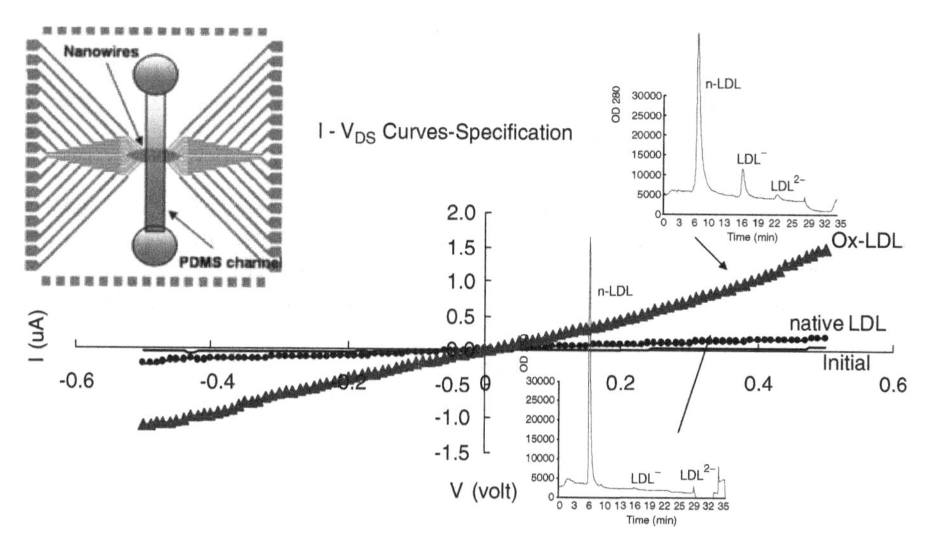

Figure 16.12. I-V_{DS} selectivity curve under dynamic condition. The presence of oxLDL (15.1%) gave rise to the highest conductivity compared with nLDL (4.4%). HPLC chromatograms of LDL were used to validate the results. Three peaks are seen, starting with the most abundant native LDL (nLDL), followed by the oxidized LDL. The top left insert shows the electrodes of source and drain in relation to the nanowires.

fluctuation in ambient temperature (Fig. 16.13). The temperature overheat ratio governs the temperature variations of the sensor

$$\alpha_T = \frac{(T - T_0)}{T_0}, \tag{16.5}$$

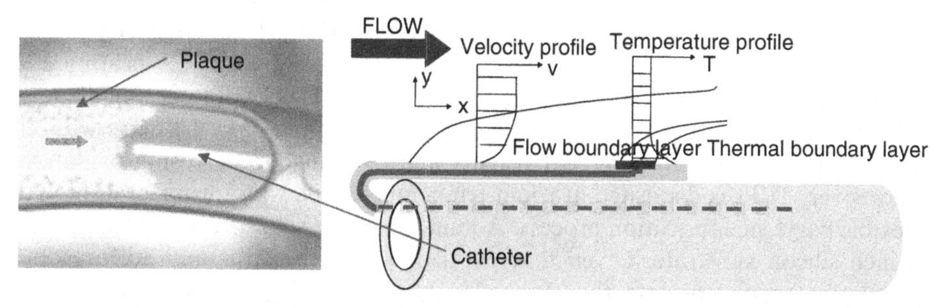

Figure 16.13. (a) A polymer-based shear stress sensor in a model of rabbit aorta. The diameter of catheter is 0.52 mm and aorta is 3.4 mm.(b) Packaging of sensors on the catheter in relation to thermal boundary layer.

where T denotes the temperature of the sensor and T_0 the ambient temperature. The relation between resistance and temperature overheat ratios is expressed as

$$\alpha_R = \frac{(R - R_0)}{R} = \alpha(T - T_0), \tag{16.6}$$

where α is temperature coefficient of resistivity or TCR. For shear stress measurement, we applied a high overheat ratio by passing higher current and by generating a "hot" sensing element to stabilize the sensor. Calibration was conducted for individual sensors to establish a relationship between heat exchange (from the heated sensing element to the flow field) and shear stress over a range of steady flow rates (Q_n) in the presence of rabbit blood flow at 37.8°C. The theoretical shear stress value corresponding to each flow rate was calculated using Equation 16.7.

$$\tau_w = \frac{6Q_n\mu}{h^2 w}, \tag{16.7}$$

where τ_w is the wall shear stress, μ is the blood viscosity, and h and w are the dimensions of the flow channel. The viscosity of the blood as a function of flow rate was measured using a viscometer. The individual calibrated sensors were then deployed into the NZW rabbit's aorta for real-time shear stress assessment.

16.4.3.1. Fabrication. Yu et al. have recently developed biocompatible, polymer-based MEMS sensors for real time measurements of shear stress and velocity with excellent spatial and temporal resolution. The micron-size and flexibility of the sensors allow the sensors to be inserted and/or attached into various anatomic structures of biological systems. The sensor is currently targeted to assess the physical parameters in the circulatory system of New Zealand White (NZW) rabbits (Fig. 16.14).

The newly developed micro polymer-based vascular sensors are composed of resistive heating and sensing elements that are encapsulated in the biocompatible polymer (parylene C). The sensor is fabricated by surface micromachining technique utilizing parylene C as microelectronic insulation. The sensor detects small temperature perturbation as fluid passed the sensing elements leading to changes in the resistance, from which shear stress is inferred. Novel biocompatible materials such as Titanium (Ti) and Platinum (Pt) are used as the heating and sensing elements which are exposed to blood flow. The Ti/Pt sensing element offers low resistance drift, large range of thermal stability, low 1/f noise without piezoresistive effect, and biocompatibility and resistance to corrosion/oxidation. Moreover, Ti/Pt is deposited at room temperature, allowing for integration with flexible parylene fabrication process. A hundred of the sensors are fabricated on a 3-inch silicon substrate. Given that parylene offers the structural stiffness and sturdiness to encapsulate the electrodes, the sensor has excellent mechanical strength and can be easily conformed to various anatomic curvatures. The resistance of the sensing element is about 1.7 kΩ, and the temperature coefficient of resistance (TCR) is at 0.11%/°C, compatible for blood rheology.

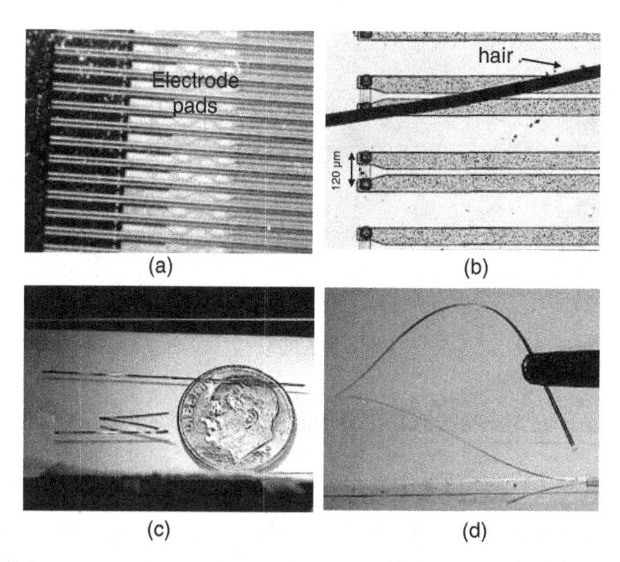

Figure 16.14. (a) An array of micro polymer sensors. (b) Magnification of individual sensors in relation to a string of hair. The sensing element is embedded in the tip of the sensor, measured at 200 μm (c) The sensor is flexible, assuming linear or zigzag fashions. (d) The sensor is folded by a tweezers without structural or functional damage. The sensor dimension is 4 cm in length, 320 μm in width and 21 μm in thickness.

The polymer-embedded sensor enables conformability to the arterial bifurcations and curvatures while retaining its mechanical strength and operational function. A flexible guide wire is cannulated through the catheter and connected to the sensing element for transmitting the signal from the vessel to the external circuitry. The micro intravascular sensors provide an entry point to link spatial and temporal variations in shear stress and the pathogenesis of coronary artery disease.

It has been also demonstrated that the polymer-based sensor is able to detect small but distinct temperature perturbation in response to the pulsatile blood flow in two specific regions of the arterial circulation, namely the abdominal aorta and aortic arch, in the New Zealand White (NZW) rabbits. Several engineering challenges have been addressed for *in vivo* investigation: (1) hemocompatibility and hemostasis of the sensor function in the rabbit blood; (2) signal-to-noise ratios and frequency responses under pulsatile arterial blood flow; and (3) novel packaging technique to transmit voltage signals to the external electronics. The overall microfabrication process is simple and compatible for both large and small scale hemodynamics.

16.4.3.2. Integrating the Sensors with Catheters. The sensors were packaged to a catheter (Fig. 16.15). The Cr/Au electrode leads were connected to a guide wire with conductive epoxy (which will cure at 90°C in about three

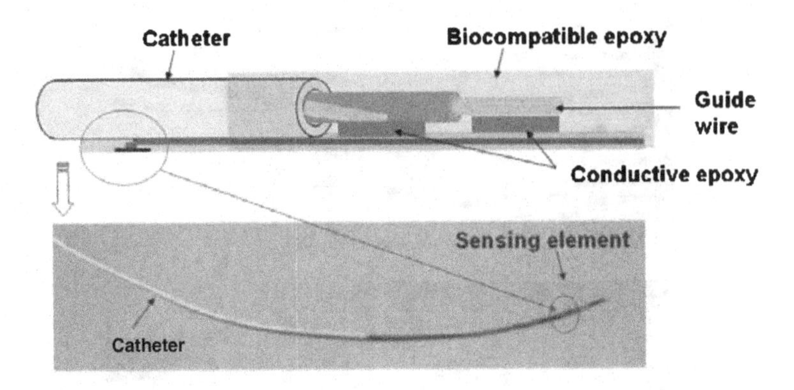

Figure 16.15. Packaging of the polymer shear stress sensor.

hours) to carry the electric signals from the arterial circulation to the external circuitry. After that the sensor was mounted to the catheter with biocompatible epoxy leaving the sensing elements facing the blood flow. This step will take one day for the epoxy to cure and complete the packaging process. The diameter of the packaged sensor is around 400 μm.

16.4.3.3. In vivo Assessment of Intravascular Shear Stress. Real time ISS measurements were acquired from the NZW rabbit's aorta, specifically the abdominal and aortic arch. Deployment of the polymer device into the rabbit's aorta was performed in compliance with the Institutional Animal Care. A pilot study has been conducted by deploying the MEMS sensors into the aortas of New Zealand White (NZW) rabbits (Fig. 16.16a). A 23-gauge hypodermic needle and a 26-gauge guide wire were introduced into the left femoral artery via a cut-down. A rabbit catheter (0.023″ID × 0.038′OD) was passed through the left femoral artery. The diameter of the sensor with guide wire was 400 μm and that of the aorta was 2.4 mm.The animal was heparinized (100 U/kg) intra-arterially. Under the fluoroscopic guidance (Phillips BV-22HQ C-arm), the catheter integrated with the MEMS intravascular device was deployed in the abdominal aorta for real-time intravascular shear stress measurements (Fig. 16.16b). The position of the catheter was visualized by fluoroscopy. The catheter was steered to a position (orientation) that the sensors were able to detect the maximal voltage output. Shown in Figure 16.16b is the real-time voltage signal output in response to ∼200 beats/min. The periodic changes reflected the respiratory variations at ∼30 breaths/min. Calibration and validation were performed to assess the non-Newtonian properties of blood flow.

16.4.4. MEMS Liquid Chromatography (LC)

In parallel to MEMS shear stress sensors and nanowire-based FET, the development of MEMS liquid chromatography (LC) provides a new venue to detect

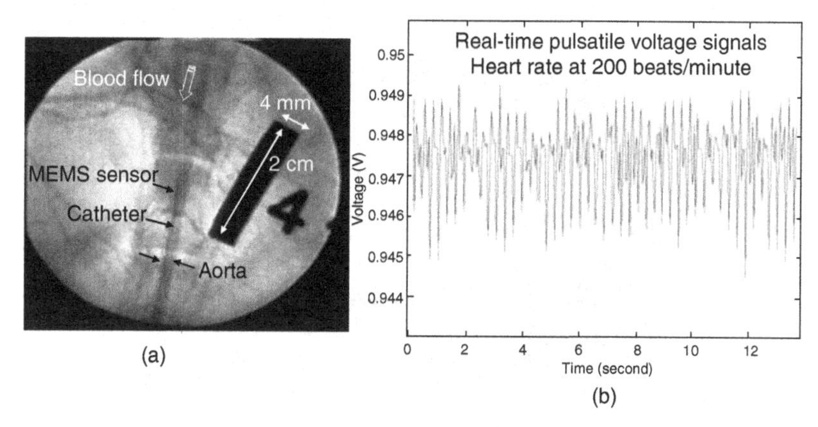

Figure 16.16. *In vivo* testing. (a) Catheter-based polymer sensor in the thoracic aorta of NZW rabbit. Contrast dye was injected to delineate the diameter of the aorta in relation to that of the catheter. (b) Voltage signal output (v) was detected by the sensor over time. The respiratory rate of the rabbit was at 30/minutes. The peaks and troughs reflect systoles and diastoles.

oxidative modification of circulating LDL. The MEMS LC as a lab-on-a-chip platform minimizes sample volume and patient discomfort, thus, providing an efficacious venue to detect clinical markers such as the levels of circulating oxidized LDL for early diagnosis of unstable angina and prevention of acute coronary syndrome.

Chromatography has made the separation of molecules possible in a fast and efficient manner. Protein molecules are separated according to their physical properties such as their size, shape, charge, hydrophobicity, and affinity for other molecules. High performance liquid chromatography (HPLC) enables the separation of molecules under high pressure in a stainless steel column filled with beads. Although liquid chromatography (LC) column is normally made of capillary tubes due to fluidics limitations, the miniaturization of the column can enhance separation performance (Fig. 16.17). When identical separation chemistry applies, separated peak width is independent of column internal diameter. The "bandwidth" of peaks is more distinct for the smaller columns with smaller beads [38]. MEMS-based LC provides high pressure compatibility of the microfluidic devices for operating HPLC [39]. The column is packed with conventional LC stationary phase support materials such as microbeads with surface functional groups. A self-aligned, channel-anchoring technique has been developed to increase the pressure rating of the Parylene microfluidic devices from 30 to at least 800 psi. MEMS-based LC also minimizes the input sample volume to less than 1 μL from the patients.

Purified LDL samples suspended in PBS solution were obtained from human blood plasma at the Atherosclerosis Research Unit at the University of Southern California. When the MEMS LC column was packed with ion-exchange

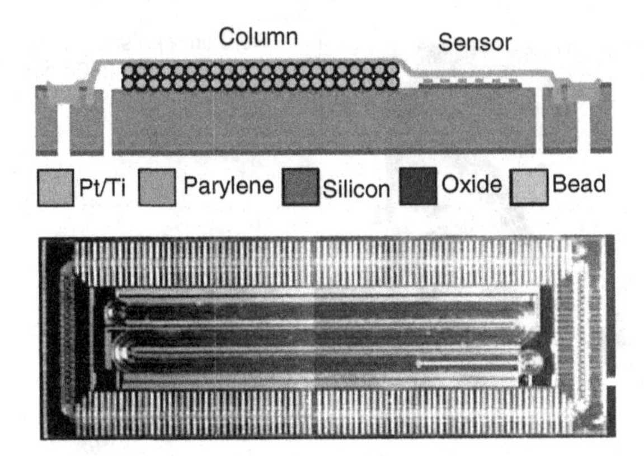

Figure 16.17. MEMS LC system. The top diagram illustrates the key components, including the column packed with beads and on-chip conductivity sensor. The photograph shows column. The column inner diameter is at 50 μm and the length at 1 cm.

chromatography beads, LDL subfractions were separated by anion exchange chromatography into native LDL (nLDL) and oxLDL in terms of LDL^- LDL^{2-}, with LDL^{2-} being more electronegative [40]. An LDL chromatogram was obtained using the MEMS LC with isocratic elution and on-chip conductivity sensing (Fig. 16.18). The result was compared with the chromatogram obtained from HPLC with salt gradient elution and UV detection (Fig. 16.19).

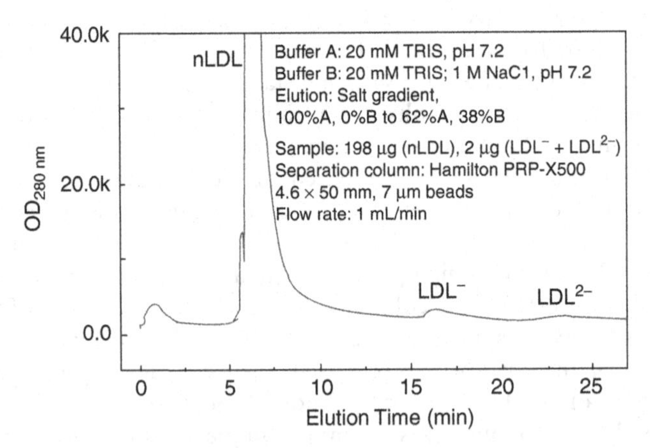

Figure 16.18. Chromatogram obtained from HPLC with the identical LDL sample and buffer solution A and B. The elution contains salt gradient from 100% A and 0% B to 62% A and 38% B. The flow rate was 1 ml/min. The identical beads used for the MEMS LC were packed the Hamilton PRP-X500 (4.6 × 50 mm) separation column.

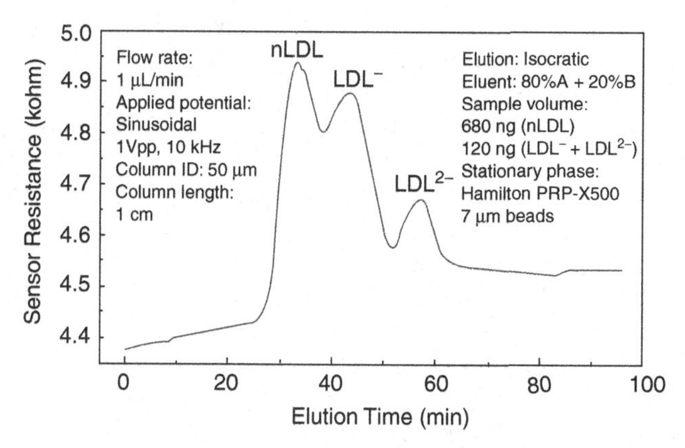

Figure 16.19. Chromatogram obtained from MEMS LC. The flow rate was controlled by allied sinusoidal potential at 1 Vpp and 10 kHz.The elution was isocratic. The eluent contain 80% of buffer A (20mM TRIS, pH 7.2) and buffer B (20 mM TRIS, I M NaCl, pH 7.2. Hamilton PRP-X500 separation column was packed with 7 μm beads. The LD sample contained 680 ng of nLDL and 120 ng of LDL^{-} and LDL^{2-} as validated by HPLC.

16.4.4.1. Sensitivity of Conductivity Detection For nLDL and oxLDL.

Since nLDL and oxLDL are both charged molecules, it makes perfect sense to use conductivity detection for sensing of those molecules. In conductivity detection, the baseline conductance signal is generated by the eluent. When sample peak enters the detector, a change of conductance ΔG can be measured. For anion exchange ion chromatography, the magnitude of this change is expressed in the following equation, assuming the samples and eluent are completely ionized [38].

$$\Delta G = \frac{1}{10^{-3} K_{cell}} (\lambda_{S^-} - \lambda_{E^-}) C_S, \tag{16.8}$$

where λ_S- and λ_E- are the equivalent conductance for the sample and eluent anions, C_S is the concentration of the sample anion, and K_{cell} is the conductivity sensor cell constant. The conductance change ratio on sample elution is usually very small, only a few percent change of the background conductivity:

$$\frac{\Delta G}{G_{Background}} = \frac{(\lambda_{S^-} - \lambda_{E^-}) C_S}{(\lambda_{E^+} + \lambda_{E^-}) C_E} \approx \frac{C_S}{C_E} \approx 1\%, \tag{16.9}$$

where C_E is the concentration of eluent anions. The sensitivity of conductivity detection is affected by the conductance noise level, sample injection volume, system dead volume and the inherent detector sensitivity. The conductance noise is mainly contributed from the temperature-fluctuation-induced background conductance change. In general, 1°C temperature change will result in 2% of conductance change. (Web site information is available from http://www.lsbu.ac.uk/water/explan4.html.) Several approaches are available to increase the sensitivity for conductivity detection. First, suppress the background conductance by using a suppressor column inserted after the separation column [38]. Second, increase the sample plug concentration by using sample preconcentration techniques (Dionex Corp., Sunnyvale, California), the use of concentrator columns in ion chromatography, 1994. Third, reduce sample injection volume and system dead volume. Fourth, reduce the sensor cell constant K_{cell} which is especially achievable using MEMS technology [39]. The state-of-the-art conductivity detection has a detection limit of 100 ppb ($\sim$10 nM) for unsuppressed detection and 10 ppb ($\sim$1 nM) for suppressed detection using a 50 μL sample injection for separation [38].

The synergism of micro- and nanocomponents for cardiovascular diagnostics includes (1) integration of sensors with the biological activities of vascular endothelium and circulating oxidized LDL from patients enrolled in the existing clinical trials at the Atherosclerosis Research Unit, and (2) a potential for investigation of *in vivo* animal models. Clinically, direct measurement of shear stress is important to predict in-stent restenosis and bypass graft occlusion. The MEMS LC as a lab-on-a-chip platform minimizes sample volume, providing an efficacious venue to detect clinical markers such as the levels of circulating oxidized LDL for early detection of unstable angina and prevention of acute coronary syndromes.

16.5. INTEGRATED MICRO- AND NANOSCALE COMPONENTS BIOFUEL CELLS

Fundamental to numerous biosensing applications are the biochemical redox reactions via micro electrodes and nanoscale molecules such as enzymes. The last decades have attested intense investigation in surface chemistry to optimize electron transport from the enzymatic reactions to the electrodes as a basis of biofuel cells. By converting chemical energy to electrical energy, biofuel cells have brought about the possibility to generate power for the micro implantable devices.

Biofuel cells consist of an anode and a cathode where the transfer of electron between the electrodes generates electrical power (Fig. 16.20). Miniaturization of biofuel cells has been materialized with microelectrodes and nanoscale particles. In an ideal microenvironment, the cell voltage is generated by virtue of the difference in the potentials between the anode, where fuel such as glucose is converted to gluconolactone, and cathode, where oxygen reacts with proton to form water. However, efficiency of power generation hinges on the kinetic electron transfer across the electrode interfaces, ohmic resistances of the electrode, and rates of electron formation by the redox reactions. Also important to the rate of cell current are the surface area of electrodes and proton transfer across the permeable membrane that separate the catholyte and anolyte compartments of the biofuel cells [41].

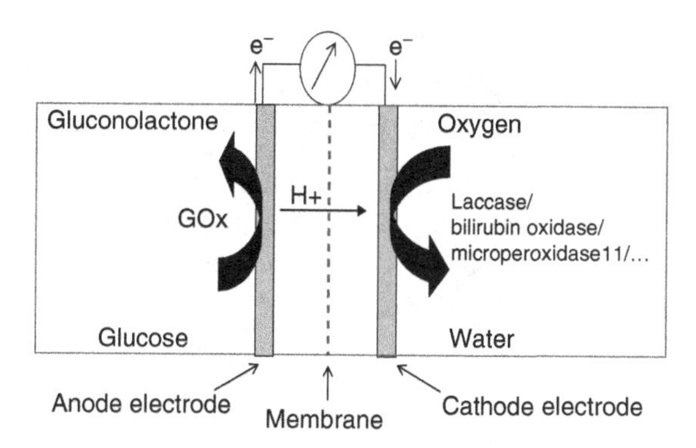

Figure 16.20. Schematics diagram of a biofuel cell where cathode and anode electrodes' surface are coated with enzymes catalyzing redox reactions.

A widely used driving force to catalyze the oxidation of (beta)-D-glucose to D-gluconic acid and hydrogen peroxide in the biofuel cells is glucose oxidase (GOx).

$$\beta\text{-D-glucose} + O_2 \xrightarrow{\text{Gox}} \text{D-glucose-1,5-lactone} + H_2O_2$$

Flavin adenine dinucleotide (FAD) is a cofactor of GOx, an apoenzyme. As an electron acceptor, FAD is reduced to $FADH_2$ and $FADH_2$ is subsequently oxidized to FAD by oxygen (O_2). This FAD/FADH redox centers are embedded in the GOx at 13 Å beneath the surface. Hence, the direct electron transfer (DET) rate between active site of GOx and electrode surface is usually slow.

To facilitate the electron transfer between the FAD site and microelectrode, researchers have engineered the positioning and alignment of electron mediators between the electrodes and redox sites to improve the output power [42]. One approach is to functionalize the nanostructure such as carbon nanotubes or gold particles with the cofactors. The apo-enzymes are positioned proximal to the cofactor-functionalized nanostructures and the electrodes. This construct provides a means to physically align the biocatalysts on the conductive surface and to connect electrically redox enzymes with electrodes (Fig. 16.20) [43].

Chemically inert and highly hydrophobic carbon nanotubes (CNT) are suitable as direct electron transfer mediators [44]. However, Yan et al. have demonstrated a miniaturized biofuel using a multiwall carbon nanotube as the anode and cathode electrodes [45]. The investigators decorate the CNTs with carboxylic acid functionality as the mediators. The flavoenzymes are covalently bound to the carboxylic functional group to facilitate the electrical contacts between the electrodes and the redox enzyme. The immobilization of CNTs to the bulk electrode has been achieved by covalent binding with a cystamine-monolayer-modified electrode (Fig. 16.21) [46].

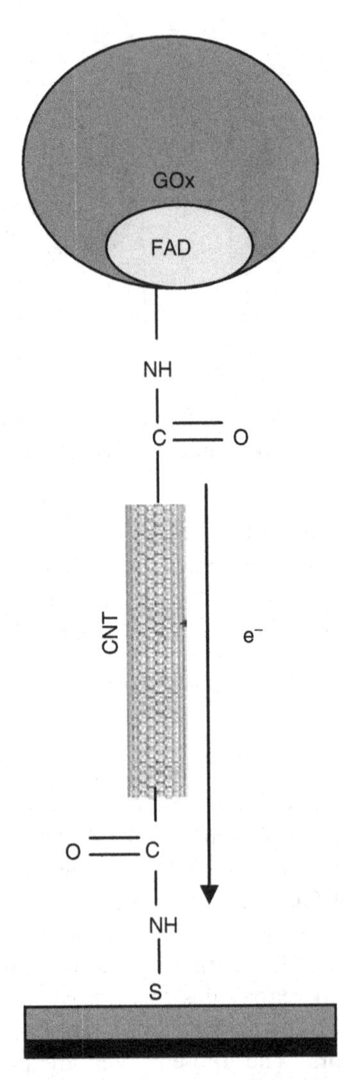

Figure 16.21. A schematic diagram of CNT mediator between GOx and anode of the biofuel cell. The cofactor, FAD, is embedded in GOx, which is linked to the mediator, a carbon nanotube (CNT). The carboxylic group on the other end is linked to the anode electrode. This construct brings the redox reactions close to the surface of electrode to facilitate electron transfer [44].

Alternatively, Katz and Willner have demonstrated gold (Au) nanoparticles as an efficient mediator for glucose sensing applications since it is based on the variations in surface plasmon resonance reflectivity angle due to steady state charging of nanoparticles [42].

16.6. CONCLUSIONS

Tremendous progress has been made in various fields of nano- and microtechnology in recent years. Simultaneously, the applications of these technologies into the biomedical domain have revolutionized the fields of BioMEMS and BioNEMS. Micro implantable devices, lab on a chip devices, and nano diagnostic methodology are likely to further bridge biology and physical sciences. Integration of micro and nano components continues to represent an engineering challenge. However, we have attested novel biochemical techniques to integrate the heterogeneous nanostructures to micro-scale components. The collaboration among scientists, engineers, and clinicians will likely generate a paradigm shift to enhance diagnosis and treatment of disease in the years to come.

REFERENCES

1. R. Feynman. *Feynman's Thesis*. Singapore: World Scientific, 2005.
2. M. Dickinson. Animal locomotion: how to walk on water. *Nature*, 424(6949): pp 621–622, 2003.
3. X. Gao and L. Jiang. Biophysics: water-repellent legs of water striders. *Nature*, 432: p 7013, 2004.
4. K. Autumn et al. Evidence for van der Waals adhesion in gecko setae. *Proceedings of the National Academy of Sciences of the United States of America*, 99(19): pp 12252–12256, 2002.
5. Y. Yokoyama, M. Takeda, T. Umemoto, and T. Ogushi. Active micro heat transport device using thermal pumping system. In: IEEE The Sixteenth Annual International Conference on Micro Electro Mechanical Systems MEMS-03, Kyoto: 2003.
6. H. L. Lai, J. Zhang, D. Li, Q. Xia, and S. Wang, Dehua. Research and development on the valve-less piezoelectric pump with unsymmetrical corrugation chamber bottom. 2006 IEEE International Conference on Information Acquisition.
7. V. I. Furdui et al. Microfabricated electrolysis pump system for isolating rare cells in blood. *Journal of Micromechanics and Microengineering*, 13(4): p S164–S170, 2003.
8. H. Y. Yu et al. Chembio extraction on a chip by nanoliter droplet ejection. *Lab on a Chip*, 5(3): p 344–349, 2005.
9. C. Y. Lee, et al. Nanoliter droplet coalescence in air by directional acoustic ejection. *Applied Physics Letters*, 89(22): 2006.
10. M. G. Lippman. Relations entre les phénomènes electriques et capillaries. *Annals of Chemistry and Physics*, 5(11): p 494–549, 1875.
11. J. Y. Chen, et al. Electrowetting in carbon nanotubes. *Science*, 310(5753): p 1480–1483, 2005.
12. H. B. Chan, et al. Quantum mechanical actuation of microelectromechanical systems by the Casimir force. *Science*, 293(5530): p 607, 2001. (Originally published in Science, 291: p 1941, Mar 2001.)
13. D. Kupiszewska and J. Mostowski. Casimir effect for dielectric plates. *Physical Review A*, 41(9): p 4636–4644, 1990.
14. Geon. Nanotechnology and Quantum Mechanics. *Arstechnica*, 2007.

15. G. K. Hansson. Inflammation, atherosclerosis, and coronary artery disease: reply. *New England Journal of Medicine*, 353(4): pp 429–430, 2005.
16. J. L. Witztum and D. Steinberg. Role of oxidized low-density-lipoprotein in atherogenesis. *Journal of Clinical Investigation*, 88(6): pp 1785–1792, 1991.
17. P. F. Davies, et al. Influence of hemodynamic forces on vascular endothelial function. *In vitro* studies of shear stress and pinocytosis in bovine aortic cells. *Journal of Clinical Investigation*, 73(4): pp 1121–1129, 1984.
18. C. F. Dewey, et al. The dynamic response of vascular endothelial cells to fluid shear stress. *Journal of Biomechanical Engineering*, 103(3): pp 177–185, 1981.
19. J. J. Chiu, et al. Effects of disturbed flow on endothelial cells. *Journal of Biomechanical Engineering*, 120(1): pp 2–8, 1998.
20. Y. C. Fung and S. Q. Liu. Elementary mechanics of the endothelium of blood vessels. *Journal of Biomechanical Engineering*, 115(1): pp 1–12, 1993.
21. N. DePaola, et al. Vascular endothelium responds to fluid shear stress gradients. *Arterioscler Thrombosis*, 12(11): pp 1254–1257, 1992. (Published erratum appears in *Arterioscler Thromb,* 13(3): p 465, Mar 1993.)
22. J. A. Frangos, et al. Steady shear and step changes in shear stimulate endothelium via independent mechanisms: superposition of transient and sustained nitric oxide production. *Biochemistry Biophysics Research Communications*, 224(3): pp 660–665, 1996.
23. G. Helmlinger, et al. Calcium responses of endothelial cell monolayers subjected to pulsatile and steady laminar flow differ. *American Journal of Physiology*, 269: pp C367–375, 1995.
24. J. R. Buchanan, et al. Relation between non-uniform hemodynamics and sites of altered permeability and lesion growth at the rabbit aorto-celiac junction. *Atherosclerosis*, 26(2): pp 215–224, 1999.
25. X. Deng, et al. Luminal surface concentration of lipid (LDL) and its effect on the wall uptake of cholesterol by canine carotid arteries. *Journal of Vascular Surgery*, 21(1): pp 135–145, 1995.
26. P. D. Henry and C. H. Chen. Inflammatory mechanisms of atheroma formation: influence of fluid mechanics and lipid-derived inflammatory mediators. *American Journal of Hypertension*, 6(11 pt. 2): pp 328S–334S, 1993.
27. T. K. Hsiai, et al. Monocyte recruitment to endothelial cells in response to oscillatory shear stress. *Faseb Journal*, 17(12): pp 1648–1657, 2003.
28. J. Hwang, et al. Pulsatile versus oscillatory shear stress regulates NADPH oxidase subunit expression: implication for native LDL oxidation. *Circulation Research*, 93(12): pp 1225–1232, 2003.
29. M. Rouhanizadeh, et al. Differentiation of oxidized low density lipoproteins by nanosensors. *Sensors and Actuators B-Chemical*, 114(2): pp 788–798, 2006.
30. M. Rouhanizadeh, et al. MEMS sensors to resolve spatial variations in shear stress in a 3D blood vessel bifurcation model. *IEEE Sensors Journal*, 6(1): pp 78–88, 2006.
31. C. Li, et al. Chemical gating of In_2O_3 nanowires by organic and biomolecules. *Applied Physics Letters*, 83(19): pp 4014–4016, 2003.
32. T. Tang, et al. Complementary response of In_2O_3 nanowires and carbon nanotubes to low-density lipoprotein chemical gating. *Applied Physics Letters*, 86(10): 2005.
33. C. Li, et al. In_2O_3 nanowires as chemical sensors. *Applied Physics Letters*, 82(10): pp 1613–1615, 2003.

34. I. Tanaka, et al. Electronic structure of indium oxide using cluster calculations. *Physical Review B*, 56(7): pp 3536–3539, 1997.
35. C. Li, et al. In_2O_3 nanowires as chemical sensors. *Applied Physics Letters*, 82: 2003.
36. M. Rouhinizadeh. et al. Applying indium oxide nanowires as sensitive and selective redox protein sensors. In: 17th IEEE Intertional Conference on Micro Electro Mechanical Systems MEMS 2004. Maastricht, The Netherlands.
37. Q. Lin, et al. Experiments and simulations of MEMS thermal sensors for wall shear-stress measurements in aerodynamic control applications. *Journal of Micromechanics and Microengineering*, 14(12): pp 1640–1649, 2004.
38. D. T. Gjerde and J. S. Fritz. Ion chromatography, 3rd ed, Weinheim: Wiley-VCH, 2000, pp 123–125.
39. Q. He, C. Pang, Y. C. Tai, and T. D. Lee. Ion liquid chromatography on-a-chip with beads-packed parylene column. In: Proceedings of The 17th IEEE International Conference on MicroElectroMechanical Systems (MEMS2004). Maastricht, The Netherlands.
40. A. Sevanian, et al. Low density lipoprotein (LDL) modification: basic concepts and relationship to atherosclerosis. *Blood Purification*, 17(2–3): pp 66–78, 1999.
41. I. Willner, et al. Biofuel cell based on glucose oxidase and microperoxidase-11 monolayer-fundionalized electrodes. *Journal of the Chemical Society-Perkin Transactions*, 2(8): pp 1817–1822, 1998.
42. B. Willner and I. Willner. Reconstituted redox enzymes on electrodes: from fundamental understanding of electron transfer at functionalized electrode interfaces to biosensor and biofuel cell applications. In: I. Willner and E. Katz (eds), *Bioelectronics*. Weinheim: Wiley-VCH. 2005. pp 35–98.
43. B. Willner, et al. Electrical contacting of redox proteins by nanotechnological means. *Current Opinion in Biotechnology*, 17(6): pp 589–596, 2006.
44. D. Ivnitski, et al. Glucose oxidase anode for biofuel cell based on direct electron transfer. *Electrochemistry Communications*, 8(8): pp 1204–1210, 2006.
45. Y. M. Yan, et al. Carbon-nanotube-based glucose/O-2 biofuel cells. *Advanced Materials*, 18(19): pp 2639, 2006.
46. P. Diao, et al. Chemically assembled single-wall carbon nanotubes and their electrochemistry. *ChemPhysChem*, 3(10): pp 898–901, 2002.

17

BIOMIMETIC CORTICAL NANOCIRCUITS

Alice C. Parker, Aaron K. Friesz, and Ko-Chung Tseng

This chapter describes a study of the feasibility of artificial brains in the future and emphasizes the necessity of using novel nanotechnologies in their implementation. The focus here is on biomimetic neural models and electronic circuits that implement those models, considering complexities in modeling biological neural tissue. Many problems present themselves when considering construction of a synthetic cortex, the most prominant being capturing biomimetic behavior; accomodating the scale of neurons, synapses, and axons; emulating the interconnectivity; and modeling the plasticity that underlies learning. Nanotechnologies offer not only the obvious advantages of scale and complexity, but exhibit the potential to support interconnectivity due to the inherent 3D nature of some nanotechnologies. Early experiments demonstrating the potential for self-assembly and reconfiguration of nanotechnologies provide some evidence that plasticity might be possible. We have designed an electronic synapse based on carbon nanotubes and elaborate on our simulation studies that demonstrate the range of behavior of the synapse. Estimates are given for the size of artificial neural systems based on CMOS technology in 2021. We predict some upper bounds on neural interconnections using CMOS nanotechnology.

17.1. INTRODUCTION AND MOTIVATION

Since the early days of vacuum tube and relay electronics, researchers have been developing electronic neurons designed to emulate neural behavior with electrical signals that mimic the measured potentials of biological neurons. However, in the past, the size and cost of available electronics made construction of complex brain-like structures infeasible due to the scale of the modeling problem.

Bio-Inspired and Nanoscale Integrated Computing. Edited by Mary Mehrnoosh Eshaghian-Wilner

As technologies become smaller and less expensive, there is the possibility of constructing neural structures on the scale of a human brain in the foreseeable future. Researchers are beginning to design, simulate, and construct cortical circuits with the goal of determining the feasibility of implementing a synthetic cortex (e.g., [1–3]). Jeff Hawkins, inventor of the PalmPilot, presents an eloquent informal discussion of the brain [4] that motivates our biomimetic assumptions. Therefore, we have started to examine the feasibility of building extremely large-scale neural systems using nanotechnology.

There is strong motivation for developing synthetic cortical circuits. We believe revolutionary changes both in technology and computing paradigms are essential in order to extend the scope of hard computational problems that can be solved. Synthetic brain structures could be used to solve some problems that have eluded conventional approaches. A synthetic cortex that could solve some difficult image understanding and speech recognition problems would be invaluable. Robots that possessed these capabilities could care for the elderly, react in situations too dangerous for humans, and provide intelligent support for routine human activities like vehicle collision avoidance [5]. Biomimetic cortical circuits could also be used as prosthetic devices [6], although there are other significant engineering challenges to such devices.

Although the timeline for large-scale biomimetic processing is some decades in the future, research on comprehensive models of large networks of cortical neurons is in the early stages (e.g., [7, 8]). A future synthetic cortex could be constructed using custom mixed-signal circuits that mimic the activites of individual neurons or simulated using parallel processing. There are fundamental challenges with both approaches, most critical of which are the issues of *scale*, *connectivity* and *plasticity*. The scale and interconnectivity of a synthetic cortex is daunting, with about 100 billion neurons, each possessing an average of 10,000 (and up to 100,000) distinct synapses [9]. The axon of each neuron fans out to around 10,000 presynaptic terminals. While some connections to each neuron originate in proximal (near) neurons, some originate in distal (far) neurons, posing interconnection problems for the candidate modeling technologies. Further, new synapses can form in a neuron in as little as an hour, expanding networks of cells and creating new networks. Such plasticity appears to be required for learning. Modeling this plasticity, coupled with nonstructural changes to the neuron as a result of learning, further complicates solutions to the synthetic cortex.

Constructing a biomimetic neuron poses additional challenges. The biological synapse has a complex physiology that we believe should be modeled. One of the complexities of neural tissue is the existence of transmitters, chemical messengers that can decrease or increase the excitability of the postsynaptic receptors to stimuli by the presynaptic cells, possibly by altering cell membrane conductance to charge-carrying ions via chemically gated ion channels. A further complication of transmitter function is via the long-term retrograde process that directly or indirectly modulates transmitter release in the presynaptic junction, a form of extremely local feedback. Transmitters acting via secondary messengers can have short- or long-term effects on synaptic junction activation. The activation probability of a given synaptic junction is up- or down-regulated by the amount and timing of presynaptic and postsynaptic activity. Neurotransmitters must be present in sufficient amounts to develop post-synaptic potentials (PSPs), and the concentration of transmitters released can affect both the height and duration of the PSP [9], phenomena that we model.

17.2. CHAPTER OVERVIEW

Our long-term objective is the development of the technology and methodology to implement a synthetic cortex. While there are hundreds of brain structures that have somewhat different anatomical characteristics [9], this research focuses on cortical structures because the density and complexity of their interconnections poses argueably the greatest neural modeling challenge. The cortex and cortical neurons exhibit many morphological variations that we will investigate. These neural circuits should accurately model inter- and intra-cellular mechanisms that neuroscientists believe are important to neural processing (e.g., the effects of neurotransmitter concentrations on cellular communication).

Future systems using these circuits could be used to construct large-scale cortical structures. Because this goal is in the distant future, while working toward the goal we find it necessary to answer to one simple question: When is technology going to progress sufficiently to be able to construct a synthetic cortex of reasonable size and cost that exhibits almost real time behavior? We believe that such a complex nanotechnological synthetic cortex will eventually be possible. This chapter describes research results in two areas:

- the design and simulation of biomimetic neural nanoelectronic circuits using future technologies, and
- the prediction of the possibilities for three-dimensional *interconnectivity* of CMOS neural circuit dies using flip-chip technology.

The emphasis in this chapter is on the first objective: circuits modeling cortical synaptic structures, designed and simulated using carbon nanotube transistor SPICE models provided by Wong [10] based on carbon nanotubes fabricated by Zhou [11, 12]. These cortical structures are being designed by us using custom biomimetic circuits, sometimes called neuromorphic circuits [1], based on carbon nanotube transistors. Carbon nanotubes are widely studied nanotechnology materials that have the potential to support three-dimensional circuit structures. The custom neural circuits that we present here use advanced nanotechnologies, are programmable at a very high level (virtually software free), integrate memory and processing, and execute in parallel in a natural manner. To date, we have designed and simulated a nanoelectronic neural synapse using carbon nanotube models, a revolutionary change to the underlying technology.

The mathematical models we are developing include prediction of the timeframe and feasibility of a synthetic cortex, prediction of the neuron size over time, and prediction of interconnection requirements.

17.2.1. Justification for the Custom Circuit Approach

Our focus on custom circuits is in contrast to computer models that simulate neural behavior using conventional multiprocessors. In recent years, several artificial brain projects have been launched that rely on computer simulations of neural behavior, hosted on multiprocessors, such as the IBM artificial brain project in cooperation with EFPL [8]. However, even on much smaller simulations such as a partial mouse brain, the IBM simulations ran an order of magnitude slower than real time [13]. Emulation can have several advantages over simulation, generally being faster, sometimes running at nearly real time for small problems. For brain emulation, speed may not be as critical since neurons are slow in comparison to electronics, but performance becomes an issue when the interconnection hardware is extensively time-shared to interconnect many parts of the brain. Of course, interprocess communication is also problematic in software brain emulations.

The challenge of using programmable processors to simulate the cortex is the scale. Although software models could solve the problems of neural interconnectivity and brain plasticity, since they can be programmed into the neural data structures, the inefficiencies of using software to model biological (continuous, analog-like) behavior requires massive supercomputing structures to simulate small networks of neurons. Even if we exploit novel technologies like carbon nanotubes and single electron transistors to implement conventional computer architectures, the problems of scale are still present. Nanotechnologies like carbon nanotubes possess physical properties that could allow significantly denser packaging in three dimensions when used in custom electronic circuits, resulting in neural structures on the same order of magnitude of volume as a human brain, dense enough to implement some limited neural computation capability.

While custom CMOS hardware solutions might be significantly smaller than software solutions per neuron modeled, synthetic cortical structures implemented using distinct CMOS electronic circuits for each synapse and neuron, sometimes called neuromorphic circuits [1], could still be quite large. Boahen has fabricated and demonstrated integrated circuits with 10,000 neurons by sharing the synaptic circuitry so that each neuron only possesses a single synaptic circuit. In order to capture the computations performed in the dendritic tree believed to be important, such sharing might need to be limited or abandoned. Assuming distinct synapse circuits for each synapse, we have predicted [2] that the CMOS technology required to implement the neurons in a synthetic human cortex in a single room is at least a decade away. This prediction did not include interconnectivity and learning structures.

17.2.2. Justification for a Radical Change in Technology

When the hardware for interconnectivity and learning are included, along with the possible end of the continued scaling of CMOS circuits,[1] a future CMOS cortex would still be very large. Interconnectivity, even with some 3D interconnection capabilities like flip-chip technology, would be a major problem, and plasticity costly to implement. Connecting 10^4 presynaptic terminals to each neuron poses technological challenges. Furthermore, the rigidity of conventional integrated circuits makes plasticity virtually impossible to implement directly. We believe that a change in technology is required to create circuits that exhibit human-like intelligence, especially via plasticity.

Carbon nanotubes may support the scale and interconnection density of a synthetic cortex. For this reason, we have begun preliminary studies into possible carbon nanotube circuits that could form the basis for a synthetic cortex. These technologies may, in the distant future, enable the construction of a reasonably sized synthetic cortex, as we predict here. We believe that with the right combination of manipulation techniques true 3D structures will be possible, thus alleviating a large portion of the interconnection and scale issues.

17.2.3. Our Modeling Approach

We have begun circuit design with the excitatory synapse [15], modeling neurotransmitter action and ion channels. The synaptic circuits we have designed translate higher-voltage short-duration signals to longer-duration signals of smaller magnitude. Their behavior is chaotic [16], with present state sensitive to initial conditions, and output behavior highly nonlinear with respect to the input behavior. We have simulated a carbon nanotube transistor circuit model of this neural synapse that captures, in a coarse manner, the actions of neurotransmitters, ion channel and ion pump mechanisms, and temporal summation of PSPs. We have chosen to focus on excitatory PSP's (EPSP's) first, and have chosen economy of size over exact replication of waveforms, to facilitate scaling to cortical-sized neural networks.

Our neural models focus on the ion channel, one fundamental unit of processing, and model the complexities described here. Ion channels are molecular structures that block or permit the flow of charge-carrying ions through the cell membrane. Ion pumps subsequently return the neuron to a resting state by restoring the concentration of ions relative to the extra-cellular fluid. Ion channels may be voltage or chemically gated. Voltage-gated channels respond to membrane potential differentials. Chemically gated ion channels open and close in response to specific chemicals called transmitters and may be directly gated (ionotropic) channels or indirectly gated (metabotropic) channels. The chemical process that causes a metabotropic channel to open is called a secondary messenger cascade.

[1] Recent Intel and IBM breakthroughs in dielectric and gate materials extend MOS scaling possibilities in the immediate future [14].

Transmitters acting via secondary messengers can have short- or long-term effects on synaptic junction activation. Altering the conductance of the membrane to ions based on potential differentials or neurotransmitter concentrations creates important feedback that is vital to the function of a neuron. Further complicating transmitter function is the long-term chemical retrograde process that directly or indirectly modulates transmitter release in the presynaptic terminal, a form of extremely local feedback.

In Section 17.4, we will describe circuit simulations of carbon nanotube neural circuits. These circuits display variations in behavior related to neurotransmitter concentrations and reuptake rates. We demonstrate, via circuit simulations, ion channel mechanisms. Finally we have produced simple prediction models of neural size, and mathematical models of neural interconnectivity in order to predict wireability of the synthetic cortex.

17.3. RELATED WORK

Related work in electronic neural modeling, nanotechnology, and measurement of brain properties is presented here.

17.3.1. Neural Modeling

Substantial research in related areas of neural modeling using electronics has been performed, although nanotechnologies are not considered. Schüffny et al. provide a good survey of research up to 1999 [17]. Many researchers focus on specific brain structures, like the retina, or applications, like image recognition. The goal of a very small number of projects is construction of an entire artificial brain or cortical columns consisting of many neurons. Some highly visible research projects [8, 18] envision synthetic structures built using general purpose processors, specialized architectures such as cellular automata [19], or asynchronous ARM processors [18]. Wells [20] proposes a neurocomputer architecture intended to solve the problems of interconnectivity, variable synaptic weights, and learning. Moravec [3] has performed optimistic predictions of when inexpensive general purpose computers will match the human brain in processing power.

The most notable research on electronic neural circuits includes Mead's artificial retina [21], and continuing work by Boahen (e.g., [1]) and others [22] on neuromorphic circuits that emulate the behavior of individual neurons. This significant body of work originated with Mahowald and Mead's pioneering research [23]. Boahen, who studied with Mead, has also concentrated on visual processing [7, 24]. Hynna and Boahen report on a circuit that generates a calcium spike with exact replication of waveforms, and describe incorporation of the calcium spike circuit in an entire neuron circuit [25]. Some mixed-signal electronic models close to biological neurons include Liu and Frenzel's spike train neuron [26], Pan's bipolar neuron [27], Wells' research (e.g., [28]), and Chua's cellular neural network (CNN) [19, 29]. Chiju et al. extends the CNN work and tests their neural model on specific applications [30]. Sato et al. [31] use stochastic logic to obtain analog behavior from digital circuits. Chen and Shi [32] use pulse-width modulation. Linares-Barranco et al. [33] describe a CMOS implementation of oscillating neurons. Fu et al. [34] present thin-film analog artificial neural networks.

The master's thesis by Chao describes a basic CMOS neuron, designed by Parker, with learning capabilities [35]. An 8-transistor CMOS synapse [36] is close in scale and nature to our current synapse circuit. Analog synapses have been reported by Pinto et al. [37] and Lee et al. [38] and a phase-lock loop synapse has been reported by Volkovskii [39]. Elias [40] has performed modeling of dendritic trees that are similar to our models. The primary difference involves our use of transistors and his use of resistors and capacitors for dendritic computations. In addition, his simple synapses involve single transistors. Noteworthy neurons capable of learning have been proposed [35, 41–47]. Koosh and Goodman [42] put a digital computer in the loop for training, control and weight updates, and the neural network is entirely analog, a style realized by several research groups.

While many neurons in the literature have some biomimetic features (e.g., [22, 26, 40, 50]), the complete range of neural variations has not been implemented in a single model or even in the variety of neuron models distributed throughout the research community. The most important research on neural interconnectivity is Boahen's neurogrid project, with roots in the pioneering research by Mahowald [23].

Commercial minimally biomimetic neural networks incorporating learning are available, and in use by the high-energy physics community. However, in contrast to our model, there is little correspondence in the majority of these models between individual circuit elements and specific physiological mechanisms in the biological neuron. The correspondence between specific biological mechanisms and circuit elements allows us to vary synapse behavior easily with control inputs. This and our choice of carbon nanotube technology differentiates us from related work.

17.3.2. Nanotechnology

While nanowires and nanotubes form the basis for a significant volume of nanotechnology research, the focus here is on the nanotechnologies incorporated into the neural circuits described in this chapter. Nanowires [11, 12, 51–55] are one-dimensional structures with diameters typically around 10λ nm and lengths up to tens of microns. Zhou's group is one of the leaders working on the synthesis and device applications of various semiconductive nanowires such as In_2O_3 (shown in Fig. 17.1a), SnO_2, ZnO, Si and Ge. These materials can be very good electrical conductors, and their conductance can be varied over many orders of magnitude by applying a gate voltage bias [55]. Zhou has demonstrated high-performance transistors based on many kinds of nanowires that can work as the active channel in artificial neurons.

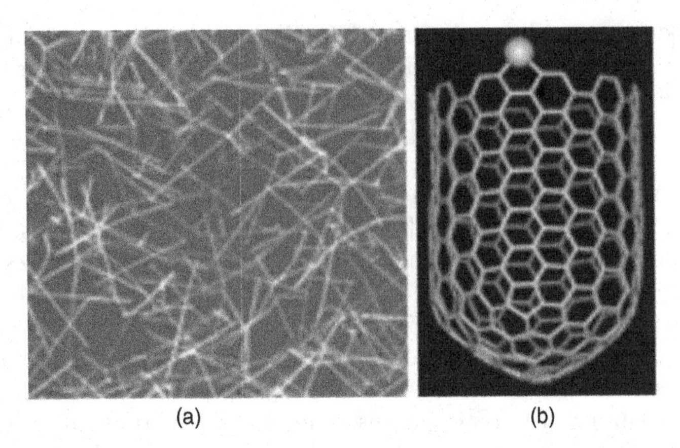

Figure 17.1. (a) SEM image of In_2O_3 nanowires. (b) Schematic diagram of a single-walled carbon nanotube.

Carbon nanotubes are sheets of graphite rolled into seamless cylinders (Fig. 17.1b) with diameters 1–2λ nm and micron scale lengths. This one-dimensional structure exhibits useful electronic and mechanical properties, behaving as metals, semimetals, or semiconductors. Carbon nanotubes can carry a current density up to 109 ampere/cm^2 and electrons can move ballistically in a carbon nanotube without any significant scattering even at room temperature. In addition, carbon nanotubes possess very high thermal conductivity and significant mechanical strength. These properties make them ideal to work as synaptic connections between artificial neurons, as they can conduct sufficient current without generating significant heat, and the high thermal conductivity and mechanical strength guarantee good heat dissipation and robust operation. In comparison, interconnects made of top-down fabricated metal electrodes may never reach a width of 1–2λ nm due to issues like the finite grain size and instability related to the grain wall migration. Following Zhou's previous work on carbon nanotube p–n junctions [56], mechanical switches, and chemical sensors [57], Zhou's group has carried out significant work, including the demonstration of the complementary nanotube inverter [58], band engineering of nanotube transistors via selected-area chemical doping [59], and template-free directional growth of nanotubes on a- and r-plane sapphire substrates [60, 61].

Single-walled carbon nanotubes avoid most of the fundamental scaling limitations of silicon devices, and appear to be an appropriate technology for a synthetic cortex. Liu, Han, and Zhou have demonstrated directional growth of high-density single-walled carbon nanotubes on a- and r-plane sapphire substrates over large areas [60, 62]. They have developed a novel nanotube-on-insulator (NOI) approach for producing high-yield nanotube devices based on aligned single-walled carbon nanotubes. Also, they have developed a way to transfer these aligned nanotube arrays to flexible substrates.

Efforts have been made in recent years on modeling CNFETs [63, 64] and CNT interconnects [65, 66] to evaluate the potential performance at the device level. Very promising single device DC performance over silicon devices has been demonstrated either by modeling or experimental data. However, the dynamic performance of a complete circuit system, consisting of more than one CNFET and interconnects, differs from that of a single device. All but one of the reported models to date used a single lumped gate capacitance and ideal ballistic model to evaluate the dynamic performance, which results in an inaccurate prediction [67, 68]. To evaluate CNFET circuit performance with improved accuracy, a CNFET device model with a more complete circuit-compatible structure and including the typical device nonidealities was constructed [10]. This recent publication presents a novel circuit-compatible compact SPICE model for short channel length (5λ nm 100λ nm), quasi-ballistic single wall carbon nanotube field-effect transistors (CNFETs). This model includes practical device nonidealities, e.g., the quantum confinement effects in both circumferential and channel length direction, the acoustical/optical phonon scattering in channel region and the resistive source/drain, as well as the real time dynamic response with a transcapacitance array. This model is valid for CNFETs for a wide diameter range and various chiralities as long as the carbon nanotube (CNT) is semiconducting.

17.3.3. Measurement and Modeling of Cortical Data

Significant cortical data has been collected by Braitenberg and mathematically categorized [69, 70]. Recent research by Stevens at the Salk Institute [71] relates the volume of an axonal arbor to the total length of the axon branches in the arbor. This type of information can be useful in predicting interconnection requirements for a synthetic cortex. Sejnowski [72, 73], also at the Salk Institute, has focused on statistics concerning neural communication.

3D biological data obtained in collaboration with Manbir Singh (USC Department of Biomedical Engineering) provides statistical information for our prediction models [74]. In the nanoelectronic approach case, Singh's three-dimensional statistics can be applied directly to provide insight into the interconnectivity problem for the synthetic cortex. We postulate that the distance presynaptic terminals are from the soma can be modeled as an exponential (Poisson) distribution, with the axon more likely to have presynaptic terminals close to the soma. Experimental results [74] support this assumption. We also postulate, similar to Steven's work at Salk, that there is a Rent's rule for regions of the cortex, relating statistically the number of axons emerging from a spherical volume of brain tissue to the volume of brain tissue enclosed.

Bailey and Hammerstrom provided early predictions on interconnection possibilities using CMOS technology [75], including an early reference to Sejnowski's research.

17.4. A CARBON NANOTUBE SYNAPSE

This section presents a novel carbon nanotube synaptic circuit, and we provide some SPICE simulation results that demonstrate the range of synaptic behaviors possible.

The carbon nanotube synaptic circuit is shown in Figure 17.2. Action potentials (see Fig. 17.3) arrive from the presynaptic neuron and terminate in our synaptic circuit. The simple piecewise action potential used here is an approximation of a biological action potential [76]. A single postsynaptic terminal is shown in this figure. An incoming action potential will cause the potential in the synaptic cleft to rise via the pull-up network. The PFET limits the peak amplitude of the synaptic cleft potential by turning off before the synaptic cleft potential rises to Vdd. The synaptic cleft potential in the electronic neuron (also shown in Fig. 17.3) models the biological release of neurotransmitters stored in the presynaptic neuron into the cleft, where they bind to receptor proteins on the recipient (postsynaptic) neuron, causing the potential across the postsynaptic neural membrane to change. Once the neurotransmitters are released from the presynaptic terminal and bound in the postsynaptic terminal, they will be cleared from the synaptic cleft by reuptake mechanisms. The presence of bound neurotransmitters is represented electronically by a potential increase in the synaptic cleft. Vesicles depleted of neurotransmitters will be replenished in the presynaptic terminal. Reuptake is modeled via the pull-down network attached to the synaptic cleft.[2] The reuptake control voltage, *R*, is an analog potential, which allows the

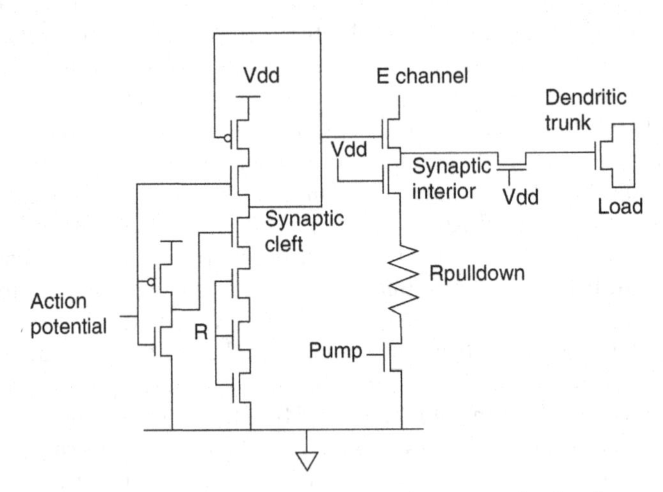

Figure 17.2. The carbon nanotube synaptic circuit.

[2] Since the SPICE model used in this study is geared toward short gate length transistors with near-ballistic transport, we use three transistors in series to model the resistance that represents the reuptake delay. The quantum contact resistances of the three transistors are added in series.

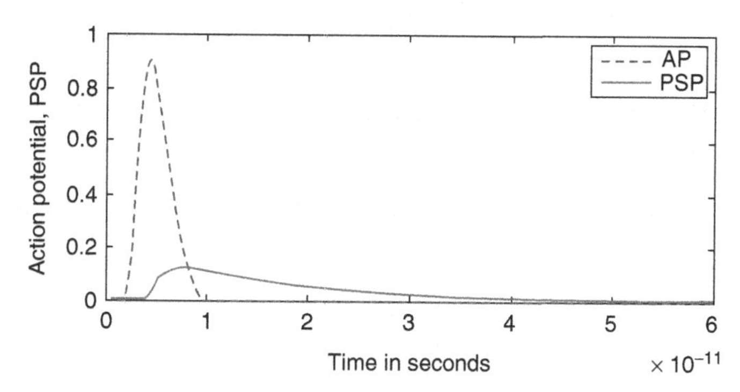

Figure 17.3. The action potential and PSP at the dendritic trunk under normal operation.

efficiency of the reuptake mechanism to be tuned. The action potential is inverted and used to block pulldown of the synaptic cleft potential to delay reuptake of the neurotransmitter at the presynaptic neuron while there is a positive action potential. Some antidepressants inhibit neurotransmitter reuptake, and this is modeled by lowering the reuptake control voltage in the synapse circuit.

The increase in potential in the synthetic synaptic cleft will temporarily turn on the transistor connected to the ion electromotive force control, *E-channel*, causing the potential at Synaptic Interior to rise. This models the change in biological membrane potential due to the increased conductance of neurotransmitter gated ion channels and the subsequent influx of charge carrying ions. A pull-down network[3] models the action of the biological ion pump. As with the reuptake mechanism the efficiency of the ion pump may be tuned via the *pump* control voltage.

The synaptic interior potential is transferred through a resistive connection to the dendritic trunk, which carries it to the cell body of the neuron. The synaptic weight control allows the importance of this synapse to be tuned, as is often done in learning algorithms. The potential on the dendrite trunk represents the postsynaptic potential (PSP). The dendrite trunk terminates on the gate of a single NFET that acts as a load for the synaptic circuit for testing purposes. This load would be replaced with a block of circuitry representing the dendritic tree structure, cell body and axon of the postsynaptic neuron when the synapse is used in a complete neuron model.

In the archetypical biological neuron we are modeling, potentials range from around −75 mV to +40 mV. with action potentials peaking around +40 mV. Since the carbon nanotubes are designed to operate with V_{dd} around 0.9 V and with 0 V. (Ground) as the lowest potential, the potentials were scaled in the synaptic circuit accordingly, with 0v. circuit potential corresponding to −75 mV biological potential and 0.9 V circuit potential corresponding to 40 mV biological

[3] Three transistors in series provide the R_{pulldown} resistance.

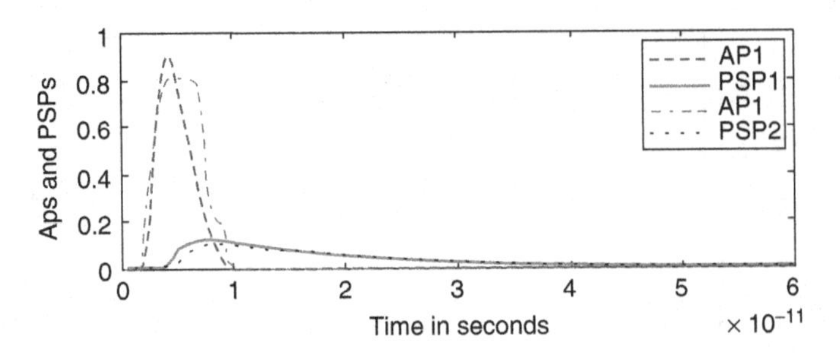

Figure 17.4. The effect of presynaptic action potential variation on the PSP.

potential. Likewise, the speed of the carbon nanotubes allowed us to scale the delays significantly, with about 1 ms in the biological neuron scaling to about 2 ps in the nanotube neuron [76].

17.4.1. Experiments with the Nanotube Synapse

We simulated the synapse with the ion channel electomotive force, E-channel, at 0.8 V, and with reuptake control, *R*, at 0.3 V. The ion pump gate voltage was set to 0.4 V representing typical operation. The postsynaptic potential appearing at the dendritic trunk is shown in Figure 17.3. This potential is approximately 14% of the action potential and the duration is about six times as long as the action potential, similar to EPSP's described in the literature [77]. While the control voltages were tuned to yield reasonable operation, the input voltage was varied in duration to remain at a maximum of 0.8 V for 2.5 ps and an acceptable PSP still resulted. This sensitivity testing revealed that the circuit is not highly sensitive to the presynaptic action potential's gross characteristics. This second action potential with increased duration and lowered magnitude is shown in Figure 17.4. The typical PSP from Figure 17.3 is reproduced here, along with the PSP resulting from the modified action potential.

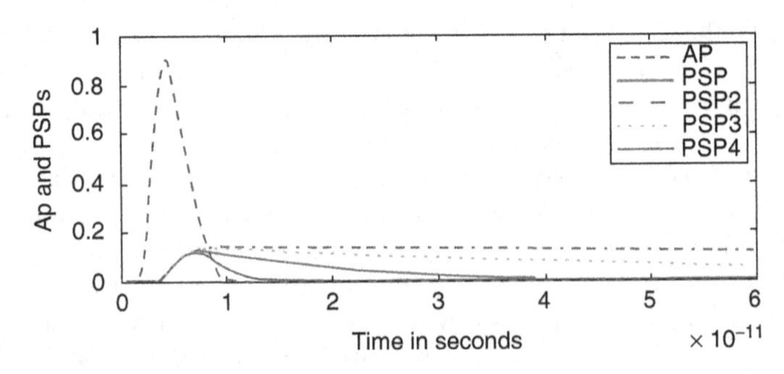

Figure 17.5. The action potential and resulting PSPs with reuptake control varying from 0 V to 0.9 V.

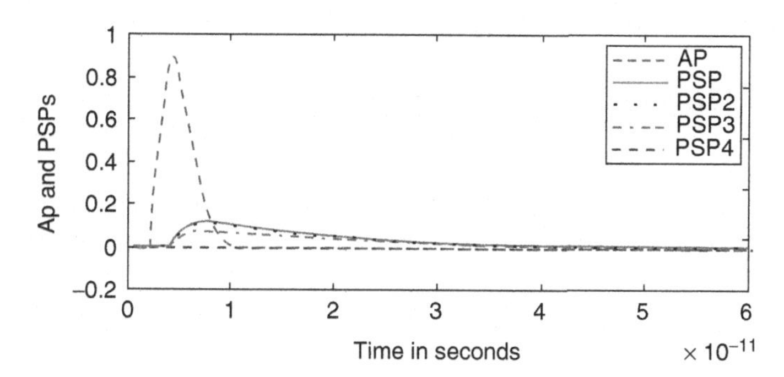

Figure 17.6. The PSPs for different values of E-channel, the ion emf control.

In a third experiment we decreased the neurotransmitter reuptake control, *R*, to slow ruptake of the neurotransmitter and increased it to speed up the reuptake. The resultant PSPs are shown in Figure 17.5, with the greatest magnitude and longest duration PSP resulting from the lowest voltage, and the shortest, lowest magnitude PSP resulting from the maximum value of R. The The ion channel control, E Channel, is held at 0.8 V, and the ion pump gate voltage at 0.4 V.

A fourth experiment shows the ability of the ion electromotive force, represented by E-channel voltage, to control the PSP that results from ion channels opening and closing. E-channel is varied from 0 V to 0.8 V, and the resulting PSPs are shown in Figure 17.6. The reuptake control, *R*, is held at 0.3 V, and the Pump Control voltage at 0.4 V. The PSPs are reduced and eventually disappear when the ion electromotive force drops to 0 V.

The final experiment (Fig. 17.7) illustrates the temporal summation of PSPs over time as a result of two successive action potentials at the same synapse.

This synapse is an example of the type of circuit that can be used to model neural physiology. There are many synaptic variations yet to be included as the research progresses.

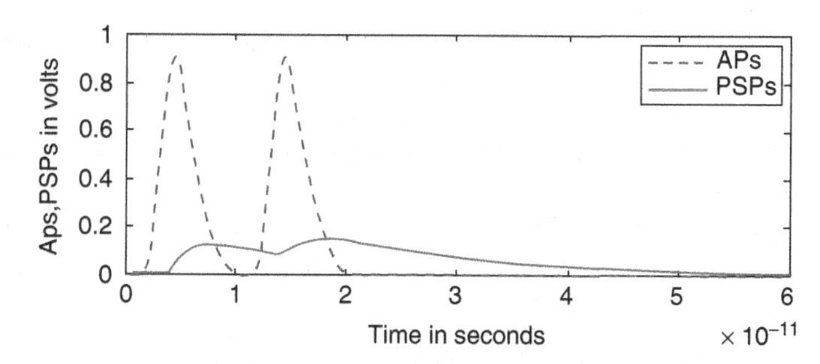

Figure 17.7. Summation of PSPs when action potentials arrive close together.

17.5. CURRENT PREDICTION RESEARCH

This section describes prediction research regarding the feasibility of a CMOS synthetic cortex [2], including flip-chip technology to gain additional vertical interconnections between two integrated circuit dies. First, we predict the size of a single neuron, and then predict the space occupied by the 100 billion neurons in the cortex. Finally we approach the problem of interconnectivity, predicting the connections possible to a CMOS core, including the use of flip-chip technology.

17.5.1. Predicted CMOS Neuron Size

We have predicted the size of a single neuron, absent any interconnections or learning capability, and projected the physical size over time until 2019 using the International Technology Roadmap for Semiconductors [78]. Our assumptions of neural size were based on electronic neurons found in the literature and our own synapse design, and we predict a significant part of an entire office being required to house the neurons in a synthetic neural cortex in 2021. We begin our CMOS prediction using typical biomimetic neural electronic circuits (the Liu-Frenzel neuron [26] and Boahen's neuron [1] are examples) and extend the prediction using our own synapse described in Section 17.4 as a measure of synaptic complexity.

The simplest possible synaptic structure is a single MOS transistor. However, additional transistors would be required to model memory, learning, neurotransmitters, the refactory period, and other phenomena that complicate the behavior of the synapse. While the biological mechanisms underlying the synaptic behavior are complex and not completely understood, we are making an assumption that charge storage can be used to model a wide range of phenomena, like the influence of neurotransmitter concentrations and the impact of learning. Additional elements can be used to model dynamic behavior such as the refactory period. Therefore, our estimate of synaptic complexity at this point is a simple order of magnitude more complex than the simplest synapse, or 10 transistors. The custom electronic synapses in the literature that do not contain significant learning mechanisms, with the exception of the research by Hynna and Boahen [25] fall well under this estimate. The prediction given here ignores dendritic computations that could contribute significantly to neural complexity [83], ignores calcium spiking mechanisms in the dendritic arbor, and ignores learning mechanisms.

Because the neuron designs we chose were fairly simplistic, we estimated that 10^7 neurons could fit onto a single die in 2019. Even today, we estimate that 10^4 neurons could fit on a single die.

Let ψ_{init} be the number of transistors per chip possible in 2006 [84], and ψ_{fin} be the number of transistors predicted to be possible per chip for 2021, 15 years from 2006. $\psi_{init} = 10^9$ transistors per chip and

$$\psi_{fin} = \psi_{init} * \chi^{M} \text{transistors/chip}, \tag{17.1}$$

where χ is the multiplicative factor, and M is the number of months between initial and final estimates.

$$Y = M/R \tag{17.2}$$

so

$$\chi^M = 2^{(M/R)} \tag{17.3}$$

where Y is the number of times the transistor count doubles, and R is the number of months elapsed for the transistor count to double. Assuming doubling in 18 months, and assuming we look 180 months (15 years) into the future from 2006, to the year 2021, the transistor counts should have doubled 10 times.

$$\chi^{180} = 2^{(180/18)} = 2^{10} = 1024 \tag{17.4}$$

$$\psi_{fin} = \psi_{init} * \chi^{180} = 10^{12}\text{transistors/chip} \tag{17.5}$$

Therefore, we predict that in 2021 there would be 10^{12} transistors/chip.

We assume that on the average, each cortical neuron has 10,000 synapses [9]. If each synapse contains 10 transistors, plus the few transistors representing the axon, the total number of transistors in a neuron is 10^5. Therefore, the number of neurons per chip can be estimated to be 10^7.

Thus, in 15 years, we will require 10^4 chips (integrated circuits) to construct an artificial brain with 100 billion neurons. Boahen predicts biomimetic chip densities will be within a factor of 10 of biological neuron density within the decade [1]; however, his estimates regarding the number of transistors per synapse and number of distinct synapses differ from ours.

There is more uncertainty when estimating system size because the future of multi-chip modules (MCMs) is less predictable. However, we will assume MCMs occupy 30% of board space. On each MCM, we assume dies occupy 70% of the space. We assume that six 12″ × 12″ boards fit vertically in a cubic foot of space. Then the system could hold 180 $chips/ft^3$. Based on these assumptions, in the year 2021, we require a space of 55.5 cubic feet to house our neural circuits, absent any room required for interconnections between neurons on chip, between boards and between racks. If the racks are 8 feet tall, then we require 6.9 square feet of floor space for the equipment. Allowing for air space around the equipment, we estimate the neurons in the artificial brain to occupy 14 sq. ft. of floor space. This is approximately the total free space available in the first author's university office.

This can be contrasted with the IBM/Swiss project that allocates only two neurons per processor, highlighting the economies of a hardware solution. Neurons with learning capabilities are significantly larger than the ones on which we based our predictions [79]. Even if we are optimistic in our predictions by two orders of magnitude, 50 times as many custom neurons could still easily fit on a single die as the software approach using current technology. Neuron size is not the limiting factor to the conventional hardware approach; interconnections prove to be a more difficult problem.

17.5.2. Predicting Neural Interconnections in CMOS

Each synaptic connection that provides input to a neuron originates in a separate, distinct neuron, requiring an average of 10,000 distinct inputs to each neuron. The address decoding required when some form of multiplexing is implemented (bus or network) makes scaling such approaches to cortical-sized neural networks difficult. Using a straightforward connection scheme, we address the question "How many distinct connections can be made to a CMOS integrated circuit area the size of an individual neuron?"

Using a pyramidal windowing structure we designed [80] (Fig. 17.8a), we made predictions of the upper bound of the number of connections possible to a core of silicon C that is a function of the silicon area and the available layers of metal. As you can see in Figure 17.8a window is nothing more than an exit for the connections from the core C to elsewhere, providing a way for us to count interconnections. In Figure 17.8a, we have several Windows (i.e., Window 1, Window 2, and Window 3) for the connections. Window 1 uses the first available metal layer for the connection. Window 2 and Window 3, respectively, use the second and third available metal layers for connection. Higher layers of metal (e.g., layer m) are used for connections through corresponding windows (e.g., Window *m*). The width of metal will increase on higher layers of metal according to the design rules. Therefore, required window size will also become larger as the level of the metal layer increases.

We extend the 1D pyramidal model to create a 2D model. The best way to assign the position of windows inside the core *C* is shown in Figure 17.8b. Window 1 is placed at the outer edges of *C* and then Window 2 and Window 3 will be placed concentrically inside, one after another, with each successive higher window placed inside the lower ones on the die. In general, the minimum metal width is usually wider than the size of contact size or via size. Therefore, the size of each side of the window is the width of the corresponding metal layer instead of the size of the contacts or via.

Connections to the core could use these windows as entries to connect to the core C. Assuming that connections to the core C can be made in an optimal fashion, the windows provide a count of the maximum possible number of connections to C.

Intuitively, the parameters in the pyramidal model are given by the design rules and process. Given a core area, more windows are possible as the required size of the windows decreases. Also, a process with more metal layers will have more possible windows, up to the limit imposed by the size of the core.

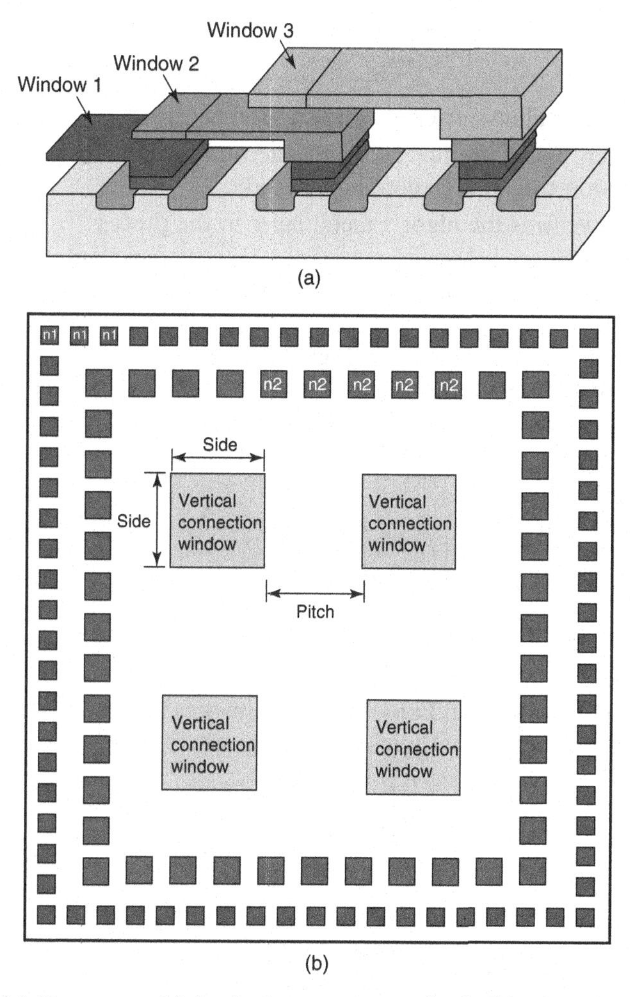

Figure 17.8. (a) The pyramidal window structure [80]. (b) The arrangement of contacts for maximum connections to a core [80].

In order to provide a mathematical pyramidal model of interconnections, we first identify parameters. We assume the following:

- The area A of the core C is $H \times H$.
- The distance from the edge of the area A to the first available layer of metal is E.
- S_i is the minimum spacing between metal i windows.
- D_i is the minimum size of metal i windows.
- L_i is the minimum spacing between the metal i windows and the metal $i+1$ windows.

All distances are in lambda (λ) design rules. The following assumptions also apply to the mathematical model:

- The process allows stacked contacts (stacked vias).
- Placement issues are not taken into account; every window is used by one connection to/from the core under study.
- Metal layer n is the highest metal layer in the process.
- The metal layers below m are used for local interconnections inside the core. The metal layers from $m+1$ to n are used to connect outside the core.

These dimensions are shown in Figure 17.9.

The number of connections possible to a core C with area $H \times H$, can be computed with Equations 17.6 to 17.8 [80]. Equation 17.6 computes the maximum number of possible connections on metal layer j.

$$C_j = 4\left\lfloor \frac{H - 2\left[E + \sum_{i=1}^{j-1}(D_i + L_i) - D_j\right]}{D_j + S_j} \right\rfloor \tag{17.6}$$

The inequality shown in Expression 17.7 ensures that there is enough space in the center of the core to continue adding rings of windows.

$$H - 2\left[E + \sum_{i=1}^{j-1}(D_i + L_i)\right] > 0 \tag{17.7}$$

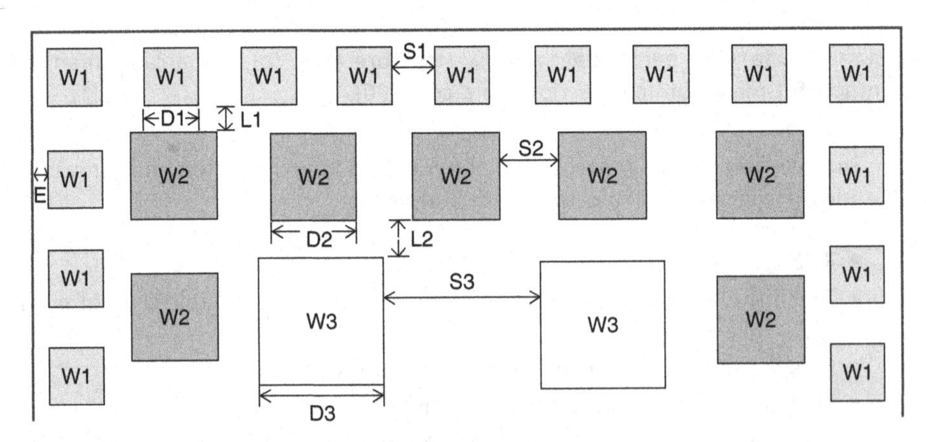

Figure 17.9. The model parameters.

Equation 17.8 sums the connections from the first available layer $m+1$ to layer j.

$$C_{total} = 4 \sum_{j=m+1}^{n} \left\lfloor \frac{H - 2\left[E + \sum_{i=1}^{j-1} (D_i + L_i) - D_j\right]}{D_j + S_j} \right\rfloor \tag{17.8}$$

A rudimentary calculation using a neuron without learning capabilities shows that less than 6000 connections can emerge from the projected area of a single neuron, 62.5 square microns in the year 2021. If each neuron has 10,000 synapses, and an axonal fanout of 10,000, this projected neural area cannot support the number of necessary point-to-point connections. Adding learning capabilities to the neuron effectively increases the size of each synapse (and ultimately the entire neuron) by a factor of 10 in the extreme case [79]. This size increase provides many more locations to place interconnections on the surface of the neuron, but the interconnections possible are limited by the number of metal layers.

Even if the die is mounted so that a second die is flipped[4] onto it, and 3D interconnections are made [81], the vertical connections between the dies are so large with current technologies that the number of additional connections is relatively small. Assuming a flip-chip design, the connections of the flip-chip design consist of two types of connections: One is the connections horizontally on the chip outside the core; another is the interconnections between two chips, one flipped. Vertical windows are used to indicate those windows for connecting to the second die. Horizontal windows are for those wires used to communicate between two cores on the same die. We apply the Pyramidal model for the outgoing connections. There are two key terms that we are going to use for this model. Figure 17.10 shows one side of a flip-chip. The outer area of the flip-chip is assigned to the horizontal windows. The central area of the flip-chip is assigned to the vertical windows. The purpose for interconnections both laterally and off-chip vertically is to maximize the number of connections to the core. In order to achieve this, the way we assign these two types of windows is to maximize the number of horizontal windows with available metal layers and utilize the remaining space for windows which are used for the vertical interconnections between two chips.

The ninth equation computes the possible intercore intrachip (horizontal) connections on available metal layers. The tenth expression, an inequality, is a constraint on the parameters, and when it holds Equation 17.11 can be applied to compute the number of vertical connections possible to the core using a flip-chip approach.

[4] Flip-chip technology implies that two dies are joined face-to-face, with vertical connections between the two dies.

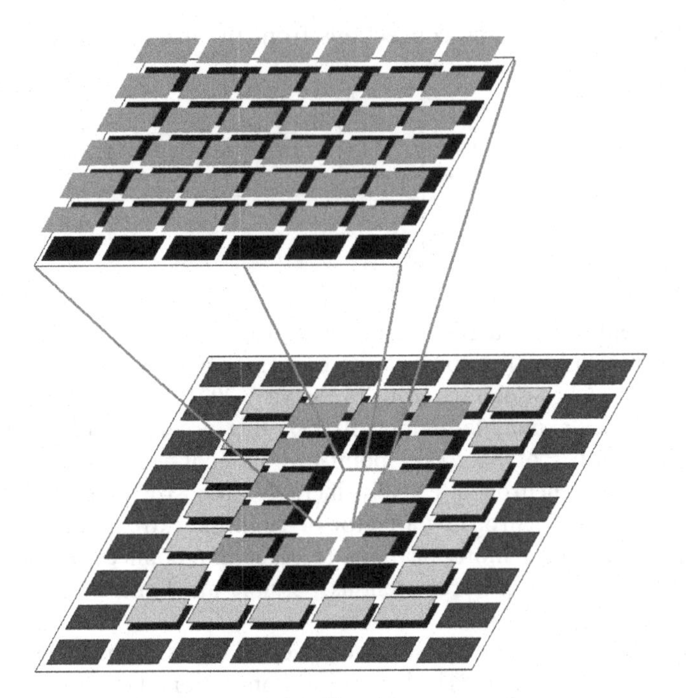

Figure 17.10. The flip-chip connections.

$$C_{\text{horizontal}} = 4 \sum_{j=m+1}^{n} \left\lfloor \frac{H - 2\left[E + \sum_{i=1}^{j-1} (D_i + L_i) - D_j\right]}{D_j + S_j} \right\rfloor \tag{17.9}$$

$$H - 2\left(E + \sum_{i=1}^{n} D_i + \sum_{i=1}^{n-1} L_i\right) \geq D_n + 2S_n \tag{17.10}$$

$$C_{\text{vertical}} = \left\lfloor \frac{H - 2\left[E + \sum_{i=1}^{n} D_i + \sum_{i=1}^{n-1} L_i\right]}{D_n + S_{in}} \right\rfloor^2 \geq 1 \tag{17.11}$$

With the 2D pyramidal model, we can estimate the maximum number of connections for each metal layer. Assuming a square core with size of 0.0004 mm^2, we use the parameters from The International Technology Roadmap for Semiconductors (ITRS) in 2006 [78]. For simplicity, we will also assume that window-to-window spacings are equal, even between layers, and the required spacing of the windows from the edges of the core is zero. According to the Technology Roadmap, the highest metal layer we could have in 2006 is metal 11. However, in order to demonstrate the relationship of the number of connections to the number of metal layers, we increase metal layers up to a maximum of 29 layers. We illustrate the possible connections for each metal layer in Figure 17.11.

As you can see in Figure 17.11, the number of connections for each layer decreases as the metal layers increase since the window size becomes larger and the remaining space for windows on each higher layer becomes smaller. Therefore, given a core area, we can determine how many more connections we can get as we add more metal layers with the 2D pyramidal model.

The utility of the flip-chip technology to enhance interconnection is decreased as the number of available metal layers increases. In order to demonstrate our model, we utilize Direct Bond Interconnect (DBI) [82] as an example. The Ziptronix DBI technology enables silicon and other technologies to be bonded and interconnected in a 3D fashion. It achieves high-density vertical interconnections without any volume exclusions (i.e., without "wasted space" for the interconnect) in the CMOS device. It supports an interconnect pitch of less than 10 microns, with typical interconnect width of 2λ μm and alignment accuracy of 1λ μm.

For our analysis, we choose the pitch of the vertical window to be $1\lambda\,\mu$. At this point, other parameters are unchanged. Figure 17.12 shows that the total outgoing number of connections to the core increases but the number of vertical interconnections between chips will decrease due to less availability of vertical connection area as we have more layers to connect horizontally (on the die) to the core.

Given the same core area, Figure 17.13 compares the number of connections of a side of a flip-chip design with that of a conventional 2D design. DBI

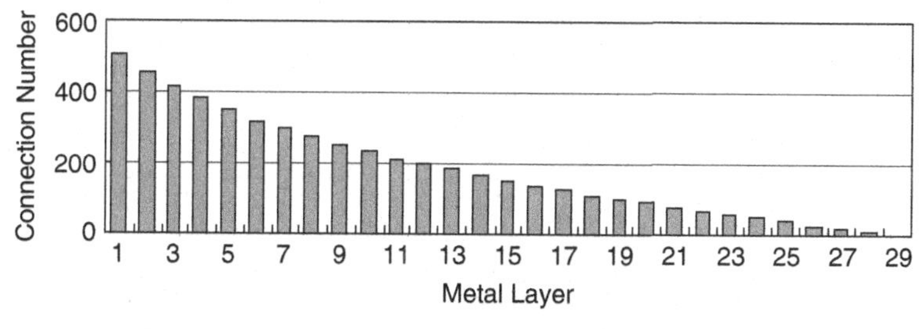

Figure 17.11. Number of connections as a function of available metal layers.

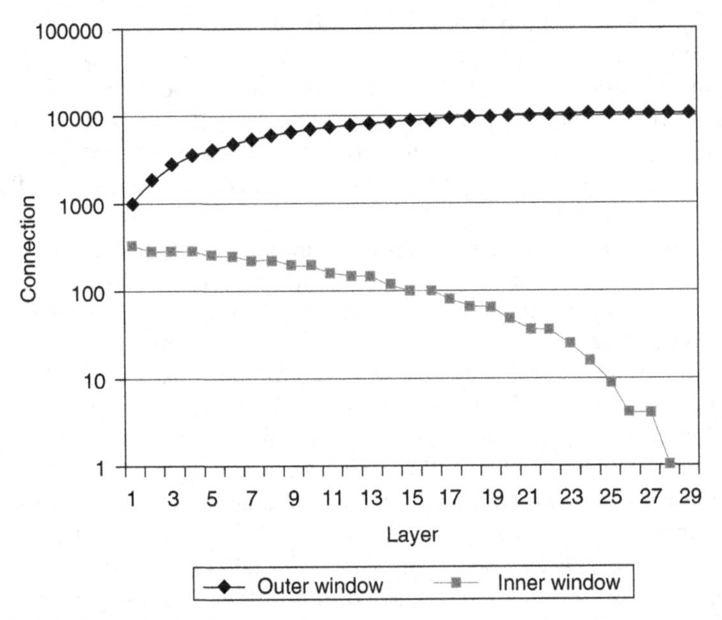

Figure 17.12. Horizontal window availability and vertical window availability versus quantity of metal layers in a log-scaled version. (MPU interconnect technology requirements in 2006.)

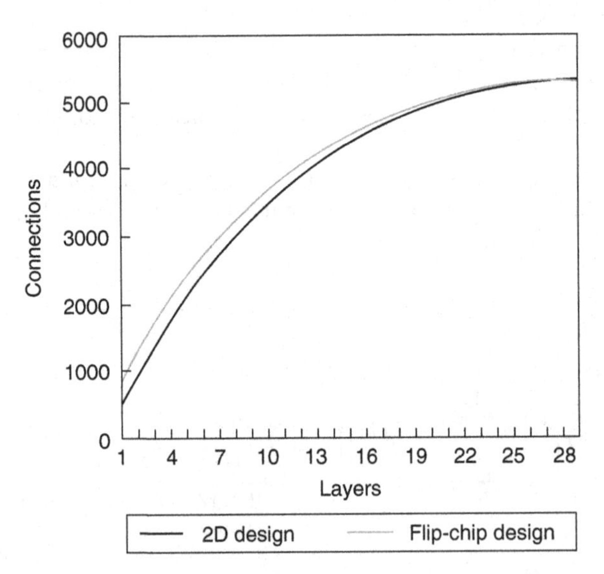

Figure 17.13. 2D design versus flip-chip design given the same core area. (MPU interconnect technology requirements in 2006.)

technology requires a very long pitch between two adjacent windows. Therefore, as the number of metal layers increases, the quantity of connections we can gain using flip-chip technology decreases. According to the technology roadmap, the maximum number of metal layers we might have in the near future is around 11 to 14 layers. Therefore, we still could obtain some amount of increased interconnections with flipped chips.

17.6. CONCLUSIONS

A carbon nanotube synapse typical of cortical synapses has been designed and simulated using SPICE. While the simulations were successful, the design of a single typical synapse is only a small step along the path to a synthetic cortex. The variations in synapses, including inhibitory synapses, will be the focus of future research. Predicting the interconnection capabilities of nanotube circuits is also important in understanding the future prospects for a synthetic cortex. Predictions of cortex size and interconnection capabilities highlight the necessity of using nanotechnology in providing a solution to the synthetic cortex design problem.

ACKNOWLEDGMENT

We acknowledge helpful suggestions of Bartlett Mel, Larry Swanson, Michel Baudry, Judith Hirsh, Norberto Grzywacz, and Gil Case regarding this research. H. S. Philip Wong, Jie Deng, Chongwu Zhou and Kyoungmin Ryu provided important research results without which the carbon nanotube synapse design could not have been possible. Jason Sanders was helpful in the understanding of the carbon nanotube models. Support for this research has been provided by the Viterbi School of Engineering at the University of Southern California.

REFERENCES

1. K. Boahen. Neuromorphic microchips. *Scientific American*, pp 56–63, May 2005.
2. A. C. Parker, A. K. Friesz, and A. Pakdaman. Towards a nanoscale artificial cortex. In: Proceedings of The 2006 International Conference on Computing in Nanotechnology (CNAN'06): June 26–29, 2006.
3. H. Moravec. When will computer hardware match the human brain? *Journal of Transhumanism*, 1: March 1998.
4. J. Hawkins and S. Blakeslee. *On Intelligence*. New York: Times Books, 2004.
5. J. Bebel, B. Raskob, A. Parker, and D. Bebel. Managing Complexity in an Autonomous Vehicle. In: Proceedings of PLAN 2006, San Diego: 2006.

6. T. W. Berger, et al. Restoring lost cognitive function. *IEEE Engineering in Medicine and Biology Magazine*, 24(5): pp 30–44, 2005.
7. J. V. Arthur and K. Boahen. Learning in silicon: timing is everything. *Advances in Neural Information Processing Systems*, 18: pp 75–82, 2006.
8. Brain Mind Institute, Ecole Polytechnique Federale de Lausanne. http://bmi.epfl.ch/.
9. G. Shepherd. Introduction to synaptic circuits. In: G. Sheperd, editor *The Synaptic Organization of the Brain*, 5th ed. New York: Oxford University Press, 2004.
10. J. Deng and H. S. P. Wong. A circuit-compatible SPICE model for enhancement mode carbon nanotube field effect transistors. Conference on Simulation of Semiconductor Devices and Processes, SISPAD 2006, Monterey, California, Sept. 6–8: pp 166–169, 2006.
11. C. Li, D. Zhang, S. Han, X. Li, T. Tang, and C. Zhou. Diameter-controlled growth of single-crystalline In_2O_3 nanowires and their electronic properties. *Advanced Materials*, 15: pp 143–145, 2003.
12. Z. Liu, D. Zhang, S. Han, C. Li, T. Tang, W. Jin, X. Liu, B. Lei, and C Zhou. Laser ablation synthesis and electronic transport studies of tin oxide nanowires. *Advanced Materials*, 15: pp 1754–1757, 2003.
13. Mouse brain simulated on computer. http://news.bbc.co.uk/2/hi/technology/6600965.stm.
14. http://www.intel.com/technology/silicon/high-k.htm.
15. A. K. Friesz, A. C. Parker, C. Zhou, K. Ryu, J. M. Sanders, H. S. Philip Wong, and J. Deng. A biomimetic carbon nanotube synapse circuit. Presented as a poster at The Biomedical Engineering Society 2007 Annual Fall Meeting, BMES 2007: Sep 2007. http://eve.usc.edu/Publications/cntfinal2.pdf/.
16. G. Chen and T. Ueta. *Chaos in Circuits and Systems*. Singapore: World Scientific, 2002.
17. Schuffny, R. et al. Hardware for neural networks. 4th International Workshop Neural Networks in Applications: Mar 1999.
18. S. Furber. http://www.cs.manchester.ac.uk/apt/people/sfurber/.
19. L. O. Chua. CNN chips crank up the computing power. *IEEE Circuits and Devices*, 12(4): pp 18–27, July 1996.
20. R. B. Wells. Preliminary discussion of the design of a large-scale general-purpose neurocomputer. MRC Institute, The University of Idaho, Nov 14, 2003. http://www.mrc.uidaho.edu/rwells/techdocs/Preliminary Discussion of the Design of a GP Neurocomputer.pdf
21. C. Mead. *Analog VLSI and Neural Systems*. Reading, MA: Addison-Wesley, 1989.
22. E. Farquhar and P. Hasler. A bio-physically inspired silicon neuron. *IEEE Transactions on Circuits and Systems*, 52(3): pp 477–488, Mar 2005.
23. M. Mahowald. VLSI analogs of neuronal visual processing: a synthesis of form and function. Ph.D. Dissertation, California Institute of Technology, Pasadena, 1992.
24. K. A. Zaghloul and K. Boahen. Optic nerve signals in a neuromorphic chip II: testing and results. *IEEE Transactions on Biomedical Engineering*, 51(4): pp 667–675, Apr 2004.
25. K. M. Hynna and K. Boahen. Neuronal ion-channel dynamics in silicon. IEEE International Symposium on Circuits and Systems, 2006 (ISCAS 2006): pp 21–24, May 2006.

26. B. Liu and J. F. Frenzel. A CMOS neuron for VLSI implementation of pulsed neural networks. In Proceedings of the 28th Annual Conference of Industrial Electronics (IECON02), Nov 5–8, Sevilla, Spain: pp 3182–3185, 2002.
27. D. Pan and B. M. Wilamowski. A VLSI implementation of mixed-signal mode bipolar neuron circuitry. International Joint Conference on Neural Networks, July 20–24 2003: Volume 2: pp 971–976.
28. B. C. Barnes, R. B. Wells, and J. F. Frenzel. PWM characteristics of a capacitor-free integrate-and-fire neuron. *IEEE Electronics Letters*, 39(16): pp 1191–1193, Aug 7 2003.
29. L. O. Chua and T. Roska. The CNN paradigm. *IEEE Transactions on Circuits and Systems I*, 40(3): pp 147–156, 1993.
30. C. Chiju, et al. Analysis and Performance of a Versatile CMOS Neural Circuit based on Multi-Nested Approach. 7th IEEE International Symposium on Signals Circuits and Systems, July, 2005: pp 417–420.
31. S. Sato, K. Nemoto, S. Akimoto, M. Kinjo, and K. Nakajima. Implementation of a new neurochip using stochastic logic. *IEEE Transactions on Neural Networks*, 14(5): pp 1122–1127, Sep 2003.
32. L. Chen and B. Shi. Building blocks for PWM VLSI neural network. 5th International Conference on Signal Processing Proceedings WCCC-ICSP 2000, (1): pp 563–566, 2000.
33. B. Linares-Barranco, E. Sanchez-Sinencio, A. Rodriguez-Vazquez, and J. L. Huertas. A CMOS implementation of Fitz Hugh–Nagumo neuron model. *IEEE Journal of Solid-State Circuits*, 26(7): pp 956–965, July 1991.
34. C. Fu, et al. A novel technology for fabricating customizable VLSI artificial neural network chips. In: International Joint Conference on Neural Networks, 1992.
35. C. Chao. Incorporation of Learning within the CMOS Neuron. M.S. Thesis, University of Southern California, July, 1990.
36. R. Z. Shi and T. Horiuchi. A Summating, Exponentially-Decaying CMOS Synapse for Spiking Neural Systems. Neural Information Processing Systems Foundation NIPS 2003.
37. R. D. Pinto, et al. Synchronous behavior of two coupled electronic neurons. *Physical Review E*, 62: pp 2644–56, 2000.
38. Y. Lee, J. Lee, Y. Kim, and J. Ayers. A low power CMOS adaptive electronic central pattern generator design. 48th Midwest Symposium on Circuits and Systems, Aug 7–10 2005, Volume 2: pp 1350–1353.
39. A Volkovskii, et al. Analog electronic model of the lobster pyloric central pattern generator. I, *Journal of Physics: Conference Series*, 23: p 4757, 2005.
40. J. G. Elias, H. H. Chu, and S. M. Meshreki. Silicon implementation of an artificial dendritic tree. International Joint Conference on Neural Networks, 1992, Volume 1: pp 154–159.
41. J. Liu and M. Brooke. Fully parallel on-chip learning hardware neural network for real-time control. Proceedings of the 1999 IEEE International Symposium on Circuits and Systems ISCAS 99, Volume 5: pp 371–374, 1999.
42. V. F. Koosh and R. Goodman. VLSI neural network with digital weights and analog multipliers. Proceedings of the 2001 IEEE International Symposium on Circuits and Systems ISCAS, May 6–9 2001. Volumes 2–3: pp 233–236.

43. W. Gerstner and W. Kistler. *Spiking Neuron Models*. Cambridge, MA: Cambridge University Press, 2002.
44. W. Mass and C. Bishop. *Pulsed Neural Networks*. Cambridge, MA: The MIT Press, 1999.
45. A. Perez-Uribe. Structure-adaptable digital neural networks. Ph.D. Thesis, EPFL, 1999.
46. E. Ros, R. Agis, R. Carrillo, and E. Ortigosa. Post-synaptic time-dependent conductance in spiking neurons: FPGA implementation of a flexible cell model. Proceedings of IWANN'03: LNCS 2687, pp 145–152, 2003.
47. A. Upegui, C. A. Pena-Reyes, and E. Sanchez. A methodology for evolving spiking neural network topologies on line using partial dynamic reconfiguration. Submitted to International Congress on Computational Intelligence (CIIC03), Medellín, Colombia.
48. X. Yao. Evolving artificial neural networks. In: Proceedings of the IEEE, 87(9): pp 1423–1447, Sep 1999.
49. L. Reyneri. *On the Performance of Pulsed and Spiking Neurons: Analog Integrated Circuits and Signal Processing*. Amsterdam: Kluwer 2002 pp 30, 101–119.
50. A. Murray, D. Corso, and L. Tarassenko. Pulse-stream VLSI neural networks mixing analog and digital techniques. *IEEE Transactions on Neural Networks*: Mar 1991.
51. M. S. Gudiksen, J. Wang, and C. M. Lieber. Synthetic control of the diameter and length of single crystal semiconductor nanowires. *Journal of Physical Chemistry B*, 105: p. 4062, 2001.
52. Y. Huang, X. Duan, Q. Wei, and C. M. Lieber. Directed assembly of one dimensional nanostructures into functional networks. *Science*, 291: p. 630, 2001.
53. N. Melosh, A. Boukai, F. Diana, B. Gerardot, A. Badolato, P. Petroff, and J. R. Heath. Ultrahigh density nanowire lattices and circuits. *Science*, 300: p. 112, 2003.
54. M. Huang, Y. Wu, H. Feick, N. Tran, E. Weber, and P. Yang. Catalytic growth of zinc oxide nanowires through vapor transport. *Advanced Materials*, 13(2): p. 113, 2001.
55. D. Zhang, C. Li, S. Han, X. Liu, T. Tang, W. Jin, and C. Zhou. Electronic transport studies of single-crystalline In_2O_3 nanowires. *Applied Physics Letters*, 82: pp 112–114, 2003.
56. C. Zhou, J. Kong, E. Yenilmez, and H. Dai. Modulated chemical doping of individual carbon nanotubes. *Science*, 290: pp 1552–1555, 2000.
57. J. Kong, N. Franklin, C. Zhou, S. Peng, K. Cho, and H Dai. Nanotube molecular wires as chemical sensors. *Science*, 287: 622–625, 2000.
58. X. Liu, R. Lee, J. Han, and C. Zhou. Carbon nanotube field-effect inverters. *Applied Physics Letters*, 79: pp 3329–3331, 2001.
59. X. Liu, Z. Luo, S. Han, T. Tang, D. Zhang, and C. Zhou. Band engineering of carbon nanotube field-effect transistors via selected area chemical gating. *Applied Physics Letters*, 86: 243501 1–3, 2005.
60. X. Liu, S. Han, and C. Zhou. Template-free directional growth of single-walled carbon nanotubes on a- and r-plane sapphire. *Journal of American Chemistry Society*, 127: pp 5294–5295, 2005.
61. X. Liu, S. Han, and C. Zhou. A novel nanotube-on-insulator (NOI) approach toward nanotube devices. *Nano Letters*: 2005. Published online.

62. X. Liu, S. Han, and C. Zhou. A novel nanotube-on-insulator (NOI) approach toward nanotube devices. *Nano Letters*, 6(3): pp 4–39, 2006.
63. K. Natori, Y. Kimura, and T. Shimizu. Characteristics of a carbon nanotube field-effect transistor analyzed as a ballistic nanowire field-effect transistor. *Journal of Applied Physics*, 97: pp 034306, 2005.
64. J. Guo, M. Lundstrom, and S. Datta. Performance projections for ballistic carbon nanotube field-effect transistors. *Applied Physics Letters*, 80: pp 3192–3194, 2002.
65. P. J. Burke. Carbon Nanotube Devices for GHz to THz Applications. Proceedings of the 2003 International Semiconductor Device Research Symposium, invited paper.
66. A. Naeemi, R. Sarvari, and J. D. Meindl. Performance comparison between carbon nanotube and copper interconnects for gigascale integration (GSI). *IEEE Electron Device Letters*, 26: pp 84–86, 2005.
67. A. Raychowdhury, S. Mukhopadhyay, and K. Roy. A circuit-compatible model of ballistic carbon nanotube field-effect transistors. *IEEE Transactions on Computer-Aided Design of Integrated Circuits and Systems*, 23: pp 1411–1420, 2004.
68. M. Cheung and C. Dwyer, Sorin. Semi-empirical SPICE Models for Carbon Nanotube FET Logic. Proceedings of the 4th IEEE Conference on Nanotechnology: 2004.
69. V. Braitenberg and A Schuz. *Cortex: Statistics and Geometry of Neuronal Connectivity*. New York: Springer, 1998.
70. V. Braitenberg. *On the Texture of Brains*. New York: Springer, 1977.
71. C. Stevens. Seminar at University of Southern California: Jan 29, 2007. http://www.salk.edu/faculty/faculty/details.php?id = 50.
72. K. Zhang and T. Sejnowski. A universal scaling law between gray matter and white matter of cerebral cortex. Proceedings of the National Academy of Sciences, 97(10): pp 5621–5626, May 9 2000.
73. S. Laughlin and T. Sejnowski. Communication in neuronal networks. *Science*, 301: pp 1870–1874, Sep 26 2003.
74. M. Singh, D. Hwang, W. Sungkarat, and K. Veera. Evaluation of MRI DTI-tractography by tract-length histogram. *Progress in Biomedical Optics and Imaging: Physiology, Function and Structure from Medical Images*, 5746(1): pp 138–147, 2005.
75. J. Bailey and D. Hammerstrom. Why VLSI implementations of associative VLCNs require connection multiplexing. IEEE International Conference on Neural Networks, Jul 1988, Volume 2: pp 173–180.
76. Y. M. Chang, D. L. Rosene, R. J. Killiany, L. A. Mangiamele, and J. I. Luebke. Increased action potential firing rates of layer 2/3 pyramidal cells in the prefrontal cortex are significantly related to cognitive performance in aged monkeys. *Cerebral Cortex Magazine*, 15(4): 409–418, 2005.
77. G. Gonzalez-Burgos, L. S. Krimer, N. N. Urban, G. Barrionuevo, and D. A. Lewis. Synaptic Efficacy during repetitive activation of excitatory inputs in primate dorso-lateral prefrontal cortex. *Cerebral Cortex Magazine*, 14(5): 2004.
78. International Technology Roadmap for Semiconductors, 2005. http://www.itrs.net/Common/2005ITRS/Interconnect2005.pdf.
79. C. Lu, B. X. Shi, and L Chen. An on-chip learning neural network with ideal neuron characteristics and learning rate adaptation. *Analog Integrated Circuits and Signal Processing*, Amsterdam: Kluwer, 31: pp 55–62, 2002.

80. C. C. Tseng. Estimation of maximum connections for CMOS neuron chip design. University of Southern California Electrical Engineering Systems Department. http://eve.usc.edu/tseng.pdf, Directed Research Report, 2005.
81. http://www.ziptronix.com/news/oct17_2005.html?pg=releases fuseaction=detail Story-ID=37
82. http://www.ziptronix.com/techno/dbi.html.
83. B. Mel and J. Schiller. On the fight between excitation and inhibition: location is everything. Science's STKE: Sep 2004.
84. Intel Corporation. http://www.intel.com/pressroom/kits/events/idfspr_2006/20060313_multicore_fact_sheet_decoder.pdf.

18

BIOMEDICAL AND BIOMEDICINE APPLICATIONS OF CNTs

Tulin Mangir

In this chapter, potential applications of carbon nanotubes (CNTs) in biomedicine and biomedical engineering are discussed. Because of their unique quasi-one-dimensional structure and fascinating mechanical and electronic properties, CNTs have captured the attention of physicists, chemists, and material scientists. Biology and medicine are rapidly emerging as new areas for the application of CNTs. Many CNT applications require handling in solution-phase; however, CNTs have proven difficult to disperse in solvents. Chemical modification of SWCNTs is often required for more versatile suspension capabilities and enablement of certain applications. This has encouraged greater exploitation of their intrinsic properties, as well as the capability to modify these properties. In particular, the functionalization of CNTs is required for their aqueous suspension and allowance for molecular interactions with biological systems. Here, we summarize the breadth of techniques that have been employed in functionalizing CNTs. We then review the potential applications in biology and medicine and commercial applications of CNTs in biomedical engineering. We describe some of our recent observations impact of CNTs on behavior of living organisms. We conclude with a discussion of issues in using CNTs in biological systems and questions of toxicity that need further exploration for full utilization of CNTs in these applications.

18.1. INTRODUCTION

Since Edison's discovery that carbon changes its resistance with pressure and a carbon filament glows when an electric current is passed through it, the unique properties of carbon have intrigued scientists. After more than a century of

Bio-Inspired and Nanoscale Integrated Computing. Edited by Mary Mehrnoosh Eshaghian-Wilner

interest, carbon has found its apogee in fullerenes and carbon nanotubes (CNTs), arguably the most promising of all nanomaterials. Because of their unique quasi-one-dimensional structure and fascinating mechanical and electronic properties, CNTs have captured the attention of physicists, chemists, and material scientists. Biology and medicine are rapidly emerging as new areas for the application of CNTs [1–8].

In the last few years, several studies have been proposed that indicate the potential applications of CNTs. The applications include the use of CNTs in energy storage and energy conversion devices, high strength composites, nano-probes and sensors, actuators, electronic devices, production of nanorods using CNTs as reacting templates, catalysis, and hydrogen storage media.

Carbon nanotubes, a form of carbon that did not exist in our environment before being manufactured, possess unique chemical, physical, optical, and magnetic properties that make them suitable for many uses in industrial products and in the field of nanotechnology, including nanomedicine. Hence, it is of the utmost importance to explore the yet almost unknown issue of toxicity of this new material. In the following sections, we will first review the functionalization that is essential for use in various applications. After that we will describe potential and commercial biomedical applications of CNTs. Carbon nanotube composite materials, such as those based on carbon nanotubes bound to nanoparticles, are suitable to tailor carbon nanotube properties for specific applications. In addition, the ability to control nanotube morphology and bead size, coupled with the versatility of chemistry, makes these structures an excellent platform for the development of biosensors (optical, magnetic, and catalytic applications).

The exploration of carbon nanotubes for biomedical applications is just underway, but the technology has significant potential. Some of the applications being investigated include tumor-targeted therapies that wipe out cancer cells without side effects; drugs and vaccines for AIDS and other infectious diseases; plaque-busting nanoparticles that clear clogged arteries; and bionanorobots and biosensors that detect and treat disease in the human body. Carbon nanotubes are also being investigated for their applications in tissue engineering and body implants [4, 6].

Cells have been shown to grow on carbon nanotubes, so they appear to not have a significant toxic effect. On the other hand, we have observed carbon nanotubes interfering with and stopping the cell division in bacteria when ingested [1, 7]. This can be interpreted in multiple ways: (1) as a toxic effect, (2) as a means for chemotherapy by stopping the division of cancer cells, and (3) as targeted drug delivery. The cells do not adhere to the carbon nanotubes; this feature has potential in applications such as coatings for prosthetics and anti-fouling coatings for ships [7].

The ability to chemically modify the sidewalls of carbon nanotubes also leads to potential in biomedical applications such as vascular stents, neuron growth, and neuron regeneration.

Fibers of pure carbon nanotubes have recently been demonstrated [6]; along with CNT composite fibers, they are undergoing rapid development. Such super strong fibers will have applications that include body and vehicle armour, transmission line cables, woven fabrics, and textiles.

There are almost certainly many unanticipated applications for this remarkable material that will come to light in the years ahead; these future applications may prove to be the most important and valuable of all. Based on our experiments, it seems that for cells to ingest the CNTs they need to be mixed in suitable liquid or some particular "nutrient" for the cells. Nanomaterials have been designed for a variety of biomedical and biotechnological applications, including bone growth, enzyme encapsulation, biosensors, and as vehicles for DNA delivery into cells [1]. Whereas nanotechnology may provide novel materials that can result in revolutionary new structures and devices, biotechnology already offers extremely sophisticated tools to precisely position molecules and assemble hierarchal structures and devices. The application of the principles of biology to nanotechnology provides a valuable route for further miniaturization and performance improvement of artificial devices. The feasibility of the bottom-up approach, which is based on molecular recognition and self-assembly properties of proteins, has already been proved in many inorganic–organic hybrid systems and devices. Nanodevices with biorecognition properties provide tools at a scale, offering a tremendous opportunity to study biochemical processes and manipulate living cells at the single molecule level. The synergetic future of nano- and biotechnologies hold great promise for further advancement in tissue engineering, prostheses, genomics, pharmacogenomics, drug delivery, surgery, and general medicine.

Due to their electrical, chemical, mechanical, and thermal properties, carbon nanotubes are one of the most promising materials for the electronics, computer, and aerospace industries. We discuss their properties in the context of future applications in biotechnology and biomedicine. The purification and chemical modification of carbon nanotubes with organic, polymeric, and biological molecules are discussed. Additionally, we review their uses in biosensors, assembly of structures and devices, scanning probe microscopy, and as substrates for neuronal growth. We note that additional toxicity studies of carbon nanotubes are necessary so that exposure guidelines and safety regulations can be established in a timely manner.

18.2. FUNCTIONALIZATION OF CARBON NANOTUBES

Many CNT applications require handling in a solution phase; however, CNTs have proven difficult to disperse in solvents. Chemical modification of single-walled carbon nanotubes (SWCNTs) is often required for more versatile suspension capabilities and the enabling of certain applications. This has encouraged greater exploitation of their intrinsic properties and the capability to modify these properties. In particular, the functionalization of CNTs is required for their aqueous suspension and to allow for molecular interactions with biological systems. In general, different noncovalent and covalent modification strategies can be used.

Noncovalent methods preserve the pristine CNT structure while covalent modification introduces structural perturbations. Details of the various functionalization techniques have been previously reviewed. Here, we summarize the breadth of techniques that have been employed in functionalizing CNTs [1, 3, 5].

18.2.1. Noncovalent Functionalization

Various noncovalent interactions such as π stacking, hydrophobic, and van der Waals interactions have allowed for the functionalization of CNTs with a wide range of molecules.

18.2.1.1. Surfactants. Surfactants have been found to associate with CNTs via van der Waals interactions through their hydrophobic chains, rendering the CNTs hydrophilic, and enabling them to disperse in aqueous environments. With this method, SWCNTs can be isolated from aggregated bundles, allowing for spectroscopic probing of individual SWCNTs. Surfactants have been used to prevent nonspecific interactions between SWCNTs and proteins. Surfactants with modified head groups have also been used to link SWCNTs to specific molecules.

18.2.1.2. Aromatic Organic Molecules. Aromatic organic molecules can associate with CNTs via π stacking interactions. In particular, modified pyrene molecules have been used to modify SWCNTs in various ways. This strategy has been used to modify SWCNT surface properties and the linking of SWCNTs to various moieties, imparting diverse functionality of the CNT complexes. For example, functionalization through modified pyrene molecules has been used to link the following: CNTs to proteins, opening the door for biological applications; gold nanoparticles, resulting in enhanced Raman signals of the CNTs; metalloporphyrins, in order to exploit the electron acceptor/donor properties of SWCNTs; and initiation sites for polymerization, allowing the ability to coat SWCNTs with polymers.

18.2.1.3. Fluorophores. Fluorophores have been shown to associate with CNTs via hydrophobic interactions, allowing for the visualization of CNTs with fluorescence microscopy. Other noncovalent means of introducing fluorescence capabilities include the addition of quantum dots (QDs) via streptavidin-conjugated QDs, where the streptavidin is absorbed on the CNT surface. Finally, fluorescent polymer wrapping also allows for the fluorescent detection of CNTs.

18.2.1.4. Polymers. Polymers interact with CNTs through several noncovalent interactions and offer a wide-range functionality, as demonstrated through the rich history of polymer science and advances in their use for drug delivery.

18.2.1.5. Lipids. Lipids comprise a class of molecules that interact with CNTs similar to surfactants. Lipids offer control of their interactions with CNTs through modification of their hydrophobic chains, while also providing versatility of the functionality that they impart through the modification of their head groups.

18.2.1.6. DNA. DNA, as a naturally occurring polymer, can also associate with carbon nanotubes through π stacking interactions formed by the aromatic nucleotide bases on the DNA molecule. DNA has proved effective in allowing for the aqueous suspension of SWCNTs; by adjusting the DNA sequence, its binding properties towards CNTs can be modified.

18.2.1.7. Proteins. Proteins have been shown to associate with CNTs predominantly through hydrophobic interactions, as well as through charge transfer interactions. This property has been exploited in the design of specific peptides for controlled interactions with CNTs.

18.2.1.8. Endohedral Functionalization. Endohedral functionalization refers to modification of the inside of CNTs in order to generate hybrid materials and utilize CNTs as "nano test-tubes" incorporating biomolecules. Endohedral growth of metallic, ionic, and semiconducting nanocrystals, as well as the incorporation of fullerenes, has been shown, and the incorporation of proteins demonstrates potential in sensing applications.

18.2.2. Covalent Functionalization

There are two main strategies for covalently functionalizing nanotubes: end and defect modification and sidewall modification. These covalent modifications arise from the difference in reactivity at the nanotube ends and sidewalls (as well as at structurally perturbed areas) and, therefore, each type of functionalization requires distinct chemical approaches.

18.2.2.1. Ends and Defects. As discussed in the chemical reactivity of CNTs, the ends are more reactive than the sidewalls. Treatment of CNTs with oxidative agents results in introducing oxygenated groups, such as carboxylic acid, ketone, alcohol, and ester, at the nanotube ends and defect sites, thereby leaving the ends open and possibly cutting and shortening CNTs. Such treatments remove amorphous carbon and metal catalyst particles and can remove smaller diameter (more reactive) tubes. Techniques have been developed to probe the degree of oxidation and the type of oxygenated groups formed. Oxidation also results in hole-doping of the CNTs causing perturbation of its electronic structure, though the perturbations are less pronounced than those produced in the case of sidewall modification as discussed below.

Through the oxidation of CNTs, various chemical species can be introduced via the covalently incorporated oxygenated species. Coordination of cadmium to oxidized multiwalled carbon nanotubes (MWCNTs) was used to grow quantum dots on the tube surfaces. Other modifications of oxidized CNTs by noncovalent interactions have also been performed through nonspecific protein binding and ionic interactions, which have coupled CNTs to crown ether.

18.2.2.2. Sidewall Functionalization. Sidewall functionalization methods can introduce higher concentrations of covalently attached functional groups onto the surface of CNTs with the trade off of significant perturbation in the electronic structure. Indeed, the loss of van Hove singularities as observed in the absorption spectrum has been used to probe the degree of sidewall modification.

Halogenation of CNTs has been achieved through the incorporation of a high density of fluorine, chlorine, and bromide atoms on tube surfaces. In particular, fluorination of CNTs is achieved through gas-phase reactions and offers a nondestructive means of high-density sidewall functionalization. The degree of fluorination can be controlled and in any case, the electronic properties of the CNT are significantly altered. Fluorination has been shown to be reversible, therefore somewhat restoring the electronic properties of the tubes. Fluorination also offers a means of introducing a wide variety of additional functional groups on the CNT surface. Chlorination and bromination of CNTs has also been demonstrated. Hydrogenation, nitration, radical addition, nucleophilic addition, electrophilic addition, diazotization, ozmylation, and ozonolysis also offer a variety of modification routes that result in a wide range of functionality. In particular, treatment with ozone in the presence of various reagents provides a means of controlling the proportions of the type of oxygenated groups introduced onto CNTs. Furthermore, CNTs oxidized with this method were coordinated with metal ions that acted as precursors for the formation of quantum dots on the CNT surface. Cycloaddition reactions of various types and have been employed in introducing a variety of functionality for biomedical applications of CNTs.

18.3. POTENTIAL APPLICATIONS OF CARBON NANOTUBES IN BIOLOGY AND MEDICINE

The unique and diverse properties of CNTs, in addition to the wide range of functionality afforded by chemical modification, allow for many exciting applications. Due to their nanometer dimensions, CNTs have the potential to interact at the cellular and molecular level.

In the last few years, several studies have been proposed that indicate potential applications of CNTs. However, this section attempts to report the existing and future applications of CNTs in the biomedical industry exclusively. The following sections detail the areas in biomedical engineering where CNTs can be potentially applied.

18.3.1. Diagnostic Tools and Devices

18.3.1.1. Radiation Oncology. The traditional method of generating X-rays comprises of a metallic filament (cathode) that acts as a source of electrons when heated resistively to a very high temperature. The accelerated electrons (that are emitted) are bombarded on a metal target (anode) to generate X-rays. The advantage associated with this method is that it works even in non-ultrahigh vacuum ambiences, which contain various gaseous molecules. But this method has several limitations: 1) a slow response time; 2) consumption of high energy; and 3) a limited lifetime. Recent research has reported that field emission, compared to thermoionic emission, is a better mechanism of extracting electrons. This is because electrons are emitted at room temperature and the output current is voltage controllable. In addition, giving the cathode the form of tips increases the local field at the tips and, as a result, the voltage necessary for electron emission is lowered. An optimal cathode material should have a high melting point, low work function, and high thermal conductivity. CNTs can be used as a cathode material for generating free flowing electrons. Electrons are readily emitted from their tips either due to oxidized tips or the curvature when a potential is applied between a CNT surface and an anode. Yue et al. [8] generated continuous and pulsed X-rays using a CNT-based field emission cathode. The field emission currents were found to follow the Fowler–Nordheim regime, where the applied voltage and constants are dependent on the cathode geometry and work function. Plotting current density against 1/E (cm/mV) yields a straight line for a current of field emission origin. Yu et al. obtained a total emission current of 28 mA from a 0.2-cm-area CNT cathode. By programming the gate voltage, pulsed X-rays with a repetition rate greater than 100 kHz was readily achieved. The X-ray intensity was sufficient to image human organs at 14 kVp and 180 mAs. Recently, a dynamic radiography system using CNT-based field emission has been proposed by Cheng et al. [9]. X-ray radiation with continuous variation of temporal resolution as short as nanoseconds can be readily generated by their system. The advantages of CNT-based X-ray devices are fast response time, fine focal spot, low power consumption, possible miniaturization, longer life, and low cost. Besides, it minimizes the need of cooling required by the conventional method. Miniaturized X-ray devices can be inserted into the body by endoscopy to deliver precise X-ray doses directly at a target area without damaging the surrounding healthy tissues, as malignant tumors are highly localized during the early stage of their development. With time, the cancer spreads to neighboring anatomic structures. Other processes such as chemotherapy and conventional radiation doses kill the cancer but may also kill healthy tissues. This is not desired from a health point of view.

18.3.1.2. Sensors. Sensors are devices that detect a change in physical quantity or event. There are many studies that have reported use of CNTs as pressure, flow, thermal, gas, and chemical and biological sensors, as mentioned at the beginning of this section. Liu and Dai [10] demonstrated that piezoresistive pressure sensors can be made with the help of CNTs. They grew SWNTs on suspended square polysilicon membranes. When uniform pressure was applied on the membranes, a change in resistance in the SWNTs was observed.

Silicon piezoresistors have the disadvantage in that their resistance is highly sensitive to variations in temperature. As CNTs have increased sensitive and a temperature coefficient almost two orders of magnitude lower than that of silicon, highly efficient pressure sensors incorporating CNTs can be fabricated. Fabrication of piezoresistive pressure sensors that incorporate CNTs can bring dramatic changes in the biomedical industry, as many piezoresistance-based diagnostic and therapeutic devices are currently in use. Pressure sensors can be used in eye surgery, hospital beds, respiratory devices, patient monitors, inhalers, and kidney dialysis machines [11, 12]. During eye surgery, fluid is removed from the eye and, if required, cleaned and replaced. Pressure sensors measure and control the vacuum that is used to remove the fluid, and provide input to the pump's electronics by measuring barometric pressure. Hospital bed mattresses for burn victims consist of pressure sensors that regulate a series of inflatable chambers. To reduce pain and promote healing, sections can be deflated under burn areas. Pressure sensors can also be used for sleep apnea (a cessation of breathing during sleep) detection. The pressure sensor monitors the changes in pressure in inflated mattresses. If no movement is found for a certain period, the sleeper is awakened by an alarm. Pressure sensing technology is used in both invasive and noninvasive blood pressure monitors. Many patients who use inhalers activate their inhalers at an inappropriate time, resulting in an insufficient dose of medication. Pressure sensors in the inhalers identify the breathing cycle and release the medication accordingly [11, 12]. During kidney dialysis, blood flows from the artery to the dialysis machine and after cleaning flows back into the vein. Waste products are removed from the blood through osmosis and move across a thin membrane into a solution that has the blood's mineral makeup [13]. Using pressure sensors, the operation of the dialysis system can be regulated by measuring the inlet and outlet pressures of both the blood and the solution. Intelligent pressure sensing systems play an important role in portable respiratory devices that consist of both diagnostic (spirometers, ergometers, and plethysmographs) and therapeutic (ventilators, humidifiers, nebulizers, and oxygen therapy equipment) equipments. They serve patients with disorders of asthma, sleep apnea, and chronic obstructive pulmonary disease. They measure pressure by known fluid dynamic principles.

In [14] CNTs are used as an immobilization matrix for the development of an amperometric biosensor. The biosensor was developed by growing aligned MWNTs on platinum (Pt) substrates. The platinum substrate served as the transduction platform for signal monitoring, whereas opening and functionalization of large CNT arrays allowed for the efficient immobilization of the model enzyme (glucose oxidase in this case). The schematic diagram of the CNT array biosensor is shown [14]. The arrays were purified by treatments with acid or air. The acid treatment resulted in the removal of impurities including amorphous carbon that occurred during the production procedure. The lengths of the nanotubes were also reduced by approximately 50%. Air oxidation resulted in the production of thinner nanotubes because of the peeling of the outer graphitic layers from the nanotubes.

CNT-based nanobiosensors may be used to detect DNA sequences in the body [15, 16]. These instruments detect a very specific piece of DNA that may be related to a particular disease. Such sensors enable detection of only a few DNA molecules that contain specific sequences; thus, the patient can be diagnosed as having specific sequences related to a cancer gene. The use of nanotube-based sensors will avoid problems associated with the current, much larger implantable sensors (which can cause inflammation) and can eliminate the need to draw and test blood samples. The devices can be administered transdermally (through the skin) avoiding the need for injections during space missions. Biosensors can also be used for glucose sensing. CNT chemical sensors for liquids can be used for blood analysis (for example, detecting sodium or finding pH value). CNTs can also be used as flow sensors.

Ghosh et al. [17] found that the flow of a liquid on bundles of SWNTs induces a voltage in the direction of flow. This finding can be used in the future in micromachines that work in a fluid environment, such as heart pacemakers that need neither heavy battery packs nor recharging. Flow sensors can also be used for precise measurements of gases utilized by respiratory apparatuses during surgery and for automatic calculation of medical treatment fees based on output data leading to reduced hospital costs and more accurate calculation. In summary, the advantages of CNT-based sensors are that they are less sensitive to variations in temperature (compared to silicon piezoresistors), which enables them perform better in many of the biomedical sensing applications mentioned above. Also, their unique flow sensing properties can be exploited in making heart pacemakers. As CNT-based sensors are smaller in size, they consume less power, avoid chances of inflammation, and eliminate the need to draw and test blood samples. They can be used in detecting very specific pieces of DNA by utilizing their electron transfer characteristics and functionalization properties. This can help in detecting important cancerous genes and biomolecules such as antibodies associated with human autoimmune diseases.

18.3.1.3. Probes. Probes are devices that are designed to investigate and obtain information on a remote or unknown region or cavity. There are many studies that have reported the use of CNTs for making probes [18–24]. CNTs are highly suitable materials for AFM probes, as the AFM-generated image is dependent upon the shape of the tip and surface structure of the sample of interest. An optimal probe should have vertical sides and a tip radius of atomic proportions. AFM tips made of silicon or silicon nitride are pyramidial in shape and have a radius of curvature around 5 nm. In comparison, nanoprobes made of CNTs have high resolution, as their cylindrical shape and small tube diameter enable imaging in narrow and deep cavities. In addition, probe tips made of CNTs have mechanical robustness and low buckling force. Low buckling force lessens the imaging force exerted on the sample and therefore can be applied for imaging soft materials such as biological samples. Besides, these factors enhance the life of probes and minimize sample damage during repeated hard crashes into substrates [25].

Nanotweezers are a type of probe that is driven by the electrostatic interaction between two nanotubes on a probe tip. Their operating principle is to balance the elastic restoring force with the electrostatic force. To fabricate nanotweezers, a tapered glass micropipette is selected. Free-standing, electrically independent metal electrodes (generally of Au) are deposited on the micropipettes. At the end, electrically controlled CNTs made from a bundle of MWNTs are attached to the independent electrodes.

Nanotube nanotweezers have several features. First, they can be used for manipulation and modification of biological systems such as DNA and structures within a cell. Second, they can be used as nanoprobes for assembling structures and will be helpful in increasing the value of measurement systems for characterization and manipulation at nanometer scale. Third, once a nanoscale object has been grasped by the nanotweezer, the object's electrical properties can be probed, as nanotube arms serve as conducting wires. Fourth, nanotweezers can be used as an electromechanical sensor for detection of pressure or viscosity of media by measuring the change in resonance frequency.

Using the developed nanotweezers, researchers Kim and Lieber [26] were able to remove a semiconductor wire 2 nm wide from a mass of entangled wires. They were also able to grasp polystyrene spheres of 500 nm clusters. Currently, research is being carried out that will grow SWNTs onto the electrodes to produce nanotweezers to grab individual macromolecules. Application of CNTs as nanoprobes for crossing the tumor but not crossing into healthy brain tissue should also be investigated, as the presence of cancer in a brain tumor may result in weakening of the blood–brain barrier.

18.3.1.4. Quantum Dots. Quantum dots are tiny, light-emitting particles that are 2–10 nm long. They are a new class of biological labels. Semiconductor quantum dots can be used for quantitative imaging and spectras copy of single cancer cells. Semiconductor quantum dots can be used for quantitative imaging and spectra for generating molecular fingerprints of individual cancer cells. This is because of their dimensional similarity with biological molecules such as nucleic acids and proteins, and their size-tunable properties. In addition, they allow longer periods of observation and do not fade when exposed to UV light. They can also be used to track many biological molecules simultaneously, as their color can be tailored by changing the size of the dot.

Researchers have shown that CNTs can form quantum dots. Coulomb blockade and a quantization of electron states were shown by their transport experiment that implied that a CNT quantum dot had been formed. Buitelaar et al. [27] showed that MWNTs can also form clean quantum dots where the level separation exceeds the charging energy. Recently, a structure has been proposed to fabricate a quantum well on the order of a few nanometers by using the electromechanical properties of SWNTs. When embedded in biological fluids and tissues, quantum dot excitation wavelengths are often quite constrained.

Therefore, excitation and emission wavelengths should be selected carefully based on the particular application [28].

18.3.2. Biopharmaceutics

18.3.2.1. Drug Delivery. Like other device industries, increasing concerns over rising health care costs have led to the development of novel drug delivery techniques that are cost effective. The important characteristic of an efficient drug delivery system is the ability to perform controlled and targeted drug delivery. For this purpose, drugs should be released at an appropriate rate, as rapid release of drugs may lead to incomplete absorption, gastrointestinal disorders, and other side effects. In addition, care needs to be taken so that the drugs will not decompose during delivery, as some drugs are toxic in nature. To this end, drugs should be encapsulated in a carrier during transit in the body while maintaining their biological and chemical properties. Also, the drug delivery material must be compatible with the drug; the delivery material should bind easily with the drug. The material should decompose at the completion of its use either by solubilizing or by elimination via excretory roots of the body. CNTs can be used as a carrier for drug delivery, as they are adept at entering the nuclei of cells. Researchers have found that functionalized CNTs can cross the cell membrane. In addition, they are of a size where cells do not recognize them as harmful intruders. Martin and Kohli [29] reported that CNTs can be used as drug delivery materials because they have larger inner volumes, as compared to the dimensions of the tube, that can be filled with the desired chemical and biological species. In addition, CNTs have distinct inner and outer surfaces that can be differentially modified for functionalization. As a result, the outer surface of CNTs can be immobilized with biocompatible materials and the inside can be filled with the desired biochemical payload. CNTs have open mouths, which make the inner surface accessible for insertion of species inside the tube. Molecular dynamics simulations show that van der Waals and hydrophobic forces are important for the insertion process. The van der Waals forces have a dominant role in the CNT species. Gao et al. [30] found that in a water-solute environment, a DNA molecule could be inserted into CNTs via an extremely rapid dynamic interaction process. Based on their studies, they suggested that the encapsulated CNT–DNA molecular complex can be applied in DNA-modulated molecular electronics, molecular sensors, electronic DNA sequencing, and gene delivery systems. Kong et al. [31] found that the encapsulation of biological molecules in CNTs can be accelerated at high temperatures. Finally, CNTs can be organically functionalized to make them soluble in organic solvents and aqueous solutions. Soluble CNTs can be coupled with amino acids and bioactive peptides for further derivatization. These drug delivery systems may form the basis for anticancer treatments, gene therapies, and vaccines in future, as the carrier can enter damaged cells and release enzymes either to initiate an autodestruct sequence of cells or to repair the cell for normal functioning. In the future, the application of CNTs as drug encapsulators for treatment of brain tumors can be investigated by determining if the drug crosses the blood–brain barrier or not. Other applications to be investigated include the use of CNTs to deliver drugs to the eye beyond the blood–retina barrier and to the central nervous system beyond the blood–brain barrier. If this is found successful, CNTs can be used for treating eye diseases and other diseases such as Alzheimer's and Parkinson's. Another interesting area invokes using CNTs as pills that contain a miniaturized video system. The swallowed pill can be used for assessing diseases in areas that are difficult to access with other techniques such as endoscopies and colonoscopies.

18.3.2.2. Drug Discovery. Traditional trial and error methods have very high lead time that takes several years for a new drug to reach the market. The critical bottlenecks in drug discovery may be overcome by using arrays of CNT sensors and current information technology solutions (such as data mining and computer-aided drug design) for identification of genes and genetic materials for drug discovery and development [32].

18.3.2.3. Implantable Materials and Devices

Implantable Nanosensors and Nanorobots. There are certain cases, such as diabetes, where regular tests by patients themselves are required to measure and control the sugar level in the body. Children and elderly patients may not be able to perform this test properly. Another similar example is regular tests of persons exposed to hazardous radiations or chemicals. The objective is the detection of the disease in its early stage so that appropriate action for higher chances of success can be taken. Implantable sensors and nanorobots can be useful in health assessment. CNT-based nanosensors have the advantages that they consume less power and are a thousand times smaller than even microelectromechanical systems (MEMS) sensors. Therefore, because of their small size and less power consumption, CNT-based nanosensors are highly suitable as implantable sensors. Implanted sensors can be used for monitoring pulse, temperature and blood glucose, and diagnosing diseases [33, 34]. Besides, CNTs can be used for repairing damaged cells or killing them by targeting tumors through chemical reactions.

Implantable nanosensors can also monitor the heart's activity level and regulate heartbeats by working with an implantable defibrillator. Possible application of CNTs for treatment of retinal diseases due to loss of photoreceptors can be investigated. One way of to compensate for the loss of photoreceptors is to bypass the destroyed photoreceptors and artificially stimulate the intact cells in the neighborhood. Another possible area related to the application of CNTs that can be investigated is cochlear implants related to hearing problems. According to Bhargava [35], implanted nanorobots can have following possible applications:

1. Cure skin diseases. A cream-containing nanorobots could remove the right amount of dead skin, remove excess oils, add missing oils, and apply the right amount of moisturizing compounds.

2. Protect the immune system by identifying unwanted bacteria and viruses and puncturing them to end their effectiveness.
3. Ensure that the right cells and supporting structures are at right place.
4. Use as a mouthwash to destroy pathogenic bacteria and lift food, plaque, or tartar from the teeth to rinse them away.

ACTUATORS. Actuators are devices that put something (such as a robot arm) in action. They do so by converting electrical energy to mechanical energy. The direct conversion of electrical energy to mechanical energy through a material response is crucial for many biomedical applications such as microsurgical devices, artificial limbs, artificial ocular muscles, or pulsating hearts in addition to robotics, optical fiber devices, and optical displays. The main technical requirements of these actuators are low weight, low maintenance voltage, large displacements, high forces, fast response, and long cycle-life [37]. Different materials that have been previously investigated for use in electrochemical actuators are ceramics (piezoelectrics), shape memory alloys, and polymers. However, they have certain limitations. For example, piezoelectrics have low stiffness and electromechanical coupling coefficients. Research reveals that CNTs can act as actuators. CNT actuators can work under physiological environment, low voltages, and temperatures as high as 350°C [36, 37–40].

Nanotube-based polymer composites have promise as possible artificial muscle devices because of their incredible strength and stiffness in addition to relatively low operating voltage. The mats of nanotubes expand and contract when operated as assembled electrodes in an electrochemical cell. When a potential is applied, the charging of the electrodes takes place; there is a linear change in CNT length because of the introduction of electronic charge on the tube and a restructuring of the double layer of charge in the double layer outside the tube. The biocompatibility, crystallinity, and morphology of the composites were evaluated using SEM, TEM, hot stage microscopy, and polarized light. Also, thermal analysis was performed. Methods of characterization included thermal analysis using thermal gravimetric analysis (TGA) and differential scanning calorimetry (DSC). The results of all these analyses were promising. Baughman et al. [36] were the first to provide evidence of the actuator property of CNTs. They used actuators based on sheets of SWNTs. The CNT electromechanical actuators (also known as artificial muscles) generated higher stresses than natural muscles and higher strains than high-modulus ferroelectrics. MWNTs are excellent candidates for electromechanical devices because of their large surface area as well as their high electrical conductivity. Gao et al. [38] were the first to show an electromechanical actuator based on MWNTs. The actuator sheets were based on arrays of parallel nonbundled MWNTs whose tubes were perpendicular to the sheet planes. The MWNTs had a length from 5 to 40 m and diameter from 10 to 60 nm. The actuators utilized the electrostatic repulsion between electrical double layers associated with parallel MWNTs. Vohrer et al. [37] developed an experimental setup for the measurement of the actuation forces and the displacement of CNT sheets for the first time. With their setup, vertical elongation or forces of bucky papers can be observed, a feature that is a prerequisite for the optimization of artificial muscles for industrial applications. The fastest actuation time observed by them was approximately 3 s. The parameters that affect the electromechanical properties of CNT sheets are the raw material used (SWNTs and MWNTs). Compared to SWNTs, MWNTs have varying features: contorted structure; different production techniques (for example, arc-discharge material shows very low actuation); purification grade (not only the amount of carbon particles but also the amount of remaining catalysts decrease the actuation time and actuation amplitude); chirality and diameter of CNTs; homogeneity of the nanotubes' distribution; alignment of the nanotubes; size and thickness of the produced bucky paper (it was observed that thinner bucky papers react faster than thicker ones at comparable thickness values); type of electrolyte (chemistry, concentration, viscosity); electrode material; surface area of electrode; arrangement of electrodes; surface electrode resistance; applied voltage; and polarity [37].

Nanofluidic Systems. If the implantable fluid injection systems are large in size, functions of surrounding tissues are adversely affected. However, tiny nanodispensing systems can dispense drugs on demand using nanofluidic systems, miniaturized pumps, and reservoirs. Currently, limited attention has been given to understand fluid mechanics at the nanoscale. As fluid mechanics at the nanoscale is in infant stage, there is a scarcity of experimental data [41]. The research so far reveals that MWNTs show great potential for use in nanofluidic devices. This can be attributed to their extremely high mechanical strength, coupled with their ability to provide a conduit for fluid transport at near-molecular length scales. Furthermore, there is a lack of defects on their inner surface. The nanodispensing systems using CNTs can be applied for chemotherapy where precise amounts of drugs are targeted directly at the tumors when the patient falls asleep. Other potential areas where fluid dispensing systems could be applied are lupus, AIDS, and diabetes [42].

18.3.2.4. Surgical Aids. Surgery using macro instruments can be cumbersome for both the surgeon and the patient. On one hand, the patient experiences severe pain, scarring, and long healing time because of large cuts; on the other hand, the surgeon requires high concentration for a long period to perform the surgery accurately. Sometimes, it may lead to surgical error due to the surgeon's fatigue. In many cases, surgical error may result because of the limited view of the organs by the surgeon. In addition, macro surgical instruments are not suitable for certain delicate cases such as surgeries related to heart, brain, eyes, and ears. One of the solutions is laparoscopic surgery, which uses a small entry port, long and narrow surgical instruments, and a rod-shaped telescope attached to a camera. However, laparoscopic surgery requires highly skilled surgeons for efficient surgery [42]. Research needs to be carried out to investigate if smart instruments (such as forceps, scalpels, and grippers with embedded sensors to provide improved functionality and real-time information) using CNTs can be developed to aid surgeons by providing specific properties of tissue to be cut and information about performance of their instruments during surgery. The usefulness of CNTs for optically guiding surgery should also be investigated. This can lead to easy removal of tumors and other diseased sites. Another option is the use of molecular nanotechnology (MNT) or nanorobotics in surgery [43]. In nanorobotics, surgeons move joystick handles to manipulate robot arms containing miniature surgical instruments at the ports. Another robot arm contains a miniature camera for a broad view of the surgical site. It results in less stress for surgeons and less pain for patients; at the same time, high precision and safety is achieved. MNT allows in vivo surgery on individual human cells. Nanorobotics-based surgery can be used for gall bladder, cardiac, prostrate, bypass, colorectal, esophageal, and gynecological surgery. However, nanorobotic systems for performing surgery require the ability to build precise structures, actuators, and motors that operate at molecular level to enable manipulation and locomotion. As nanotweezers (that can be used for manipulation and modification of biological systems such as structures within a cell) have already been created using CNTs, they have the potential to be used in medical nanorobotics. Besides, Cumings and Zettl [44] have demonstrated that nested CNTs can make exceptionally low friction nanobearings. These nanobearings can be used in many surgical tools. Therefore, research can be extended to investigate the application of CNTs in other surgical tools.

18.3.2.5. Tissue Engineering and Implantable Devices. The final areas of medicine that could be impacted by carbon nanotubes are tissue repair and implantable devices. In the repair of damaged tissue, a simple transplant often does not restore the functionality of that tissue. Cells respond to their physical and chemical environment, and tissue generation requires support from the extracellular matrix for proper cell differentiation and mediation of interactions between cells such that a particular functionality is achieved. The engineered generation of functional tissue as such requires scaffolds, and carbon nanotubes are emerging as promising materials in this regard. Scaffolds have been constructed using CNTs and have been further functionalized in order to promote proper cell growth and tissue formation. Collagen–SWCNT composites have been used as scaffolds for seeding smooth muscle cells. Neurons were shown to grow on unmodified MWCNT surfaces and further studies looked into modification of these surfaces, introducing cellular growth signals to promote specific growth. These studies pointed toward the possibility of using CNT scaffolds for the development of neural prostheses as implantable devices for the regeneration of neuronal tissue. Research is also being conducted using CNT scaffolds for the regeneration of visual tissue. Retinal pigment epithelial cells have been shown to grow well on these surfaces, and the scaffolds showed potential in subretinal implantation. Other work in this area used vertically aligned MWCNTs on electrode surfaces in order to exploit their electrical properties for potential implantable retinal prosthesis. Studies looking into the growth of osteoblasts on as-produced and functionalized MWCNTs and SWCNTs have determined variations in differentiation due to functionalization and show promise for bone graft materials [2, 4, 45].

18.4. COMMERCIAL BIOMEDICAL APPLICATIONS OF CARBON NANOTUBES

The unique and diverse properties of CNTs, in addition to the wide range of functionality afforded by chemical modification, allows for many exciting applications. Due to their nanometer dimensions, CNTs have the potential to interact at the cellular and molecular level. This characteristic is just beginning to be exploited and has already shown great promise in biological and medical applications. At the time of this writing, only a few companies have begun to invest in commercializing carbon nanotubes for different medical applications [1–5].

18.4.1. Drug and Gene Delivery

The global drug delivery market is expected to grow substantially in the coming years. Developments in nanotechnology-based drug delivery are expected to have significant impacts, enhancing the capabilities of current drugs and creating whole new therapeutics. It is estimated that the global market share for nanoparticle-based therapeutics is expected to grow to from 0.9% to 5.2% by 2012, comprising a $4.8 billion industry. Carbon nanotubes are among the promising new nanoparticles that can be used for drug and gene delivery.

Since the first demonstration of the functionalized SWCNT translocation of the cell membrane and the ability to penetrate the cell nucleus to deliver a bioactive peptide, [46] several CNT-based drug and gene delivery applications have been developed. In the initial study, cycloaddition techniques were used to covalently modify the sidewalls of CNTs. Cationically modified SWCNTs and MWCNTs have been used for electrostatic association with plasmid DNA that was subsequently delivered into cells. Double functionalization of SWCNTs for delivery of multiple therapeutics has shown promise, as well as similar applications achieving dual functionality through sidewall cyclo-adition and simultaneous end oxidation routes to introduce drug delivery and fluorescent probing functionalities [47].

During an *in vivo* study, covalently functionalized SWCNTs with a neutralizing B cell epitope from the foot-and-mouth disease virus were administered into mice. Immunization of the mice with the SWCNT complexes induced a strong immune response from the functionalized peptide while immunogenicity was not observed from the nanotubes themselves.

A more recent *in vivo* study covalently functionalized SWCNTs with carborane (a cluster compound of boron and carbon atoms) and administered the complex intravenously into mice with implanted EMT6 tumor cells for the purpose of boron neutron capture therapy [48]. Analysis of the distribution of the SWCNT complex in tumor, blood, lung, liver, and spleen samples found that the complexes were tumor specific. SWCNTs have also been used to deliver RNA interference [49]. A recent study employed short interfering RNA (siRNA), which is being developed as therapy to inhibit expression of certain genes. Noncovalently modified SWCNTs with phospholipids were linked through disulfide bonds to siRNA as well as DNA. Upon endocytosis of the SWCNT complexes, the disulfide bonds were cleaved by natural cellular mechanisms and the delivered cargo was able to take effect. This group also covalently attached proteins to SWCNTs through end and defect oxidized modification and conducted a detailed study of the internalization mechanism of SWCNTs, concluding that it occurs through an endocytosis process. In addition, the group showed that noncovalent absorption of proteins onto end and defect oxidized SWCNTs serves as a general intracellular protein transporter for proteins ≤ 80 kDa [50].

Another unique means of CNT-based gene delivery was achieved through "nanotube spearing" where as-grown CNTs with carbon encapsulated nickel (catalyst) particles remaining at their tips were used to magnetically drive plasmid DNA modified CNTs into cells. The group succeeded in transfecting cells that had only previously been transfected with viral vectors [51].

18.4.2. Hyperthermia Therapy

In addition to delivering therapeutics, carbon nanotubes are a prime candidate for hyperthermia therapy. Hyperthermia therapy is the use of heat to damage or kill cancer cells. SWCNTs have been shown to absorb NIR radiation and generate heat while tissue is transparent in this region. This property has been exploited through noncovalent functionalization of SWCNTs with oligo nucleotides and phospholipids containing folic acid that targets folate receptors on cancer cells, thereby targeting the SWCNT complex to these cells. Researchers have shown that the SWCNT complexes selectively target the folate receptor containing cells and are internalized, while cells without these receptors are avoided [52].

Irradiation of the tissue with an 808 nm laser (GaAs) causes the SWCNTs to heat up. The SWCNT complexes, which remain in endosomes after cellular uptake, are released upon pulsed radiation by the laser source, allowing the complexes to enter the cell nucleus. Prolonged irradiation generates enough heat to destroy the targeted cells while the cells without internalized SWCNTs are unharmed.

18.4.3. Biomedical Imaging

In addition to their potential as therapeutics or delivery vehicles for therapeutics, nanotubes might also be used for imaging purposes. New imaging capabilities based on nanotubes could be harnessed in biomedical research, clinical diagnosis, and new multifunctional therapeutics. It is likely that future therapeutics will not only have the capabilities to target specific tissues or cell, but they will also have imaging agents that will enable clinicians to ensure that the therapeutics have hit their targets. Because nanotubes have unique optical properties and can be linked to other nanostructures that have unique optical properties, they could play an important role in this new class of therapeutics.

As previously discussed, SWCNTs intrinsically emit in the NIR region [53]. Most biological material is fairly transparent in this spectral range, and very little endogenous fluorescence is generated in the tissue. In addition, emission from SWCNTs can penetrate though relatively thick tissue. Furthermore, CNTs do not blink and photo bleach. The first example of the use of nanotubes as imaging agents involved surfactant encapsulated SWCNTs that were internalized by live phagocytic cells [54]. The samples were excited at 660 nm and emission from the CNTs was detected at 1125–1600 nm, producing NIR images of SWCNT within the cells. More recently, a similar study with 3T3 fibroblast and myoblast stem cells used SWCNTs functionalized with DNA; the Raman and fluorescence spectra inside the cells was monitored for a period of three months. The group demonstrated unprecedented stability of CNTs as fluorescent probes; spectral changes in the CNT emission over the period of incubation were also noticed. The observed behavior changes were attributed to changes in the local environment surrounding the nanotubes. Thus, it may even be possible to use nanotubes for real-time, *in vivo* imaging of changes taking place inside of cells.

In addition to their intrinsic fluorescent properties, it is also possible to fluorescently label CNTs such that they emit in the visible spectrum and can therefore be imaged through more conventional fluorescence microscopy. Many of the CNT-based drug and gene delivery applications mentioned above have specifically introduced fluorophores onto the CNTs in order to determine their internalization through confocal fluorescence microscopy [55]. Also, conventional fluorophores have been found to associate nonspecifically with MWCNTs through hydrophobic interactions; this has allowed for the fluorescent imaging of MWCNTs. Fluorescent polymers have been used as a noncovalent means of introducing fluorescence onto CNTs as well. Quantum dots, another emerging nanomaterial with promising applications in biomedical applications, have been conjugated to CNTs and have even been grown from functional groups attached to their surfaces. Although devices comprised of nanotubes and quantum dots have been developed primarily for electronic and photonic applications in mind, new work is focused on using such devices as intracellular fluorescent probes [56].

18.4.4. Biosensors for New Diagnostics

While the future of medical treatment lies in targeted therapeutics, the future of clinical diagnosis is likely to involve a shift to early detection of disease. Today, when a patient observes symptoms of a possible disease, she visits the doctor, and a sample of body fluid is taken and sent out for testing. The results are generally not available for days. Further, because existing tests are largely based on systemic physiological changes rather than local or cellular changes, disease at an early stage often goes undetected. The development of new, minute and ultrasensitive biosensors could lead to personalized diagnostic kits that allow patients to determine their state of health from the comfort of their own homes. By placing a drop of bodily fluid on a small, disposable chip, patients may be able to identify the early onset of disease. The market potential for health care devices with these characteristics is undoubtedly tremendous.

The capability of specifically attaching (and preventing the attachment of) biomolecules to CNTs has provided the possibility of using nanotubes as the biosensors for future diagnostic products. There are two main types of CNT sensors: electronic sensors and optical sensors.

18.4.4.1. Electronic Sensors. The introduction of electronic biosensors would provide several advantages over existing biosensors, which are primarily based on different optical techniques. Electronic sensors do not require complex reagents and expensive equipment, as in the case of optical-based detection systems, and have the added advantage of versatility of implementation due to being an electronic device. Additionally, electronic detectors could be more sensitive and accurate than optical systems. There are two different electronic sensors that are being developed using nanotubes.

Field effect transistors (FETs) are constructed by connecting the source and drain of the FET with a single semiconducting SWCNT that forms the conducting channel [57–59]. The device detects changes in the electrical characteristics of the FET resulting from molecules interacting with the nanotube. As chemical sensors, these devices were initially shown to be capable of single molecule detection both in air and in liquid environments. Through the ability of CNTs to interact with biomolecules and by taking advantage of functionalization schemes, these devices have also been employed in biosensing. Carbon nanotube FETs (NTFETs) exhibit unparalleled sensitivities as compared to conventional FET devices while having advantages over optical techniques with comparable sensitivity. NTFETS have since been used to probe the mechanism of CNT–protein surface interactions, in the detection of receptor–ligand interactions through the use of biomolecule-funtionalized CNTs, in probing enzymatic reactions, and in the detection of viruses. Another class of NTFET devices employs a network of SWCNTs forming a CNT film as the conducting channel. The network architecture allows for better reproducibility and eased manufacturability, but is less sensitive than a single nanotube device. CNT network FETs have been integrated with cell membranes. Further research will focus on sensing capabilities in physiological serum [59]. The low limit of detection of NTFETs is characteristic of CNTs responding to the binding of any molecule. Applications in complex mediums such as serum will require stringent prevention of nonspecific binding through robust functionalization schemes.

Amperometric sensors relate the current induced in a particular reaction to the concentration of the species involved in the reaction. CNT-based sensors employing dense, vertically aligned, and biofunctionalized SWCNTs have been developed for probing biomolecule-induced charge transfer [64]. These devices take advantage of the directional conduction pathway of CNTs by attaching the base of the tubes to an electrode with analyte interactions occurring at their functionalized tips. Like field effect transistor sensors, further advances for amperometric sensors are needed to prevent nonspecific binding while maintaining high specificity at the molecular level.

18.4.4.2. Optical Sensors. NIR optical sensors based on CNTs take advantage of perturbations in the nanotube's optical properties due to their surface modification. For example, a group [65] developed a CNT-based probe to detect DNA hybridization. Single-stranded DNA was noncovalently absorbed onto SWCNTs, and the hybridization with its complementary DNA on the nanotube surface was detected through monitoring alterations in the NIR band gap fluorescence of the tubes. This same group has also developed an *in vivo* optical sensor for monitoring glucose levels using the optical properties of CNTs [66] by noncovalently modifying SWCNTs affecting the optical properties of the CNT. In this way, the NIR fluorescence can be coupled to glucose concentration. The SWCNT complex was encapsulated into a 200-μm-diameter dialysis capillary to retain the complex while allowing the analyte to diffuse through the dialysis membrane. This device was then imaged through a human epidermal tissue sample and exhibited a limit of detection of 34.7 μM and sensitivities within the range of blood glucose regulations of diabetic patients.

18.5. RESEARCH TOOLS FOR DISCOVERING NEW DRUGS

The costs of bringing a new drug to market using existing methods such as combinatorial chemistry are substantial. Current techniques in drug discovery require considerable quantities of proteins of interest in order to meet the limits of detection of the probing devices employed. Additionally, existing tools do not provide sensitive enough data about binding events between proteins and other biological molecules. Beyond the diagnostic applications outlined above, CNT-based biosensors could also be widely employed in drug discovery. CNT electrical and optical detection systems might be used to characterize induced molecular interactions. Perhaps the most powerful new tool for drug discovery will emerge from CNTs incorporated onto the tips of atomic force microscopes (AFM). Nanotube-based AFM tips could enable researchers to prove an array of target proteins for specific interactions at the single-molecule level.

CNTs have been incorporated onto the tips of conventional probes for atomic force microscopes, greatly enhancing this already powerful tool [67]. The high aspect ratio and small radius of CNTs offers enhanced resolution. The flexibility of nanotubes makes these tips robust and sensitive to fragile biological samples. Additionally, beyond topographical imaging, the rich functionalization chemistry of CNTs offers the potential for probing a variety of different biological molecules. End oxidized MWCNTs, with diameters ranging from 15 to 50 nm, were shown to achieve less than 3 nm resolution due to their functionalization. Chemical modification of CNT tips has allowed for applications such as conducting measurements as a function of pH, imaging based on specific molecular interactions, and measuring single-molecule protein–protein interactions. An extension of this technology combines a functionalized SWCNT probe with specifically probing single proteins labeled for single-molecule fluorescence studies. In addition, researchers can electrically couple the SWCNT probe to the protein in order to induce conformational changes within the protein that can be detected by AFM and fluorescence measurements.

18.6. ISSUES IN THE BIOLOGICAL APPLICATIONS

18.6.1. Challenges

In a very short duration, CNTs appear to be the frontrunner that has the potential to dominate the biomedical research. However, challenges remain that need to be addressed before the full potential of CNTs for biomedical applications can be realized. For example, there is a lack of detailed understanding of the growth mechanism of CNTs. First, an efficient growth approach to structurally perfect nanotubes at large scales is currently not available. Second, it is difficult to grow defect-free nanotubes continuously to macroscopic lengths. Third, control over nanotubes' growth on surfaces is required in order to obtain large-scale ordered nanowire structures. Fourth, controlling the chirality of SWNTs by any existing growth method is very difficult. Also, the above-mentioned limitations result in high cost of production for pure and uncontaminated nanotubes with uniform characteristics. In short, the optimization of production parameters and the control on the growth of nanotubes is to be mastered. In addition to the challenges at the fabrication level, the low dimensional geometry results in structural instability, which is an important issue for the mechanical application of CNTs, as larger strains are prone to buckling, kink forming, and collapse. Besides, the toxicology of CNTs is not well understood. Dagani [90] compiled a report on the adverse effects of potential drug-carrying nanoparticles at the blood–brain barrier from the national meeting of the American Chemical Society. On the basis of their experiments, researchers suggested that CNTs possess health risks. They reasoned that humans can potentially be exposed to CNTs by inhalation because unprocessed CNTs are lightweight and, therefore, can become airborne. If CNTs reach the lung, they can agglomerate and fill the air passages, which may lead to suffocation. This warrants an in-depth study about the toxicology of CNTs to come up with a final conclusion with respect to their acceptance by the human immune system. Finally, the time from proof of concept in the laboratory of the CNT-based devices to the commercial marketplace should be reduced as the competition from other novel materials and technologies continue to emerge.

18.6.2. Toxicity

18.6.2.1. Occupational Exposure: Unmodified CNTs. To date, most studies regarding carbon nanotube toxicity have focused on potential hazards associated with carbon nanotube production, with the primary risk being CNT inhalation and epidermal exposure [94–101]. Because the production of CNTs includes the use of metal catalysts such as iron and nickel, toxic effects from these metal particles should also taken into consideration. Preliminary *in vivo* studies in rats and mice revealed the development of pulmonary epethilial granulomas and mortality from intratracheally–instilled, unpurified SWCNTs. The extent of toxicity due to catalyst particles is unclear from this rat study although in the mouse study granulomas were observed regardless of metal content from various SWCNT samples. However, these studies used excessive amounts of SWCNTs, administered in a nonphysiological manner, thus making it difficult to interpret results [97]. In the rat study, mortality was attributed to blockage of the upper airways by the instillate and not inherently by SWCNTs. Mortality in the mouse study was suggested to have been caused from the known toxicity of residual catalyst particles in the sample.

In vitro studies have also been conducted using human keratinocyte epidermis cells and alveolar macrophage cells (these being the first line of defense within the lung against foreign particulates). The keratinocyte study specifically addressed the toxic effect of ferrous iron from residual catalyst particles, citing its effect of catalyzing the decomposition of hydrogen peroxide and lipid peroxides, thus resulting in the generation of free radical species that enhance oxidative stress. In the macrophage study, high levels of toxicity were observed although (again) the extent of toxicity attributed to catalyst content is unknown.

The toxicity of MWCNTs has also been studied *in vitro* on keratinocytes and was found to elicit an inflammatory effect but no severe toxicity [99]. In fact, the discussed SWCNT studies on alveolar macrophage cells compared the toxicity of SWCNTs to MWCNTs and found the latter to be less toxic. A more recent study compared the *in vivo* and *in vitro* effects of SWCNTs on mice and mice cells. The study used purified SWCNT samples with less catalyst content than was present in previous studies, and the method of SWCNT administration was refined through the use of pharyngeal aspiration, as well as consideration of relevant dose as compared to the permissible exposure limits (PEL) for graphene particles established by the United States Operational Safety and Health Administration. The *in vivo* study revealed distinct effects due to aggregated and dispersed SWCNT particles where aggregates induced granulomas due to macrophage accumulation, while dispersed particles resulted in a fibrogenic response. Furthermore, exposure to SWCNTs affected pulmonary function by inhibiting bacterial clearance. The *in vitro* study was conducted on macrophage cells and found that the cells exhibited changes in gene expression but no severe toxic effects.

From examining the degree of airborne particulate generation by simulating a SWCNT manufacturing environment and from the *in vivo* results to date, it has been determined that particulate release from SWCNT manufacturing can produce dangerous exposure levels. However, the administration of SWCNTs in these *in vivo* experiments has not simulated physiological conditions (the most accurate being the sudden dose of total PEL for 20 work days while bypassing the nose and with the possibility of the incorporation of nonrespirable particles).

From *in vitro* studies at the cellular level for epidermal and pulmonary cells, it is difficult to determine proper dosage and infer the *in vivo* effects since cells are isolated from physiological interactions. Yet, it is clear that SWCNTs have the potential to elicit pulmonary damage. The potential risks are gradually becoming understood and will allow for the development of appropriate precautions.

18.6.2.2. Cell Behavior After Ingesting CNTs. In our experiments, we have observed that the CNTs interfere and stop cell division in bacteria after ingesting CNTs [7]. The picture was taken after filtering the bacteria. In the SEM figure (Fig. 18.1), the bacteria changes its shape due to the CNT now inside of it, leaving a thin shell supported only by the CNT forming a film inside of it. Since one of the nutrients of Pseudomonas is iron, the bacteria started to look for the only supply of food available in its environment, the CNTs, which contain iron particles. Since the pseudomonas has iron receptors, those absorbed the CNTs in order to extract the iron from them. In Figure 18.1, it can be observed how the mitosis of the bacteria is being stopped by the amount of CNT ingested for the bacteria interfering in the cytokinesis process of the bacterial division. One can observe how the division was blocked because of CNT. Hence, the implications in the bio-realm may suggest a way to further stop cell division of unwanted cells. Furthermore, we will be studying the implications of this particular process for both drug delivery, as well prevention and shrinking of unwanted tissues.

A closer look at the bacteria shell in Figure 18.2 reveals the amount of CNT inside the bacteria and the blockage of the cytokinesis process for the CNT.

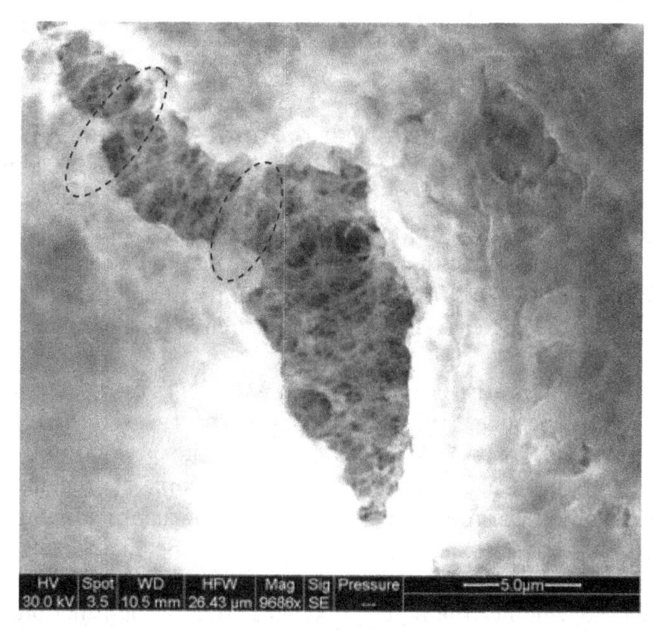

Figure 18.1. SEM image of pseudomonad after CNT ingestion.

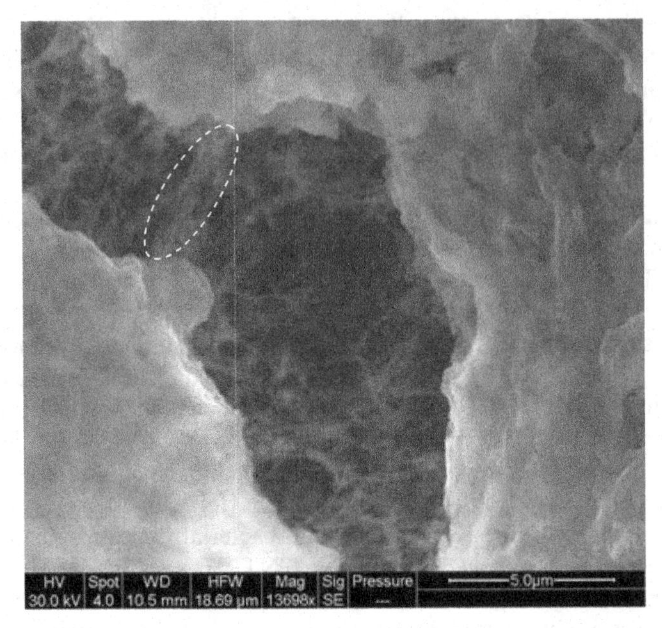

Figure 18.2. A closer look of block cell division.

As stated before, this was a surprising finding of this approach creating thin films. Another issue is the manipulation of the bacteria shell containing the thin film inside, i.e., the handling of the CNT thin film created inside the bacteria.

18.6.2.3. Clinically Administered Exposure—Modified CNTs. The potential of CNT toxicity in terms of medical applications only becomes relevant in terms of functionalized CNTs and their resulting properties. As stated before, impeding of the cell cytokinesis was a surprising finding of this approach to creating thin films. Another issue is the manipulation of the bacteria shell containing the CNT thin film inside, i.e., the handling of the CNT thin film created inside the bacteria. This will be researched further by characterization studies.

An *in vitro* study of pristine and oxidized MWCNTs on T lymphocytes found significant, dose-dependent toxicity that was most severe for the oxidized MWCNTs. Like SWCNTs, MWCNTs are known to be internalized into cells. Furthermore, pristine CNTs tend to form macroscopic aggregates in solution, while oxidation allows for finer dispersion and homogeneous aqueous suspension. Thus, a factor affecting toxicity in this case could include the fact that higher concentrations of oxidized MWCNTs are able to interact with the T lymphocytes, making it unclear whether the higher toxicity is due to the functional groups on oxidized MWCNTs.

A recent study probing the effects of CNT size on toxicity conducted *in vitro* and *in vivo* experiments with 220-nm and 850-nm oxidized MWCNTs. As opposed to the previously mentioned study, an order of magnitude lower MWCNT concentration was used to test toxicity against human acute monocytic leukemia cells *in vitro*. These studies observed an induced inflammation-associated gene expression with no size correlation *in vitro*, while *in vivo*, subcutaneous administration of the larger MWCNTs generated a heightened inflammatory response. Increased inflammation was associated with an inability of the macrophages to effectively engulf the larger MWCNTs.

More recently, an *in vitro* study was conducted on human dermal fibroblasts involving sidewall functionalized SWCNTs with phenyl-SO_3H, phenyl-SO_3Na, and phenyl-$(COOH)_2$ groups, as well as unmodified SWCNTs dispersed in the block-copolymer Pluronic F108. Sidewall functionalized SWCNTs contained varying degrees of functionalization density, and toxicity substantially decreased with increasing sidewall functionalization density. Furthermore, in all cases, sidewall functionalization resulted in less toxicity than the surfactant coated SWCNTs.

All of the *in vitro* studies mentioned in the above discussion of CNT applications found low or no toxicity for the doses of functionalized CNTs that were employed. *In vitro* studies offer valuable insight for specific cellular level toxicity and therapeutic properties of functionalized CNTs; however, whole animal *in vivo* studies are crucial to determine targeting specificity as well as potential side effects.

A recent *in vivo* animal study employed ammonium and indium-labeled dimethylentriamine pentaacetic functionalized SWCNTs and MWCNTs administered intravenously into mice. The data showed that the SWCNTs were not retained in any reticuloendothelial organs and were completely cleared from the animals through the renal excretion route without any toxic side effects. The development of blood-compatible CNTs has been reported recently. MWCNTs were functionalized with glycosaminoglycan (GAG), which is commonly found on the luminal surface of the endothelium. Further developments involving intravenous administration of carbon nanotubes are on the horizon.

Several reviews regarding nanoparticle toxicity are emerging that offer comparisons between clinically relevant nanoparticles. CNTs offer unique properties as therapeutic materials over other nanoparticles, and it is reasonable to expect that their toxicity can be effectively managed.

18.7. CONCLUSIONS

Although there may be a long process of getting some of the above-mentioned biomedical products to the market, the research on some of these aspects are already on the floor. In future, many of these developed devices could save millions of lives, and therefore gives companies and institutions the incentives to hasten their research.

This chapter provides the most contemporary overview possible of synthesis, properties, and potential biomedical applications of CNTs through recent examples. The exceptional physical, mechanical, and electronic properties of CNTs allow them to be used in sensors, probes, actuators, nano-electronic devices, and drug delivery systems within biomedical applications. With the increasing interest shown by the nanotechnology research community in this field, it is expected that plenty of applications of CNTs will be explored in future. At the same time, it is believed that the continued development and application of CNTs can enhance the practice of biomedical industries. However, amidst all the hope and hype, CNTs have yet to cross many technological hurdles in order to fulfill their potential as the most preferred material for biomedical applications. Hopefully, the descriptions provided and references to the literature therein will allow researchers to develop new applications besides proposing improvement in the current application areas. Almost certainly many unanticipated applications for this remarkable material will come to light in the years ahead and may prove to be the most important and valuable of all.

Carbon nanotubes, a form of carbon that did not exist in our environment before being manufactured, possess unique chemical, physical, optical, and magnetic properties, which make them suitable for many uses in industrial products and in the field of nanotechnology, including nanomedicine. Hence, it is of the uttermost importance to explore the yet almost unknown issue of the toxicity of this new material. The control of the nanotube morphology and the bead size, coupled with the versatility of silica chemistry, makes these structures an excellent platform for the development of biosensors (optical, magnetic and catalytic applications). When comparing the toxicity of pristine and oxidized multiwalled carbon nanotubes on human T cells–which would be among the first exposed cell types upon intravenous administration of carbon nanotubes in therapeutic and diagnostic nanodevices–results suggest that carbon nanotubes indeed can be very toxic and induce massive loss of cell viability through programmed cell death at sufficiently high concentrations (>1 ng/cell). The cytotoxicity of carbon nanotubes does depend on many other factors than concentration, including their physical form, diameter, length, and the nature of attached molecules or nanomaterials. Carbon black, for instance, is less toxic than pristine CNTs (which shows the relevance of structure and topology); oxidized CNTs are more toxic than pristine CNTs. We conclude that careful toxicity studies need to be undertaken particularly in conjunction with nanomedical applications of carbon nanotubes.

REFERENCES

1. E. Bekyarova, et al. Applications of carbon nanotubes in biotechnology and biomedicine. *Journal of Biomedical Nanotechnology*, 1: pp 3–17, 2005.
2. G. E. Park and T. J. Webster. A review of nanotechnology for the development of better orthopedic implants. *Journal of Biomedical Nanotechnology*, 1: pp 18–29, 2005.

3. N. Sinha and J. T. Yeow. Carbon nanotubes for biomedical applications. *IEEE Transactions on Nanobioscience*, 4(2): pp. 180–195, June 2005.
4. B. S. Harrison and A. Atala. Carbon nanotube applications for tissue engineering. *Biomaterials*, 28(2): pp 344–353, 2007.
5. D. A. Rey, C. A. Batt, and J. C. Miller. Carbon nanotubes in biomedical applications. *Nanotechnology Law and Business*: p 263, Sep 2003.
6. R. H. Baughman. Putting a new spin on carbon nanotubes. *Science*, 290: p 1310, 2000.
7. T. Mangir, J. Chaves, and S. Chaves. Carbon nanotube/bacteria interface: implications for bioapplications of carbon nanotubes. Submitted to NSTI 2008.
8. G. Z. Yue, Q. Qiu, B. Gao, Y. Cheng, J. Zhang, H. Shimoda, S. Chang, J. P. Lu, and O. Zhou. Generation of continuous and pulsed diagnostic imaging x-ray radiation using a carbon-nanotube-based field-emission cathode. *Applied Physics Letters*, 81(2): pp 355–357, 2002.
9. Y. Cheng, J. Zhang, Y. Z. Lee, B. Gao, S. Dike, W. Lin, J. P. Lu, and O. Zhou. Dynamic radiography using a carbon-nanotube-based fieldemission x-ray source. *Review of Scientific Instruments*, 75(10): pp 3264–3267, 2004.
10. J. Liu and H. Dai. Design, fabrication, and testing of piezoresistive pressure sensors using carbon nanotubes. 2002. http://www.nnf.cornell.edu/2002reu/Liu.pdf.
11. M. Romero, R. Figueroa, and C. Madden. Pressure sensing systems for medical devices. *Medical Device and Diagnostic Industry Magazine*, 2000. http://www.devicelink.com/mddi/archive/00/10/004.html.
12. H. Joseph, B. Swafford, and S. Terry. MEMS in the medical world. *Sensor Magazine*, 14: pp 47–51, Apr 1997.
13. K. Sakai. Artificial kidney engineering-dialysis membrane and dialyzer for blood purification. *Journal of Chemical Engineering of Japan*, 30(4): pp 587–599, 1997.
14. S. Sotiropoulou and N. A. Chaniotakis. Carbon nanotube array-based biosensor. *Analytical and Bioanalytical Chemistry*, 375: pp 103–105, 2003.
15. J. Wang, G. Liu, and M. R. Jan. Ultrasensitive electrical biosensing of proteins and DNA: carbon-nanotube derived amplification of the recognition and transduction events. *Journal of American Chemistry Society*, 126: pp 3010–3011, 2004.
16. Y. Xu, Y. Jiang, H. Cai, P. G. He, and Y. Z. Fang. Electrochemical impedance detection of DNA hybridization based on the formation of M-DNA on ploypyrrole/carbon nanotube modified electrode. *Analytica Chimica Acta*, 516: pp 19–27, 2004.
17. S. Ghoh, A. K. Sood, and N. Kumar. Carbon nanotube flow sensors. *Science*, 299(5609): pp 1042–1044, 2003.
18. K. Moloni, A. Lal, and M. Lagally. Sharpened carbon nanotube probes. *Proceedings of SPIE*, 4098: pp 76–83, 2000.
19. Y. Nakayama, H. Nishijima, S. Akita, K. I. Hohmura, S. H. Yoshimura, and K. Takeyasu. Microprocess for fabricating carbon nanotube probes of a scanning probe microscope. *Journal of Vacuum Science and Technology B: Microelectronics and Nanometer Structures*, 18(2): pp 661–664, 2000.
20. R. M. D. Stevens, N. A. Frederick, B. L. Smith, D. E. Morse, G. D. Stucky, and P. K. Hansma. Carbon nanotubes as probes for atomic force microscopy. *Nanotechnology*, 11(1): pp 1–5, 2000.
21. C. V. Nguyen, K. J. Chao, R. M. D. Stevens, L. Delzeit, A. Cassell, J. Han, and M. Meyyappan. Carbon nanotube tip probes: stability and lateral resolution in

scanning probe microscopy and application to surface science in semiconductors. *Nanotechnology*, 12(3): pp 363–367, 2001.

22. Y. N. Emirov, M. Beerbom, D. A. Walters, Z. F. Ren, Z. P. Huang, B. B. Rossie, and R. Schlaf. Making carbon nanotube probes for high aspect ratio scanning probe metrology. *Proceedings of the SPIE*, 5038 I: pp 493–495, 2003.
23. R. M. D. Stevens, C. V. Nguyen, and M. Meyyappan. Carbon nanotube scanning probe for imaging in aqueous environment. *IEEE Transactions on Nanobioscience*, 3(1): pp 56–60, Mar 2004.
24. C. V. Nguyen, C. So, R. M. D. Stevens, Y. Li, L. Delzeit, P. Sarrazin, and M. Meyyappan. High lateral resolution imaging with sharpened tip of multi walled carbon nanotube probe. *Journal of Physical Chemistry B*, 108(9): pp 2816–2821, 2004.
25. R. H. Baughman, A. A. Zakhidov, and W. A. de Heer. Carbon nanotubes: the route toward applications. *Science*, 297(5582).
26. P. Kim and C. M. Lieber. Nanotube nanotweezers. *Science*, 286: pp 2148–2150, 1999.
27. M. R. Buitelaar, A. Bachtold, T. Nussbaumer, M. Iqbal, and C. Schonenberger. Multiwall carbon nanotubes as quantum dots. *Physical Review Letters*, 88(15): p. 156801, 2002.
28. Y. T. Lim, S. Kim, A. Nakayama, N. E. Stott, M. G. Bawendi, and J. V. Frangioni. Selection of quantum dot wavelengths for biomedical assays and imaging. *Molecular Imaging*, 2(1): pp 50–64, 2003.
29. C. R. Martin and P. Kohli. The emerging field of nanotube biotechnology. *Nature Reviews Drug Discovery*, 2: pp 29–37, 2003.
30. H. Gao, Y. Kong, D. Cui, and C. Z. Ozkan. Spontaneous insertion of DNA oligonucleotides into carbon nanotubes. *Nano Letters*, 3(4): pp 471–473, 2003.
31. Y. Kong, D. Cui, C. S. Ozkan, and H. Gao. Modeling carbon nanotube based bio-nano systems: a molecular dynamics study. In: Proceedings of Materials Research Society Symposium, 773: pp 111–116, 2003.
32. W. L. Jorgensen. The many roles of computation in drug discovery. *Science*, 303: pp 1813–1818, 2004.
33. M. C. Shults, R. K. Rhodes, S. J. Updike, B. J. Gilligan, and W. N. Reining. A telemetry-instrumentation system for monitoring multiple subcutaneously implanted glucose sensors. *IEEE Transactions on Biomedical Engineering*, 41(10): pp 937–942, Oct 1994.
34. R. Shandas and C. Lanning. Development and validation of implantable sensors for monitoring function of prosthetic heart valves: *in vitro* studies. *Medical and Biological Engineering and Computing*, 41(4): pp 416–424, 2003.
35. A. Bhargava. 1999. Nanorobots: Medicine of the future. http://www.ewh.ieee.org/r10/Bombay/news3/page4.html.
36. R. H. Baughman, C. Cui, A. A. Zakhidov, Z. Iqbal, J. N. Barisci, G. M. Spinks, G. G. Wallace, A. Azzoldi, D. De Rossi, A. G. Rinzler, O. Jaschinski, S. Roth, and M. Kertesz. Carbon nanotube actuators. *Science*, 284: pp 1340–1344, 1999.
37. U. Vohrer, I. Kolaric, M. H. Haque, S. Roth, and U. Detlaff-Weglikowska. Carbon nanotube sheets for use as artificial muscles. *Carbon*, 42: pp 1159–1164, 2004.
38. M. Gao, L. Dai, R. H. Baughman, G. M. Spinks, and G. G. Wallace. Electrochemical properties of aligned nanotube arrays: basis of new electromechanical actuators. *Proceedings of SPIE*, 3987: pp 18–24, 2000.

39. J. Fraysse, A. I. Minett, O. Jaschinski, G. S. Duesberg, and S. Roth. Carbon nanotubes acting like actuators. *Carbon*, 40: pp 1735–1739, 2002.
40. H. P. Monner, S. Muhle, and P. Wierach. Carbon nanotubes as actuators in smart structures. *Proceedings of SPIE*, 5053: pp 138–146, 2003.
41. Y. Gogotsi, J. A. Libera, A. G. Yazicioglu, and C. M. Megaridis. *In situ* multiphase fluid experiments in hydrothermal carbon nanotubes. *Applied Physics Letters*, 79(7): pp 1021–1023, 2001.
42. Briefing Paper: Institute of Neurosciences, Mental Health and Addiction. CIHR. Feb 2003. http://www.regenerativemedicine.ca/nanomed/Nanomedicine%20Taxonomy%20(Feb%202003).pdf
43. K. E. Drexler. *Nanosystems: Molecular Machinery, Manufacturing, and Computation-* New York: Wiley, 1992.
44. J. Cumings and A. Zettle. Low-friction nanoscale linear bearing realized from multiwall carbon nanotubes. *Science*, 289: pp 602–604, 2000.
45. M. A. Correa-Duarte, N. Wagner, J. Rojas-Chapana, C. Morsczeck, M. Thie, and M. Giersig. Fabrication and biocompatibility of carbon nanotube-based 3D networks as scaffolds for cell seeding and growth. (Letter): pp 2233–2236.
46. D. Pantarotto, et al. Translocation of bioactive peptides across cell membranes by carbon nanotubes. *Chemical Communications:* pp 16–17, 2004.
47. W. Wu, et al. Targeted delivery of amphotericin B to cells by using functionalized carbon nanotubes. *Angewandte Chemie International Edition*, 44: pp 6358–6362, 2009.
48. Z. Yinghuai, et al. Substituted carborane-appended water-soluble single-wall carbon nanotubes: new approach to boron neutron capture therapy drug delivery. *Journal of American Chemistry Society* SOC: pp 9875–9880, 2005.
49. Q. Lu, et al. RNA polymer translocation with single-walled carbon nanotubes. *Nano Letters*, 6: pp 2473–2477, 2004.
50. N. W. S. Kam, et al. Carbon nanotubes as intracellular transporters for proteins and DNA: an investigation of the uptake mechanism and pathway. *Angewandte Chemie International Edition*, 45: pp 577–581, 2006.
51. D. Cai, et al. Highly efficient molecular delivery into mammalian cells using carbon nanotube spearing. *Natural Methologies*, 2: pp 449–454, 2005.
52. N. W. S. Kam, et al. Carbon nanotubes as multifunctional biological transporters and near-infrared agents for selective cancer cell destruction. Proceedings of the National Academy of Sciences, 102: pp 11600–11605, 2005.
53. R. B. Weisman, et al. Fluorescence spectroscopy of single-walled carbon nanotubes in aqueous suspension. *Applied Physics A: Materials Science and Processing*, 78: pp 1111–1116, 2004; D. A. Tsyboulski, et al. Versatile visualization of individual single-walled carbon nanotubes with near-infrared fluorescence microscopy. *Nano Letters*, 5: pp 975–979, 2005.
54. K. König. Multiphoton microscopy in life sciences. *Journal of Microscopy*, 200: pp 83–104, 2000.
55. D. R. Larson, et al. Water-soluble quantum dots for multiphoton fluorescence imaging *in vivo*. *Science*, 300: pp 1434–1436, 2003.
56. M. Bottini, et al. Full-length single-walled carbon nanotubes decorated with streptavidin conjugated quantum dots as multivalent intracellular fluorescent nanoprobes. *Biomacromolecules*: 2006.

57. S. J. Tans, et al. Room-temperature transistor based on a single carbon nanotube. *Nature*, 393: pp 49–52, 1998.
58. E. Artukovic, et al. Transparent and flexible carbon nanotube transistors. *Nano Letters*, 5: pp 757–760, 2005.
59. K. Bradley, et al. Integration of cell membranes and nanotube transistors. *Nano Letters*, 5, 841–845, 2005.
60. A. Star, et al. Electronic detection of specific protein binding using nanotube fet devices. *Nano Letters*, 3: pp 459–463, 2003.
61. K. Bradley, et al. Charge transfer from adsorbed proteins. *Nano Letters*, 4: 253–256, 2004.
62. R. J. Chen, et al. Noncovalent functionalization of carbon nanotubes for highly specific electronic biosensors. Proceedings of the National Academy of Sciences, 100: pp 4984–4989, 2003.
63. R. Singh, et al. Binding and condensation of plasmid DNA onto functionalized carbon nanotubes: toward the construction of nanotube-based gene delivery vectors. *Journal of American Chemistry Society:* pp 4388–4396, 2005.
64. X. Yu, et al. Protein immunosensor using single-wall carbon nanotube forests with electrochemical detection of enzyme labels. *Molecular Biosystems*, 1: pp 70–78, 2005. J. J. Gooding, et al. Protein electrochemistry using aligned carbon nanotube arrays. *Journal of American Chemistry Society*, 125: pp 9006–9007, 2003.
65. E. S. Jeng, et al. Detection of DNA hybridization using the near-infrared band-gap fluorescence of single- walled carbon nanotubes. *Nano Letters*, 6, 371–375, 2006.
66. P. W. Barone, et al. Near-infrared optical sensors based on single-walled carbon nanotubes. *Nature*, 4: pp 86–92, 2005.
67. See H. Dai, et al. Nanotubes as nanoprobes in scanning probe microscopy. *Nature*, 384: pp 147–150, 1996; S. S. Wong, et al. Carbon nanotube tips: high-resolution probes for imaging biological systems. *Journal of American Chemistry Society*, 120: pp 603–604, 1998.
68. E. Lee and M. J. Chung. Carbon nanotube-tipped microcantilever arrays for imaging, testing, and 3D nanomanipulation: design and control. 4th IEEE Conference, Aug 16–19, 2004. *Nanotechnology:* pp 32–34, 2004.
69. E. Lee. Unified design control for biological applications of carbon nanotube-tipped microcantilever, 5th IEEE Conference, July 11–15, 2005. *Nanotechnology*: pp 733–737, 2005.
70. J. D. Gong, G. Jensen, A. Bhirde, Y. Xin Yu, B. Munge, V. Patel, N. K. Nyon, J. Silvio Gutkind, F. Papadimitrakopoulos, and J. F. Rusling. Wearable and Implantable Body Sensor Networks, BSN 2006. Single-walled carbon-nanotube forest immunosensor for amplified detection of cancer biomarkers. International Workshop on April 3–5, 2006.
71. H. Xu, G. Li, M. Fu, Y. Wang, and J. Liu. Engineering in Medicine and Biology Society, 2005. IEEE-EMBS 2005. 27th Annual International Conference. Study of carbon nanotube modified biosensors for monitoring uric acid and total cholesterol in blood: pp 255–258.
72. J. D. Yantzi and J. T. W. Yeow. Carbon nanotube enhanced pulsed electric field electroporation for biomedical applications. Mechatronics and Automation, 2005 IEEE International Conference, Volume 4, July 29–Aug 1. 2005: pp 1872–1877.

73. K. Teker, E. Wickstrom, and B. Panchapakesan. Biomolecular tuning of electronic transport properties of carbon nanotubes via antibody functionalization. *Sensors Journal*, 6(6): pp 1422–1428, Dec 2006.
74. G. Ruffini, et al. Engineering in Medicine and Biology Society, 2006 (EMBS '06) 28th Annual International Conference of the IEEE, Aug 2006. ENOBIO: First Tests of a Dry Electrophysiology Electrode using Carbon Nanotubes.
75. J. Tkac, J. Whittaker, T. Ruzgas, and S. O. Tautgirdas. The use of single walled carbon nanotubes dispersed in a chitosan matrix for preparation of a galactose biosensor. *Biosensors and Bioelectronic*, 22(8): pp 1820–1824, Mar 15, 2007.
76. F. Qu, et al. Electrochemical biosensing utilizing synergic action of carbon nanotubes and platinum nanowires prepared by template synthesis. *Biosensors and Bioelectronics*, 22(8): pp 1749–1755, Mar 15, 2007.
77. R. Yang, et al. Single-walled carbon nanotubes-mediated *in vivo* and *in vitro* delivery of siRNA into antigen-presenting cells. *Gene Therapy*, 13(24): pp 1714–1723, 2006.
78. Yu, Xin, et al. Carbon nanotube amplification strategies for highly sensitive immunodetection of cancer biomarkers. *Journal of the American Chemical Society*, 128(34): pp 11199–11205, 2006.
79. J. Meng, et al. Using single-walled carbon nanotubes nonwoven films as scaffolds to enhance long-term cell proliferation *in vitro*. *Journal of Biomedical Materials Research*, 79A(2): pp 298–306, 2006.
80. A. G. Cuenca, et al. Emerging implications of nanotechnology on cancer diagnostics and therapeutics. *Cancer*, 107(3): pp 459–466, 2006.
81. K. Donaldson, et al. Carbon nanotubes: a review of their properties in relation to pulmonary toxicology and workplace safety. *Toxicological Sciences*, 92(1): pp 5–22; SN 1096–6080, 2006.
82. A. G. Cuenca, et al. Emerging implications of nanotechnology on cancer diagnostics and therapeutics. *Cancer*, 107(3): pp. 459–466, 2006.
83. X. Chen, et al. Interfacing carbon nanotubes with living cells. *Journal of the American Chemical Society*, 128(19): pp 6292–6293, 2006.
84. K. Balasubramanian and M. Burghard. Biosensors based on carbon nanotubes. *Analytical and Bioanalytical Chemistry*, 385(3): pp 452–468, 2006.
85. C. Klumpp, et al. Functionalized carbon nanotubes as emerging nanovectors for the delivery of therapeutics. *Biochimica et Biophysica Acta*, 1758(3): pp 404–412, 2006.
86. A. K. Wanekaya, et al. Nanowire-based electrochemical biosensors. *Electroanalysis*, 18(6): pp 533–550.
87. X. Shi, et al. Injectable nanocomposites of single-walled carbon nanotubes and biodegradable polymers for bone tissue engineering. *Biomacromolecules*, 7(7): pp 2237–2242, 2006.
88. B. Marrs, et al. Augmentation of acrylic bone cement with multiwall carbon nanotubes. *Journal of Biomedical Materials Research*, 77A(2): pp 269–276, 2006.
89. J. Riu, A. Maroto, F. Rius, and T. Xavier. Nanosensors in environmental analysis, *Nanowerk News*, 69(2): pp 288–301, 2006.
90. G. Gruner. Carbon nanotube transistors for biosensing applications. *Analytical and Bioanalytical Chemistry*, 384(2): pp 322–335, 2006.
91. R. Dagani. Nanomaterials: safe or unsafe? *Chemical and Engineering News*, 81(17): pp 30–33, 2003.

92. J. S. Tsuji, et al. Research strategies for safety evaluation of nanomaterials part IV: risk assessment of nanoparticles. *Toxicological Sciences*, 89(1): pp 42–50, 2006.
93. T. Mangir, J. Chaves, and S. Chaves. 2007. A novel technique for purification and segregation of CNTs For nanoscale thin film studies. NSTI 2007, Santa Clara, California, May 21–24.
94. For a detailed review and analysis of the different toxicity studies, see: J. C. Miller, et al. *The Handbook of Nanotechnology Business, Policy, and Intellectual Property Law*, pp 51–64, 2004.
95. D. B. Warheit, et al. Comparative pulmonary toxicity assessment of single-wall carbon nanotubes in rats. *Toxicological Sciences*, 7: pp 117–125, 2004.
96. C. W. Lam, et al. Pulmonary toxicity of single-wall carbon nanotubes in mice 7 and 90 days after intratracheal instillation. *Toxicological Sciences*, 77: pp 126–134, 2004.
97. P. H. M. Hoet, et al. Nanoparticles: known and unknown health risks. *Journal of Nanobiotechnology*, 2: 2004.
98. O. Oberdörster, et al. Nanotoxicology: an emerging discipline evolving from studies of ultrafine particles. *Environmental Health Perspectives:* pp 823–839, 2005.
99. A. A. Shvedova, et al. Exposure to carbon nanotube material: assessment of nanotube cytotoxicity using human keratinocyte cells. *Journal of Toxicology and Environmental Health, Part A: Current Issues*, 66: pp 1909–1926, 2003.
100. G. Jia, et al. Cytotoxicity of carbon nanomaterials: single-wall nanotube, multi-wall nanotube, and fullerene. *Environmental Science and Technology*, 39: pp 1378–1383, 2005.
101. N. Monteiro-Riviere, et al. Multi-walled carbon nanotube interactions with human epidermal keratinocytes. *Toxicology Letters*, 155: pp 377–384, 2005.

19

NANOSCALE IMAGE PROCESSING

Mary Mehrnoosh Eshaghian-Wilner and Shiva Navab

This chapter has two main sections. The first section concentrates on image processing applications of nanoscale spin-wave architectures. More specifically, how three image processing problems, namely, labeling, finding the convex hull, and finding the nearest neighbor problem, can be solved on a spin-wave reconfigurable mesh. The second section explains how to implement one of the most widely used operations in image processing: discrete Fourier transform (DFT). How the DFT spin-wave module can be optimized to implement a spin-wave fast Fourier transform (FFT) module is also demonstrated.

19.1. IMAGE PROCESSING ALGORITHMS

Over the past few decades, several mesh-based parallel architectures such as mesh-connected computer, mesh-of-trees, pyramid, mesh with multiple buses, reconfigurable meshes, systolic meshes, and optical meshes have been considered for performing low and intermediate-level computer vision tasks [1–20]. This is due to the fact that a two dimensional image can be mapped in a straightforward fashion onto a two-dimensional mesh. In particular, reconfigurable meshes have been shown to be attractive computational engines due to the flexibility that the reconfigurable bus offers. Here, we use the nanoscale reconfigurable mesh architecture that is interconnected with ferromagnetic spin-wave buses, presented in Chapter 7. As described in that chapter, the nanoscale spin-wave-based reconfigurable mesh, while requiring the same number of switches as standard reconfigurable meshes, is capable of simultaneously transmitting N waves on each

The authors of this chapter are listed alphabetically.

Bio-Inspired and Nanoscale Integrated Computing. Edited by Mary Mehrnoosh Eshaghian-Wilner

of the spin-wave paths. So, this architecture represents a new approach for designing nanometer-scale computational structures, thus aiming to preserve the main advantages of wave-based computers [21–24].

In this section, we study different image processing algorithms on the spin-wave reconfigurable mesh architecture [22]. We first show that given an $N^{1/2} \times N^{1/2}$ image, using a $N \times N$ spin-wave reconfigurable mesh, in $O(1)$ time, the figures can be labeled and their nearest neighbor and convex hull can be found. This is due to the fact that the superposition characteristic of spin waves allows concurrent writes if all the requesting processors write a "1." We then present the same algorithms for "multiple" figures and show that they can be solved in $O(\log N)$ time, considering the fact that multiple spin waves at different frequencies can be transmitted on each of the spin-wave buses.

The input to our algorithms is a digitized (black and white) image with processor (i,j) storing the pixel (i,j), $0 \leq i,j \leq N-1$, in the plane, where the black pixels are 1-valued and white pixels are 0-valued. Figures correspond to connected 1's (black pixels) in the image.

19.1.1. Labeling Problem

An early step in image processing is the identification of the figures in the image. Figures correspond to connected 1's (black pixels) in the image. Since this architecture allows concurrent writes all the 1's are connected in $O(1)$ time. Each black node (1-valued) broadcasts 1 on the four buses connected to it, so that all the black pixels simultaneously connect to each of their 1-valued neighbors.

As mentioned above, in a 0/1 picture, the connected 1's are said to form a figure. An $N \times N$ digitized image may contain more than one region of black pixels. The labeling problem is to identify to which figure each '1' belongs and to label each figure with a unique ID.

Here we present three variations of the labeling problem. First, we show a simple algorithm that labels $O(N)$ figures in an $N \times N$ image in $O(\log N)$ time using an $N \times N$ spin-wave reconfigurable mesh. Next, we present the same algorithm for a smaller image size and prove that on an $N^{1/2} \times N^{1/2}$ image, the figures can be labeled in $O(1)$ time. Finally, we show that the figures of an $N \times N$ image can be labeled in $O(N)$ steps in a systolic fashion, where each systolic cycle takes just $O(1)$ time.

Algorithm 19.1.1

Given an $N \times N$ image, using an $N \times N$ reconfigurable mesh with spin-wave buses, $O(N)$ figures can be labeled in $O(\log N)$ time.

We would like to label each figure with a unique identification number (ID). Associated with each PE is an ID, and we choose the largest ID in each figure to be the label of that figure. First, each processing element (PE) checks its most significant bit (MSB). If its MSB is 1, the PE broadcast 1 to all the nodes connected to it. If its MSB is 0, on the other hand, the PE listens to the channel and if there is a 1 on the bus, that PE gets disabled. At the next step, the nodes check their next most significant bit, and this procedure is repeated for all the bits. At the end only one node has not been disabled. That node broadcasts its ID to all its connected nodes as their label. □

Algorithm 19.1.2

Given an $N^{1/2} \times N^{1/2}$ image, using an $N \times N$ array of processing elements with spin-wave buses, each figure can be labeled in $O(1)$ time.

Similar to Algorithm 19.1.1, associated with each PE is an ID, and we choose the largest ID in each figure to be the label of that figure. Since the size of the figure is smaller than the size of the reconfigurable mesh, we can efficiently find the maximum value among the processor IDs in each figure. The image size is $N^{1/2} \times N^{1/2}$, so the number of processor IDs is N at the maximum. We copy each of these N values to all the nodes in one row. Now each of the N columns contains the N different values. In each of these columns (column j) all the nodes compare their value to the value in row j. Only in one column, assume in column c, there is no node with a value greater than the one in row c. Therefore, the value in row c and column c is the maximum value. Since all these comparisons are done simultaneously, the time complexity is $O(1)$. □

Algorithm 19.1.3

Given an $N \times N$ image, using an $N \times N$ reconfigurable mesh with spin-wave buses, in a systolic fashion with inputs coming from left and leaving from the right, each figure can be labeled with $O(1)$ steps per systolic cycle.

In a systolic reconfigurable mesh (SRM), as described in [19], data is input into the reconfigurable mesh and eventually output from the other side, one column of data per unit of time. A systolic reconfigurable mesh (SRM) of size 16 is shown in Figure 19.1. Each generic processor contains a switch that is under local control and can be used to configure the four bus lines in any of 15 possible combinations. Without loss of generality, assume that data is input from the left

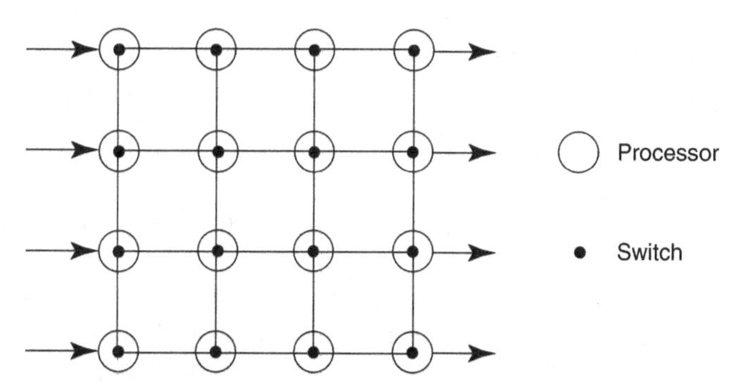

Figure 19.1. A systolic reconfigurable mesh.

side of the systolic reconfigurable mesh, one column of data per unit time, and eventually output from the right side, one column per unit time.

When designing algorithms for meshes or reconfigurable meshes, it is typically assumed that the data already resides in the processors. When designing algorithms for the systolic reconfigurable mesh, however, this assumption no longer holds, as the time to perform input and output of data is considered in the complexity analysis of an algorithm.

In this algorithm, as explained in [23] similar to the algorithm presented in [19], we represent the component label will as <CL,CR,T>, where CL is the index of the leftmost column containing a pixel of the component; CR is the index of the rightmost column containing a pixel of the component; and T is used to break ties in the case of distinct figures with the same CL and CR values. CL is computed for every pixel during the input and preprocessing phase of the algorithm, while CR is computed for every pixel during the output and post-processing phase of the algorithm. To distinguish figures that occupy the same leftmost and rightmost column of the image, a tie-breaking scheme will be used to determine T during the output and post-processing step. Each input and preprocessing cycle consists of the following:

1. In lockstep, shift the image to the right while inputting the next column of data into the first column of the SRM.
2. Initialize the component labels for all pixels now in column one of the SRM to <0,0,0>.
3. All processors that currently hold a black pixel of the image connect their bus to all neighboring pixels that also maintain black pixels. The result is that there is a sub-bus over all figures with respect to the restriction of the image that has thus far been inputted.
4. Exploiting the concurrent write capability, all processors in the first column of the SRM now broadcast their column (with respect to the entire image) label. Note that during cycle c, this is column $n-c+1$. All processors receiving such a value store this in CL, potentially replacing a previous value.

After n of these simple input and preprocessing cycles, all pixels know CL, the leftmost column of any pixel in their connected component. Both CR and T will be determined during the output and postprocessing phase. Each output and post-processing cycle is concerned with those processors maintaining pixels in the last column of the SRM that have not previously received their final component labels. For such processors, the output and post-processing cycle consists of the following:

1. Similar to step 3 of the preprocessing phase, the nodes form a sub-bus over all figures.
2. All processors in the last column of the SRM now broadcast the column (with respect to the entire image) label of their pixel. All the processors that have not previously set the value of CR store it in CR.

3. Unfortunately, multiple unique figures may now have the same component labels in terms of CL and CR. The labels are disambiguated by choosing the row index of the topmost node with the column index CR to be the third label component, T, of each figure. This label component is found on the last column of the systolic reconfigurable mesh, as the image is piped out.
   (a) All the boundary black nodes should now form a connected sub-bus. For that, each node checks its neighbors and decide if it is a boundary node (has at least a white neighbor). All the boundary nodes then connect to their neighboring boundary nodes, forming a sub-bus.
   (b) Next the local topmost nodes on the column CR are found by having each node send a signal downwards in that column. This is done by first turning off (disconnecting the channel) all the switches above each node to force the signal downwards. Then all nodes simultaneously send out a signal downwards and turn all switches on (reconnecting the channel), thereby lettings all signals pass through the bus. Each node that receives a signal will become disabled. As the result, the only active nodes are the local topmost nodes that have not received a signal.
   (c) If there is only one local topmost node in any figure, that will be the topmost node chosen as the third label component. However, there might be more than one local topmost node in a figure (in dented figures). To solve this issue, each local bottommost node is used to disable the local topmost node below it. The procedure of finding local bottommost nodes is quite similar to the one finding the local topmost nodes. Next, the switches at the topmost nodes are set so that they won't let the signals go pass through them. To prevent the last bottommost node to disable the first topmost node, the bottommost nodes broadcast their row index. The immediate topmost node after each bottommost node receives the row index and compares it to its own.

       If the signal is from a node with a row index higher than its own, that topmost node gets disabled. At the end of this step only the "top" topmost node is not disabled, and the row index of that node is used to label Figure 19.2, as explained in step (d). An example is shown in Figure 19.2 where first the local topmost and local bottommost nodes are marked, and next the local topmost nodes are disabled by the bottommost nodes above them.
   (d) The switches are set such that all the black nodes in each figure are connected. The "top" topmost node broadcasts its row index to all the nodes. Each black pixel which has not previously set its third label component T saves the broadcasted value; otherwise, it won't overwrite the existing label.
4. In lockstep, shift the image to the right, so as to output the next column of data (pixel and label information).

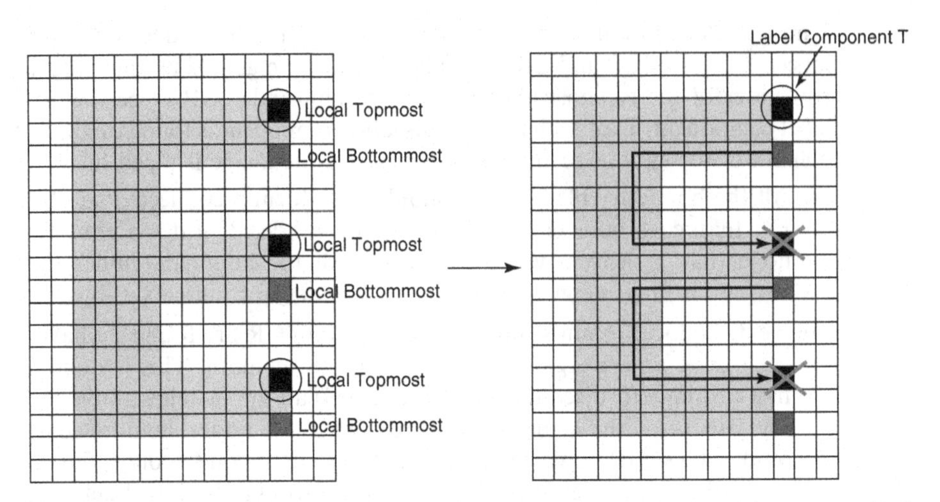

Figure 19.2. An example for finding T label component in Algorithm 19.1.3 [23].

It should be noted that there is no static stage in this algorithm; it is done in $O(1)$ time per systolic cycle, which is the fastest solution for this problem. □

19.1.2. Convex Hull Problem

Now we consider the convexity problem. We use the following definition of convexity: A set of PEs is said to be convex if and only if the corresponding set of integer lattice points is convex. Given a set S of PEs, the convex hull of S, denoted Hull (S), is the smallest convex set of PEs containing S.

In the following, first we show that if the size of the image is the same as that of the reconfigurable mesh, the convex hull of a single figure can be found in $O(\log N)$ time. Then we show that if the size of the reconfigurable mesh is N times larger than the size of the image, the above algorithm time complexity becomes $O(1)$. We also show that the convex hull of multiple (up to N) figures can be found in $O(\log N)$.

Algorithm 19.1.4

Given an $N \times N$ image, using an $N \times N$ Spin-Wave reconfigurable mesh, the convex hull of a single figure can be found in $O(\log N)$ time.

This algorithm operates in two steps. First the rightmost (RM) and the leftmost (LM) black pixels (1-valued) in each row are found. Then it will be verified if these points are extreme points or not.

Step 1: Fist all the left and right switches of each node are tuned on, so that all the channels between the nodes in the same row are open. Then each black node sends out a 1 signal on the bus and turns off its left-hand side switch, thereby closing that

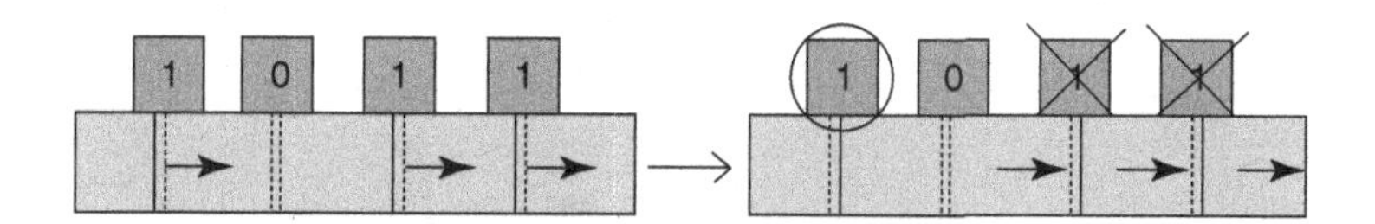

Figure 19.3. Example of finding the leftmost black node.

section of the channel and not letting the waves pass though from right to left. Closing that particular section belonging to the node will still allow the node to receive signals, but it will not let signals travel beyond (to the left) of that node. If at least one black pixel exists on a row, all the nodes on that row (except the LM node) will receive a 1. An example is shown in Figure 19.3. Now that LM is identified in each row, we do the same to find the RM node. The only difference is to turn the right-hand side switch and not let the waves go from left to right. Since setting all the switches in all the rows is done in parallel, finding RM and LM takes $O(1)$ time.

Step 2: The verification is done by having the ith column compute the enclosing angle made by the 1's in the image with the RM of the ith row (RM_i). This angle is defined to be the smallest angle Φi such that all the 1's are enclosed inside the region X RM_i Y as shown in Figure 19.6. For this, first all the RMs broadcast their address to all the 1's in their row. Now all the columns have a copy of all RM nodes. Each RM_i sends its address to all the black nodes in the ith column, and all these nodes compute their angle with $RM_{i.}$ We need to find the maximum value (angle) of the nodes above RM_i and below it. RM_i is an extreme point if and only if the sum of these two angles is less than 180°. This step is repeated for LM_i. The maximum angle value is found in $O(\log N)$ time, therefore, the total time complexity of finding the convex hull of a figure will be $O(\log N)$. The $O(1)$ time variation of this algorithm for a smaller image size is been presented in the following algorithm.

Algorithm 19.1.5

Given an $N^{1/2} \times N^{1/2}$ image, using an $N \times N$ Spin-Wave reconfigurable mesh, the convex hull of a single figure can be found in $O(1)$ time.

This algorithm is similar to Algorithm 19.1.4, except the process of finding the maximum angle. Here the image size is $N^{1/2} \times N^{1/2}$, so we need $N^{1/2}$ columns to be assigned to each RM, as explained in the proof sketch of Algorithm 19.1.2, in the previous section. We assign column (i* $N^{1/2}$) to RMi, and use the columns in between the columns assign to RMi and RM$i+1$ to find the maximum value of the angles computed on column (i* $N^{1/2}$). In this case, we have $N^{1/2}$ column to find the maximum of $N^{1/2}$ values. We copy these $N^{1/2}$ values to each of the $N^{1/2}$ columns by broadcasting column (i* $N^{1/2}$) values to its adjacent $N^{1/2}$ columns. In each column $j, 0 \le j \le N-1$, all the nodes compare their value to the value in row j. Only in one

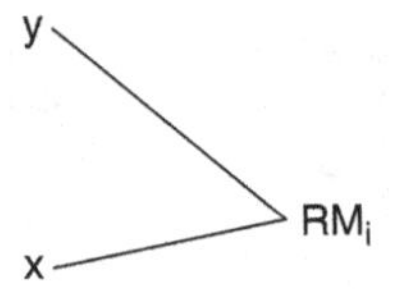

Figure 19.4. Enclosing Angle X RM_i Y [23].

column, assume in column c, no node has a value greater than the one in row c. Thus the element in row and column c is the maximum angle. Since all these comparisons are performed simultaneously, the time complexity is $O(1)$.

Algorithm 19.1.6

The convex hull of multiple (N) figures in a given $N \times N$ image using an $N \times N$ spin-wave reconfigurable mesh can be found in $O(\log N)$ time.

To find the convex hull of multiple (N) figures in an image with size of N, using a spin-wave reconfigurable mesh of size $N \times N$, we can simulate each step of the algorithm explained in [25] in constant time. So the total time complexity will be $O(\log N)$. □

19.1.3. The Nearest Neighbor Problem

In this problem we would like to find the closest neighbor of a specific figure. In the following algorithm, first it is shown that for an image size equal to the size of the spin-wave reconfigurable mesh, for a single figure, its nearest neighbor can be found in $O(\log N)$ time. The same problem is next solved in $O(1)$ time for another scenario where the size of the reconfigurable mesh is N times larger than the size of the image.

Algorithm 19.1.7

Given an $N \times N$ image, using an $N \times N$ spin-wave reconfigurable mesh, the nearest neighbor of a single figure can be found in $O(\log N)$ time.

In this stage, it is assumed that the figures are labeled and their convex hulls are found. The time complexity of labeling and finding the convex hull are $O(\log N)$, as mentioned in the previous sections. First we set up the switches so that all nodes in the whole image are connected to each other. To find the neatest neighbor of a figure, we should find the closest 1 (belonging to another figure) to all the extreme points of this figure. To do so, each extreme point of the figure broadcasts its address, as well as its ID, to all four directions. As soon as a black node from another figure (with a different ID) receives this broadcasted signal, it computes its distance to that node. Next, it sends this computed distance back to the original node. It also blocks the channel by turning off its switch and does not let the signal pass though.

In a similar fashion, we check the top, bottom, right, and left neighbors, but we may miss some closer neighbors that are located in a diagonal position from the original figure. To address this issue, we need to revise the algorithm in the following way. First, all the black nodes (including nonextreme points) broadcast their address and ID to their neighbors. When a black pixel receives this data, it blocks it. All black (and white) nodes compute their distance from the closest black pixel belonging to a different figure. All the intermediate white nodes which have received the broadcasted address find their distance to the nearest black node (with a different ID). On each side of the extreme point, the minimum distance of all these nodes is found in a binary tree fashion. In other words, each two adjacent nodes repeatedly find their minimum which will be compared to the minimum of their adjacent pair. As a result, the nearest neighbor in each direction is found in $O(\log N)$ time. The extreme point then finds the minimum distance between these four distances and knows which figure, with which ID, is the closest neighbor to it. All the extreme points know their closest neighbor in $O(\log N)$ time. The next step will be finding the minimum number among the data of all the extreme points, which takes $O(\log N)$ time as well. □

Algorithm 19.1.8

Given an $N^{1/2} \times N^{1/2}$ image, using an $N \times N$ spin-wave reconfigurable mesh, the nearest neighbor of a single figure can be found in $O(1)$ time.

This is quite similar to the algorithm presented in Algorithm 19.1.7. Here, since the image size is an order of N smaller than the size of the reconfigurable mesh, we can find the maximum and minimum of numbers in $O(1)$ using the technique discussed in Algorithm 2 in the previous section. The time complexity of most of the steps of the nearest neighbor algorithm explained above are $O(1)$, except labeling and finding the minimum number among the extreme points' ID which are $O(\log N)$. As we showed previously, for a $N^{1/2} \times N^{1/2}$ image, labeling can be done in $O(1)$ time, and similarly minimum distance is found in $O(1)$ time. □

Algorithm 19.1.9

The nearest neighbor of each of the multiple (N) figures in a given an $N \times N$ image using an $N \times N$ reconfigurable mesh with spin-wave buses can be found in $O(\log N)$ time.

As shown in Algorithm 19.1.7, the nearest neighbor of each single figure is found in $O(\log N)$ time. The nearest neighbor of all figures can be found within each connected region simultaneously. Therefore, the nearest neighbor of all figures is found in $O(\log N)$. □

19.2. DISCRETE FOURIER TRANSFORM

One of the most important operations in image processing is discrete Fourier transform (DFT) and its optimized version, fast Fourier transform (FFT). Many mathematicians, engineers, and applied scientists have tried to efficiently implement DFT due to its broad range of applications in not only image processing but also other digital signal processing applications. The DFT can calculate a signal's frequency spectrum that allows a direct examination of information encoded in the frequency, phase, and amplitude of the component sinusoids. Moreover, the DFT can find a system's frequency response from the system's impulse response, and vice versa [26]. A class of particularly efficient algorithms for computing DFT is called fast Fourier transform (FFT) [27].

DFT (and FFT) can be used in spectral estimation [28], interpolation [29], multichannel carrier modulation (MCM) such as orthogonal frequency division multiplexing (OFDM) [30], object recognition [31], frequency analysis [32], speech coding [33], harmonic analysis [34], digital audio compression [35], channel separation and combination [36], watermarking [37], and bit reversal algorithms [38]. They also have several applications in spectral analysis, convolution, and correlation [39, 40]. DFT (and FFT) can be used in designing several tools such as Dolby AC-2 and AC-3 audio coders [41], MPEG decoders [42, 43], and digital filters such as low-pass (LPF), band-pass (BPF), high-pass (HPF), generalized cepstrum, homomorphic filtering [44], nonlinear [45], and adaptive filters [46]. One of the applications of both nanoscale DFT and FFT modules can be in image processing used in medical imaging [54–56].

The DFT is identical to samples of the Fourier transform at equally spaced frequencies, $W_k = 2\pi k/N$. Several methods have been developed for computing values of the DFT. A class of particularly efficient algorithms for computing DFT is called fast Fourier transform (FFT) [27]. In this section, after a theoretical discussion of DFT and FFT, we first show how to implement a module that computes DFT coefficients at the nanoscale, and afterwards, we discuss how this solution is optimized by using FFT techniques.

The DFT and FFT architectures can be implemented on a simplified version of a spin-wave reconfigurable mesh [57]. In implementation of DFT and FFT all the processing nodes of a reconfigurable mesh are used; however, no spin-wave switches are required to realize the DFT/FFT computation. In Section 19.2.1 we show how DFT and FFT are implemented on a reconfigurable mesh.

19.2.1. Discrete Fourier Transform (DFT) and Fast Fourier Transform (FFT)

The computational problem for the DFT is to compute the sequence $\{X(k)\}$ of N complex-valued numbers given another sequence of data $\{x(n)\}$ of length N, according to Equation 19.1 [27]

$$X(k) = \sum_{n=0}^{N-1} x(n) W_N^{kn}, \quad 0 \le k \le N-1 \qquad (19.1)$$

$$W_N = e^{-j2\pi/N}$$

where $X(k)$ represents the Fourier coefficients of $x(n)$. In general, the data sequence $x(n)$ is also assumed to be complex valued. Similarly, the inverse discrete Fourier transform (IDFT) becomes

$$x(n) = \frac{1}{N} \sum_{k=0}^{N-1} X(k) W_N^{-nk}, \quad 0 \le n \le N-1. \tag{19.2}$$

Since DFT and IDFT involve the same type of computations, our discussion of efficient computational algorithms for the DFT applies to the efficient computation of the IDFT as well.

We observe that for each value of k, direct computation of $X(k)$ involves N complex multiplications ($4N$ real multiplications) and $N-1$ complex additions ($4N-2$ real additions). Consequently, to compute all N values of the DFT requires $4N^2$ complex multiplications and $4N^2-2N$ complex additions.

Direct computation of the DFT is inefficient primarily because it does not exploit the symmetry and periodicity properties of the phase factor W_N. In particular, these two properties are

$$\begin{aligned} &\text{Symmetry property} : W_N^{k+N/2} = -W_N^k \\ &\text{Periodicity property} : W_N^{k+N} = W_N^k \end{aligned} \tag{19.3}$$

The computationally efficient algorithms described in this section, known collectively as fast Fourier transform (FFT) algorithms, exploit these two basic properties of the phase factor.

FFT algorithms are based on the fundamental principle of decomposing the computation of the discrete Fourier transform of a sequence of length N into successively smaller discrete Fourier transforms. The manner in which this principle is implemented leads to a variety of different algorithms. The algorithm we describe here, called decimation-in-time, derives its name from the fact that in the process of arranging the computation into smaller transformations, the sequence $x(n)$ (time sequence) is decomposed into successively smaller subsequences. This decomposition can be performed in several ways such as radix-2, radix-4, and Split-radix. In the following passage, we explain the radix-2 FFT algorithm.

We split the N-point data sequence into two $N/2$-point data sequences $f_1(n)$ and $f_2(n)$, corresponding to the even-numbered and odd-numbered samples of $x(n)$, respectively; that is

$$\begin{aligned} f_1(n) &= x(2n) \\ f_2(n) &= x(2n+1), \quad n = 0, 1, \ldots, N/2-1 \end{aligned}. \tag{19.4}$$

Thus $f_1(n)$ and $f_2(n)$ are obtained by decimating $x(n)$ by a factor of 2, and hence the resulting FFT algorithm is called a *decimation-in-time algorithm*. Now the N-point DFT can be expressed in terms of the DFT's of the decimated sequences shown in Equation 19.5.

$$\begin{aligned} X(k) &= \sum_{n=0}^{N-1} x(n) W_N^{kn}, \quad k = 0, 1, \ldots, N-1 \\ &= \sum_{n \text{ even}} x(n) W_N^{kn} + \sum_{n \text{ odd}} x(n) W_N^{kn} \\ &= \sum_{m=0}^{(N/2)-1} x(2m) W_N^{2mk} + \sum_{m=0}^{(N/2)-1} x(2m+1) W_N^{k(2m+1)} \end{aligned} \tag{19.5}$$

But $W_N^2 = W_{N/2}$. With this substitution, Equation 19.5 can be expressed as

$$\begin{aligned} X(k) &= \sum_{m=0}^{(N/2)-1} f_1(m) W_{N/2}^{km} + W_N^k \sum_{m=0}^{(N/2)-1} f_2(m) W_{N/2}^{km}, \\ &= F_1(k) + W_N^k F_2(k), \quad k = 0, 1, \ldots, N-1 \end{aligned} \tag{19.6}$$

where $F_1(k)$ and $F_2(k)$ are the $N/2$-point DFTs of the sequences $f_1(m)$ and $f_2(m)$, respectively.

Since $F_1(k)$ and $F_2(k)$ are periodic, with period $N/2$, we have $F_1(k+N/2) = F_1(k)$ and $F_2(k+N/2) = F_2(k)$. In addition, the factor $W_N^{k+N/2} = -W_N^k$. Hence Equation 19.6 (called butterfly computation) may be expressed as

$$\begin{aligned} X(k) &= F_1(k) + W_N^k F_2(k), \quad k = 0, 1, \ldots, \frac{N}{2} - 1 \\ X\left(k + \frac{N}{2}\right) &= F_1(k) - W_N^k F_2(k), \quad k = 0, 1, \ldots, \frac{N}{2} - 1 \end{aligned} \tag{19.7}$$

This method is illustrated in Figure 19.5.

We observe that the direct computation of $F_1(k)$ requires $(N/2)^2$ complex multiplications. The same applies to the computation of $F_2(k)$. Furthermore, $N/2$ additional complex multiplications are required to compute $W_N^k F_2(k)$. Hence, the computation of $X(k)$ requires $2(N/2)^2 + N/2 = N^2/2 + N/2$ complex multiplications. This first step results in a reduction of the number of multiplications from N^2 to $N^2/2 + N/2$, which is about a factor of 2 for N large. Note that these results are based on radix-2 decomposition; however it can be repeated for $N/4$, $N/8$, ...1 points to implement radix-4, radix-8, radix-2^v FFT.

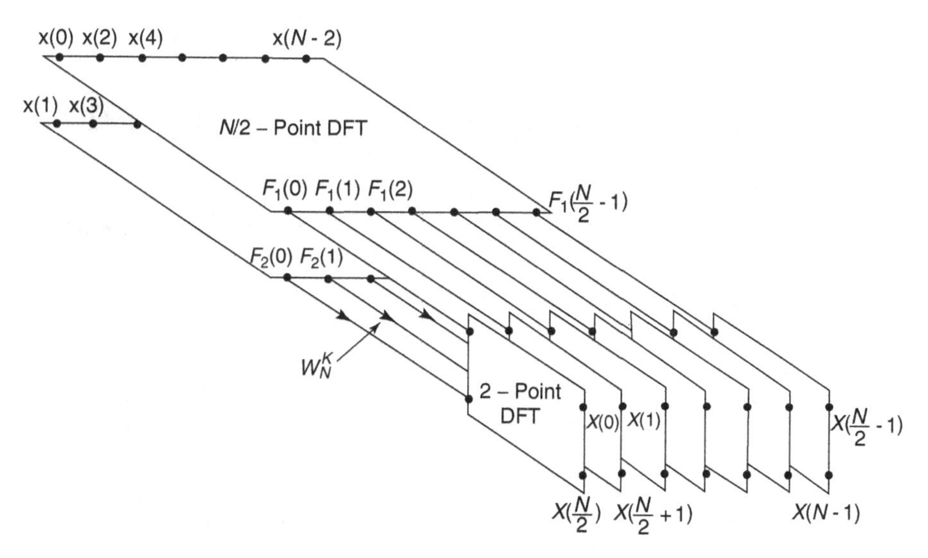

Figure 19.5. The radix-2 decimation-in-time FFT algorithm [27].

19.2.2. Implementation of the DFT and FFT

In this section, we first show how to compute DFT coefficients using a simplified version of nanoscale spin-wave reconfigurable mesh architecture, and afterwards, we discuss how this solution can be optimized using FFT techniques.

The DFT module consists of an $N \times N$ array of input plus N output computing nodes. In this architecture, in each row, one of the N Fourier coefficients is computed. In other words, the kth row computes $X[k]$ for $k = 0, \ldots, N-1$. As mentioned previously, the DFT coefficients are computed as

$$
\begin{aligned}
X(k) &= \sum_{x=0}^{N-1} x(n) W_N^{kn}, \quad 0 \leq k \leq N \\
W_N &= e^{-j2\pi/N}
\end{aligned}
\tag{19.8}
$$

First the W_N^k coefficients are mapped to nodes in row k and column N, as shown in the structure of the DFT module in Figure 19.6. The multiplication of W_N^k and the inputs, $x[k]$, is done in the processing nodes, while the summation is performed in the spin-wave bus by employing the superposition property of the waves.

Note that all the W_N^k coefficients are complex numbers. The multiplications of the complex numbers are performed in the processing nodes as four real number multiplications. The addition of the complex numbers on the spin-wave bus is performed in two steps: summation of the real parts and summation of the imaginary parts. After the multiplications are done in the processing nodes, they send the "real" part of the product on the bus in the first step and the "imaginary"

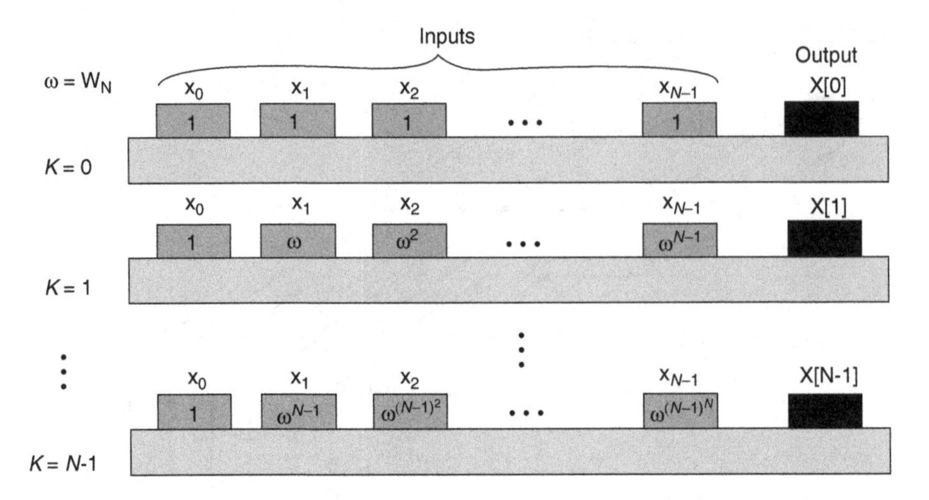

Figure 19.6. Nanoscale spin-wave DFT module.

part on the second step. Therefore, the outputs receive the real and imaginary parts of the $X[k]$ in two steps. The time complexity of this process depends on the time complexity of the multiplication, since the summation in done in $O(1)$ time. Multiplication of two numbers can be performed in constant time using a nanoscale spin-wave multiplication module presented in Chapter 9. Therefore the time-complexity of computing the DFT coefficients is $O(1)$.

The IDFT module can be implemented in a similar fashion as the DFT module. The only difference is in the coefficients that are stored in each processing node. In a DFT module the coefficient stored in the node in row k and column N is W_N^k, while in an IDFT module, this coefficient is $1/N \cdot W_N^{-k}$.

The design of the DFT module, as shown in Figure 19.2, requires N^2 processing nodes. In the following we show that the radix-2 decimation-in-time FFT algorithm can be implemented by $N^2/2$ processing nodes. As discussed in the previous section, the outputs of the radix-2 decimation-in-time FFT algorithm are computed as

$$\begin{aligned} X(k) &= F_1(k) + W_N^k F_2(k), \quad k = 0, 1, \ldots, \frac{N}{2} - 1 \\ X\left(k + \frac{N}{2}\right) &= F_1(k) + W_N^k F_2(k), \quad k = 0, 1, \ldots, \frac{N}{2} - 1 \end{aligned} \quad (19.9)$$

From Equation 19.9, we can see that by finding $F_1(k)$ and $F_2(k)$ for $N/2$ (odd and even) points, two Fourier coefficients, namely $X[k]$ and $X[k+N/2]$, can be computed at the same time. By observing this fact, we implement the FFT module as an array of $N/2 \times N$ input and N output computing nodes. In other words, the design of the DFT module is improved by eliminating half of the processor rows

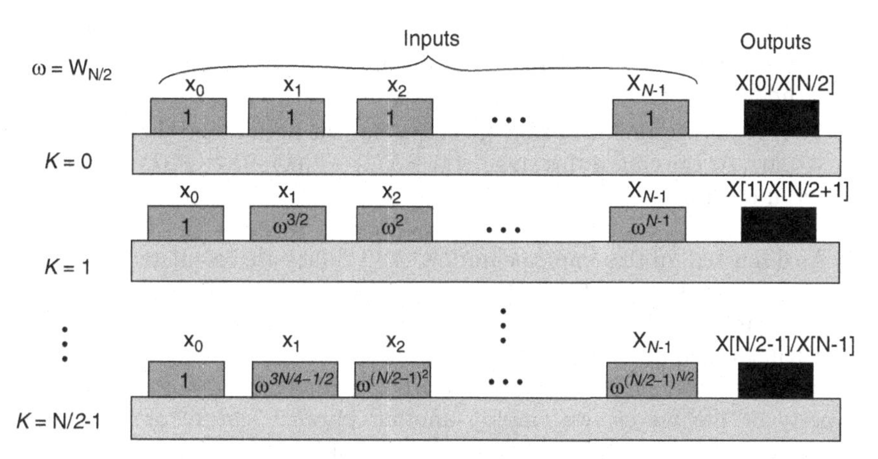

Figure 19.7. Nanoscale spin-wave FFT module (two-step output method).

or $N^2/2$ processing nodes. We present two methods for implementing the FFT module. The structure of the first implementation of the spin-wave FFT module is shown in Figure 19.7.

In this implementation, the W_N^k coefficients are stored in the respective nodes to which the inputs $x[k]$ are mapped. The multiplication of these two complex values is performed in the processing nodes and the summation in the spin-wave bus. Note that for simplifying the equations, W_N^k term was factored in the odd-indexed summation. At this step however, for more efficiently implementing the calculation of $W_N^k \cdot F_2(k)$, we multiply each of the coefficients by the factor W_N^k instead of first finding $F_2(k)$ and then multiplying it by W_N^k. So the coefficient that is multiplied by $x[k]$ is $W_N^k \cdot \mathrm{W}_{N/2}^k$:

$$W_N^k \cdot W_{N/2}^k = e^{-j2\pi k/N} \cdot e^{-j4\pi k/N} = e^{-j4\pi k/N} = W_N^{3k} = W_{N/2}^{3k/2} \tag{19.10}$$

The FFT operation is implemented in a similar fashion as the DFT in the following steps:

1. First the coefficients are multiplied by the inputs in all the nodes. As explained previously the coefficients in the odd-indexed nodes in each row contain the extra factor of W_N^k.
2. All the nodes send the "real" part of their data on the spin-wave bus. In other words, even-indexed nodes send $Re[F_1(k)]$, and the odd-indexed nodes send the $Re[W_N^k \cdot F_2(k)]$. As a result, $X[k]$ receives the superposition of these two values: $Re[F_1(k) + W_N^k \cdot F_2(k)]$.
3. This step is quite similar to the previous step, except that the nodes send their "imaginary" parts. Therefore, the superposed result would be $\mathrm{Im}[F_1(k) + W_N^k \cdot F_2(k)]$. At this stage, both the real and imaginary parts of the complex value $[F_1(k) + W_N^k \cdot F_2(k)]$ has been found at row k which is equal to $X[k]$.

4. This step is the same as step 2, except that the odd-indexed nodes send a wave with $W_N^k \cdot F_2(k)$ amplitude on opposite phase. Therefore, the output node receives $Re[F_1(k) - W_N^k \cdot F_2(k)]$.
5. This step is similar to step 4, except that the nodes send their "imaginary" part. At the end of this stage, $X[k+N/2] = F_1(k) - W_N^k \cdot F_2(k)$ is found at the output node.

As discussed, in this implementation, $X[k]$ values are calculated for $k=0, \ldots, N/2-1$ at one step and for $k=N/2, \ldots, N-1$ in another step. In another implementation of FFT module, explained in the following, all the N Fourier coefficients are computed simultaneously.

In the second implementation of FFT module, besides the superposition property of the waves, we employ another parallel feature of the spin-wave architectures, i.e., the ability of transmitting multiple waves at different frequencies. The structure of this module is shown in Figure 19.8.

As illustrated, this architecture includes $N/2$ extra output nodes to compute all the N outputs simultaneously. The $X[k]$ values for $k = 0, \ldots, N/2-1$ are computed in the first output column, called out1, while $X[k]$ values for $k = N/2, \ldots, N-1$ are computed in the second column, called out2. These values are computed in the following steps:

1. First the coefficients are multiplied by the inputs in all the nodes (odd-indexed nodes contain the extra factor of W_N^k).
2. At this step, the receiving frequency of the out1 nodes is tuned on f_{even} and out2 on f_{odd}. The even-indexed input nodes broadcast the "real" part of their data on f_{even} while the odd-indexed nodes broadcast the "real" part of their data on f_{odd}. As a result of the superposition of the spin-waves, $Re[F_1(k)]$ is detected in out1 and $\text{Re}[W_N^k \cdot F_2(k)]$ in out2. Since the odd- and

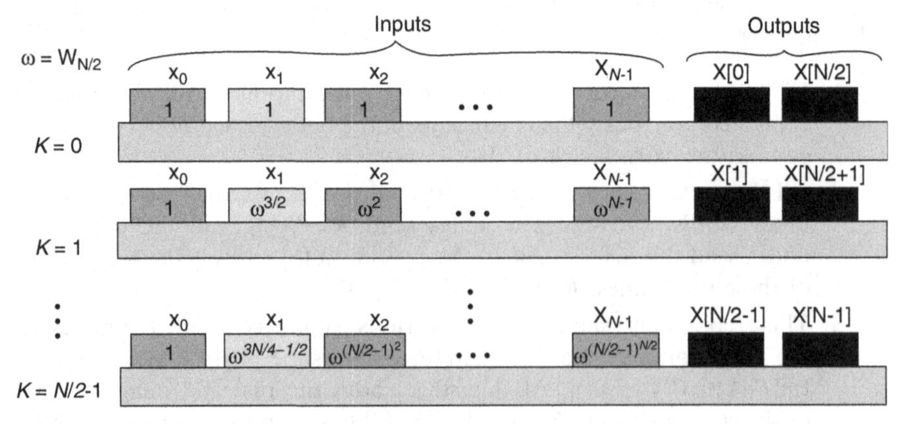

Figure 19.8. Nanoscale spin-wave FFT module (odd-even frequency split method).

even-indexed nodes send their data on two different frequencies, there is no interference between the two.

3. This step is similar to step 1, except that the nodes send the "imaginary" part of their data. The frequency tunings remains the same as previous step; therefore, out1 nodes receive $\mathrm{Im}[F_1(k)]$, while out2 nodes receive $\mathrm{Im}[W_N^k \cdot F_2(k)]$. At this stage, out1 nodes contain the complex value of $F_1(k)$, and out2, $W_N^k \cdot F_2(k)$.
4. At this step, out1 and out2 switch frequencies. The receiving frequency of the out1 nodes is tuned on f_{odd} and out2 on f_{even} and step 2 is repeated. As a result the $\mathrm{Re}[F_1(k)]$ is received by out2 and $\mathrm{Re}[W_N^k \cdot F_2(k)]$ by out1.
5. This step is similar to step 4, except that the imaginary parts are broadcasted. At this stage each of out1 and out2 contain both the complex values of $W_N^k \cdot F_2(k)$ and $F_1(k)$.
6. Out1 nodes compute and output $X[k] = F_1(k) + W_N^k \cdot F_2(k)$ while out2 nodes compute and output $X[k+N/2] = F_1(k) - W_N^k \cdot F_2(\mathrm{k})$ for $k=0, \ldots, N/2-1$. These addition and subtraction of complex numbers include two addition and subtraction for the real and imaginary parts.

The time complexity of both methods of implementing FFT, similar to DFT, is $O(1)$. As mentioned previously, one of the applications of Fourier transform is in designing digital filters. In digital filter design problems, the filter coefficients should be found based on a filter's desired frequency response. The filter's frequency response can be simplified to the form of the Fourier transform of the filter coefficients. Therefore, the filter coefficients can be calculated simply by taking the inverse Fourier transform of the desired frequency response. Finding the inverse Fourier transform using spin-wave architectures is possible in a similar fashion to finding discrete Fourier transform, discussed in this section.

19.3. CONCLUSIONS

This chapter showed how image processing tasks can be performed very fast on nanoscale spin-wave architectures. Solutions to three image processing problems, namely, labeling, finding the convex hull, and finding the nearest neighbor problem, were presented. In addition, efficient implementations of spin-wave modules for computing discrete Fourier transform (DFT) and fast Fourier transform (FFT) were shown.

REFERENCES

1. H. M. Alnuweiri. Constant-time parallel algorithms for image labeling on a reconfigurable network of processors. *IEEE Transactions on Parallel and Distributed Systems*, 5(3): pp 320–326, Mar 1994.

2. D. Chin, J. Passe, F. Bernard, H. Taylor, and S. Knight. The Princeton Engine: A real-time video system simulator. *IEEE Transactions on Consumer Electronics*, 32(2): pp 285–297, 1988.
3. R. Cypher, J. L. C. Sanz, and L. Snyder. EREW PRAM and mesh connected computer algorithms for image component labeling. Proceedings of IEEE Computer Society Workshop on Computer Architecture for Pattern Analysis and Machine Intelligence: pp 122–128, 1987.
4. M. J. B. Duff. CLIP4. *In Special Computer Architectures for Pattern Processing*, K. S. Fu and T. Ichikawa, (eds), Elkins Park, PA: Franklin Book Company: pp 65–86, 1982.
5. M. M. Eshaghian-Wilner. Parallel algorithms for image processing on OMC. *IEEE Transaction on Computers*, 40(7): 827–833, 1991.
6. M. M. Eshaghian-Wilner, K. Kim, G. Nash, and D. B. Shu. Implementation and application of a gated connection network in image understanding architecture In: H. Li and Q. Stout (eds), *Reconfigurable Massively Parallel Computers*. Upper Saddle River, NJ: Prentice Hall, 1991.
7. T. Fountain, K. Matthews, and M. Duff. The CLIP7A image processor. *IEEE Pattern Analysis and Machine Intelligence*, 10(3): pp 310–319, 1988.
8. D. W. Hammerstrom and D. P. Lulich. Image processing using one-dimensional processor arrays. *IEEE Proceedings*, 84(7): pp 1005–1018, 1996.
9. M. C. Herbordt, C. C. Weems, and M. J. Scudder. Non-uniform region processing on SIMD arrays using the coterie network. *Machine Vision and Applications*, 2(5): pp 105–125, 1992.
10. J. W. Jang, H. Park, and V. K. Prasanna. A fast algorithm for computing a histogram on reconfigurable mesh. *IEEE Transactions on Pattern Analysis and Machine Intelligence*, 17(97): p 106, 1995.
11. J. F. Jenq and S. Sahni. Histogramming on a reconfigurable mesh computer In: Proceedings of the 6th International Parallel Processing Symposium, California: Beverly Hills, pp 425–432, 1992.
12. R. Miller, V. K. Prasanna Kumar, D. I. Reisis, and Q. F. Stout. Parallel computations on reconfigurable meshes. *IEEE Transactions on Computers*, 42: pp 678–692, June 1993.
13. R. Miller and Q. F. Stout. Geometric algorithms for digitized pictures on a mesh-connected computer. *IEEE Transactions on Pattern Analysis and Machine Intelligence*, 7: pp 216–228, 1985.
14. R. Miller and Q. F. Stout. Efficient parallel convex hull algorithms. *IEEE Transactions on Computers, C*-37: pp 1605–1618, 1988.
15. V. K. Prasanna Kumar and M. M. Eshaghian. Parallel geometric algorithm for digitized pictures on mesh of trees. In: Proceedings of IEEE International Conference on Parallel Processing, 1986.
16. Y. C. Chen and W. T. Chen. Efficient median finding and its application to two-variable linear programming on mesh-connected computers with multiple broadcasting. *Journal of Parallel and Distributed Computing*, 15: pp 79–84, 1992.
17. D. Bhagavathi, H. Gurla, S. Olariu, J. L. Schwing, and I. Stojmenovic. Time-optimal visibility-related algorithms on meshes with multiple broadcasting. *IEEE Transactions on Parallel and Distributed Systems*, 6: 687–703, 1995.
18. R. Miller, V. Prasanna Kumar D. Reisis, and Q. Stout. Meshes with reconfigurable buses. In: Proceedings of the 5 MIT Conference on Advanced Research in VLSI, Boston, 1988.

19. M. M. Eshaghian-Wilner, and R. Miller. The systolic reconfigurable mesh. *Journal of Parallel Processing Letters*, 14(3–4): pp 337–350, Sep–Dec 2004.
20. C. C. Weem, S. P. Levitan, A. R. Hanson, E. M. Riseman, J. G. Nash, and D. B. Shu. The image understanding architecture. COINS Tech. Rep. University of Massachusetts: pp 87–76, 1987.
21. D. W. Goodwin. Towards an optical computer. *Engineering*, 208(5399): pp 423–424, 1969.
22. M. M. Eshaghian-Wilner, S. Navab, A. Khitun, and K. L. Wang. Constant-time image processing on spin-wave nanoarchitectures. *Physica Status Solidi a*, 204(6): pp 1931–1936, June 2007.
23. M. M. Eshaghian-Wilner, S. Navab, A. Khitun, and K. L. Wang. The spin-wave reconfigurable mesh and labeling problem. *Journal on Emerging Technologies in Computing Systems*, 3(2 no. 5): July 2007© 2007, ACM, Inc. Used by permission.
24. T. Roska. AnaLogic wave computers-wave-type algorithms: canonical description computer classes, and computational complexity. The 2001 IEEE International Symposium on Circuits and Systems (Cat. No. 01CH37196). In: *ISCAS 2001 Part 2*. Piscataway NJ: IEEE Press pp 41–42, 2001.
25. M. M. Eshaghian-Wilner and L. Hai. Application-specific design of the optical communication topology in ORM. International Symposium on Automotive Technology and Automation, Florence, Italy, June 1997. In: Proceedings of Supercomputing Applications in the Transportation Industries.
26. S. W. Smith. *The Scientist and Engineer's Engineer's Guide to Digital Signal Processing*, San Diego: California Technical Publishing, 1997.
27. http://www.cmlab.csie.ntu.edu.tw/cml/dsp/training/coding/transform/index.html.
28. P. T. Gough. A fast spectral estimation algorithm based on FFT. *IEEE Transactions on Signal Processing*, 42: pp 1317–1322, June 1994.
29. S. D. Stearn and R. A. David. Decimation and Interpolation Routines. In: *Signal Processing Algorithms*, Englewood Cliffs, NJ: Prentice-Hall. 1988.
30. C. Wu. Digital television terrestrial broadcasting. *IEEE Communications Magazine*, 32(5): pp 46–52, May 1994.
31. M. Perry. Using 2-D FFTs for object recognition. ICSPAT, DSP World Expo, Dallas, Texas: pp 1043–1048, Oct 1994.
32. R. R. Holdrich. Frequency analysis of non-stationary signals using time frequency mapping of the DFT magnitudes. ICSPAT, DSP World Expo, Dallas, Texas: Oct 1994. (www.icspat.com)
33. A. S. Spanias. Speech coding: a tutorial review. *Proceedings of the IEEE*, 82: pp 1542–1582, Oct 1994. (FFT-based RELP vocoder)
34. F. J. Harris. On the use of windows for harmonic analysis with the discrete Fourier transform. *Proceedings of the IEEE*, 66: pp 51–83, Jan 1978.
35. Digital audio compression (AC-3). ATSC standard: Dec 20 1995. (www.atsc.org)
36. M. Zhao. Channel separation and combination using fast Fourier transform. ICSPAT 97, San Diego, Sept 1997.
37. X. Kang et al. A DWT-DFT composite watermarking scheme robust to both affine transform and JPEG compression. *IEEE Transactions on Circuits and Systems for Video Technology*, 13: pp 776–786, Aug 2003.
38. J. M. Rius and R. De Povrata-Doria. New FFT bit reversal algorithm. *IEEE Transactions on Signal Processing*, 43: pp 991–994, Apr 1995.

39. S. D. Stearn and R. A. David. FFT convolution and correlation. *In: Signal Processing Algorithms*, Englewood Cliffs NJ: Prentice-Hall, 1988.
40. G. Dovel. FFT analyzers make spectrum analysis a snap. *EDN*: pp 149–155, Jan 1989.
41. G. Davison, L. Fielder, and M. Antill. Low-complexity transform coder for satellite link applications. 89th AES convention, Los Angeles, Sept 1990. (www.aes.org)
42. K. Brandengurg, et al. The ISO/MPEG audio codec: A generic standard for coding of high quality digital audio. 92nd AES convention, Vienna, Austria, 1992: preprint 3336. (www.aes.org)
43. D. Pan. An overview of the MPEG/audio compression algorithm. IS&T/SPIE Symposium on Electronic Imaging: Science and Technology, San Jose, California, Feb 1994. 2187, pp 260–273.
44. A. K. Jain. Filtering LPF BPF HPF Generalized cepstrum and homomorphic filtering. *Fundamentals of Digital Image Processing*. Englewood Cliffs, NJ: Prentice-Hall, 1988.
45. K. O. Egiazarian, et al. Nonlinear filters based on ordering by FFT structure. Photonics West, IS&T/SPIE Symposium on Electronic Imaging: Science & Technology, 2662, San Jose, Feb 1996.
46. J. J Shynk. Frequency domain and multirate adaptive filtering. IEEE Signal Processing Magazine, 9: pp 14–37, Jan 1992.
47. J. O. Smith III. *Introduction to Digital Filters with Audio Applications*, Palo Alto, CA: Stanford University Press, Aug 2006. (http://ccrma.stanford.edu/~jos/filters/)
48. I. Barhumi, G. Leus, and M. Moonen. Time-varying FIR equalization of doubly selective channels. In: IEEE International Conference on Communications, ICC, Anchorage, Alaska, May 2003.
49. M. Bi, S. H. Ong, and Y. H. Ang. Integer-modulated FIR filter banks for image compression. *IEEE Transactions on Circuits And Systems For Video Technology*, 8(8): pp 923–927, 1998.
50. A. Zergaïnoh, P. Duhamel, and J. P. Vidal. Efficient implementation methodology of fast FIR filtering algorithms on DSP. *Journal of VLSI Signal Processing Systems Archive*, 16(1): pp 81–103, May 1997.
51. J. C. Kuo III. et al. VLSI design of a variable-length FFT/IFFT processor for OFDM-based communication systems. *Journal on Applied Signal Processing*, 2003(13): Dec 2003.
52. G. Zhong, F. Xu, A. Wilson Jr. An energy-efficient reconfigurable FFT/IFFT processor based on a multiprocessor ring. EUSIPCO, Vienna, Austria, Sep 2004.
53. K. Mahoratna, E. Grass, and U. Jagdhold. A novel 64-point FFT/IFFT processor for IEEE 802.11(A) standard. In: IEEE ICASSP, II: pp 321–324, 2003.
54. http://en.wikibooks.org/wiki/Basic_Physics_of_Nuclear_Medicine/FourierFourier_Methods# Filtering_the_FourierFourier_Spectrum.
55. http://www.med.harvard.edu/AANLIB/hms1.html.
56. http://homepages.inf.ed.ac.uk/rbf/HIPR2/fourierFourier.htm.
57. S. Navab. Nanoscale digital signal processing architectures. 2007 World Congress in Computer Science, Computer Engineering, and Applied Computing WORLDCOMP'07. In: Proceedings of the 2007 International Conference on Embedded Systems and Applications (ESA'07): pp 141–147.

20

CONCLUDING REMARKS AT THE BEGINNING OF A NEW COMPUTING ERA

Varun Bhojwani, Stephen Chu, Mary Mehrnoosh Eshaghian-Wilner, Shawn Singh, and Chun Wing Yip

This chapter briefly summarizes the topics that were discussed in each chapter. In addition to the work of the scientists presented in this book, several other interesting projects in nanocomputing currently are taking place globally and are not included in this volume. Table 20.1 lists the names of some of those excluded groups. Additionally, Table 20.2 lists some of the public agencies and private companies currently offering funding for nanocomputing. Nanocomputing as a new area of research will involve many inventions that would need to be patented. As such, a brief discussion of patent issues is included. Finally, the chapter ends with with a discussion of social and economic impacts of nanocomputing.

20.1. MATERIALS PRESENTED IN THIS BOOK

Chapter 1 briefly discussed the beginning of the nanocomputing era, brought about by the fact that integrated circuits have been constantly decreasing in size for 50 years. *Computing* was defined as the representation and manipulation of abstract information; some historical context of the microcomputing era was also discussed. The limitations of traditional microcomputers motivates a vast amount of research effort to develop new nanoscale devices and paradigms, including molecular devices, quantum dots, tunneling devices, spin devices, cellular logic, wave computing, DNA computing, quantum computing, and more. Two major fields that greatly benefit from nanocomputing are biology and neurology.

Bio-Inspired and Nanoscale Integrated Computing. Edited by Mary Mehrnoosh Eshaghian-Wilner

TABLE 20.1. List of Some Current Nanocomputing Research Centers and Researchers

Research center	Director/Leader/People
Arizona Institute for Nano-Electronics (AINE) Web page: http://www.asu.edu/aine/	Dr. Stephen Goodnick, Director
Ball State University, Center for Computational Nanoscience Web page: http://www.bsu.edu/ccn/	Dr. Yong Joe, Director
California Institute of Technology—Pierce Lab Web page: http://www.piercelab.caltech.edu/	Niles A. Pierce, Professor
Caltech Kavli Nanoscience Institute (KNI) Web page: http://kni.caltech.edu/	Axel Scherer, Director
Columbia University—Nanoscale Science and Engineering Center Web page: http://www.cise.columbia.edu/nsec/	Ronald Breslow, Tony Heinz, James Yardley
Georgia Institute of Technology—Microelectronics Research Center (MiRC) Web page: http://www.mirc.gatech.edu/	Dr. James D. Meindl, Director
HP Quantum Science Research (QSR) Web page: http://www.hpl.hp.com/research/qsr	Stan Williams, HP Senior Fellow, Director
Institute for Nanoelectronics and Computing (INAC) Web page: http://www.inac.purdue.edu/home	Prof. Supriyo Datta, Director
Kavli Institute at Cornell for Nanoscale Science Web page: http://www.research.cornell.edu/KIC/	Dr. Robert C. Richardson, Director
MITRE Nanosystems Group Web page: http://www.mitre.org/tech/nanotech/ourwork/index.html	Mr. Alfred Grasso, President and Chief Executive Officer
NASA Ames Center for Nanotechnology Web page: http://www.ipt.arc.nasa.gov/index.html	M. Meyyappan, Director
National Center for Design of Biomimetic Nanoconductors Web page: http://www.nanoconductor.org	Eric Jakobsson, Director
National Composite Center (NCC) Web page: http://www.compositecenter.org	Louis Luedtke, President & CEO
NDSU Center for Nanoscale Science and Engineering Web page: http://www.ndsu.edu/cnse/	Gregory J. McCarthy, Ph.D. and Director
Northwestern University—Center for Quantum Devices Web page: http://cqd.eecs.northwestern.edu	Walter P. Murphy, Professor and Director
University of Notre Dame Center for Nano Science and Technology Web page: http://www.nd.edu/~ndnano/	Wolfgang Porod, Professor and Director
USC Nanotechnology Research Laboratory Web page: http://nanolab.usc.edu/	Dr. Chongwu Zhou
Western Michigan University Nanotechnology Computation and Research Center (NRCC) Web page: http://www.wmich.edu/nrcc/start.htm	Ekkehard Sinn, Professor

TABLE 20.2. List of Some Public and Private Agencies Funding Nanocomputing

Private companies	Public agencies
IBM	National Science Foundation
Intel	Department of Defense
Hewlett-Packard	Department of Energy
Lucent Technologies	National Institute of Health
Motorola	NASA
NEC	CIA

The remaining materials of the book where presented in two parts. In Part I, which included Chapter 2 to Chapter 10, various types of technologies that could be used in the design of nanoscale devices and paradigms were presented. In the second part, which included Chapter 11 to Chapter 19, a set of bio-inspired models and applications were presented.

20.1.1. Nanoscale Integrated Circuits

Chapter 2 gave an introductory overview of the ideas and techniques used to develop nanoscale devices. It described how resonant tunneling diodes, single-electron transistors, and quantum dots operate. Carbon nanotubes were used as an example of how new materials can provide creative solutions to overcome difficulties that arise at this small scale. It gave a very brief, high-level introduction to quantum mechanics, discussing the Schrödinger equation and the many-body problem, which is very useful for modeling and understanding how nanoscale devices work. Several representative methods for simulation of quantum behaviors were described: the Hartree–Fock approximation, density functional theory, tight-binding, the molecular dynamics approach, Monte Carlo techniques, and the use of Green's functions. More information about the author and his research can be found at the Microsystems and Nanotechnology Group at the University of British Columbia, http://www.mina.ubc.ca/.

In addition to such devices, nanoscale computing offers revolutionary paradigms of computation. Chapter 3 gave a history and discussion of quantum computing. Quantum computing is a paradigm of computation that uses quantum mechanics directly as the abstraction of computation. The chapter began by discussing reversibility, quantum Turing machines, other quantum automata models, and quantum gates and circuits. It then discussed quantum algorithms that have been developed for various applications, including compression, cryptography, Fourier transforms, factoring numbers, and search problems. It also addressed the issue of error correction due to decoherence and the potential of using distributed entanglement for a quantum computer. A variety of ways to use physics to realize quantum computing was given, and the energy and volume bounds for quantum computing were discussed. It turns out that volume bounds are challenging to determine, but understanding these bounds has implications on the benefits of quantum computation versus classical computation. More information about this author's research, including work on quantum computing, may be found at the home page of Professor John H. Reif at Duke University, http://www.cs.duke.edu/~reif/.

Chapter 4 described quantum-dot cellular automata (QCA), a paradigm that uses ground-state computation instead of switching technology. In this paradigm, QCA cells are used for both the logic function itself and interconnections between functions. It first introduced the operation of a single QCA cell and discussed that kink energy is an important parameter that makes the physical simulation of QCA tractable. It then discussed several issues of designing QCA circuits, specifically, the challenges of clocking and wire crossing. The chapter discussed the simulation tool, QCADesigner. Throughout the chapter, implementations of simple Boolean logic circuits, adders, multipliers, and even a simple processor, were shown with QCA. Finally, how QCA devices can be implemented using quantum dots, magnetic nanoparticles, or molecules was discussed. More information about the authors' affiliations and research can be found at the Microsystems and Nanotechnology Group at the University of British Columba, http://www.mina.ubc.ca/, and ATIPS Labs, http://www.atips.ca/.

This book then looked at how the dielectrophoretic effect can be exploited to move nanoscale structures–Chapter 5, which discussed the use of dielectrophoresis for assembly, reconfiguration, and disassembly of a nanoscale device. It also described circuit-level architectures that could be used to create crossbars and field-programmable gates. This effect is also envisioned to be used for fault tolerance, where the functionality to replace faulty circuits using dielectrophoresis could potentially be contained in a chip's packaging. More information about this and the author's other research may be found at the home page of Dr. Alexander D. Wissner-Gross, http://www.alexwg.org/index.html.

Data storage is also an important aspect of nanocomputing. Chapter 6 studied multilevel magnetic recording, a data storage technique that can attain storage densities of more than 10 terabits/in^2, and possibly up to 100 terabits/in^2. The chapter illustrated that the superparamagnetic limit is a fundamental limitation to improving traditional magnetic storage density, even with recent perpendicular recording techniques. It described how multilevel magnetic recording would work, and compared it to existing multilevel optical recording techniques. It also discussed how data would need to be encoded to use a multilevel magnetic storage system. More information about the authors and their research can be found at the Center for Nanomagnetic Systems at the University of Houston, http://www.uh.edu/cns/, and the Center for 3D Electronics at the University of California, Riverside, http://c3de.ee.ucr.edu/.

The use of spin-waves for nanoscale computation was presented in Chapter 7. Chapter 7 first described how spin waves propagate. Then, it discussed three main architectures that benefit from spin waves: reconfigurable mesh, crossbar, and fully interconnected cluster. Because of wave properties, these three architectures can transmit multiple signals simultaneously, perform concurrent writes, and process digitally or in analog. The chapter also discussed a hierarchical architecture that overcomes the short attenuation length of spin waves. More information about this research can be found at the Bio-Inspired and Nanoscale Integrated Computing Research Group at the University of California, Los Angeles, http://www.seas.ucla.edu/~eshaghia/lab_web/index.htm, and the Center on Functional Engineered Nano Architectronics, http://www.fena.org/.

Chapter 8 discussed parallel techniques that exploit the wave properties of the spin-wave architectures discussed in the previous chapter. It discussed a set of general algorithmic techniques, including finding the maximum/minimum, first/last of a list, prefix sum, and shifting elements of a list. It also discussed routing through the three architectures: reconfigurable mesh, crossbar, and fully interconnected cluster. The fault tolerance feature of routing with spin waves was also described. More information about this research can be found at the Bio-inspired and Nano-scale Integrated Computing Research Group at the University of California, Los Angeles, http://www.seas.ucla.edu/~eshaghia/lab_web/index.htm.

After the introduction to spin waves and its parallel techniques in previous chapters, Chapter 9 showed how a variety of fundamental operations would work using spin waves. The chapter first discussed the operation of digital-to-analog conversion, used in many of the following operations. It then described full adders, multipliers, basic logic gates, decoders, encoders, multiplexers, and demultiplexers, priority encoders, and shifters. Being able to describe this variety of functions shows that nearly any logic function, which would be a combination of these basic operations, can be realized using spin waves. More information about this research can be found at the Bio-Inspired and Nano-scale Integrated Computing Research Group in the University of California, Los Angeles, http://www.seas.ucla.edu/~eshaghia/lab_web/index.htm.

To conclude the discussion of nanoscale integrated circuits, Chapter 10 explored fault tolerant computing. The chapter first discussed problems that arise from consistenly shrinking the feature size of devices—nanoscale devices will inevitably have permanent defects as well as transient errors. It described three major ways to perform reliable computations despite these faults: structural redundancy, information redundancy, and reconfigurability. It also covered a variety of ways to evaluate the reliability of fault-tolerant devices. More information about this research can be found at the Formal Engineering Research with Models, Abstractions and Transformations Lab at Virginia Tech, http://fermat.ece.vt.edu/.

20.1.2. Bio-Inspired Models and Applications

In Part II, we turned our attention to biologically inspired models and applications of nanotechnology. Chapter 11 covered a variety of approaches for molecular devices. The chapter first described some approaches to creating molecular switches and memory devices. It then proceeded to discuss circuit and architecture level molecular integration, particularly electrostatic architectures (e.g., quantum-dot cellular automata) and nanoscale crossbar architectures. It also described the NanoCell, an interesting point in the design space of molecular devices that offers fault tolerance and reprogrammability. Finally, the chapter discussed some interesting ways that molecules can be used to enhance existing silicon-based technology, such as being able to adjust transistor characteristics more reliably and smaller scale than bulk doping. More information about the authors and their research can be found at the Rice Efficient Computing Group at Rice University, http://www.ruf.rice.edu/~mobile/, and the Tour Group at Rice University, http://www.ruf.rice.edu/~kekule/.

After the discussion of building and characterizing molecular devices, in Chapter 12, self assembly approaches were described to create highly regular arrays of metal ions. There are several future-looking uses for this self assembly approach: "ion dots" which are a smaller analog to quantum dots, cellular automata which benefits from regular, locally interconnected machines, and molecules used as qubits. It also reviewed the self-assembly of carbon nanotubes (CNTs) and their use as an interface between nanoscopic molecular devices and macroscopic environment. More information about the author's research can be found at the Ruben-Group, http://www.ruben-group.de/.

Chapter 13 overviewed DNA nanostructures and self-assembly of these nanostructures. It first reminded the reader of the basic properties of DNA, and explained that DNA is a good choice for nanostructures because its properties and usage are well understood. It then discussed the pioneering work by Adleman, as well as the flurry of more recent work that effectively demonstrated self-assembly of DNA tiles and lattices. It also discussed error correction schemes that are necessary when working with DNA self assembly. More information about the author and his research may be found at the home page of Professor John H. Reif at Duke University, http://www.cs.duke.edu/~reif/.

After the discussion of the DNA nanostructures, Chapter 14 showed how to form partial-order multiple-sequence alignment graphs (PO-MSAG) on a spin-wave reconfigurable mesh architecture as compared to a traditional VLSI reconfigurable mesh architecture. In addition to understanding how the algorithm can be realized on reconfigurable meshes, it is interesting to see that the spin wave architecture performs algorithmically faster, because wave communications can be parallelized more easily than wired communications. The chapter showed that given N distinct variables in the sequence, the spin-wave model can form a graph in $O(1)$ time, while the VLSI architecture requires $O(N)$ time. More information about this research can be found at the Bio-Inspired and Nano-Scale Integrated Computing Research Group in the University of California, Los Angeles, http://www.seas.ucla.edu/~eshaghia/lab_web/index.htm.

One major application of nanocomputing and nanotechnology will be nanoscale autonomous robots. Chapter 15 discussed what it will take to progress towards medical nanorobots. It first described how such nanorobots may be assembled, constructed of diamondoid structures. It then described four major classes of medical nanorobots proposed in literature: respirocytes, microbivores, clottocytes, and chromallocytes. Throughout the chapter, the required resources for each robot were enumerated, and then these requirements were generalized: a typical nanorobot would require the ability to perform pumping, sensing, energy control, communication, navigation, manipulation, and other related actions. The chapter also considered control protocols that would be needed to ensure nanorobots work safely and effectively. More information about the author and this research can be found at the Institute for Molecular Manufacturing, http://www.imm.org/.

Chapter 16 introduced several heterogeneous NEMS (NanoElectroMechanical Systems) and MEMS (MicroElectroMechanical Systems) devices for biomedical diagnostics. It discussed the bio-inspired use of surface (or interface) tension and the electrowetting force to drive fluid through spaces as small as a carbon nanotube. It then discussed the Casmir effect, and noted its significance for understanding NEMS components. The role, creation, and integration of nanoscale biosensors in medicine was emphasized. Finally, the chapter discussed the operation of nanoscale biofuel cells. More information about this research may be found at the home page of Professor Tzung Hsiai at University of Southern California, http://bme.usc.edu/directory/faculty/primary-faculty/tzung-k-hsiai/.

Then, in Chapter 17, the possibility of implementing artificial brains was discussed. It first motivated the idea of creating an artificial brain and why it should use custom, carbon nanotube circuits. Specifically, carbon nanotubes are a promising way to achieve three-dimensional interconnections and a small enough size for large-scale artificial brains. It then described how carbon nanotubes can be used to emulate a synapse, and showed SPICE simulations of such a device. Finally, it discussed the feasibility of creating a complex, highly interconnected system of neurons, with respect to future technology sizes and interconnection complexity. More information about the authors and their research may be found at the home page of Professor Alice Parker at University of Southern California, http://ceng.usc.edu/~parker/.

Chapter 18 discussed the use of carbon nanotubes for biomedical applications. An essential step to allowing carbon nanotubes to interact with biological systems is functionalization, the process of supplementing the carbon nanotubes with molecules of particular functional groups. It gave a brief overview of functionalization techniques, including non-covalent (structure preserving) and covalent (structure modifying) functionalization. It then described many proposed uses of carbon nanotubes, for applications such as miniaturized x-ray or radiation devices, sensors, probes, drug delivery vehicles, implants, actuators, and more. Some of the many challenges of using nanotubes were discussed, such as fabrication reliability and toxicity. More information about the author may be found at the home page of Professor Tulin Mangir at California State University, Long Beach, http://www.csulb.edu/projects/npu/Tmangir_page/index.htm.

Last but not least, image processing applications of nanoscale spin-wave architectures were discussed in Chapter 19. It was shown how the concurrent write feature of the spin wave architecture can achieve constant time complexity with a large spin-wave reconfigurable mesh for three common problems: labeling, convex hull, and nearest neighbor search. It then discussed spin wave algorithms for the discrete Fourier transform, a fundamental computational tool used in nearly all fields of engineering. The chapter also discussed an implementation of the Fast Fourier Transform. More information about this research can be found at the Bio-Inspired and Nano-Scale Integrated Computing Research Group in the University of California, Los Angeles, http://www.seas.ucla.edu/~eshaghia/lab_web/index.htm.

20.2. NANOCOMPUTING RESEARCH AND FUNDING

The previous section reviewed the work that was presented in this book. As noted earlier, there are many other interesting projects in nanocomputing that have not been included in this volume. This section first presents Table 20.1 which lists some of the nanocomputing research groups whose work has not been included in this book. A more comprehensive list can be found in [1]. Next, this section presents a reference to the funding sources available for nanocomputing.

Having gone through some of the types of research currently taking place in nanocomputing, Table 20.2 lists some of the funding sources that are available for nanocomputing.

20.3. PATENTING ISSUES IN NANOCOMPUTING

Nanotechnology thrives on the creativity and profound insight—the intellectual achievements —of researchers all over the world. The ideas that researchers create are the breadbasket of the nanotechnology industry, bringing research progress into practical products. *Patents* are a prime way to protect ideas and to encourage researchers to invest their intellectual efforts for the progress of technology.

According to Wikipedia, "*A patent is a set of exclusive rights granted by a state to an inventor or his assignee for a fixed period of time in exchange for a disclosure of an invention.*" The exclusive right granted to a patentee in most countries is the right to prevent or exclude others from making, using, selling, offering to sell or importing the invention.

Three requirements that an invention must satisfy to obtain a patent are the invention must be: (1) novel, (2) useful, and (3) nonobvious. More specifically, it must be novel, useful, and nonobvious to a person with expertise in the topic that the patent applies to. Furthermore, a patent must contain enough information so that anyone skilled in the field would be able to easily implement the invention. The patent should use language that is as clear and unambiguous as possible, explicitly defining how terminology should be interpreted.

The first issue to consider when patenting nanotechnology is that inventions need to be filed, prosecuted, examined, and defended by individuals who are knowledge in the field. However, with nanotechnology simultaneously being a new field and a vast, multidisciplinary field, there may not be enough professionals with adequate skill, who can file, prosecute, examine and defend nano inventions.

This may become a major problem when the number of nanotechnology patent applications begins to flourish.

A second problem specific to nanotechnology related inventions is that many inventions may simply be smaller versions of something that already exists. This is an interesting topic open to debate. Some may argue that as such, some nanotechnologies may not necessarily constitute an invention, because they may be an obvious and smaller version of what is already known at a bigger scale. In these cases, it may be useful to consider whether the *process* used to create the nanoscale object is worth patenting. The nanoscale version could be substantially different in terms of fabrication, since physics operates very differently at the molecular level of fabrication. Such differences could fulfill the requirement for nonobviousness of an invention.

Additionally, when deciding to patent an invention, one must consider the fact that the protection offered by a patent issued in the United States would be limited, of course, to the United States. While an inventor would ideally like to protect his invention internationally, that is not necessarily the best option. For one thing, obtaining patents is a costly matter, with many fees involved, such as attorney fees, translation fees for foreign countries, and more. An inventor should weigh such costs against what he could earn with the patents. In some case, it may be best to use the *Patent Cooperation Treaty*, or PCT. Additional information about such issues may be obtained by contacting the United State Patent and Trademark Office at www.uspto.gov, or your local intellectural property law firms.

20.4. SOCIO-ECONOMIC EFFECTS OF NANOCOMPUTING

When it comes to exploiting the economic and social potentials of nanocomputing, scientists and engineers can have high expectations. For one, the present era of nanocomputing offers extraordinary opportunities for inventors. Some of the early inventions in this field could become embedded in future technologies, which will then vastly increase the value of such inventions. Furthermore, since nanocomputing is multidisciplinary in nature, a discovery in this area could have several applications in various fields, and thereby making nanocomputing inventions even more valuable. For example, think of a miniature robot that works as a surgeon in the human brain or body, correcting problems that are ailing its host. Such an advanced nanorobot is one of the promises that nanocomputing can make to the field of neuroscience, and to the world as well.

In recent years, nanotechnology has given the world new ideas for a brighter future and its subsequent emergence is increasingly difficult to ignore. The raw potential in this new subject has even drawn the attention of many policy makers and social scientists. In addition to changing medical fields such as neuroscience, the presence of nanotechnology will leave shockwaves in the social and economic realms.

Nanocomputing as a subarea of nanotechnology comes with the foreseeable advantage of being environmentally friendly. As nanocomputing advances, electronic components such as transistors will decrease in size, more efficiently using construction materials for computing equipment. Electronic devices will not only be smaller, but will also aspire to a higher level of energy efficiency. Furthermore, improved electronic components such as transistors may result in the production of ecofriendly, cheaper, and faster electricity. Consistent with these predictions, nanocomputing can promote reduced levels of industrial and fossil-fuel waste.

Nanocomputing has the potential to impact the education system globally. As a very novel and sophisticated new area of study, nanocomputing would require highly skilled and educated individuals in both industry and research. An area of such caliber could increase the demand for skilled labor, which may promote the developing of education to match the demand. The education system may witness changes, and consequently more highly educated and adept workers would need to be produced globally. New and updated chapters may be added to books used in science courses, making students more knowledgeable and competent in nanocomputing.

Another way that nanocomputing could help the world in the future is through the opportunities it may present to the world's workforce. Nanocomputing can make the implementation of many new technologies and inventions possible. The factories where these inventions and technologies are to be implemented could have provisions for many new jobs around the world—potentially reducing unemployment and spurring economic growth. According to World Bank estimates, as of 2001 1.1 billion people globally live below the poverty line, many earning less than $1 a day. Multiple new jobs can help alleviate this situation. And if this results in decreased poverty, other related problems such as malnutrition and homelessness could also be positively affected. Individuals earning more per day could have better chances of providing themselves with basic living essentials.

The impact of nanocomputing and nanotechnology could possibly be greater than that of the Industrial Revolution. If a nation's progress is dependent upon its economy, nanocomputing may become a major factor in determining that economy, which ultimately determines the nation's power and influence. The National Science Foundation has predicted a global market for products and services made possible by nanotechnology of up to $1 trillion by the end of 2015, and certainly a significant amount will be invested in nanocomputing itself. According to economists, this is an opportunity for economic equality and healthy competition among the countries of the world. The potentially giant market for nanoproducts could create millions of employment opportunities around the world. The jobs may demand skilled labor and promote education in the developing countries, and hence promote economic growth.

Due to its high economic potential, much funding is being generated for cutting-edge research in nanocomputing. The U.S. funding for general nanotechnology research has quadrupled from $270 million in 2000 to $1.08 billion in the current year, and approximately 4000 government research projects are underway.

With all the benefits that this new emerging discipline offers, nanocomputing could promise individuals around the world a better future. An overall boost to global society and the world economy can be predicted from the pledges of this discipline. Nanotechnology as a whole, with its perks and benefits definitely has the potential—directly or indirectly—to help resolve some of the world's economic and social problems. It has the capability to enhance other fields globally, such as the education system, nanomedicine, and the world workforce. Nanocomputing, therefore, certainly has the potential to bring in a positive wave of change among the lives of people everywhere.

20.5. CONCLUSIONS

This chapter briefly reviewed the main topics discussed in this book. Additionally, a listing of other research centers involved in nanocomputing research, whose work were not included in this volume was given. We discussed the importance of obtaining patents for nanocomputing inventions, and presented a basic introduction to patents, patenting issues and requirements for obtaining patents. To help readers be aware of various agencies currently funding nanocomputing research, we presented a listing of both public and private funding agencies. Lastly, we introduced a discussion on socio-economic effects of nanocomputing. In summary, nanocomputing is a very promising field with many challenges and issues that need to be addressed over the coming decades to make it truly viable. Motivated by results presented in this book, we envision that the greatest applications of nanocomputing are going to be in medicine and biological sciences, especially in building biomedical and biomimetic integrated circuits and devices.

REFERENCES

1. Nanowerk, Nanotechnology Directory. http://www.nanowerk.com/nanotechnology/research/nanotechnology_links.php.
2. United States Patent and Trademark Office. http://www.uspto.gov/.
3. Bawa, Raj, "Nanotechnology Patenting in the US," Nanotechnology Law and Business, Volume 1, Issue 1, Article 5, 2004.
4. Uhlir, Nikolas J. "Throwing a Wrench in the System: Size-Dependent Properties Inherency, and Nanotech Patent Applications," The Federal Circuit Bar Journal, 16(3): pp 327–354, 2007.
5. Chen Lucian C. and Douglas Sharrott. "Patenting Nanotech Inventions," Industrial Biotenchnology, 1(3): pp 153–155, 2005.
6. Baluch, Andrew, Radomsky, Leon, and Maebius, Stephen. "In re Kumar: The First Nanotech Patent Case in the Federal Circuit," Nanotechnology Law and Business, 2(4): pp 342–346, 2005.

INDEX

Bio-Inspired and Nanoscale Integrated Computing. Edited by Mary Mehrnoosh Eshaghian-Wilner

T

V

W

Y

Z

Printed in the United States
By Bookmasters